ADVANCED ENVIRONMENTAL MONITORING

Advanced Environmental Monitoring

Edited by

Young J. Kim

Gwangju Institute of Science and Technology (GIST),
Gwangju,
Korea

and

Ulrich Platt

University of Heidelberg,
Heidelberg,
Germany

Springer

A C.I.P. Catalogue record for this book is available from the Library of Congress.

ISBN 978-1-4020-6363-3 (HB)
ISBN 978-1-4020-6364-0 (e-book)

Published by Springer,
P.O. Box 17, 3300 AA Dordrecht, The Netherlands.

www.springer.com

Printed on acid-free paper

Contents

Contributors xi

Preface xxi

Section 1 Atmospheric Environmental Monitoring

Chapter 1 **Air Pollution Monitoring Systems—Past–Present–Future** 3
U. Platt

Chapter 2 **Radial Plume Mapping: A US EPA Test Method for Area and Fugitive Source Emission Monitoring Using Optical Remote Sensing** 21
Ram A. Hashmonay, Ravi M. Varma, Mark T. Modrak, Robert H. Kagann, Robin R. Segall, and Patrick D. Sullivan

Chapter 3 **MAX-DOAS Measurements of ClO, SO_2 and NO_2 in the Mid-Latitude Coastal Boundary Layer and a Power Plant Plume** 37
Chulkyu Lee, Young J. Kim, Hanlim Lee, and Byeong C. Choi

Chapter 4 **Laser Based Chemical Sensor Technology: Recent Advances and Applications** 50
Frank K. Tittel, Yury A. Bakhirkin, Robert F. Curl, Anatoliy A. Kosterev, Matthew R. McCurdy, Stephen G. So, and Gerard Wysocki

Chapter 5 **Atmospheric Monitoring With Chemical Ionisation Reaction Time-of-Flight Mass Spectrometry (CIR-TOF-MS) and Future Developments: Hadamard Transform Mass Spectrometry** 64
Kevin P. Wyche, Christopher Whyte, Robert S. Blake, Rebecca L. Cordell, Kerry A. Willis, Andrew M. Ellis, and Paul S. Monks

Chapter 6 **Continuous Monitoring and the Source Identification of Carbon Dioxide at Three Sites in Northeast Asia During 2004–2005** 77
Fenji Jin, Sungki Jung, Jooll Kim, K.-R. Kim, T. Chen, Donghao Li, Y.-A. Piao, Y.-Y. Fang, Q.-F. Yin, and Donkoo Lee

Chapter 7 **Aircraft Measurements of Long-Range Trans-Boundary Air Pollutants over Yellow Sea** 90
Sung-Nam Oh, Jun-Seok Cha, Dong-Won Lee, and Jin-Su Choi

Chapter 8 **Optical Remote Sensing for Characterizing the Spatial Distribution of Stack Emissions** 107
Michel Grutter, Roberto Basaldud, Edgar Flores, and Roland Harig

Section 2 Atmospheric Environmental Monitoring

Chapter 9 **Mass Transport of Background Asian Dust Revealed by Balloon-Borne Measurement: Dust Particles Transported during Calm Periods by Westerly from Taklamakan Desert** 121
Y. Iwasaka, J.M. Li, G.-Y. Shi, Y.S. Kim, A. Matsuki, D. Trochkine, M. Yamada, D. Zhang, Z. Shen, and C.S. Hong

Chapter 10 **Identifying Atmospheric Aerosols with Polarization Lidar** 136
Kenneth Sassen

Chapter 11 **A Novel Method to Quantify Fugitive Dust Emissions Using Optical Remote Sensing** 143
Ravi M. Varma, Ram A. Hashmonay, Ke Du, Mark J. Rood, Byung J. Kim, and Michael R. Kemme

Chapter 12 **Raman Lidar for Monitoring of Aerosol Pollution in the Free Troposphere** 155
Detlef Müller, Ina Mattis, Albert Ansmann, Ulla Wandinger, and Dietrich Althausen

Chapter 13 **An Innovative Approach to Optical Measurement of Atmospheric Aerosols—Determination of the Size and the Complex Refractive Index of Single Aerosol Particles** 167
Wladyslaw W. Szymanski, Artur Golczewski, Attila Nagy, Peter Gál, and Aladar Czitrovszky

Chapter 14 **Remote Sensing of Aerosols by Sunphotometer and Lidar Techniques** 179
Anna M. Tafuro, F. De Tomasi, and Maria R. Perrone

Chapter 15 **Retrieval of Particulate Matter from MERIS Observations** 190
Wolfgang von Hoyningen-Huene, Alexander Kokhanovsky, and John P. Burrows

Chapter 16 **Bioaerosol Standoff Monitoring Using Intensified Range-Gated Laser-Induced Fluorescence Spectroscopy** 203
Sylvie Buteau, Jean-R. Simard, Pierre Lahaie, Gilles Roy, Pierre Mathieu, Bernard Déry, Jim Ho, and John McFee

Chapter 17 **MODIS 500 × 500-m^2 Resolution Aerosol Optical Thickness Retrieval and Its Application for Air Quality Monitoring** 217
Kwon H. Lee, Dong H. Lee, Young J. Kim, and Jhoon Kim

Section 3 Contaminant-Control Process Monitoring

Chapter 18 **Aquatic Colloids: Provenance, Characterization and Significance to Environmental Monitoring** 233
Jae-Il Kim

Chapter 19 **Progress in Earthworm Ecotoxicology** 248
Byung-Tae Lee, Kyung-Hee Shin, Ju-Yong Kim, and Kyoung-Woong Kim

Chapter 20 **Differentiating Effluent Organic Matter (EfOM) from Natural Organic Matter (NOM): Impact of EfOM on Drinking Water Sources** 259
Seong-Nam Nam, Stuart W. Krasner, and Gary L. Amy

Chapter 21 **An Advanced Monitoring and Control System for Optimization of the Ozone-AOP (Advanced Oxidation Process) for the Treatment of Drinking Water** 271
Joon-Wun Kang, Byung Soo Oh, Sang Yeon Park, Tae-Mun Hwang, Hyun Je Oh, and Youn Kyoo Choung

Chapter 22 **Monitoring of Dissolved Organic Carbon (DOC) in a Water Treatment Process by UV-Laser Induced Fluorescence** 282
Uwe Wachsmuth, Matthias Niederkrüger, Gerd Marowsky, Norbert Konradt, and Hans-Peter Rohns

Section 4 Biosensors, Bioanalytical and Biomonitoring Systems

Chapter 23 **Biosensors for Environmental and Human Health** 297
Peter-D. Hansen

Chapter 24 **Biological Toxicity Testing of Heavy Metals and Environmental Samples Using Fluorescence-Based Oxygen Sensing and Respirometry** 312
Alice Zitova, Fiach C. O'Mahony, Maud Cross, John Davenport, and Dmitri B. Papkovsky

Chapter 25 **Omics Tools for Environmental Monitoring of Chemicals, Radiation, and Physical Stresses in *Saccharomyces cerevisiae*** 325
Yoshihide Tanaka, Tetsuji Higashi, Randeep Rakwal, Junko Shibato, Emiko Kitagawa, Satomi Murata, Shin-ichi Wakida, and Hitoshi Iwahashi

Chapter 26 **Gene Expression Characteristics in the Japanese Medaka (*Oryzias latipes*) Liver after Exposure to Endocrine Disrupting Chemicals** 338
Han Na Kim, Kyeong Seo Park, Sung Kyu Lee, and Man Bock Gu

Chapter 27 **Optical Detection of Pathogens using Protein Chip** 348
Jeong-Woo Choi and Byung-Keun Oh

Chapter 28 **Expression Analysis of Sex-Specific and Endocrine-Disruptors-Responsive Genes in Japanese Medaka, *Oryzias latipes*, using Oligonucleotide Microarrays** 363
Katsuyuki Kishi, Emiko Kitagawa, Hitoshi Iwahashi, Tomotaka Ippongi, Hiroshi Kawauchi, Keisuke Nakazono, Masato Inoue, Hiroyoshi Ohba, and Yasuyuki Hayashi

Chapter 29 **Assessment of the Hazard Potential of Environmental Chemicals by Quantifying Fish Behaviour** 376
Daniela Baganz and Georg Staaks

Chapter 30 **Biomonitoring Studies Performed with European Eel Populations from the Estuaries of Minho, Lima and Douro Rivers (NW Portugal)** 390
Carlos Gravato, Melissa Faria, Anabela Alves, Joana Santos, and Lúcia Guilhermino

Chapter 31 **In Vitro Testing of Inhalable Fly Ash at the Air Liquid Interface** 402
Sonja Mülhopt, Hanns-Rudolf Paur, Silvia Diabaté, and Harald F. Krug

List of Abbreviations 415

Index 416

Contributors

Dietrich Althausen, Leibniz Institute for Tropospheric Research, Permoserstraße 15, 04318 Leipzig, Germany

Anabela Alves, CIMAR-LA/CIIMAR – Centro Interdisciplinar de Investigação Marinha e Ambiental, Laboratório de Ecotoxicologia, Universidade do Porto, Rua dos Bragas, 177, 4050-123 Porto, Portugal.

Gary L. Amy, UNESCO-IHE Institute for Water Education, Delft, the Netherlands, g.amy@unesco-ihe.org

Albert Ansmann, Leibniz Institute for Tropospheric Research, Permoserstraße 15, 04318 Leipzig, Germany

Daniela Baganz, Department of Biology and Ecology of Fishes, Leibniz-Institute of Freshwater Ecology and Inland Fisheries, Berlin, Germany
and
Leibniz-Institute of Freshwater Ecology and Inland Fisheries, Forschungsverbund Berlin e.V., Müggelseedamm 310, 12587 Berlin, baganz@igb-berlin.de

Yury A. Bakhirkin, Rice University, Electrical and Computer Engineering Department, MS-366, 6100 Main St., Houston, TX 77005, USA

Roberto Basaldud, Centro de Ciencias de la Atmósfera, Universidad Nacional Autónoma de Mexico, 05410 México D.F. México

Robert S. Blake, Department of Chemistry, University of Leicester, Leicester, UK

John P. Burrows, University of Bremen, Institute of Environmental Physics, Otto-Hahn-Allee 1, D-28334 Bremen, Germany

Sylvie Buteau, Defence R & D Canada Valcartier, 2459 Boul. Pie-XI Nord, Québec, QC, Canada, G3J 1X5, sylvie.buteau@drdc-rddc.gc.ca

Jun-Seok Cha, Global Environment Research Center, National Institute of Environment Research, Environmental Research Complex, Gyeongseo-dong, Seo-gu, Inchon 404-708, Korea

T. Chen, Yanbian University, Yanji, Jilin, China

Byeong C. Choi, Meteorological Research Institute, 460-18 Sindaebang-dong, Dongjak-gu, Seoul 156-720, Republic of Korea

Jeong-Woo Choi, Department of Chemical and Biomolecular Engineering, Sogang University, #1 Shinsu-dong, Mapo-gu, Seoul 121-742, Korea
and
Interdisciplinary Program of Integrated Biotechnology, Sogang University, #1 Shinsu-dong, Mapo-gu, Seoul 121-742, Korea, jwchoi@sogang.ac.kr

Jin-Su Choi, Global Environment Research Center, National Institute of Environment Research, Environmental Research Complex, Gyeongseo-dong, Seo-gu, Inchon 404-708, Korea

Youn Kyoo Choung, School of Civil & Environmental Engineering, Yonsei University, Seoul, Korea

Rebecca L. Cordell, Department of Chemistry, University of Leicester, Leicester, UK

Maud Cross, Zoology Ecology and Plants Science Department, University College Cork, Distillery Fields, North Mall, Cork, Ireland

Robert F. Curl, Rice University, Electrical and Computer Engineering Department, MS-366, 6100 Main St., Houston, TX 77005, USA

Aladar Czitrovszky, Research Institute for Solid State Physics and Optics, Department of Laser Applications, Hungarian Academy of Science, H-1525 Budapest, P.O. Box 49, Hungary

John Davenport, Zoology Ecology and Plants Science Department, University College Cork, Distillery Fields, North Mall, Cork, Ireland

Bernard Déry, Defence R & D Canada Valcartier, 2459 Boul. Pie-XI Nord, Québec, QC, Canada, G3J 1X5

Silvia Diabaté, Forschungszentrum Karlsruhe, Institute for Toxicology and Genetics, Hermann-von-Helmholtz-Platz 1, 76344 Eggenstein – Leopoldshafen, Germany

Ke Du, Department of Civil & Environmental Engineering, University of Illinois at Urbana-Champaign, 205 N. Mathews Ave., Urbana, IL 61801, USA

Andrew M. Ellis, Department of Chemistry, University of Leicester, Leicester, UK

Y.-Y. Fang, Yanbian University, Yanji, Jilin, China

Melissa Faria, CIMAR-LA/CIIMAR – Centro Interdisciplinar de Investigação Marinha e Ambiental, Laboratório de Ecotoxicologia, Universidade do Porto, Rua dos Bragas, 177, 4050-123 Porto, Portugal

Edgar Flores, Centro de Ciencias de la Atmósfera, Universidad Nacional Autónoma de Mexico, 05410 México D.F. México

Peter Gál, Research Institute for Solid State Physics and Optics, Department of Laser Applications, Hungarian Academy of Science, H-1525 Budapest, P.O. Box 49, Hungary

Artur Golczewski, Faculty of Physics, University of Vienna, Boltzmanngasse 5, A-1090 Vienna, Austria

Carlos Gravato, CIMAR-LA/CIIMAR – Centro Interdisciplinar de Investigação Marinha e Ambiental, Laboratório de Ecotoxicologia, Universidade do Porto, Rua dos Bragas, 177, 4050-123 Porto, Portugal.
and
Departamento de Biologia, Universidade de Aveiro, 3810-193 Aveiro, Portugal. gravatoc@ciimar.up.pt

Michel Grutter, Centro de Ciencias de la Atmósfera, Universidad Nacional Autónoma de Mexico, 05410 México D.F. México, grutter@servidor.unam.mx

Man Bock Gu, School of Life Sciences and Biotechnology, Korea University, Seoul 136-701, Korea, mbgu@korea.ac.kr

Lúcia Guilhermino, CIMAR-LA/CIIMAR – Centro Interdisciplinar de Investigação Marinha e Ambiental, Laboratório de Ecotoxicologia, Universidade do Porto, Rua dos Bragas, 177, 4050-123 Porto, Portugal
and
ICBAS – Instituto de Ciências Biomédicas de Abel Salazar, Universidade do Porto, Departamento de Estudos de Populações, Laboratório de Ecotoxicologia, Largo Professor Abel Salazar 2, 4099-003, Porto, Portugal

Peter-D. Hansen, Technische Universität Berlin, Faculty VI, Department of Ecotoxicology, Franklin Strasse 29 (OE4), D-10587 Berlin, Germany, pd.hansen@tu-berlin.de

Roland Harig, Institut für Messtechnik, Technische Universität Hamburg-Harburg, 21079 Hamburg, Germany

Ram A. Hashmonay, ARCADIS, 4915 Prospectus Drive Suite F, Durham, NC 27713, USA, rhashmonay@arcadis-us.com

Yasuyuki Hayashi, GeneFrontier Corp., Nihonbashi Kayabacho 3-2-10, Chuo-ku, Tokyo, 103-0025, Japan

Tetsuji Higashi, Human Stress Signal Research Center (HSS), National Institute of Advanced Industrial Science and Technology (AIST), 1-8-31 Midorigaoka, Ikeda, Osaka 563-8577, Japan

Jim Ho, Defence R & D Canada Suffield, Box 4000, Medicine Hat, AB, Canada, T1A 8K6

C. S. Hong, Institute of Nature and Environmental Technology, Kanazawa University, Kanazawa, Japan

Tae-Mun Hwang, Korea Institute of Construction Technology, 2311 Daehwa-Dong, Ilsan-gu, Kyonggi-do, Korea (411–712)

Masato Inoue, GeneFrontier Corp., Nihonbashi Kayabacho 3-2-10, Chuo-ku, Tokyo, 103-0025, Japan

Tomotaka Ippongi, GeneFrontier Corp., Nihonbashi Kayabacho 3-2-10, Chuo-ku, Tokyo, 103-0025, Japan

Rudolf Irmscher, Stadtwerke Düsseldorf AG, Qualitätsüberwachung Wasser (OE 423), Postfach 101136, 40002 Düsseldorf, Germany

Hitoshi Iwahashi, Human Stress Signal Research Center (HSS), National Institute of Advanced Industrial Science and Technology (AIST), Tsukuba West, 16-1 Onogawa, Tsukuba 305-8569, Japan, hitoshi.iwahashi@aist.go.jp

Y. Iwasaka, Institute of Nature and Environmental Technology, Kanazawa University, Kanazawa, Japan, kosa@t.kanazawa-u.ac.jp

Fenji Jin, School of Earth and Environmental Science, Seoul National University, Seoul, Korea

Sungki Jung, School of Earth and Environmental Science, Seoul National University, Seoul, Korea

Robert H. Kagann, ARCADIS, 4915 Prospectus Drive Suite F, Durham, NC 27713, USA

Joon-Wun Kang, Department of Environmental Engineering, YIEST, Yonsei University at Wonju, 234, Maeji, Wonju, Korea (220–710), jwk@yonsei.ac.kr

Hiroshi Kawauchi, GeneFrontier Corp., Nihonbashi Kayabacho 3-2-10, Chuo-ku, Tokyo, 103-0025, Japan

Michael R. Kemme, U.S. Army ERDC – CERL, 2902 Farber Drive, Champaign, IL 61822 USA

Byung J. Kim, U.S. Army ERDC – CERL, 2902 Farber Drive, Champaign, IL 61822 USA

Han Na Kim, National Research Laboratory on Environmental Biotechnology, Gwangju Institute of Science and Technology (GIST), Gwangju 500-712, Korea

Jae-Il Kim, Institut für Nukleare Entsorgung (INE), Forschungszentrum Karlsruhe (FZK), 76021 Karlsruhe, Germany, jikim@t-online.de

Jhoon Kim, Department of Atmospheric Sciences, Yonsei University, Shinchondong 134, Seodaemun-gu, Seoul 120-749, Republic of Korea

Jooll Kim, School of Earth and Environmental Science, Seoul National University, Seoul, Korea

Ju-Yong Kim, Department of Environmental Science and Engineering, Gwangju Institute of Science and Technology (GIST), Gwangju 500-712, Republic of Korea

K.-R. Kim, School of Earth and Environmental Science, Seoul National University, Seoul, Korea, krkim@snu.ac.kr

Kyoung-Woong Kim, Department of Environmental Science and Engineering, Gwangju Institute of Science and Technology (GIST), Gwangju 500-712, Republic of Korea, kwkim@gist.ac.kr

Y.S. Kim, Institute of Nature and Environmental Technology, Kanazawa University, Kanazawa, Japan
and
Now: Institute of Environmental and Industrial Medicine, Hanyang University, Seoul, Korea

Young J. Kim, Advanced Environmental Monitoring Research Center (ADEMRC), Department of Environmental Science and Engineering, Gwangju Institute of Science and Technology (GIST), 1 Oryong-dong, Buk-gu, Gwangju 500-712, Republic of Korea, yjkim@gist.ac.kr

Katsuyuki Kishi, Japan Pulp & Paper Research Institute, Inc., Tokodai 5-13-11, Tsukuba, Ibaraki, 300-2635, Japan, kishi@jpri.co.jp

Emiko Kitagawa, Human Stress Signal Research Center (HSS), National Institute of Advanced Industrial Science and Technology (AIST), Tsukuba West, 16-1 Onogawa, Tsukuba, Ibaraki 305-8569, Japan

Alexander Kokhanovsky, University of Bremen, Institute of Environmental Physics, Otto-Hahn-Allee 1, D-28334 Bremen, Germany

Norbert Konradt, Stadtwerke Düsseldorf AG, Qualitätsüberwachung Wasser (OE 423), Postfach 101136, 40002 Düsseldorf, Germany

Anatoliy A. Kosterev, Rice University, Electrical and Computer Engineering Department, MS-366, 6100 Main St., Houston, TX 77005, USA

Stuart W. Krasner, Metropolitan Water District of Southern California, La Verne, California USA

Harald F. Krug, Forschungszentrum Karlsruhe, Institute for Toxicology and Genetics, Hermann-von-Helmholtz-Platz 1, 76344 Eggenstein – Leopoldshafen, Germany

Pierre Lahaie, Defence R & D Canada Valcartier, 2459 Boul. Pie-XI Nord, Québec, QC, Canada, G3J 1X5

Byung-Tae Lee, Department of Environmental Science and Engineering, Gwangju Institute of Science and Technology (GIST), Gwangju 500-712, Republic of
Korea

Chulkyu Lee, Advanced Environmental Monitoring Research Center (ADEM-RC), Department of Environmental Science and Engineering, Gwangju

Institute of Science and Technology (GIST), 1 Oryong-dong, Buk-gu, Gwangju 500-712, Republic of Korea
and
Now at Institute of Environmental Physics and Remote Sensing, University of Bremen, Atto-Hahn-Allee 1, D-28334, Bremen, Germany, cklee79@gmail.com

Dong H. Lee, Advanced Environmental Monitoring Research Center (ADEMRC), Gwangju Institute of Science & Technology (GIST), 1 Oryong-dong, Buk-gu, Gwangju 500-712, Republic of Korea

Dong-Won Lee, Global Environment Research Center, National Institute of Environment Research, Environmental Research Complex, Gyeongseo-dong, Seo-gu, Inchon 404-708, Korea

Donkoo Lee, College of Agriculture and Life Sciences, Seoul National University, Seoul, Korea

Hanlim Lee, Advanced Environmental Monitoring Research Center (ADEMRC), Department of Environmental Science and Engineering, Gwangju Institute of Science and Technology (GIST), 1 Oryong-dong, Buk-gu, Gwangju 500-712, Republic of Korea

Kwon H. Lee, Advanced Environmental Monitoring Research Center (ADEMRC), Gwangju Institute of Science & Technology (GIST), 1 Oryong-dong, Buk-gu, Gwangju 500-712, Republic of Korea

Sung Kyu Lee, Environmental Toxicology Devision, Korea Institute of Toxicology, 100 Jangdong, Yuseong, Daejeon, 305-343, Korea

Donghao Li, Yanbian University, Yanji, Jilin, China

J.M. Li, Graduate School of Environmental Studies, Nagoya University, Nagoya, Japan

Gerd Marowsky, Laser-Laboratorium Göttingen e.V., Hans-Adolf-Krebs-Weg 1, 37077 Göttingen, Germany

Pierre Mathieu, Defence R & D Canada Valcartier, 2459 Boul. Pie-XI Nord, Québec, QC, Canada, G3J 1X5

A. Matsuki, Institute of Nature and Environmental Technology, Kanazawa University, Kanazawa, Japan
and
Now: Laboratorire de Meteorologie Physique, Universite Blaise Pascal, Aubie re CEDEX, France

Ina Mattis, Leibniz Institute for Tropospheric Research, Permoserstraße 15, 04318 Leipzig, Germany

Matthew R. McCurdy, Rice University, Electrical and Computer Engineering Department, MS-366, 6100 Main St., Houston, TX 77005, USA

John McFee, Defence R & D Canada Suffield, Box 4000, Medicine Hat, AB, Canada, T1A 8K6

Mark T. Modrak, ARCADIS, 4915 Prospectus Drive Suite F, Durham, NC 27713, USA

Paul S. Monks, Department of Chemistry, University of Leicester, Leicester, UK p.s.monks@le.ac.uk

Sonja Mülhopt, Forschungszentrum Karlsruhe, Institute for Technical Chemistry, Thermal Waste Treatment Division, Hermann-von-Helmholtz-Platz 1, 76344 Eggenstein – Leopoldshafen, Germany, muelhopt@itc-tab.fzk.de

Detlef Müller, Leibniz Institute for Tropospheric Research, Permoserstraße 15, 04318 Leipzig, Germany, detlef@tropos.de

Satomi Murata, Human Stress Signal Research Center (HSS), National Institute of Advanced Industrial Science and Technology (AIST), Tsukuba West, 16-1 Onogawa, Tsukuba, Ibaraki 305-8569, Japan

Attila Nagy, Research Institute for Solid State Physics and Optics, Department of Laser Applications, Hungarian Academy of Science, H-1525 Budapest, P.O. Box 49, Hungary

Keisuke Nakazono, GeneFrontier Corp., Nihonbashi Kayabacho 3-2-10, Chuo-ku, Tokyo, 103-0025, Japan

Seong-Nam Nam, Civil and Environmental Engineering, University of Colorado, Boulder, Colorado USA

Matthias Niederkrüger, Laser-Laboratorium Göttingen e.V., Hans-Adolf-Krebs-Weg 1, 37077 Göttingen, Germany

Byung-Keun Oh, Department of Chemical and Biomolecular Engineering, Sogang University, #1 Shinsu-dong, Mapo-gu, Seoul 121-742, Korea
and
Interdisciplinary Program of Integrated Biotechnology, Sogang University, #1 Shinsu-dong, Mapo-gu, Seoul 121-742, Korea

Byung Soo Oh, Department of Environmental Engineering, YIEST, Yonsei University at Wonju, 234, Maeji, Wonju, KOREA (220-710)

Hyun Je Oh, Korea Institute of Construction Technology, 2311 Daehwa-Dong, Ilsan-gu, Kyonggi-do, Korea (411-712)

Sung-Nam Oh, Meteorological Research Institute (METRI), Korea Meteorological Administration (KMA), 460-18 Shindaebang-dong, Dongjak-gu, Seoul 156-720, Korea, snoh@metri.re.kr

Hiroyoshi Ohba, GeneFrontier Corp., Nihonbashi Kayabacho 3-2-10, Chuo-ku, Tokyo, 103-0025, Japan

Fiach C. O'Mahony, Biochemistry Department & ABCRF, University College Cork, Cavanagh Pharmacy Building, Cork, Ireland

Dmitri B. Papkovsky, Biochemistry Department & ABCRF, University College Cork, Cavanagh Pharmacy Building, Cork, Ireland
and
Luxcel Biosciences Ltd., Suite 332, BioTransfer Unit, BioInnovation Centre, UCC, Cork, Ireland, d.papkovsky@ucc.ie

Kyeong Seo Park, National Research Laboratory on Environmental Biotechnology, Gwangju Institute of Science and Technology (GIST), Gwangju 500-712, Korea

Sang Yeon Park, Department of Environmental Engineering, YIEST, Yonsei University at Wonju, 234, Maeji, Wonju, Korea (220-710)

Hanns-Rudolf Paur, Forschungszentrum Karlsruhe, Institute for Technical Chemistry, Thermal Waste Treatment Division, Hermann-von-Helmholtz-Platz 1, 76344 Eggenstein – Leopoldshafen, Germany

Maria R. Perrone, CNISM, Dipartimento di Fisica, Università di Lecce, via per Arnesano, Lecce, Italy

Y.-A. Piao, Yanbian University, Yanji, Jilin, China

U. Platt, Institute of Environmental Physics, University of Heidelberg, INF 229, D-69120 Heidelberg, ulrich.platt@iup.uni-heidelberg.de

Randeep Rakwal, Human Stress Signal Research Center (HSS), National Institute of Advanced Industrial Science and Technology (AIST), Tsukuba West, 16-1 Onogawa, Tsukuba, Ibaraki 305-8569, Japan

Hans-Peter Rohns, Stadtwerke Düsseldorf AG, Qualitätsüberwachung Wasser (OE 423), Postfach 101136, 40002 Düsseldorf, Germany

Mark J. Rood, Department of Civil & Environmental Engineering, University of Illinois at Urbana-Champaign, 205 N. Mathews Ave., Urbana, IL 61801, USA

Gilles Roy, Defence R & D Canada Valcartier, 2459 Boul. Pie-XI Nord, Québec, QC, Canada, G3J 1X5

Joana Santos, Laboratório de Ecotoxicologia, Universidade do Porto, Rua dos Bragas, 177, 4050-123 Porto, Portugal

Kenneth Sassen, Geophysical Institute, University of Alaska Fairbanks, 903 Koyukuk Drive, Fairbanks, Alaska 99775 USA, ksassen@gi.alaska.edu

Robin R. Segall, Emission Measurement Center (E143-02), Office of Air Quality Planning and Standards, US Environmental Protection Agency, Research Triangle Park, NC 27711

Z. Shen, Cold and Arid Regions Environmental and Engineering Research Institute, Chinese Academy of Science, Lanzhou, China

G.-Y. Shi, Institute of Atmospheric Physics, Chinese Academy of Science, Beijing, China

Junko Shibato, Human Stress Signal Research Center (HSS), National Institute of Advanced Industrial Science and Technology (AIST), Tsukuba West, 16-1 Onogawa, Tsukuba, Ibaraki 305-8569, Japan

Kyung-Hee Shin, Department of Environmental Science and Engineering, Gwangju Institute of Science and Technology (GIST), Gwangju 500-712, Republic of Korea

Jean-R. Simard, Defence R & D Canada Valcartier, 2459 Boul. Pie-XI Nord, Québec, QC, Canada, G3J 1X5

Stephen G. So, Rice University, Electrical and Computer Engineering Department, MS-366, 6100 Main St., Houston, TX 77005, USA

Georg Staaks, Department of Biology and Ecology of Fishes, Leibniz-Institute of Freshwater Ecology and Inland Fisheries, Berlin, Germany

Patrick D. Sullivan, Air Force Research Laboratory, Air Expeditionary Forces Technologies Division (AFRL/MLQF), 139 Barnes Drive, Suite 2, Tyndall AFB, FL 32403

Wladyslaw W. Szymanski, Faculty of Physics, University of Vienna, Boltzmanngasse 5, A-1090 Vienna, Austria, w.szym@univie.ac.at

Anna M. Tafuro, CNISM, Dipartimento di Fisica, Università di Lecce, via per Arnesano, Lecce, Italy, anna.tafuro@le.infn.it

Yoshihide Tanaka, Human Stress Signal Research Center (HSS), National Institute of Advanced Industrial Science and Technology (AIST), 1-8-31 Midorigaoka, Ikeda, Osaka 563-8577, Japan

Frank K. Tittel, Rice University, Electrical and Computer Engineering Department, MS-366, 6100 Main St., Houston, TX 77005, USA, fkt@rice.edu

F. De Tomasi, CNISM, Dipartimento di Fisica, Università di Lecce, via per Arnesano, Lecce, Italy

D. Trochkine, Institute of Nature and Environmental Technology, Kanazawa University, Kanazawa, Japan
and
Now: Institute for Water and Environmental Problems, Siberian Branch of Russian Academy of Science, Barnaul, Russia

Ravi M. Varma, ARCADIS, 4915 Prospectus Drive Suite F, Durham, NC 27713, USA
and
Department of Physics, National University of Ireland, University College Cork, Cork, Ireland, r.varma@ucc.ie

Wolfgang von Hoyningen-Huene, University of Bremen, Institute of Environmental Physics, Otto-Hahn-Allee 1, D-28334 Bremen, Germany, hoyning@iup.physik.uni-bremen.de

Uwe Wachsmuth, Laser-Laboratorium Göttingen GmbH, Hans-Adolf-Krebs-Weg 1, 37077 Göttingen, Germany, uwachsm@llg.gwdg.de

Shin-ichi Wakida, Human Stress Signal Research Center (HSS), National Institute of Advanced Industrial Science and Technology (AIST), 1-8-31 Midorigaoka, Ikeda, Osaka 563-8577, Japan

Ulla Wandinger, Leibniz Institute for Tropospheric Research, Permoserstraße 15, 04318 Leipzig, Germany

Christopher Whyte, Department of Chemistry, University of Leicester, Leicester, UK

Kerry A. Willis, Department of Chemistry, University of Leicester, Leicester, UK

Kevin P. Wyche, Department of Chemistry, University of Leicester, Leicester, UK

Gerard Wysocki, Rice University, Electrical and Computer Engineering Department, MS-366, 6100 Main St., Houston, TX 77005, USA

M. Yamada, Institute of Nature and Environmental Technology, Kanazawa University, Kanazawa, Japan

Q.-F. Yin, Huaiyin Teacher's College, Huaian, Jiangsu, China

D. Zhang, Faculty of Environmental and Symbiotic Sciences, Prefectural University of Kumamoto, Kumamoto, Japan

Alice Zitova, Biochemistry Department & ABCRF, University College Cork, Cavanagh Pharmacy Building, Cork, Ireland

Preface

We are facing increasing environmental concerns associated with water, air, and soil pollution as well as climate change induced by human activities. Therefore accurate assessment of the state of the environment is a prerequisite for undertaking any course of action towards improvement. In particular, development of new environmental monitoring technologies for the detection of hazardous pollutants and environmental change has become increasingly important to scientists and to regulatory agencies. In recent years there has been much progress in the field of environmental monitoring research, resulting in the development of more accurate, fast, compound-specific, convenient, and cost-effective techniques by integrating emerging technologies from various disciplines.

This book is a result of the *6th International Symposium on Advanced Environmental Monitoring*, organized by ADvanced Environmental Monitoring Center (ADEMRC), Gwangju Institute of Science and Technology (GIST), Korea and held in Heidelberg, Germany on June, 27–30, 2006. It presents recent advances in the research and development of forthcoming technologies, as well as in field applications in advanced environmental monitoring. It is our hope that the papers presented in this book will provide a glimpse of how cutting-edge technologies involving monitoring of pollutants, determination of environmental status, and the detection and quantification of toxicity are being developed and applied in the field.

We give many thanks to all authors for their participation and contributions and to the reviewers for their goodwill in providing a rapid turnover of the manuscripts and the critical comments necessary for ensuring the quality of this publication. We gratefully acknowledge Dr. Paul Roos, Editorial Director, and Betty van Herk of Springer for their continuing support and cooperation in making this book a reality. Members of the symposium organizing committee deserve the most credit for the success of the symposium and their critical suggestions for collection of the manuscripts. This symposium was supported in part by the Korea Science and Engineering Foundation (KOSEF) through the Advanced Environmental Monitoring Research Center at Gwangju Institute of Science and Technology.

April 2007

Young J. Kim
Editor
Director, Advanced Environmental
Monitoring Research Center (ADEMRC)
Professor, Dept. of Environmental
Science and Engineering
Gwangju Institute of Science
and Technology (GIST)
1 Oryong-dong, Buk-gu
Gwangju 500-712, Republic of Korea
E-mail: yjkim@gist.ac.kr

Ulrich Platt
Editor
Professor and Director
Institute of Environmental
Physics (IUP)
University of Heidelberg
Im Neuenheimer Feld 229
D-69120 Heidelberg, Germany
E-mail: ulrich.platt@iup.uni-heidelberg.de

Section 1
Atmospheric Environmental Monitoring

Chapter 1
Air Pollution Monitoring Systems—Past–Present–Future

U. Platt

Abstract Measurements of trace gas concentrations and other parameters like photolysis frequencies are a crucial tool for air pollution monitoring and the investigation of processes in the atmosphere. However, the determination of atmospheric trace gas concentrations constitutes a technological challenge, since extreme sensitivity (mixing ratios as low as 10^{-13}) is desired simultaneously with high specificity i.e. the molecule of interest usually must be detected in the presence of a large excess of other species. In addition, spatially resolved measurements are becoming increasingly important.

Today none of the existing measurement techniques meets all above requirements for trace gas measurements in the atmosphere. Therefore, a comprehensive arsenal of different techniques has been developed. Besides a large number of special techniques (like the ubiquitous short-path UV absorption for O_3 measurement) universal methods gain interest, due to their economy and relative ease of use. In particular, a single instrument can register a large number of different trace species.

The different types of requirements and the various techniques are discussed; special emphasis is given to spectroscopic methods, which play a large and growing role in atmospheric chemistry research. For instance, only spectroscopic methods allow remote sensing and spatially resolved determination of trace gas concentrations e.g. from space-borne platforms. Today many varieties of spectroscopic methods are in use (e.g. tunable diode laser- and Fourier-transform spectroscopy). The basic properties and recent applications of this technique are presented using differential optical absorption spectroscopy (DOAS) as an example. Future requirements and expected developments are discussed.

Keywords: Air pollution monitoring, trace gas, DOAS, spectroscopy, remote sensing

Institute of Environmental Physics, University of Heidelberg, INF 229, D-69120 Heidelberg
Tel: 49 6221 546339, Fax: 49 6221 546405

Y.J. Kim and U. Platt (eds.), *Advanced Environmental Monitoring*, 3–20.

1.1 Introduction

Measurements of trace gas and aerosol concentrations (and other quantities like the intensity of the radiation field in the atmosphere) are experimental prerequisites for pollution monitoring and the understanding of the underlying physicochemical processes in the earth's atmosphere (Roscoe and Clemitshaw 1997; Platt 1991, 1999; Clemitshaw 2004). At the same time the determination of trace gas concentrations in the atmosphere is a challenge for the analytical techniques employed in several respects.

First, the technique must be very sensitive to detect the species under consideration at ambient concentration levels. This can be a very demanding criterion, since, for instance, species present at mixing ratios ranging from as low as 0.1 ppt (1 ppt corresponds to a mixing ratio of 1 pmol of trace gas per mole of air or a mixing ratio of 10^{-12}, equivalent to about 2.4×10^7 molecules/cm^3) to several ppb (1 ppb corresponds to 1 nmol mol^{-1} or a mixing ratios of 10^{-9}) can still have a significant influence on the chemical processes in the atmosphere (Perner et al. 1987). Thus, detection limits from below 0.1 ppt up to the lower ppb-range are usually required, depending on the application.

Second, it is equally important for the measurement techniques to be specific, which means, that the result of the measurement of a particular species must neither be positively nor negatively influenced by any other trace species simultaneously present in the probed volume of air. Given the large number of different molecules present at the ppt and ppb level, even in clean air, this is not a trivial condition.

Third, the technique must allow sufficient precision and calibration to be feasible.

In most practical applications, there are other requirements, including spatial coverage, time resolution, properties like simplicity of design and use of the instruments, a capability of real-time operation (as opposed to taking samples for later analysis), and the possibility of unattended operation. Other factors to be considered are weight, portability, and dependence of the measurement on ambient conditions.

To date no single measurement technique can fulfil all the diverse requirements for trace gas measurements in the atmosphere. Therefore, specialised techniques or variants of techniques have been developed, which are tailored to the various measurement tasks occurring in atmospheric research, pollution control, and monitoring of atmospheric change:

1. Long-term observations are aimed at monitoring gradual changes in atmospheric parameters, e.g. its trace gas composition. Typical examples are
    - Trends of greenhouse gases like CO_2, CH_4, N_2O, or CFM's
    - Stratospheric ozone
    - Change of stratospheric chemistry (e.g. realised in the Network for the Detection of Atmospheric Composition Change, NDACC)
    - The temporal evolution in the abundance of species supplying halogens to the stratosphere (e.g. CFC-and HCFC-species)
    - Trend of the tropospheric ozone mixing ratio as routinely monitored by the Global Atmospheric Watch (GAW) programme.

In this context the so-called 'operator dilemma' should be noted: the measurement of a particular set of species over an extended period is frequently not considered a scientific challenge; on the other hand, the success of the data series hinges on the very careful execution of the measurements. Here the psychological side of the project may be as critical as the technology.

2. Regional and episodic studies seek to investigate causes, extent, and consequences of regional events like air pollution episodes or boundary layer ozone depletion events (Barrie et al. 1988). While routine monitoring is an issue many fundamental questions can only be investigated by observations made on a regional scale. Typical measurements tasks in this context are
   - Monitoring of air pollutants (like O_3, SO_2, NO, NO_2, hydrocarbons)
   - Investigation of urban plume evolution (e.g. with respect to O_3 formation downwind of source regions)
   - Mapping of continental plumes
   - Observation of the Antarctic Stratospheric Ozone Hole
   - Polar boundary-layer ozone loss events (the 'tropospheric ozone hole', (Platt and Lehrer 1997)
3. Investigation of fast in situ (photo) chemistry allows to neglect the effect of transport, in particular this is true for the following systems:

Free-radical (e.g. OH, HO_2, BrO) photochemistry, where the lifetime of the reactive species is of the order of seconds (OH: below 1 s, HO_2: from <1 s at high NO_X levels to ≈200 s at zero NO_X, BrO: ≈100 s).

'Smog'-chamber (today frequently called reaction chamber or photoreactor) studies allow to suppress transport. However, care has to be taken to avoid artefacts which may arise from chemical processes at the chamber walls.

Today, atmospheric chemistry has a comprehensive arsenal of measurement techniques at its disposal; Table 1.1 gives an overview of the techniques available for a series of key species relevant for studies of atmospheric chemistry. Among a large number of specialised techniques (such as the gas-phase chemiluminescence detection of NO) universal techniques are of great interest, due to their relative simplicity and economy.

1.2 Measurement Techniques by Broad Categories

In this section we group the available techniques according to a series of broad criteria. According to the remarks above the degree of specialisation is of importance and we may distinguish between

- Specialised techniques, where one instrument measures a single species ('box per species')
- Universal techniques, where a single instrument can determine a large set of species

Examples for specialised techniques include gas-phase chemiluminescence detection of NO (Drummond et al. 1985) or short-path UV absorption detection of ozone using a

Table 1.1 Overview of species of relevance to atmospheric chemistry research and measurement techniques

Species	UV/ vis	FT-IR	TDLS (IR)	GC	MS (CIMS)	Fluorescence Chemolum- inescence	Other
NO	O	O	O			+	
NO_2	+	O	+			+	MI-ESR[a]
NO_3	+						MI-ESR, LIF
HNO_2	+						Denuder
HNO_3		O	O				Denuder
OH	+				O		LIF
HO_2/RO_2			?		+		LIF, Ch. A[b]
H_2O_2		O	+			+	
O_3	+	O	O			O	Electrochemistry[c]
HCHO	+	O	+				Derivat.[d]
RCHO							Derivat.[d]
Alkanes				+			
Olefins				+	O		
Aromatic	+			+	O		
CO			O	+		+	
DMS				+			
SO_2	+		O			+	
N_2O			+	+			
CFC's		+		+			
HX[e]		+					Wet chemistry
XO[e]	+				O	+	
HOX[e]			?				

Symbols denote: well measurable (+), measurable (O), not measurable (empty field)
UV/vis UV/visible spectroscopy, FT-IR Fourier-transform IR Spectroscopy, TDLS tunable diode laser spectroscopy, GC gas chromatography, MS (CIMS) mass spectrometry (chemical ionisation mass spectrometry)
[a] Matrix isolation-electron spin resonance
[b] Chemical amplifier
[c] Electrochemical cell
[d] Derivatisation + HPLC
[e] X = Halogen atom (F, Cl, Br, I)

strong emission line of mercury near 253.7 nm (Proffitt and McLaughlin 1983). On the other hand, universal techniques include gas-chromatography or spectroscopic methods.

Another fundamental property of instruments is the spatial range of the measurements, usually expressed in terms of

- In situ measurements
- Remote sensing measurements

While in situ measurements come close to the ideal to determine trace gas concentrations at a 'point' in space, which is usually very close to the instrument, remote sensing techniques usually average the trace gas concentration over a relatively large volume of air, thus providing more representative measurements.

In addition remote sensing techniques allow observations from a (large) distance, perhaps as far as from a satellite instrument in the earth's orbit. Present remote sensing techniques always rely on the sensing of electromagnetic radiation, i.e. they are spectroscopic methods.

Further criteria are the capability of techniques to perform spatially resolved measurements:

- Volume integrated measurements
- Spatially resolved measurements

Also to be considered is the degree of redundancy in the measurements, i.e. the result of a measurement can be

- Just a number, i.e. the mixing ratio or concentration of a trace gas
- Redundant data, for instance the strength of several absorption lines

Examples of instruments belonging to either category include

- Gas-chromatography (universal technique, in situ, redundant data)
- Optical spectroscopy (universal technique, in situ and remote sensing, redundant data)
- Mass spectrometry (MS, redundant data)
- 'Any other (in situ) technique', where the most commonly employed principles include

Chemiluminescence (e.g. for the detection of NO or O_3, usually no redundant data)

Chemical amplifiers for the detection of peroxy radicals (Cantrell et al. 1993; Clemitshaw et al. 1997)
Electrochemical techniques
Matrix isolation–electron spin resonance (MI–ESR) (Mihelcic et al. 1985)
Derivatisation + HPLC (e.g. for the determination of carbonyls, (Lowe and Schmidt 1983)
Bubbler + wet chemistry or ion chromatography (IC), (in situ, usually no redundant data)

In this context, spectroscopic techniques are a promising variety: these techniques are highly sensitive, very specific, universally useable, provide absolute results, and have the potential for remote sensing. It is, therefore, not surprising, that spectroscopic techniques assume a unique role among the many methods, which are in use today. In the following section we will focus further on spectroscopic techniques.

1.3 Selection Criteria for Spectroscopic Air Monitoring Techniques

For a particular application, the selection of a specific spectroscopic technique will be based on the particular requirements as outlined above: which species are to be measured, is the simultaneous determination of several species with the same technique necessary, what is the required accuracy, time resolution, and spatial resolution? Also to be considered are logistic requirements like power consumption,

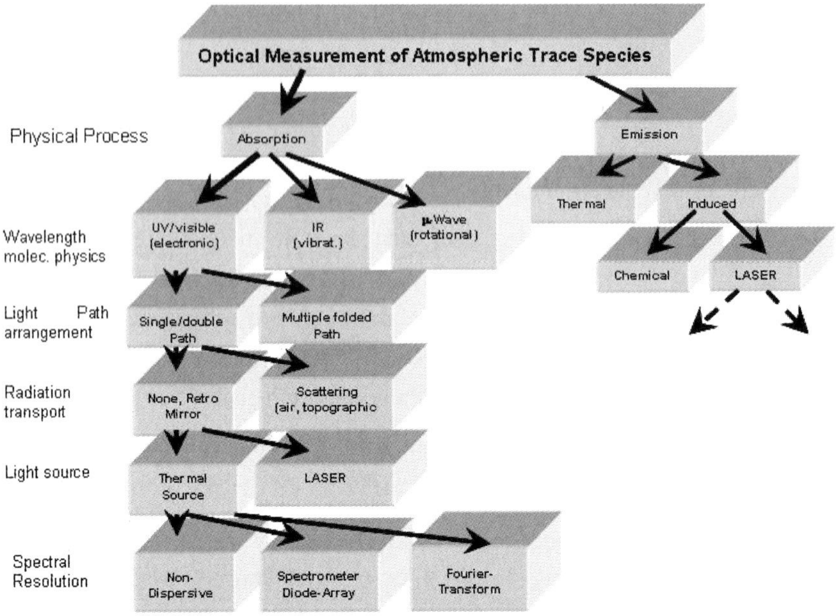

Fig. 1.1 Spectroscopic techniques are used in for the measurement of atmospheric trace species and parameters in a large number of variants. Shown here is a 'family tree' of spectroscopic techniques

mounting of light sources or retro-reflectors (see below), or accommodation of the instrument on mobile platform.

Important technical criteria of spectroscopic instruments are the wavelength region used (see Fig. 1.1), the physical principle (i.e. absorption- or emission spectroscopy), the arrangement of the light path, or the type of light source used. The following spectroscopic techniques are presently employed to measure atmospheric trace gases (see also Sigrist 1994; Clemitshaw 2004; Platt and Stutz 2007):

- Tunable diode laser spectroscopy (TDLS)
- Photo acoustic spectroscopy (PAS)
- Light detection and ranging (LIDAR)
- Differential absorption LIDAR (DIAL)
- Laser-induced fluorescence (LIF)
- Differential optical absorption spectroscopy (DOAS)
- Cavity-ringdown spectroscopy (CRDS)

1.4 The Principle of Absorption Spectroscopy

This universal spectroscopic technique makes use of the absorption of electromagnetic radiation by matter (Fig. 1.2). Quantitatively, the absorption of radiation is expressed by Lambert–Beers law:

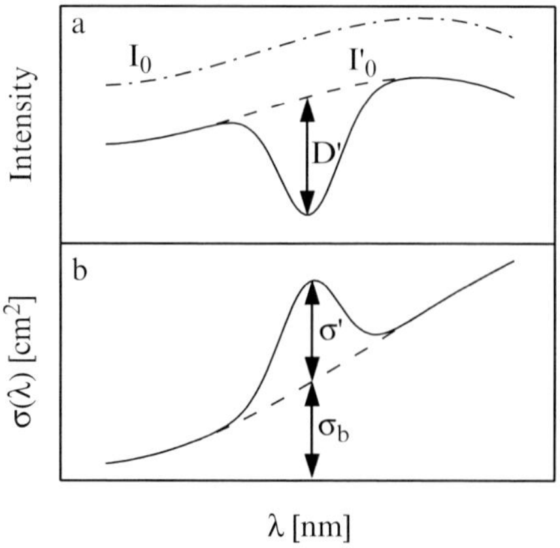

Fig. 1.2 DOAS principle: the trace gas concentrations are calculated (Stutz and Platt 1996) from the amplitude of absorption structures, e.g. from differences of the absorption in the centre of an absorption band (or line) and the spectral range between bands

$$I(\lambda) = I(\lambda) \cdot \exp\left(-S \cdot \sigma(\lambda)\right) \tag{1}$$

where $\sigma(\lambda)$ denotes the absorption cross section at the wavelength λ, $I_0(\lambda)$ is the initial intensity emitted by some suitable source of radiation, and $I(\lambda)$ is the radiation intensity after passing through a layer column density S of the trace gas (what is D in Eq. 1?)

$$S = \int_0^L c(s)\,ds = \bar{c} \cdot L \tag{2}$$

where the species to be measured is present at a concentration (or number density) $c(s)$, which may change along the light path (with $\bar{c}$ being the average) and thus vary with s, while L denotes the total length of the light path. The absorption cross section, $\sigma(\lambda)$ is a characteristic property of any species, it can be measured in the laboratory, while the determination of the light path length, L is usually trivial in the case of active instruments, but requires radiation transport calculations for passive measurements (Platt and Stutz 2007). Once those quantities are known the path averaged trace gas concentration $\bar{c}$ and/or the column density S can be calculated from the measured spectrum $I(\lambda)$ according to Eqs. 1 and 2. For measurements, where the light path is in the atmosphere there will be usually more than one absorbing species present, thus a more comprehensive description of atmospheric absorption can be expressed as

$$I(\lambda) = I(\lambda) \cdot \exp\left(-L\left[\sum_i \bar{c}_i \cdot \sigma(\lambda) + \sigma_{R0} \cdot \lambda^{-4} \cdot c_{air} + \sigma_{M0} \cdot \lambda^{-n} \cdot c_{aer}\right]\right) \tag{3}$$

The quantities $I(\lambda)$ and $I_0(\lambda)$ have the meaning as defined in Eq. 1, $\sigma_i(\lambda)$ and c_i denote the absorption cross-section and the concentration of the ith species, and c_{air} the concentration of air molecules (2.4×10^{19} cm^{-3} at 20°C, 1 atm). The expressions $\sigma_{R0}(\lambda)\,\lambda^{-4}$ and $\sigma_{M0}\,\lambda^{-n}$ describe the effective wavelength dependence of the Rayleigh- and Mie- extinction, respectively (with n in the range of 1–3, depending on the aerosol size distribution), while c_{aer} denotes the average aerosol number density. At long wavelengths, i.e. in the microwave or infrared spectroscopy, Rayleigh- and Mie- scattering from aerosol are usually of minor importance (however, Mie-scattering in clouds can play a role). However, scattering processes can not be neglected in the UV/visible part of the spectrum. Grouped by wavelength (see Fig. 1.1), the techniques can be categorised as follows:

1.4.1 Microwave Spectroscopy

This spectral range is presently not used for tropospheric measurements, due to the relatively large pressure broadening of the lines and would require measurements at reduced pressure. However, spectroscopy of thermally emitted microwave radiation is an established technology to study stratospheric ozone and ClO (DeZafra et al. 1995; Janssen 1993).

1.4.2 IR Spectroscopy

Infrared spectroscopy is a technique in use for several decades, initially developed for the detection of atmospheric CO_2 by non dispersive instruments (URAS). More modern instruments are based on Fourier transform techniques to measure HNO_3, CH_2O, HCOOH, H_2O_2, and many other species in km-path lengths multiple-reflection cells (Pitts et al. 1977; Tuazon et al. 1980; Galle et al. 1994). The sensitivity is in the low ppb-range. Thus these instruments appear to be best suited for studies of polluted air. The technique can be applied in two modes of operation: (1) active operation, where an artificial light source is used (Pitts et al. 1977; Tuazon et al. 1980) or (2) passive operation using the thermal emission from the trace gases under consideration (Fischer et al. 1983; Clarmann et al. 1995).

In recent years, tunable diode laser spectrometers (TDLS) were developed to become field-usable instruments, successfully employed to measure HNO_3, NO, NO_2, CH_2O, H_2O_2 at sub-ppb levels (Harris et al. 1989; Schiff et al. 1990; Sigrist 1994; Tittel et al. 2003; Clemitshaw 2004). In the usual arrangement, the merit of TDLS coupled to a multipass gas cell lies in the mobility and sensitivity of instrument allowing concentration measurements on board of ships and aircraft. Limitations are due to the need to operate at low pressures (in most applications),thus introducing possible losses at the walls of the closed measurementcell. Furthermore the present diode-laser technology still remains complex.

1.5 Differential UV/Visible Absorption Spectroscopy

In the ultraviolet and visible wavelength ranges, electronic transitions of the trace gas molecules (or atoms) are observed. Like the other absorption–spectroscopic techniques, DOAS makes use of the characteristic absorption features of trace gas molecules along a path of known length in the open atmosphere. Thereby the problem of determining the true intensity $I_0(\lambda)$, as would be received from the light source in the absence of any extinction is solved by measuring the 'differential' absorption. It is defined as the part of the total absorption of any molecule 'rapidly' varying with wavelength and is readily observable as will be shown below. Accordingly, the absorption cross section of a given molecule (numbered *i*) is split into two portions:

$$\sigma_i(\lambda)=\sigma_i'(\lambda)+\sigma_{i0}(\lambda) \tag{4}$$

Where σ_{i0} varies only 'slowly' (i.e. essentially monotonously) with the wavelength λ, for instance describing a general 'slope', (e.g. Rayleigh- and Mie- scattering) while $\sigma_i(\lambda)$ shows rapid variations with λ, for instance due to an absorption line (see Fig. 1.2). The meaning of 'rapid' and 'slow' variation of the absorption cross section as a function of wavelength is, of course, a question of the observed wavelength interval and the width of the absorption bands to be detected. After introduction of Eq. 4 into Eq. 3, we obtain

$$\begin{aligned} I(\lambda)=I(\lambda)\cdot\exp\left(-L\left[\sum_i \bar{c}_i\cdot\sigma'(\lambda)\right]\right) \\ \cdot\exp\left(-L\left[\sum_i \bar{c}_i\cdot\sigma_0(\lambda)+\varepsilon_R(\lambda)+\varepsilon_M(\lambda)\right]\right)\cdot A(\lambda) \end{aligned} \tag{5}$$

where the first exponential function describes the effect of the structured 'differential' absorption of trace species, while the second exponential constitutes the slowly varying absorption of atmospheric trace gases as well as the influence of Rayleigh- and Mie-scattering (described by the wavelength-dependent extinction coefficients ε_R and ε_M) and the (slowly) wavelength-dependent transmission of the optical system used (summarised in the attenuation factor $A[\lambda]$).

Atmospheric trace gas concentrations are then calculated from the first exponential term in Eq. 5 using least squares fitting procedures as outlined by (Stutz and Platt 1997; Platt and Stutz 1996). The second exponential in Eq. 5 describing rather continuous extinction is usually neglected. Obviously DOAS can only measure species with reasonably narrow absorption features. Thus continuous absorptions of trace gases will be neglected by DOAS. On the other hand, DOAS is insensitive to extinction processes, which vary only monotonously with wavelength, like Mie-scattering by aerosol-, dust- or haze particles. Likewise slow variations in the spectral intensity of the light source or in the transmission of the optical system (telescope, spectrometer etc.) are also essentially eliminated.

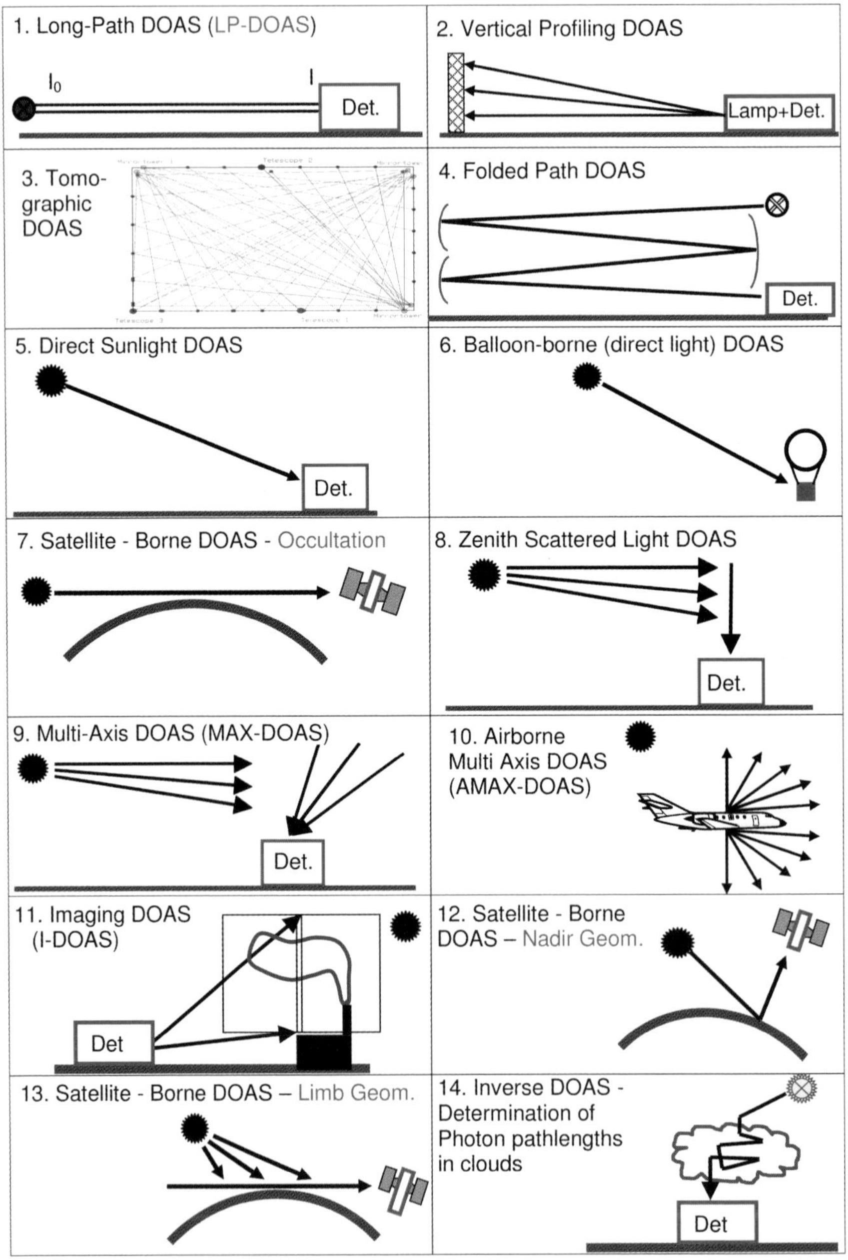

Fig. 1.3 The DOAS principle can be applied in a several light path arrangements and observation modes using artificial (arc lamps, incandescent lamps, or lasers, 1–4) as well as natural (sunlight or starlight; 5–14) light sources. Either the (light path averaged) trace gas concentration (1–4), the trace gas column density (5–13), or the length of the light path (e.g. in clouds, 14) can be determined

The DOAS principle (Platt 1994) has been applied in a wide variety of light path arrangements and observation modes as sketched in Fig. 1.3. The strength of DOAS lies in the absence of wall losses, good specificity, and the potential for real-time measurements. In particular, the first property makes spectroscopic techniques especially well suited for the detection of unstable species like OH radicals (Dorn et al. 1988; Brauers et al. 1996) or nitrate radicals (Platt et al. 1984; Platt and Janssen 1996; Allan et al. 1999).

Limitations of systems using a separate light source and receiving system are due to logistic requirements (the need for electric power at two sites separated by several kilometres, but in sight of each other) in the case of unfolded path arrangements (Fig. 1.3), also conditions of poor atmospheric visibility can make measurements with this technique difficult.

LIDAR techniques, on the other hand, combine the absence of wall losses and good specificity with fewer logistic requirements and the capability to make range-resolved measurements (while the above systems can only make point or path-averaged measurements). Unfortunately, this advantage is usually obtained at the expense of sensitivity.

1.6 Sample Applications of DOAS

DOAS applications encompass studies in urban air, measurements in rural areas, observations in the background troposphere as well as investigations of the distribution of stratospheric ozone and species leading to its destruction.

Using the DOAS technique, numerous new results of atmospheric chemistry could be obtained. For instance, the atmospheric concentration of several free radicals (such as OH, NO_3, BrO, and IO.) was determined. Further modern applications of the DOAS technique include the determination of the concentration of aromatic hydrocarbons and their degradation products in urban air (Etzkorn et al. 1999; Kurtenbach et al. 2002).

A growing field of DOAS application is the observation of trace gas concentration from space as implemented in the GOME instrument on ERS-2 (Burrows et al. 1999) and the SCIAMACHY sensor launched on ENVISAT in 2002 (Borrell et al. 2003). In addition, geometric light path lengths in clouds or haze could be determined (Noël et al. 1999).

An important early result obtained with DOAS was the first unambiguous detection of nitrous acid (HONO, [Perner and Platt 1979]) in urban air. Nitrous acid is produced from NO_2 and water at various types of surfaces. While many subsequent DOAS investigations confirmed that HONO levels rarely exceed 5% of the NO_2 it is, nevertheless, significant for atmospheric chemistry since its photolysis (HONO + $h\nu \rightarrow$ OH + NO) leads to the production of OH radicals, which in turn initiate most chemical degradation process of air pollutants.

The detection of OH radicals presented a major challenge for DOAS for a long time, since (daytime) atmospheric OH levels are of the order of 10^6 molecules/cm^3 (roughly 0.04 ppt). After early attempts (Perner et al. 1976) steady progress is being made (Platt et al. 1988; Dorn et al. 1988, 1996; Brauers et al. 1996).

Another radical species, the nitrate radical (NO_3,) was also discovered in the troposphere (Platt et al. 1980; Noxon et al. 1980; Allan et al. 1999) by DOAS techniques. Nitrate radicals (NO_3) are formed via oxidation of NO_2 by ozone ($NO_2 + O_3 \rightarrow NO_3 + O_2$). The nitrate radical is a strong oxidant initiating the degradation of many (unsaturated) hydrocarbons. Some of the oxidation products lead to the formation of organic peroxy- and HO_2 radicals, which in turn can yield OH radicals. Since the formation of NO_3 does not require sunlight, this OH source will also be active at night time.

Among the first indications for the involvement of chlorine species in the formation of the Antarctic ozone hole was the detection of OClO by ZSL-DOAS (Solomon et al. 1987). Also BrO could since be detected in stratospheric air (Solomon et al. 1989; Wahner et al. 1990; Richter et al. 1998; Otten et al. 1998).

Recently reactive halogen species could also be found in the troposphere: bromine monoxide (BrO) and ClO (Platt and Hausmann 1994; Tuckermann et al. 1997; Stutz et al. 2002), IO (Alicke et al. 1998; Allan et al. 2000; Saiz-Lopez et al. 2004) by using ground-based DOAS systems with artificial light sources.

The distribution of stratospheric BrO (Richter et al. 1998; Hegels et al. 1998) was mapped by satellite-borne DOAS (GOME instrument). However, also successful detection and mapping of tropospheric species by GOME was demonstrated in the cases of NO_2 (Leue et al. 1998; Richter and Burrows 2002; Beirle 2004), CH_2O, SO_2 (Eisinger and Burrows 1998; Khokhar et al. 2005), and BrO (Richter et al. 1998; Wagner and Platt 1998). Like in the stratosphere, halogen monoxide radicals lead to very efficient, catalytic ozone destruction. A very spectacular phenomenon caused by BrO (and possibly ClO) is the complete, episodic destruction of boundary layer ozone during polar spring (the 'polar tropospheric ozone hole') (Platt and Lehrer 1997). In addition the ability to map the global NO_2 distribution allows determining human activities (e.g. industrial and traffic related) as well as the extent of biomass burning. A sample distribution of tropospheric NO_2 is shown in Fig. 1.4.

By 'reversing' the usual DOAS approach (i.e. instead determining an unknown trace gas concentration at known light path length L, an unknown L is derived from the absorption of an absorber with known concentration) the average lengths of photon paths in clouds (see Fig. 1.3-1.4) could be determined by making use of the known concentrations of oxygen (O_2)-, tropospheric ozone-, or oxygen dimers $(O_2)_2$ (Erle et al. 1995; Wagner et al. 1998). By analysing the absorption of individual rotational lines (e.g. of the O_2 a-band around 765 nm) it is possible to infer not only the average photon-path length in clouds but also

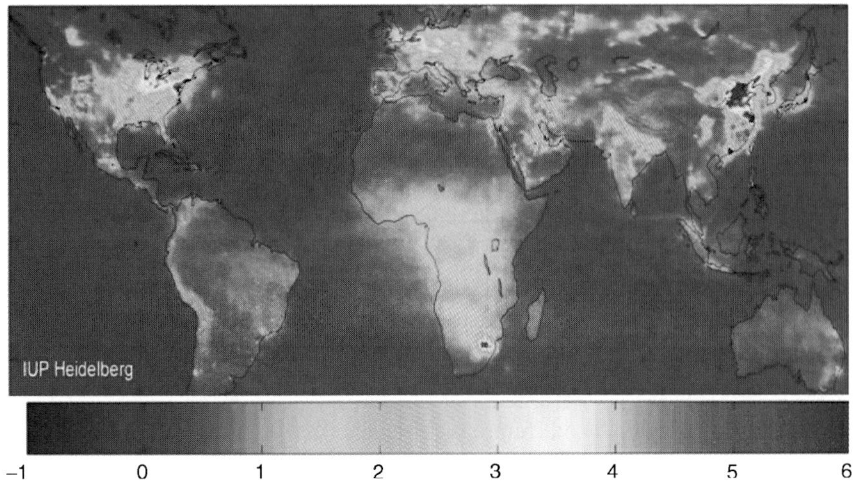

Fig. 1.4 The global NO_2 distribution (in units of 10^{15} molecules/cm^2), determined by the (SCIAMACHY) on the ENVISAT satellite. The data represent the tropospheric fraction of the total NO_2-column only, they are averaged over the period of Jan 2003–June 2004 (Beirle 2004). The industrial centres in Europe, North America, and Asia are clearly visible. Biomass burning plumes in equatorial America and Africa are less pronounced in the yearly average, but show up clearly in the 'burning seasons'

moments of its distribution (Pfeilsticker et al. 1998). These data give new insight into the internal structure and properties of the radiation field inside clouds.

1.7 Measurement Techniques: Tomorrow

At this point in time, it is interesting to speculate about the future of the measurement techniques. This can be approached from two directions: (1) by extrapolation, i.e. by extending present trends in instrumental design and development and (2) starting from the opposite direction by analysing the measurement requirements that might arise in future atmospheric chemistry research. Following the first approach evolutionary improvements are likely:

1. What can we expect?

Miniaturisation of instrumentation is a foreseeable trend, in particular in the electronics for the instruments, but we will also see applications of micro-mechanical devices, which are presently being developed. Further trends include:

Wider application of gas-chromatography (quadrupole) mass-spectrometry (GC–MS)

The following are the improvements in optical spectroscopy:

- Innovative passive DOAS spectrometers (e.g. multi axis-DOAS, MAX-DOAS [Hönninger et al. 2004]) or topographic target light scattering DOAS, ToTaL-DOAS (Frins et al. 2006)
- More compact DOAS instruments
- Application of tomographic techniques to determine the spatial distribution of trace gases (Hashmonay et al. 1999; Hartl et al. 2006; Pundt et al. 2005)
- Long-path infrared spectroscopy (LP-IR)
- Application of TDLS (mid-IR, near IR, UV for OH?)
- 'White light' LIDAR (South et al. 1998)
- Miniaturised, automated gas-chromatographs

2. What do we actually need (future requirements)?

Future research will require the study of new species with more compact, more universal instruments, which can be more readily calibrated. In particular, the measurement techniques for many free radicals are still not satisfactory (e.g. for RO_2 radicals or halogen radicals) or too difficult to use for routine measurements. In addition, modern chemistry—transport models cannot be tested because there are simply no techniques to observe the two- and three-dimensional distributions of trace gases on regional- or global scales. Thus a short list of requirements would include

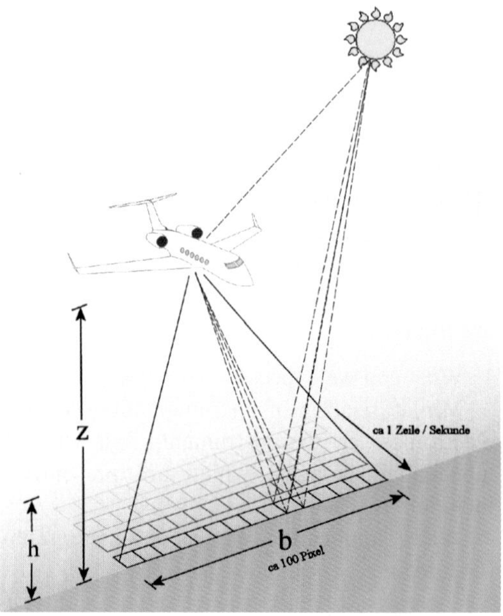

Fig. 1.5 Determination of 2D trace gas column density distributions (e.g. of NO_2, SO_2, CH_2O) in 'stripes' (≈10 km width) along the flight track

- Techniques for continuous hydrocarbon (VOC) measurements
- Instruments allowing detection of NO at mixing ratios <1 ppt
- Simple ozone monitoring instruments
- Simple sensors for the free radicals RO_2, HO_2, and OH.
- Techniques for the determination of the isotopic composition (with respect to e.g. $^{18}O/^{16}O$, D/H, $^{14}C/^{12}C$) of trace gases at ambient levels
- Measurement techniques for the sub-ppt detection of reactive halogen species (X, XO, OXO, HOX, where X = Cl, Br, I)
- Techniques that allow the mapping of two- and three-dimensional distributions of trace gases at high spatial resolution.
- Remote sensing (satellite-based) instruments which allow the global observation of trace gas distributions

In summary at present it can be said that the future development of atmospheric monitoring and research will to a large extent depend on the progress in instrumentation development. The recent years largely saw much evolutionary development but also new principles. (e.g. CRDS or MAX-DOAS).

Acknowledgements I would like to thank Christoph Kern for kindly providing Fig. 1.5. Also I am thankful to an anonymous reviewer for making helpful suggestions to improve the presentation of the manuscript.

References

Alicke B., Hebestreit K., Stutz J., and Platt U. (1998), Detection of iodine by DOAS in the marine boundary layer, *Nature*, 397, 572–573.

Allan B.J., Carlslaw N., Coe H., Burgess R.A., and Plane J.M.C. (1999), Observation of the nitrate radical in the marine boundary layer, *J. Atm. Chem.*, 33, 129–154.

Allan B.J., McFiggans G., and Plane J.M.C. Coe H. (2000), Observation of iodine monoxide in the remote marine boundary layer, *J. Geophys. Res.*, 105, 14363–14369.

Barrie L.A., Bottenheim J.W., Schnell R.C., Crutzen P.J., and Rasmussen R.A. (1988), Ozone destruction and photochemical reactions at polar sunrise in the lower Arctic atmosphere, *Nature*, 334, 138–141.

Beirle S. (2004), Estimating source strengths and lifetime of nitrogen oxides from satellite data. Dissertation, University of Heidelberg, Germany.

Borrell P., Burrows J.P., Richter A., Platt U., and Wagner T. (2003), New directions: New developments in satellite capabilities for probing the chemistry of the troposphere, *Atmos. Environ.*, 37, 2567–2570.

Brauers T., Aschmutat U., Brandenburger U., Dorn H. P., Hausmann M., Heßling M., Hofzumahaus A., Holland F., Plass-Dülmer C., and Ehhalt D. H. (1996), Intercomparison of tropospheric OH radical measurements by multiple folded long-path laser absorption and laser induced fluorescence, *Geophys. Res. Lett.*, 23, 2545–2548.

Burrows J.P., Weber M., Buchwitz M., Rozanov V., Ladstätter-Weißenmayer A., Richter A., DeBeek R., Hoogen R., Bramstedt K., Eichmann K.-U., Eisinger M., and Perner D. (1999), The global ozone monitoring experiment (GOME): Mission concept and first scientific results, *J. Atmos. Sci.*, 56, 151–171.

Cantrell C.A., Shetter R.E., Lind J.A., Mcdaniel A.H., Calvert J.G., Parrish D.D., Fehsenfeld F.C., Buhr M.P., and Trainer M. (1993), An improved chemical amplifier technique for peroxy radical measurements, *J. Geophys. Res.*, 98, 2897–2909.

Clarmann von T., Linden A., Oelhaf H., Fischer H., Friedlvallon F., Piesch C., and Seefeldner M. (1995), Determination of the stratospheric organic chlorine budget in the spring arctic vortex from MIPAS B limb emission spectra and air sampling, *J. Geophys. Res.*, 100, 13979–13997.

Clemitshaw K.C., Carpenter L.J., Penkett S.A., and Jenkin M.E. (1997), A calibrated peroxy radical chemical amplifier for ground-based tropospheric measurements, *J. Geophys. Res.*, 102, 25405–25416.

Clemitshaw K.C. (2004), A review of instrumentation and measurement techniques for ground-based and airborne field studies of gas-phase tropospheric chemistry, critical reviews in environmental science and technology, 34:1–108, *Taylor & Francis Inc.*, ISSN: 1064–3389, doi: 10.1080/10643380490265117.

DeZafra R.L., Reeves J.M., and Shindell D.T. (1995), Chlorine monoxide in the Antarctic spring vortex 1. Evolution of midday vertical profiles over McMurdo station, 1993, *J. Geophys. Res.*, 100, 13999–14007.

Dorn H.P., Callies J., Platt U., and Ehhalt D.H. (1988), Measurement of tropospheric OH concentrations by laser long-path absorption spectroscopy, *Tellus*, 40B, 437–445.

Dorn H.P., Brandenburger U., Brauers T., Hausmann M., and Ehhalt D.H. (1996), In- situ detection of tropospheric OH radicals by folded long-path laser absorption. Results from the POPCORN field campaign in August 1994, *Geophys. Res. Lett.*, 23, 2537–2540.

Drummond J.W., Volz A., and Ehhalt D.H. (1985), An optimized chemiluminescence detector for tropospheric NO measurements, *J. Atmos. Chem.*, 2, 287–306.

Eisinger M. and Burrows J.P. (1998), Tropospheric sulfur dioxide observed by the ERS-2 GOME instrument, *Res. Lett.*, 25, 4177–4180.

Erle F., Pfeilsticker K., and Platt U. (1995), On the influence of tropospheric clouds on zenith-scattered-light measurements of stratospheric species, *Geophys. Res. Lett.*, 22, 2725–2728.

Etzkorn T., Klotz B., Sörensen S., Patroescu I.V., Barnes I., Becker K.H., and Platt U. (1999), Gas-phase absorption cross sections of 24 monocyclic aromatic hydrocarbons in the uv and ir spectral ranges, *Atmos. Environ.*, 33, 525–540.

Fischer H., Fergg F., Oelhaf H., Rabus D., Voelker W., and Burkert P. (1983), Simultaneous detection of trace constituents in the middle atmosphere with a small He-cooled, high resolution Michelson interferometer (MIPAS), *Phys. Atmos.*, 56, 260–275.

Frins E., Bobrowski N., Platt U., and Wagner T. (2006), Tomographic MAX-DOAS observations of sun illuminated targets: A new technique providing well defined absorption paths in the boundary layer, *Appl. Optics*, 45(24), 6227–6240.

Galle B., Klemedtsson L., and Griffith D.W. (1994), Application of an FTIR for measurements of N_2O fluxes using micrometeorological methods, an ultralarge chamber system and conventional field chambers, *J. Geophys. Res.*, 99, 16575–16583.

Hartl A., Song B.C., and Pundt I. (2006), 2-D reconstruction of atmospheric concentration peaks from horizontal long path DOAS tomographic measurements: Parameterisation and geometry within a discrete approach, *Atmos. Chem. Phys.*, 6, 847–861.

Hashmonay R.A., Yost M.G., and Wu C.-F. (1999), Computed tomography of air pollutants using radial scanning path-integrated optical remote sensing, *Atmos. Environ.*, 33, 267–274.

Harris G.W., Mackay G.I., Iguchi T., Mayne L.K., and Schiff H.I. (1989), Measurements of formaldehyde in the troposphere by tunable diode laser absorption spectroscopy, *J. Atmos. Chem.*, 8, 119–137.

Hegels E., Crutzen P.J., Klüpfel T., Perner D., and Burrows P.J. (1998), Global distribution of atmospheric bromine monoxide from GOME on earth-observing satellite ERS 2, *Geophys. Res. Lett.*, 25, 3127–3130.

Hönninger G., Friedeburg C.V., and Platt U. (2004), Multi axis differential absorption spectroscopy (MAX-DOAS), *Atmos. Chem. Phys.*, 4, 231–254.

Janssen M.A. (1993), *An Introduction to the Passive Remote Atmospheric Remote Sensing by Microwave Radiometry*, John Wiley & Sons Inc., New York, pp.1–36.
Khokhar M.F., Frankenberg C., Van Roozendael M., Beirle S., Kühl S. Richter A., Platt U., and Wagner T. (2005), Satellite observations of atmospheric so_2 from volcanic eruptions during the time period of 1996 to 2002, *Advances in Space Res.*, 36 (5), 879–887.
Kurtenbach R., Ackermann R., Becker K.H., Geyer A., Gomes J.A.G., Lörzer J.C., Platt U., and Wiesen P. (2002), Verification of the contribution of vehicular traffic to the total NMVOC emissions in Germany and the importance of NO_3 chemistry in the city air, *J. Atmos. Chem.*, 42, 395–411.
Leue C., Wenig M., Platt U. (1998), Retrieval of atmospheric trace gas concentrations, *Handbook of Computer Vision and Applications,* Vol. III, Systems and Applications, B. Jähne, H. Haußecker, P. Geisler, Eds., Academic Press, San Diego, CA, 783–805.
Lowe D.C. and Schmidt U. (1983), Formaldehyde (HCHO) measurements in the non urban atmosphere. *J. Geophys. Res.*, 88, 10844–10858.
Noël S., Bovensmann H., Burrows J.P., Frerick J., Chance K.V., and Goede A.H.P. (1999), Global atmospheric monitoring with SCIAMACHY, *Phys. Chem. Earth (C)*, 24 (5), 427–434.
Mihelcic D., Muesgen P., and Ehhalt D.H. (1985), An improved method of measuring tropospheric NO_2 and RO_2 by matrix isolation and electron spin resonance, *J. Atmos. Chem.*, 3, 341–361.
Noxon J.F., Norton R.B., and Marovich E. (1980), NO_3 in the troposphere, *Geophys. Res. Lett.*, 7, 125–128.
Otten C., Ferlemann F., Platt U., Wagner T., and Pfeilsticker K. (1998), Ground-based DOAS UV/visible measurements at Kiruna (Sweden) during the SESAME winters 1993/94 and 1994/95, *J. Atmos. Chem.*, 30, 141–162.
Perner D., Ehhalt D. H., Pätz H.W., Platt U., Röth E. P., and Volz A. (1976), OH-radicals in the lower troposphere, *Geophys. Res. Lett.*, 3, 466–468.
Perner D. and Platt U. (1979), Detection of nitrous acid in the atmosphere by differential optical absorption, *Geophys. Res. Lett.*, 6, 917–920.
Perner D., Platt U., Trainer M., Hübler G., Drummond J.W., Junkermann W., Rudolph J., Schubert B., Volz A., Ehhalt D.H., Rumpel K.J., and Helas G. (1987), Measurement of tropospheric OH concentrations: A comparison of field data with model predictions, *J. Atmos. Chem.*, 5, 185–216.
Pfeilsticker K., Erle F., Funk O., Veitel H., and Platt U. (1998), First geometrical path lengths probability density function derivation of the skylight from spectroscopically highly resolved oxygen a-band observations. 1. Measurement technique, atmospheric observations and model calculations, *J. Geophys. Res.*, 103, 11483–11504.
Pitts J.N., Finlayson B.J., and Winer A.M. (1977), Optical systems unravel smog chemistry, *Environ. Sci. Technol.*, 11, 568–573.
Platt U., Perner D., Harris G.W., Winer A.M., and Pitts J.N. (1980), Detection of NO_3 in the polluted troposphere by differential optical absorption, *Geophys. Res. Lett.*, 7, 89–92.
Platt U., Winer A.M., Biermann H.W., Atkinson R., and Pitts J.N. (1984), Measurement of nitrate radical concentrations in continental air, *Environ. Sci. Technol.*, 18, 365–369.
Platt U., Rateike M., Junkermann W., Rudolph J., and Ehhalt D.H. (1988), New tropospheric OH measurements, *J. Geophys. Res.*, 93, 5159–5166.
Platt U. (1991), Spectroscopic measurement of free radicals (OH, NO_3) in the atmosphere, Fresenius *J. Anal. Chem.*, 340, 633–637.
Platt U. (1994), Differential optical absorption spectroscopy (DOAS). Air monitoring by spectroscopic techniques, M.W. Sigrist (Ed.), *Chemical Analysis Series*, Vol. 127, John Wiley & Sons Inc., New York, pp. 27–84.
Platt U. and Hausmann M. (1994), Spectroscopic measurement of the free radicals NO_3, BrO, IO, and OH in the troposphere, *Res. Chem. Intermed.*, 20, 557–578.
Platt U. and Lehrer E., Eds. (1997), *Arctic tropospheric ozone chemistry*, ARCTOC, Contract No. EV5V-CT93-0318 (DTEF), European Commission Air pollution research report No. 64, EUR 17783 EN July 1997, Office for Official Publications of the European Communities, Luxembourg, ISBN 92-828-2350-4.

Platt U. and Janssen C. (1996), Observation and role of the free radicals NO_3, ClO, BrO and IO in the troposphere, *Faraday Discuss*, 100, 175–198.
Platt U. (1999), Modern methods of the measurement of atmospheric trace gases, *J. Phys. Chem. Chem. Phys. PCCP*, 1, 5409–5415.
Platt U. and Stutz J. (2007), Differential Optical Absorption spectroscopy, Principles and Applications, *Springer*. (Heidelberg), in press.
Proffitt M.H. and McLaughlin R.J. (1983), Fast response dual-beam UV-absorption photometer suitable for use on stratospheric balloons, *Rev. Sci. Instrum.*, 54, 1719–1728.
Pundt I., Mettendorf K.-U., Laepple T., Knab V., Xie P., Lösch J., Friedeburg C.V., Platt U., and Wagner T. (2005), Measurements of trace gas distributions by long-path DOAS-tomography during the motorway campaign BAB II: Experimental setup and results for NO_2, (BAB II special issue) *Atmos. Environ.*, 39, 5, 967–975, doi:10.1016/j.atmosenv.2004.07.035.
Richter A., Wittrock F., Eisinger M., and Burrows J.P. (1998), GOME observation of tropospheric BrO in northern hemispheric spring and summer 1997, *Geophys. Res. Lett.*, 25, 2683–2686.
Richter A. and Burrows J.P. (2002), Retrieval of tropospheric NO_2 from GOME measurements, *Adv. Space Res.*, 29(11), 1673–1683.
Roscoe H.K. and Clemitshaw K.C. (1997), Measurement techniques in gas-phase tropospheric chemistry: A selective view of the past, present, and future, *Science*, 276, 1065–1072.
Saiz-Lopez A., Plane J.M.C., and Shillito J.A. (2004), Bromine oxide in the mid-latitude marine boundary layer, *Geophys. Res. Lett.*, 31, L03111, doi: 10.1029/2003GL018956.
Schiff H.I., Karecki D.R., Harris G.W., Hastie D.R., and Mackay G.I. (1990), A tunable diode laser system for aircraft measurements of trace gases, *J. Geophys. Res.*, 95, 10147–10154.
Sigrist M.W. (Ed.) (1994), Air monitoring by spectroscopic techniques, *Chemical Analysis Series*, Vol. 127, John Wiley & Sons, Inc.
Solomon S., Mount G., Sanders R.W., and Schmeltekopf A. (1987), Visible spectroscopy at McMurdo station, Antarctica: 2. Observation of OClO, *J. Geophys. Res.*, 92, 8329–8338.
Solomon S., Sanders R.W., Carroll M.A., and Schmeltekopf A.L. (1989), Visible and near-ultraviolet spectroscopy at McMurdo station, Antarctica: 5. Observations of the diurnal variations of BrO and OClO, *J. Geophys. Res.*, 94, 11393–11403.
South A.M., Povey I.M., and Jones R.L. (1998), Broadband lidar measurements of tropospheric water vapour profiles, *J. Geophys. Res.*, 103, 31191–31202.
Stutz J. and Platt U. (1996), Numerical analysis and error estimation of differential optical absorption spectroscopy measurements with least squares methods, *Appl. Optics*, 35, 6041–6053.
Stutz J., Ackerman R., Fast J.D., and Barrie L. (2002), Atmospheric reactive chlorine and bromine at the Great Salt lake, Utah, *Geophys. Res. Lett.*, 29, No. 10, 10.1029/2002GL014812, 18-1–18-4
Tittel F.K., Richter D., and Fried A. (2003), Mid-infrared laser applications in spectroscopy. Solid state mid-infrared laser sources, I.T Sorokina, and K.L. Vodopyanov (Eds), *Topics Appl. Phys.*, Springer Verlag, Berlin, Heidelberg, 89, 445–510.
Tuazon E.C., Winer A.M., Graham R.A., and Pitts J.N. (1980), Atmospheric measurements of trace pollutants by kilometer-pathlength FT-IR spectroscopy, *Environ. Sci. Technol.*, 10, 259–299.
Tuckermann M., Ackermann R., Gölz C., Lorenzen-Schmidt H., Senne T., Stutz J., Trost B., Unold W., and Platt U. (1997), DOAS-Observation of halogen radical-catalyzed Arctic boundary layer ozone destruction during the ARCTOC-campaigns 1995 and 1996 in Ny-Alesund, Spitsbergen, *Tellus*, 49B, 533–555.
Wagner T. and Platt U. (1998), Observation of tropospheric BrO from the GOME satellite, *Nature*, 395, 486–490.
Wagner T., Erle F., Marquard L., Otten C., Pfeilsticker K., Senne T.H., Stutz J., and Platt U. (1998), Cloudy sky photon path lengths as derived from DOAS observations, *J. Geophys. Res.*, 103, 25,307–325.
Wahner A., Callies J., Dorn H.-P., Platt U., and Schiller C. (1990), Near UV atmospheric absorption measurements of column abundances during airborne arctic stratospheric expedition, Jan-Feb. 1989: 3. BrO observations, *Geophys. Res. Lett.*, 17, 517–520.

Chapter 2
Radial Plume Mapping: A US EPA Test Method for Area and Fugitive Source Emission Monitoring Using Optical Remote Sensing

Ram A. Hashmonay[1], Ravi M. Varma[1], Mark T. Modrak[1], Robert H. Kagann[1], Robin R. Segall[2], and Patrick D. Sullivan[3]

Abstract This paper describes the recently developed United States Environmental Protection Agency (US EPA) test method that provides the user with unique methodologies for characterizing gaseous emissions from non-point pollutant sources. The radial plume mapping (RPM) methodology uses an open-path, path-integrated optical remote sensing (PI-ORS) system in multiple beam configurations to directly identify emission "hot spots" and measure emission fluxes. The RPM methodology has been well developed, evaluated, demonstrated, and peer reviewed. Scanning the PI-ORS system in a horizontal plane (horizontal RPM) can be used to locate hot spots of fugitive emission at ground level, while scanning in a vertical plane downwind of the area source (vertical RPM), coupled with wind measurement, can be used to measure emission fluxes. Also, scanning along a line-of-sight such as an industrial fenceline (one-dimensional RPM) can be used to profile pollutant concentrations downwind from a fugitive source. In this paper, the EPA test method is discussed, with particular reference to the RPM methodology, its applicability, limitations, and validation.

Keywords: Area fugitive emission sources, open-path fourier transform infrared (FTIR), open-path tunable diode laser absorption spectroscopy (TDLAS), optical remote sensing (ORS), radial plume mapping (RPM)

2.1 Introduction

Optical remote sensing (ORS) is a powerful technique for measuring air contaminant emissions from fugitive area sources (Walmsley and O'Connor 1996; Hashmonay and Yost 1999a; Gronlund et al. 2005). Under the auspices of

[1] *ARCADIS, 4915 Prospectus Drive Suite F, Durham, NC 27713, USA*

[2] *Emission Measurement Center (E143-02), Office of Air Quality Planning and Standards, US Environmental Protection Agency, Research Triangle Park, NC 27711*

[3] *Air Force Research Laboratory, Air Expeditionary Forces Technologies Division (AFRL/MLQF), 139 Barnes Drive, Suite 2, Tyndall AFB, FL 32403*

Y.J. Kim and U. Platt (eds.), *Advanced Environmental Monitoring*, 21–36.

the US Department of Defense's (DoD) Environmental Security Technology Certification Program (ESTCP) and the US Environmental Protection Agency (EPA), a radial plume mapping (RPM) methodology to directly characterize gaseous emissions from area sources has been demonstrated and validated, and a protocol has been developed and peer reviewed. This EPA "other test method" was made available for use on the US EPA website in July 2006.[1] The RPM-based methodologies use ORS techniques to collect path-integrated concentration (PIC) data from multiple beam paths in a plane and combine these with optimization algorithms to map the field of concentration across the plume of contaminant (Hashmonay et al. 1999; Hashmonay et al. 2001).

This test method currently describes three methodologies, each for a specific use. The horizontal radial plume mapping (HRPM) methodology was designed to map pollutant concentrations in a horizontal plane. This methodology is used to locate hot spots close to the ground. The vertical radial plume mapping (VRPM) methodology was designed to measure mass flux of pollutants through a vertical plane downwind from an emission source. VRPM utilizes multiple non-intersecting beam paths in a vertical plane downwind from the emission source to obtain a mass-equivalent plume map. This map, in conjunction with wind speed and direction, is used to obtain the flux of pollutants through the vertical plane. The measured flux is then used to estimate the emission rate of the upwind source being characterized. The one-dimensional (1D) RPM methodology (1D-RPM) was designed to profile pollutant concentrations along a line-of-sight (e.g., along an industrial site fenceline). The peak concentration position along the line-of-sight can be correlated with wind direction to estimate the location of an upwind fugitive emission source. The methodologies are independent of the particular PI-ORS system used to generate the PIC data.

Any scanning PI-ORS system that can provide PIC data may be considered for the purposes of the methodologies described in this test method and may include the following: open-path Fourier transform infrared (OP-FTIR) spectroscopy, ultraviolet differential optical absorption spectroscopy (UV-DOAS), open-path tunable diode laser absorption spectroscopy (TDLAS), and path-integrated differential absorption LIDAR* (PI-DIAL) (*LIDAR—light detection and ranging). The choice of instrument must be made based on its performance relative to the data quality objectives of the study. The OP-FTIR and UV-DOAS technologies are widely used due to their capability of simultaneous chemical detection for a large number of gas species of environmental interest. However, when only a few gas species are of interest, it may be more beneficial to employ other PI-ORS instrumentation, such as the TDLAS or PI-DIAL.

[1] See www.epa.gov/ttn/emc/tmethods.html. The "other test methods" category of the EPA Emission Measurement Center website includes test methods which have not yet been subject to the Federal rulemaking process. Each of these methods, as well as the available technical documentation supporting them, have been reviewed by the Emission Measurement Center staff and have been found to be potentially useful to the emission measurement community.

2.2 Methodologies

The RPM methodologies were validated by measuring emission fluxes from controlled gas releases simulating area sources. Examples from the validation experiments are presented and discussed. This method was applied and demonstrated at various sites for measuring emission fluxes (Modrak et al. 2004; Thoma et al. 2005). Typically, PIC data are collected over time, completing many cycles through the defined beams of each configuration. Because data are acquired sequentially, a moving average is required to reduce errors that originate from temporal variability. Typically, a moving average with a grouping of three cycles is sufficient to provide stable results. Once the PIC for all beam paths are averaged with the predetermined grouping of cycles for the gas species of interest, the RPM calculations make use of the information to reconstruct a plume map or profile.

2.2.1 HRPM Methodology—Hot Spot Source Location

The HRPM methodology is used to locate the source of fugitive emissions or hot spots. A rectangular area (which may be a square) is defined around the ground location where the suspected gaseous emissions are originating. Ideally, the HRPM configuration will cover the entire suspected source area; however, this may be prevented by equipment limitations or site conditions. Larger areas may need to be divided into smaller sections and studied separately.

The PI-ORS instrument is typically placed at the origin (in the first quadrant of the Cartesian convention) of the rectangular area to be measured (see Fig. 2.1). Once the HRPM measurement area and the number of path determining components (PDCs) have been determined the area is divided into smaller rectangular areas called pixels. The total number of pixels required is smaller or equal to the total number of beam paths. The methodologies here are not instrument specific. For ease of presentation, PDC is used to denote the component on the other end of the optical path from the PI-ORS instrument. Depending on the instrument selected, this could be a source, detector, mirror, or other reflecting object.

In Fig. 2.1, the survey area is divided into nine pixels of equal size. It should be noted that the survey area may be irregular in size, so that the resulting pixel grid is asymmetric (e.g., 2 by 4 pixels, 3 by 5 pixels, etc.). Each pixel will have at least one optical beam path that terminates within its boundaries at a PDC (a retro-reflecting mirror for most ORS instruments). This geometry maximizes the spread of the optical beams inside the area of emissions by passing one optical beam through the center of each pixel. It is recommended that a meteorological station is set up as part of the HRPM configuration. Although this data is not required for reconstructing the hot spot source location, wind speed and direction data may be helpful in interpreting the results. Dwelling time per PDC is determined by (1) the specific project goals,

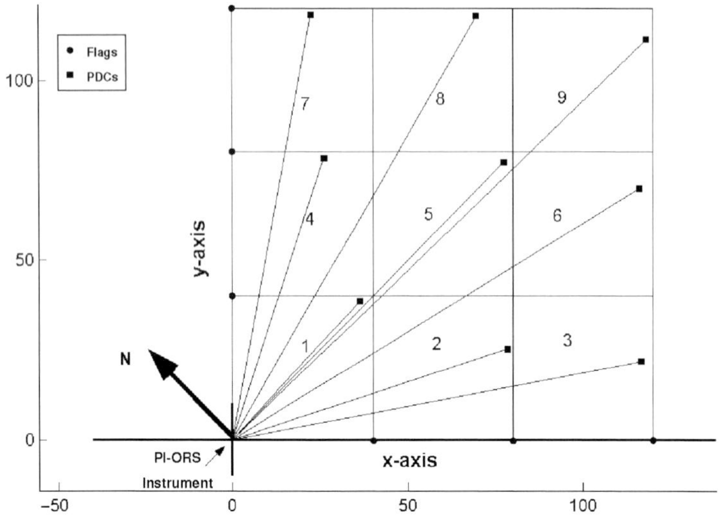

Fig. 2.1 Example of an HRPM configuration setup

and (2) the PI-ORS instrument-specific detection limits of the expected target gases. A recommended range for dwelling time per PDC is 10–60 s.

Once the PIC for all beam paths are averaged with the predetermined grouping of cycles for the gas species of interest, the HRPM calculations make use of the information to reconstruct a plume map over the area of interest. Average concentrations for each pixel can be obtained by applying an iterative algebraic deconvolution algorithm. The measured PIC, as a function of the field of concentration, is given by

$$PIC_k = \sum_m K_{km} c_m \tag{1}$$

where

K = a kernel matrix that incorporates the specific beam geometry with the pixel dimensions;

k = the number index for the beam paths;

m = the number index for the pixels; and

c = the average concentration in the mth pixel.

Each value in the kernel matrix K is the length of the kth beam within the mth pixel; therefore, the matrix is specific to the beam geometry. The HRPM procedure solves for the average concentrations (one for each pixel) by applying non-negative least squares (NNLS) algorithm (Lawson et al. 1995). The HRPM procedure multiplies the resulting column vector of averaged concentration by the matrix K to yield the end vector of predicted PIC data.

The second stage of the plume reconstruction involves interpolation among the reconstructed pixel's average concentration, providing a peak concentration not limited to the center of the pixels. A triangle-based cubic interpolation procedure (in Cartesian coordinates) is currently used in the HRPM procedure (Barber et al. 1996). The HRPM procedure provides a plume map and calculates the location of the peak concentrations. It is up to the user to interpret this information, taking into consideration site-specific characteristics such as obstructions or terrain complexities, to determine the actual location of the hot spot.

2.2.2 VRPM Methodology—Estimation of Emission Rate

The VRPM methodology is used to estimate the rate of gaseous emissions from an area fugitive source. A vertical scanning plane, downwind of the source, is used to directly measure the gaseous flux. Two different beam configurations of the VRPM methodology are recommended: the five-beam (or more) and the three-beam VRPM configuration. Figure 2.2 illustrates the setup for these two VRPM beam configurations. In the five-beam (or more) configuration, the ORS instrument sequentially scans over five PDCs. Three PDCs are along the ground-level

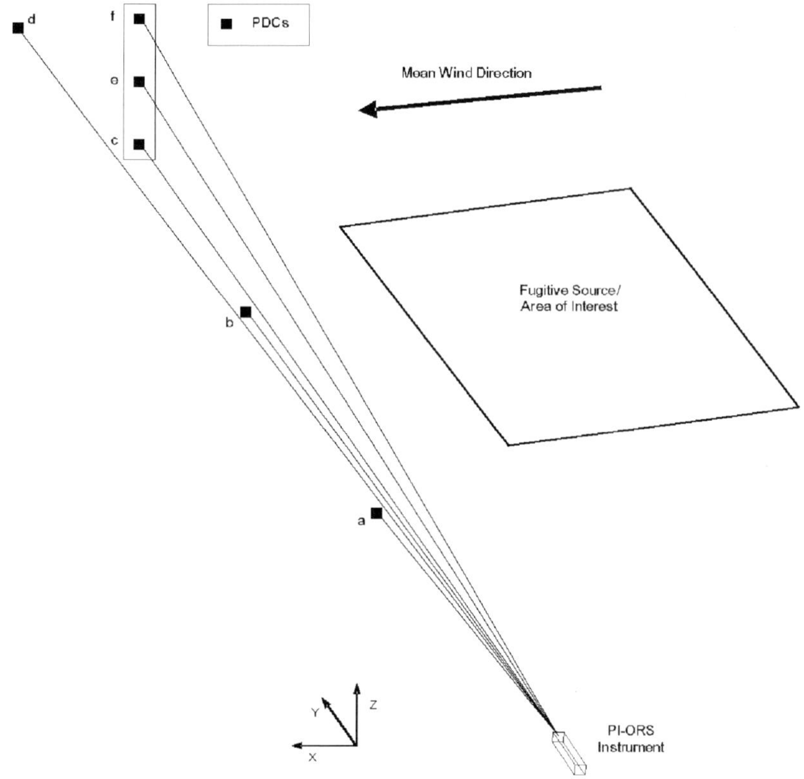

Fig. 2.2 Example of a VRPM configuration setup

crosswind direction (beams a, b, and c in Fig. 2.2), and the other two are elevated on a vertical structure (beams e and f in Fig. 2.2). The additional beam (d) in Fig. 2.2 is for the six-beam configuration, which provides better spatial definition of the plume in the crosswind direction. In the three-beam configuration, the ORS instrument sequentially scans over three PDCs. Only one beam is along the ground level (beam c or d in Fig. 2.2) and the other two are elevated on a vertical structure (beams e and f in Fig. 2.2).

A two-phase smooth basis function minimization (SBFM) approach is applied where there are three or more beams along the ground level (five-beam, or more, configuration). In the two-phase SBFM approach, a one-dimensional SBFM (1D-SBFM) reconstruction procedure is first applied in order to reconstruct the smoothed ground level and crosswind concentration profile. The reconstructed parameters are then substituted into the bivariate Gaussian function when applying a two-dimensional SBFM (2D-SBFM) procedure.

A 1D-SBFM reconstruction is applied to the ground level segmented beam paths (Fig. 2.2) of the same beam geometry to find the crosswind concentration profile. A univariate Gaussian function is fitted to measured PIC ground-level values.

The error function for the minimization procedure is the sum of squared errors (SSE) function and is defined in the 1D-SBFM approach as follows:

$$SSE(B_j, m_{y_j}, \sigma_{y_j}) = \sum_i \left(PIC_i - \sum_j \frac{B_j}{\sqrt{2\pi}\sigma_{y_j}} \int_0^{r_i} \exp\left[-\frac{1}{2}\left(\frac{m_{y_j} - r}{\sigma_{y_j}} \right)^2 \right] dr \right)^2 \quad (2)$$

where

B = equal to the area under the one-dimensional Gaussian distribution (integrated concentration);

r_i = the pathlength of the ith beam;

m_y = the mean (peak location);

σ_y = the standard deviation of the jth Gaussian function; and

PIC_i = the measured PIC value of the ith path

The SSE function is minimized using the simplex minimization procedure to solve for the unknown parameters (Press et al. 1992). When there are more than three beams at the ground level, two Gaussian functions are fitted to retrieve skewed and sometimes bi-modal concentration profiles. This is the reason for the index j in Eq. 2.

Once the one-dimensional phase is completed, the two-dimensional phase of the two-phase process is applied. To derive the bivariate Gaussian function used in the second phase, it is convenient to express the generic bivariate function G in polar coordinates r and θ:

$$G(r,\theta) = \frac{A}{2\pi\sigma_y\sigma_z\sqrt{1-\rho_{12}^2}} \exp\left\{ -\frac{1}{2(1-\rho_{12}^2)} \left[\frac{(r\cdot\cos\theta - m_y)^2}{\sigma_y^2} - \frac{2\rho_{12}(r\cdot\cos\theta - m_y)(r\cdot\sin\theta - m_z)}{\sigma_y\sigma_z} + \frac{(r\cdot\sin\theta - m_z)^2}{\sigma_z^2} \right] \right\} \quad (3)$$

The bivariate Gaussian has six unknown independent parameters:

A = normalizing coefficient which adjusts for the peak value of the bivariate surface;

ρ_{12} = correlation coefficient which defines the direction of the distribution-independent variations in relation to the Cartesian directions y and z ($\rho_{12} = 0$ means that the distribution variations overlap the Cartesian coordinates);

m_y and m_z = peak locations in Cartesian coordinates; and

σ_y and σ_z = standard deviations in Cartesian coordinates.

Six independent beam paths are sufficient to determine one bivariate Gaussian that has six independent unknown parameters. Some reasonable assumptions are made when applying the VRPM methodology to this problem to reduce the number of unknown parameters. The first is setting the correlation parameter ρ_{12} equal to zero. This assumes that the reconstructed bivariate Gaussian is limited only to changes in the vertical and crosswind directions. Secondly, when ground level emissions are known to exist, the ground level PIC is expected to be the largest of the vertical beams. Therefore, the peak location in the vertical direction can be fixed to the ground level. In the above ground-level scenario, Eq. 3 reduces into Eq. 4:

$$G(r,\theta) = \frac{A}{2\pi\sigma_y\sigma_z} \exp\left\{-\frac{1}{2}\left[\frac{(r\cdot\cos\theta - m_y)^2}{\sigma_y^2} + \frac{(r\cdot\sin\theta)^2}{\sigma_z^2}\right]\right\} \tag{4}$$

The standard deviation and peak location retrieved in the 1D-SBFM procedure are substituted in Eq. 4 to yield

$$G(A,\sigma_z) = \frac{A}{2\pi\sigma_{y-1D}\sigma_z} \exp\left\{-\frac{1}{2}\left[\frac{(r\cdot\cos\theta - m_{y-1D})^2}{\sigma_{y-1D}^2} + \frac{(r\cdot\sin\theta)^2}{\sigma_z^2}\right]\right\} \tag{5}$$

where

$\sigma_{y\text{-}1D}$ = standard deviation along the crosswind direction (found in the 1D-SBFM procedure);

$m_{y\text{-}1D}$ = peak location along the crosswind direction (found in the 1D-SBFM procedure);

A and σ_z are the unknown parameters to be retrieved in the second phase of the fitting procedure. An error function (SSE) for minimization is defined for this phase in a similar manner. The SSE function for the second phase is defined as

$$SSE(A,\sigma_z) = \sum_i \left(PIC_i - \int_0^{r_i} G(r_i,\theta_i,A,\sigma_z)\,dr \right)^2 \tag{6}$$

The SSE function is minimized using the simplex method to solve for the two unknown parameters.

When the VRPM configuration consists only of three-beam paths—one at the ground level and the other two elevated—the one-dimensional phase can be skipped, assuming that the plume is very wide. In this scenario, peak location can be arbitrarily assigned to be in the middle of the configuration. Therefore, the three-beam VRPM configuration is most suitable for area sources or for sources with a series of point and fugitive sources that are known to be distributed across the upwind area. In this case, the bivariate Gaussian has the same two unknown parameters as in the second phase (Eq. 5), but information about the plume width or location is not known. The standard deviation in the crosswind direction is typically assumed to be about ten times that of the ground level beam path (length of vertical plane). If r_1 represents the length of the vertical plane, the bivariate Gaussian would be as follows:

$$G(A,\sigma_z) = \frac{A}{2\pi(10r_1)\sigma_z}\exp\left\{-\frac{1}{2}\left[\frac{(r\cdot\cos\theta - \frac{1}{2}r_1)^2}{(10r_1)^2} + \frac{(r\cdot\sin\theta)^2}{\sigma_z^2}\right]\right\} \tag{7}$$

This process is for determining the vertical gradient in concentration. It allows an accurate integration of concentrations across the vertical plane as the long-beam ground-level PIC provides a direct integration of concentration at the lowest level.

Once the parameters of the function are found for a specific run, the VRPM procedure calculates the concentration values for every square elementary unit in a vertical plane. Then, the VRPM procedure integrates the values, incorporating wind speed data at each height level to compute the flux. The concentration values are converted from parts per million by volume (ppmv) to grams per cubic meter (g/m^3), taking into consideration the molecular weight of the target gas. This enables the direct calculation of the flux in grams per second (g/s), using wind speed data in meters per second (m/s).

As described in earlier studies (Hashmonay et al. 2001), the concordance correlation factor (CCF) was used to represent the level of fit for the reconstruction in the path-integrated domain (predicted versus measured PIC). CCF is defined as the product of two components:

$$\mathrm{CCF} = rA \tag{8}$$

where

r = the Pearson correlation coefficient and

A = a correction factor for the shift in population and location.

This shift is a function of the relationship between the averages and standard deviations of the measured and predicted PIC vectors:

$$A = \left[\frac{1}{2}\left(\frac{\sigma_{PIC_P}}{\sigma_{PIC_M}} + \frac{\sigma_{PIC_M}}{\sigma_{PIC_P}} + \left(\frac{\overline{PIC_P} - \overline{PIC_M}}{\sqrt{\sigma_{PIC_P}\sigma_{PIC_M}}}\right)^2\right)\right]^{-1} \tag{9}$$

where

σ_{PIC_P} = standard deviation of the predicted PIC vector;

σ_{PIC_M} = standard deviation of the measured PIC vector;

$\overline{PIC_P}$ = the mean of the predicted PIC vector; and

$\overline{PIC_P}$ = the mean of the measured PIC vector.

The Pearson correlation coefficient is a good indicator of the quality of fit to the Gaussian mathematical function. In this procedure, typically an r close to 1 will be followed by an A very close to 1. This means that the averages and standard deviations in the two concentration vectors are very similar and the mass is conserved (good flux value). However, when a poor CCF is reported (CCF <0.80) at the end of the fitting procedure it does not directly mean that the mass is not conserved. It could be a case where only a poor fit to the Gaussian function occurred if the correction factor A was still very close to 1 ($A > 0.90$). However, when both r and A are low one can assume that the flux calculation is inaccurate.

2.2.3 1D-RPM Methodology—Line-of-Sight Profile Concentrations and Upwind Source Location Estimation

The 1D-RPM methodology is used to profile pollutant concentrations along a line-of-sight downwind of a fugitive emission source. This pollutant concentration profile can be combined with wind data to estimate the location of an upwind source, when applicable (Hashmonay and Yost 1999b). The scanning PI-ORS instrument and three or more PDC are placed in a crosswind direction along a line, such as an industrial site fenceline, and PIC measurements are made. The 1D-RPM configuration setup is shown in Fig. 2.3. A minimum of three PDCs are needed, but four to six are recommended for this configuration. The additional PDCs provide a more detailed concentration profile. PDCs should be placed on the line-of-sight, with an equal distance between each subsequent PDC, if possible (see Fig. 2.3).

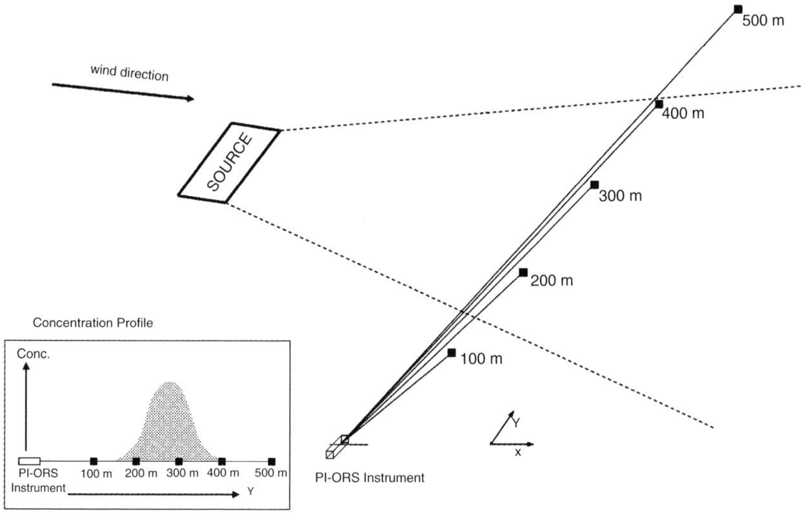

Fig. 2.3 Example of a 1D-RPM configuration setup

Once the PIC for all beam paths are averaged with the predetermined grouping of cycles for the gas species of interest, the 1D-RPM calculations make use of the information to reconstruct a plume concentration profile along the measurement line-of-sight. Similar to the case of VRPM, the 1D-RPM calculations utilize the 1D-SBFM to reconstruct a mass-equivalent plume concentration profile along the line-of-sight measurement.

The 1D-RPM procedure fits a univariate Gaussian function to measured PIC ground-level values. The error function for the minimization procedure is the SSE function, and is defined in Eq. 2. The 1D-RPM procedure reconstructs the plume profile along the measurement line-of-site and notes the peak location. Over time, as the wind direction fluctuates, different peak locations are reconstructed from the PIC measurements as illustrated in Fig. 2.4. Each time a peak location is noted, a source projection line is drawn for each peak location. This is done by calculating a line equation through the peak location, with the same orientation as the averaged wind direction for the same measurement time interval. Ideally, for a stationary point source, all source projection lines drawn over time should intersect at a point upwind of the measurement line in the vicinity of the real emission source location. Calculating the density of lines per unit area upwind from the measurement plane, the most likely location of the source can be estimated as the region of the maximal line density.

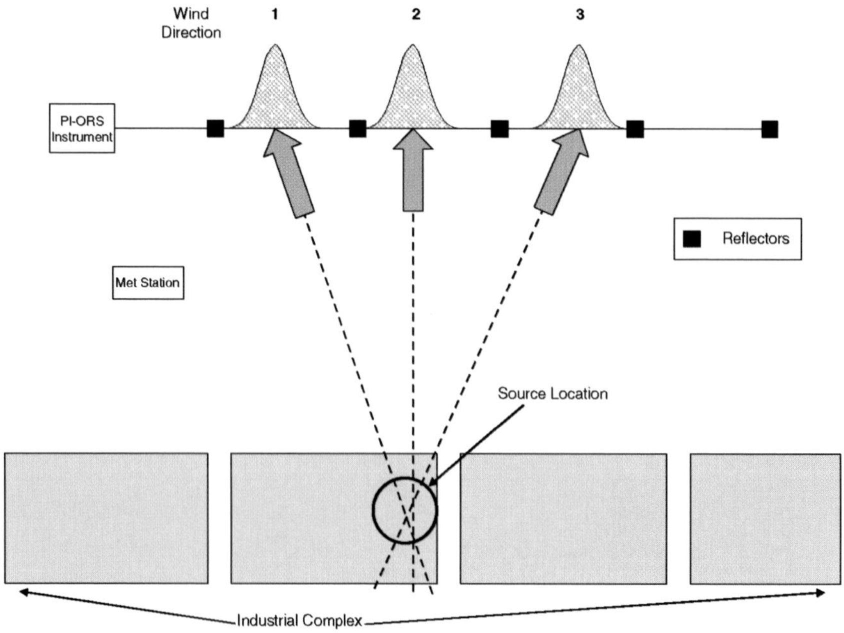

Fig. 2.4 An Example of a 1D-RPM fenceline monitoring setup

2.3 Example of Methodologies Validation using OP-FTIR

During the Fall of 2002 and Spring of 2003, the methodologies described in the EPA protocol were evaluated using the controlled release of verification gases and a scanning monostatic OP-FTIR instrument. The experiments were performed at the Duke Forest Facility of Duke University, located in Chapel Hill, North Carolina. A nearly flat and square area of 14,400 m^2 was selected for the study. The verification gases released during the experiments included ethylene and acetylene. This example briefly describes the preliminary results obtained from this controlled demonstration study using the HRPM and VRPM methodologies. This study was supported by the ESTCP of the DoD and the Emissions Monitoring Center (EMC) of the EPA.

In this study, an OP-FTIR mounted on a scanner and an array of mirrors (PDCs) formed multiple, non-intercepting beam paths. Nine mirrors were used for the HRPM experiment, and five were used for the VRPM experiment. The OP-FTIR was scanned sequentially from mirror to mirror, acquiring spectra from each beam path. The OP-FTIR dwelling time on each mirror was 30 s for the HRPM experiment and 60 s for the VRPM experiment.

2.3.1 *HRPM Results*

Several verification gases were released at random locations and spectral data were collected by sequentially scanning the OP-FTIR from mirror to mirror. This spectral data were used to derive PIC data along each beam path, which were then input into the RPM software to construct a surface concentration contour map. The hot spot location was determined by the area of highest concentrations in the contour map. This location was then compared to the actual release location of the verification gas.

For this example, acetylene was released and data were collected for 18 cycles. Fig. 2.5 shows the concentration contour map constructed by averaging PIC data from all 18 cycles. The open circle indicates the actual release location, and the closed circle indicates the hot spot location determined by the RPM software. The dislocation distance (distance between the actual and determined release location) is shown with a double-headed arrow. In this case, the dislocation distance was 17 m.

2.3.2 *VRPM Results*

Several verification gases were released in an “H” pattern upwind of the VRPM configuration to simulate an area source in Duke Forest. The verification gas cylinders were weighed before and after each release to calculate the actual release rate (in g/s). Spectral data were collected by sequentially scanning the OP-FTIR from

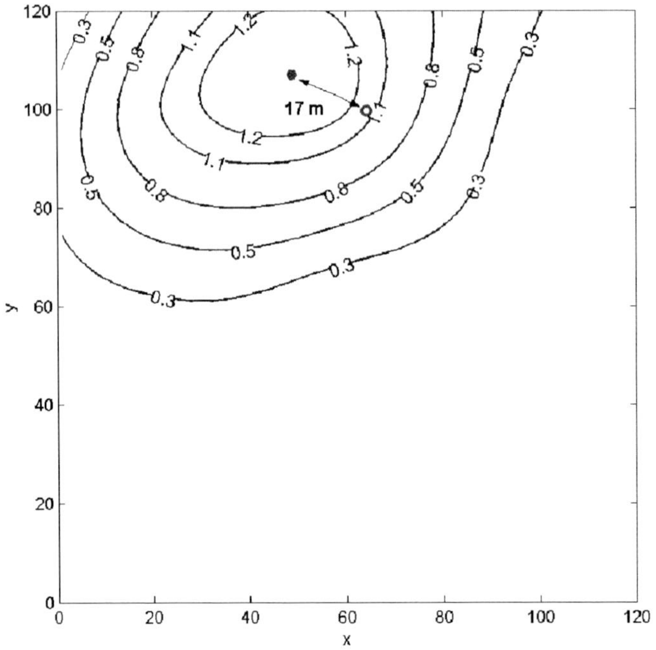

Fig. 2.5 Actual versus determined source location for the acetylene release

mirror to mirror. These data were used to derive PIC data along each beam path, which were used to reconstruct the plume map in the vertical plane (see Fig. 2.6). Wind data were simultaneously collected at two heights (2 and 10 m), and were incorporated with the plane-integrated concentration data to calculate the emission flux through the VRPM configuration.

For this example, ethylene was released and data were collected for 12 cycles. Fig. 2.6 shows the reconstructed mass-equivalent plume map constructed by averaging PIC data from all 12 cycles. A total of 1.0 lb of ethylene was released over a period of 68 min, which resulted in an average release rate of 0.11 g/s. The average calculated emission flux was 0.10 g/s.

Wind measurements were interpolated for every 2 m between the 1 and 9 m levels (wind data remain constant beyond 10 m). The wind direction is measured clockwise from an axis perpendicular to the VRPM configuration. In this convention, 0° is perpendicular to the vertical plane.

A moving average is used in the calculation of average values to show temporal variability in the measurements. A moving average involves averaging values from several different consecutive cycles. For example, a data set may be using a moving average with a group size of four consecutive cycles, where concentration and wind values from cycles one to four, and two to five and so on are averaged together prior to application of the VRPM algorithm.

The following example explores the dependence of the results on the group size of the moving average, and is illustrated in Fig. 2.7. To assess the accuracy of the reconstruction for each moving average group, the CCF is computed. It is apparent in Fig. 2.7 that as the group size of the moving averages increase, the standard deviation decreases (smaller error bar). The CCF value increases for larger group sizes of moving averages (indicating a better fit for the measured data), but levels off at a group size of about three. Therefore, a moving average group of three is recommended as a starting point for the VRPM methodology.

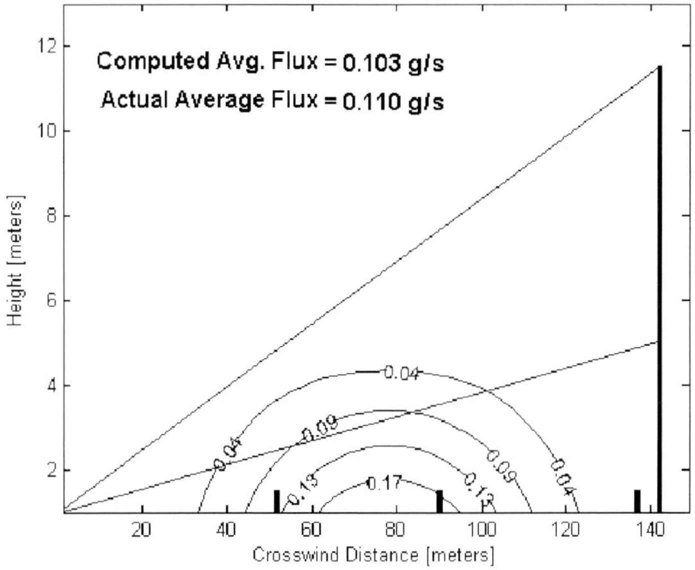

Fig. 2.6 Flux measurement for ethylene

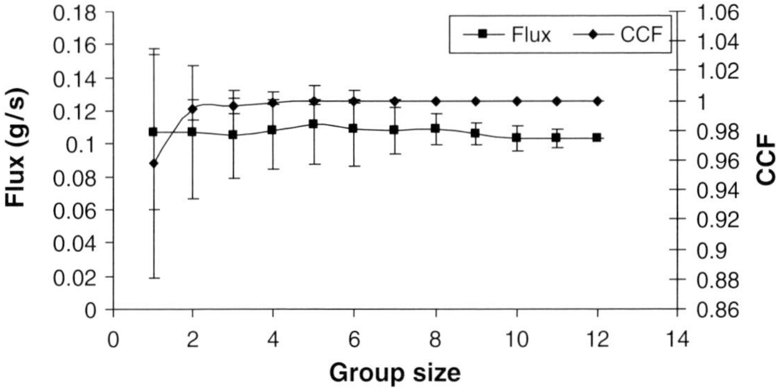

Fig. 2.7 Dependence of flux estimation on moving average group size

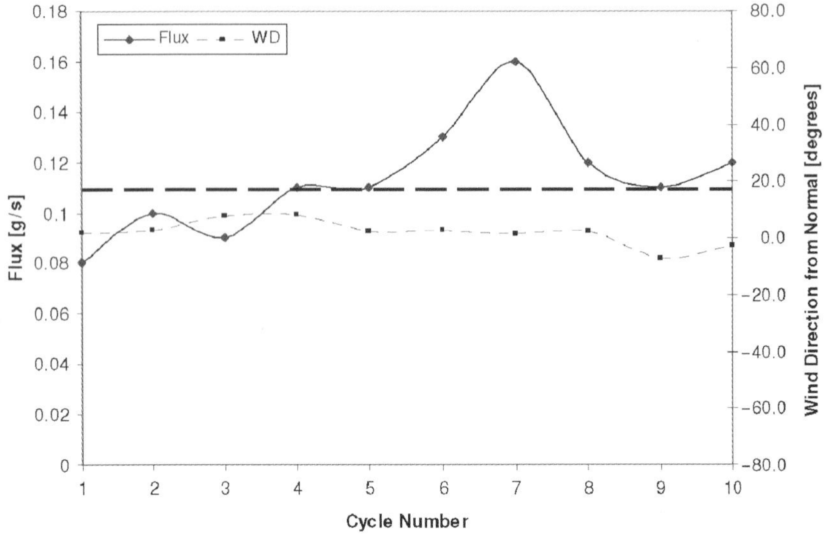

——— Indicates the actual release rate

Fig. 2.8 Time series of ethylene emission flux estimation

The total emission flux calculation depends on the wind direction perpendicular to the VRPM configuration. A time plot of the flux calculation for the ethylene verification release, using a moving average with a group size of 3, is shown in Fig. 2.8. The dashed horizontal line corresponds to the actual mass released; the dashed plot indicates the measured wind direction from perpendicular to the VRPM configuration; and the solid plot indicates the calculated emission flux values.

2.3.3 Horizontal Plume Capture Determination

The example below of flux data from the Duke Forest study provides guidance for determining the range of wind directions that will ensure complete capture of the plume (in the horizontal direction) by the VRPM configuration. The VRPM flux results are representative of the source emission rate only during periods that the prevailing wind directions meet the established criterion.

The determination of the acceptable wind criterion is done by analyzing a data set of calculated flux values, with the corresponding wind direction, at the time of the flux calculation. The data set should consist of data collected over a wide range of wind directions on both sides of the perpendicular direction to the VRPM configuration. A plot is created of the calculated flux values as a function of the angle of the prevailing wind direction (from perpendicular to the VRPM configuration) of all runs completed. Because these runs were collected over a period of 6 months, the flux values were normalized to the maximum flux value in each run.

Figure 2.9 presents a plot of this data set. The figure shows the calculated normalized flux values and the corresponding wind direction (from the normal to the

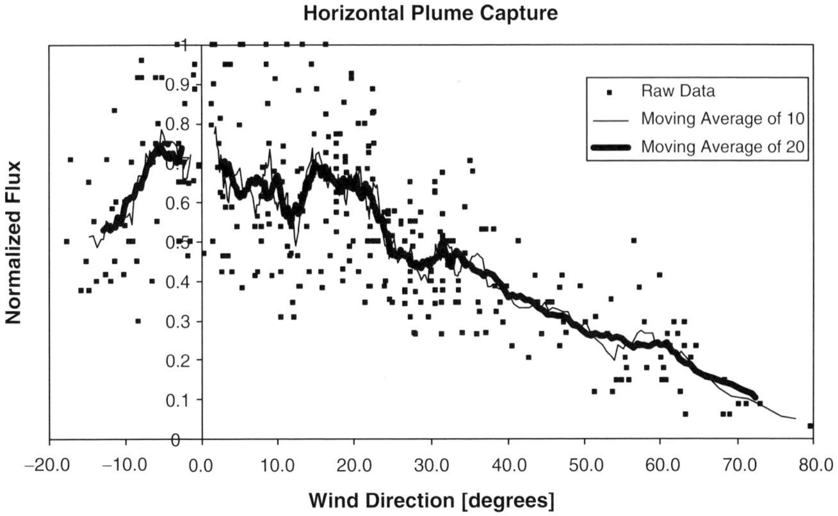

Fig. 2.9 Plot of calculated flux values and corresponding wind direction during the time of measurements

VRPM configuration during the time of the measurements). The figure shows that the raw flux values steadily increase, up to a certain point, as the prevailing wind direction becomes closer to perpendicular to the configuration. However, over a certain range of wind directions, there appears to be no relationship between calculated flux values and prevailing wind direction. This range is approximately between $-10°$ and $+25°$ for the Duke Forest validation study. This is the approximate range of acceptable wind directions, as the variations in flux values in this range are due to variations in the source strength, and not due to changes in wind direction. It is recommended that these flux plots be smoothed out with a moving average approach to retrieve a more definite and accurate cutoff of the wind criterion. The lined curves in Fig. 2.9 show examples of such moving averages, one with a grouping of ten data values (thin line) and the other with a grouping of 20 data values (thicker line). These examples of moving averages clearly define the range of wind directions where flux values are independent from the wind direction values.

2.4 Conclusion

The US EPA "other test method 10 (OTM-10)" described in this paper is a result of an extensive research, validation, and demonstration effort. As shown in the example above, the validation results successfully met quality objectives for all the three methodologies. For the HRPM validation the distance of the reconstructed "hot spot" location from the actual release location was within 10% of the rectangle diagonal in all completed runs (total of nine random release locations). For the VRPM validation, the average flux for all runs was within 25% of the actual release rate (when wind direction met the plume capture criterion) for eight completed runs (four 5-beam configurations and four 3-beam configurations, with the three recommended dwelling times of 10 s, 30 s, and 60 s on each mirror). A detailed description of the VRPM is in preparation for peer-reviewed publication (Hashmonay et al. 2007). Validation data for the 1D-RPM methodology is reported in Hashmonay and Yost (1999b).

The choice of PI-ORS system to be used for the collection of measurement data (and subsequent calculation of PIC) is left to the discretion of the user, and should be dependent on the compounds of interest and the purpose of the study. Basic knowledge of a PI-ORS system and the ability to obtain quality PIC data is assumed. The user must be capable of using commercial software to utilize the RPM-related procedures and algorithms explained in this methodology.

References

Barber C.B., Dobkin D.P., and Huhdanpaa H.T. (1996, December), The Quickhull Algorithm for Convex Hulls, *ACM T. Math Software,* 22(4), 469–483.

Gronlund R., Sjoholm M., Weibring P., Edner H., and Svanberg S. (2005), Elemental Mercury Emissions from Chlor-Alkali Plants Measured by LIDAR Techniques, *Atmos. Environ.*, 39, 7474–7480.

Hashmonay R.A. and Yost M.G. (1999a), Innovative Approach for Estimating Gaseous Fugitive Fluxes Using Computed Tomography and Remote Optical Sensing Techniques, *J. Air Waste Manage. Assoc.*, 49, 966–972.

Hashmonay R.A. and Yost M.G. (1999b), Localizing Gaseous Fugitive Emission Sources by Combining Real-Time Remote Optical Sensing and Wind Data, J. *Air Waste Manage. Assoc.*, 49, 1374–1379.

Hashmonay R.A., Yost M.G., and Wu C.F. (1999), Computed Tomography of Air Pollutants Using Radial Scanning Path-Integrated Optical Remote Sensing, *Atmos. Environ.*, 33(2), 267–274.

Hashmonay R.A., Natschke D.F., Wagoner K., Harris D.B., Thompson E.L., and Yost M.G. (2001), Field Evaluation of a Method for Estimating Gaseous Fluxes from Area Sources Using Open-Path Fourier Transform Infrared, *Environ. Sci. Technol.*, 35, 2309–2313.

Hashmonay R.A., Varma R.M., Modrak M.T., Kagann R.H., Egler, K.D., Sullivan, P.D., and Segall R.R. (2007), Validation of Vertical Radial Plume Mapping Methodology for Estimating Gaseous Emission Rates from Area Sources, submitted to *J. Air Waste Manage. Assoc.*

Lawson C.L. and Janson R.J. (1995), Solving Least Squares Problems, (*In Soc. Ind. Appl. Math.,* Philadelphia, chapter 23, pp. 158–165.

Modrak M.T., Hashmonay R.A., and Kagann R.H. (2004, January), *Measurement of Fugitive Emissions at a Region I Landfill.* U.S. Environmental Protection Agency, Research and Development, EPA-600/R-04-001.

Press, W.H., Teukolsky, S.A., Vetterling, W.T., and Flannery, B.P. (1992). *Numerical Recipes in FORTRAN*, 2nd ed. (Cambridge University Press: Cambridge, MA)

Thoma E.D., Shores R.C., Thompson E.L., Harris D.B., Thornloe S.A., Varma R.M., Hashmonay R.A., Modrak M.T., Natschke D.F., and Gamble H.A. (2005), Open Path Tunable Diode Laser Absorption Spectroscopy for Acquisition of Fugitive Emission Flux Data, J. *Air Waste Manage. Assoc.*, *55*, 658–668.

Walmsley H.L. and O'Connor S.J. (1996), *The Use of Differential Absorption LIDAR to Measure Atmospheric Emission Rates at Industrial Facilities.* In Proceedings of the A&WMA International Conference on Optical Sensing for Environmental and Process Monitoring, Dallas, TX, p. 127.

Chapter 3
MAX-DOAS Measurements of ClO, SO_2 and NO_2 in the Mid-Latitude Coastal Boundary Layer and a Power Plant Plume

Chulkyu Lee[1,2], Young J. Kim[1], Hanlim Lee[1], and Byeong C. Choi[3]

Abstract Remote sensing techniques have been preferred for measurements of atmospheric trace gases because they allow direct measurement without pre- and/or post-treatment in the laboratory. UV–visible absorption measurement techniques have been used for ground-based remote sensing of atmospheric trace species. The multi-axis differential optical absorption spectroscopy (MAX-DOAS) technique, one of the remote sensing techniques for air quality measurement, uses scattered sunlight as a light source and measures it at various elevation angles by sequential scanning with a stepper motor. Ground-based MAX-DOAS measurements were carried out to investigate ClO, SO_2 and NO_2 levels in the mid-latitude coastal boundary layer from 27 May to 9 June, 2005, and SO_2 and NO_2 levels in fossil fuel power plant plumes from 10 to 14 January 2004. MAX-DOAS data were analyzed to identify and quantify ClO, SO_2 and NO_2 by utilizing their specific structured absorption features in the UV region. Differential slant column densities (dSCDs) for ClO, SO_2 and NO_2 were as high as 7.3×10^{14}, 2.4×10^{16} and 6.7×10^{16} molecules/cm^2 (with mean dSCDs of 2.3×10^{14}, 8.0×10^{15} and 1.2×10^{16} molecules/cm^2), respectively, at a 3° elevation angle in the coastal boundary layer during the measurement period. Based on the assumption that the trace gases were well mixed in the 1 km height of the boundary layer, estimates of the mean mixing ratios of ClO, SO_2 and NO_2 during the measurement period were 8.4 (±4.3), 296 (±233) and 305 (±284) pptv, respectively. MAX-DOAS measurement of the power plant plumes involved making vertical scans through multiple elevation angles perpendicular to the plume dispersion direction to yield cross-sectional distributions of ClO, SO_2 and NO_2 in the plume in terms of SCDs. Mixing ratios based on the estimated cross-sections of the plumes were 15.5 (ClO), 354 (SO_2) and 210 (NO_2) ppbv in the plumes of the fossil fuel power plant.

---

[1] *Advanced Environmental Monitoring Research Center (ADEMRC), Department of Environmental Science and Engineering, Gwangju Institute of Science and Technology (GIST), 1 Oryong-dong, Buk-gu, Gwangju 500-712, Republic of Korea*

[2] *Now at Institute of Environmental Physics and Remote Sensing, University of Bremen, Otto-Hahn-Allee 1, D-28334, Bremen, Germany*

[3] *Meteorological Research Institute, 460-18 Sindaebang-dong, Dongjak-gu, Seoul 156-720, Republic of Korea*

Y.J. Kim and U. Platt (eds.), *Advanced Environmental Monitoring*, 37–49.

Keywords: Air pollution, chlorine monoxide, DOAS, remote sensing

3.1 Introduction

Since the discovery of the ozone hole over Antarctica, halogen oxides have been of great environmental concern due to their ability to cause ozone depletion in the atmosphere (Farman et al., 1985; McElroy et al., 1986; Solomon et al., 1986). The importance of halogen chemistry for the tropospheric ozone budget was acknowledged following investigations of ozone depletion in the polar boundary layer (McElroy et al., 1999; Leser et al., 2003; Salawitch et al., 2006). Processes releasing halogens into the atmosphere and their effects on atmospheric chemistry have been the subject of a number of laboratory, field and modeling investigations (Schall and Heumann 1993; Mozurkewich 1995; Wayne et al., 1995; Sander and Crutzen 1996; von Glasow, et al., 2002; N. Bobrowski et al., 2003; Lee et al., 2005b; Salawitch 2006).

It is known that chlorine monoxide (ClO) plays a key role in processes leading to ozone loss in the troposphere and stratosphere. Gas-phase chlorine for ClO production in coastal areas is mostly released from sea salt through heterogeneous reactions (Fan and Jacob 1992; Sander and Crutzen 1996; Vogt et al., 1996). Enhanced levels of ClO in the boundary layer associated with ozone destruction have been observed over salt lakes (Stutz et al., 2002), in the polar boundary layer (Tuckermann et al., 1997), and in volcanic plumes (Lee et al., 2005b). However, there is little information on ClO and its effects on ozone depletion in the mid-latitude coastal boundary layer. Although other halogen species (e.g. BrO and IO) have been measured in coastal areas (Leser et al., 2003; Saiz-Lopez and Plane 2004; Saiz-Lopez et al., 2004), few direct measurements of ClO have been reported.

Anthropogenic SO_2 and NO_2 emissions are chemically converted to sulfuric and nitric acids in the atmosphere in the gaseous and aqueous phases. When these acids precipitate, damage is caused to buildings and ecosystems, particularly in areas where soils lack sufficient alkalinity to buffer these acids. The related formation of sulfate aerosols can increase the incidence of human respiratory ailments and associated mortality. Most measurements of these species emitted from the stack of an industrial facility such as a power plant can be performed using direct sampling methods, in situ monitoring techniques, or remote sensing techniques. The direct sampling methods are based on the collection of air samples by an appropriate container, with subsequent analysis in the laboratory. These techniques usually offer high sensitivity and selectivity, but have disadvantages in requiring real-time and continuous monitoring. Problems may also arise owing to alterations of gas composition caused by adsorption and desorption processes at the inner surface of the collecting container. In situ real-time monitoring techniques are often less sensitive and selective than direct sampling methods, though they offer the advantage of real-time measurements. Furthermore, direct sampling and in situ sampling around the outlet of an industrial stack is often impractical and hazardous.

Ultraviolet–visible absorption spectroscopy of remote sensing techniques has mostly been applied to the measurement of atmospheric trace species. The strength of this technique lies in the absence of wall losses, good specificity, and the potential for real-time measurements. In particular, the first property makes spectroscopic techniques especially well-suited for the detection of unstable species like OH radicals (Dorn et al., 1988), nitrate radicals (Platt et al., 1981) and halogen radicals (Tuckermann et al., 1997; Wagner et al., 1998; Stutz et al., 2002; Saiz-Lopez and Plane 2004; Saiz-Lopez et al., 2004). Since the 1970s, Differential optical absorption spectroscopy (DOAS) has been considered a powerful tool for the detection of atmospheric trace species on a real-time basis. The DOAS technique can gather sensitive measurements of several trace species simultaneously (Platt et al., 1981; Hausmann and Platt 1994; Tuckermann et al., 1997; Wagner et al., 1998; Hebestreit et al., 1999; Stutz et al., 2002; Saiz-Lopez and Plane 2004; Saiz-Lopez et al., 2004; Lee et al., 2005b).

Scattered sunlight in the near ultraviolet–visible spectral ranges has been employed for the analysis of atmospheric trace species by ground-based DOAS. This application is also called 'passive' absorption spectroscopy in contrast to 'active' spectroscopy using artificial light sources. Scattered sunlight DOAS measurements yield slant column densities of the respective absorbers. Most measurements have been conducted with zenith-looking instruments because the radiative transfer modeling necessary for the determination of vertical column densities is best understood for zenith-scattered sunlight. The multi-axis differential optical absorption spectroscopy (MAX-DOAS) technique is a passive DOAS technique using scattered sunlight from several viewing directions in addition to conventional zenith pointing. This ground-based instrument receives scattered sunlight from different viewing directions with several telescopes or by changing the viewing direction of a single telescope, thus collecting spatial information of atmospheric trace species. Measurement using low telescope elevation angles emphasizes the absorption path in the lowermost atmospheric layers and exhibits enhanced sensitivity for absorbers in the boundary layer, while photons received from the zenith-looking telescope have traveled a relatively long path in the stratosphere and a comparatively short path in the troposphere.

MAX-DOAS measurements were carried out to investigate ClO, SO_2 and NO_2 levels in the mid-latitude coastal boundary layer and in plumes of a fossil fuel power plant. Results concerning the levels of ClO, SO_2 and NO_2 and their diurnal variations in the mid-latitude coastal boundary layer are presented here. We also provide data on the levels and distribution of ClO, SO_2 and NO_2 in the plumes of a fossil fuel plant.

3.2 Measurement

Our sequential MAX-DOAS system essentially consists of a small aluminium box containing a miniature spectrograph and a telescope. The miniature spectrograph (OceanOptics USB2000, cross Czerny-Turner type, $1/f = 4$)

consists of a grating (2,400 grooves/mm) yielding spectral coverage between 289 and 431 nm (at a 0.7 nm FWHM spectral resolution) and a CCD detector (2,048 pixels with a 14 μm center-to-center spacing) (Lee et al., 2005b). The MAX-DOAS box was attached directly to a stepper motor, allowing sequential measurement of scattered sunlight at various elevation angles between 0° and 90° above the horizon. Dark current and offset signals were also recorded during the nighttime. The MAX-DOAS system was operated using the DOASIS software developed by the Institute of Environmental Physics, University of Heidelberg, Germany.

MAX-DOAS measurements of ClO, SO_2 and NO_2 in the mid-latitude coastal boundary layer were carried out at the Korea Global Atmosphere Watch Observatory (KGAWO) (36.54° N, 127.12° E) located on Anmyeon island off the west coast of Korea. Several resorts are located on Anmyeon island, and there is increased traffic around nearby resorts on weekends. The KGAWO station is 230 m from the west coast of Anmyeon island, and has occasionally been affected by the long-range transport of pollutants from the Asian continent. MAX-DOAS measurements were made on the roof of the KGAWO building (43 m above sea level) during the daytime from 27 May to 9 June 2005. The viewing azimuth angle of the MAX-DOAS telescope was 340° (0° indicates a true north direction), looking to the sea. Scattered sunlight signals were recorded at telescope elevation angles of 3, 6, 10, 20 and 90°. Each measurement sequence took about 10–20 min. Measured MAX-DOAS spectra were analyzed to identify and quantify levels of ClO, SO_2 and NO_2 using their specific structured absorption features in the ultraviolet region (Platt 1994). In situ gas analyzers measuring SO_2, O_3 and NO_x were also installed 10 m above the ground at KGAWO. Meteorological data that included wind speed and direction were also collected by an automatic weather station (AWS) installed 40 m above ground level at KGAWO. Measurements of SO_2 and NO_2 in the plumes emitted from a power plant were made from 10 to 14 January 2004 at a fossil fuel power plant in the west of Korea. Telescope elevation angles of 2, 5, 10, 20, 30, 40, 50, 60, 70, 80 and 90° above the horizon were used to scan the plume, resulting in cross-sectional measurements of the plume. Each measurement sequence of the plume cross-section took about 15–25 min. Mercury lamp line peaks were also recorded on 29 May and 9 June 2006 to check for any spectral shifts of the spectrograph and ensure its wavelength calibration.

Spectra taken by the MAX-DOAS system were calibrated using mercury lamp line peaks corrected for dark current and electric offset signal by subtracting the dark current and offset signals recorded. Then the calibrated DOAS spectra were calibrated again by fitting them to the solar reference file. Slant column densities (SCDs) of ClO, SO_2 and NO_2 were derived from the calibrated DOAS spectra using the WinDOAS V2.10 software package (Van Roozendael and Fayt 2001). Scattered sunlight is highly structured due to solar Fraunhofer lines. These structures were removed by including the Fraunhofer reference spectrum (FRS) in the fitting procedure. The scattered sunlight spectrum taken at an elevation angle of 90° (zenith direction) around noon (local time) on a cloudless day was used as the FRS, where only negligible background trace gas absorption was present. To compensate for the ring effect (Fish and Jones 1995; Chance and Spurr 1997), which represents the filling-in (reduction of the observed optical densities) of Fraunhofer lines, a 'ring spectrum' was calculated from the FRS. In addition to a polynomial order fit to remove broad band structures, the FRS, ring spectrum and literature sources of absorption cross-section spectra were simultaneously fitted to the scattered sunlight spectra using the nonlinear least square method (Stutz and Platt 1996). High-resolution reference absorption spectra were obtained from previous studies (Greenblatt et al., 1990; Simon et al., 1990; Vandaele et al., 1997; Wilmouth et al., 1999; Meller and Moortgat 2000; Bogumil et al., 2003). The specifications for the evaluation of ClO, SO_2 and NO_2 are summarized in Table 3.1. Figure 3.1 shows an example of a MAX-DOAS spectrum evaluation for ClO in the coastal boundary layer. All reference absorption cross-section spectra used for the evaluation of ClO, SO_2 and NO_2 were convoluted with the instrumental function to match the spectral resolution of the MAX-DOAS system used in this study.

For MAX-DOAS measurements in the coastal area, the retrieved SCDs from MAX-DOAS spectra recorded at a 90° elevation angle were subtracted from those obtained at other elevation angles during each scanning sequence. This procedure yields differential SCDs (DSCDs) (= SCD[α, θ] – SCD[90°, θ]), where α is an elevation angle and θ is a solar zenith angle, and removes absorptions by trace gases in the stratosphere (Leser et al., 2003; Friedeburg et al., 2005).

Table 3.1 Specifications for the MAX-DOAS spectrum evaluation for ClO, NO_2 and SO_2

Molecule	Wavelength range, nm	Polynomial order	Cross-sections included in the fitting procedure
ClO	302.5–316	3	ClO[a], BrO[b], SO_2[c], HCHO[d], NO_2[e], O_3[f], Ring[g], FRS[h]
SO_2	303.5–316	3	ClO[a], SO_2[c], BrO[b], NO_2[e], O_3[f], Ring[b], FRS[c]
NO_2	399–418	3	NO_2[e], O_4[i], O_3[f], Ring[g], FRS[h]

The cross-sections of NO_2, O_3 and SO_2 were I_0-corrected (Aliwel et al., 2002)
[a] ClO at 300 K (Simon et al., 1990)
[b] BrO at 298 K (Wilmouth et al., 1999)
[c] SO_2 at 293 K (Bogumil et al., 293)
[d] HCHO at 298 K (Meller and Moortgat 2000)
[e] NO_2 at 294 K (Vandaele et al., 1997)
[f] Two O_3 cross-sections obtained at 223 K and 241 K were included in the fitting routine.
[g] Ring spectrum
[h] Fraunhofer reference spectrum
[i] O_4 at 296 K (Greenblatt et al., 1990)

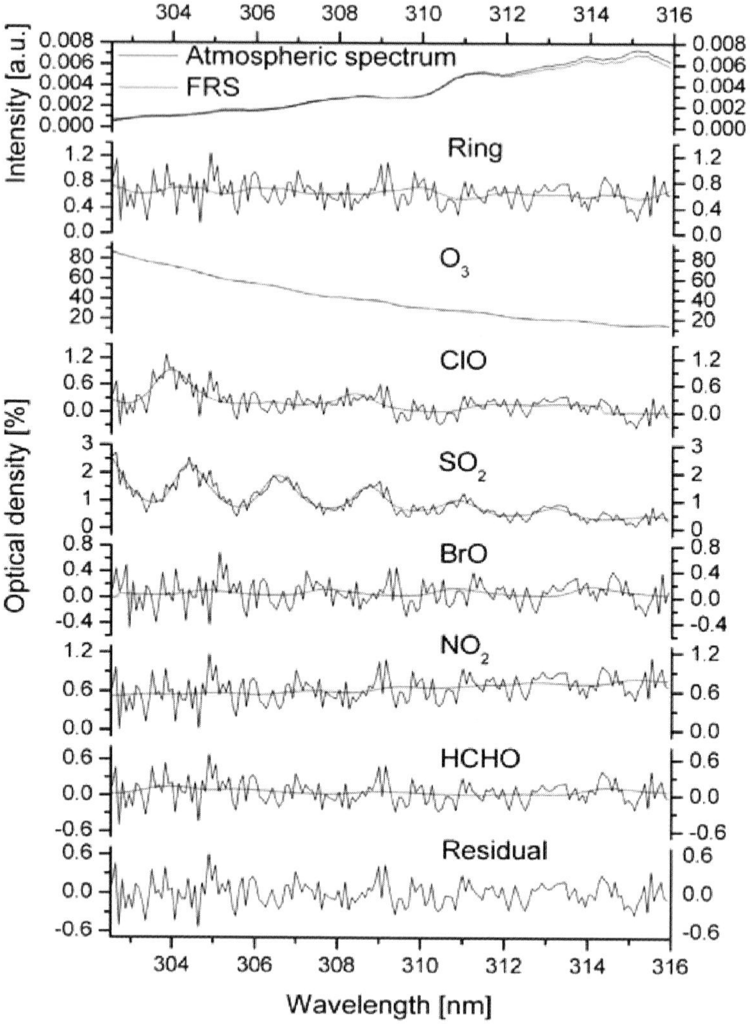

Fig. 3.1 An evaluation example of a ClO SCD from the MAX-DOAS spectrum taken at a 3° elevation angle at 16:12 on 5 June 2005 (LT)

3.3 Results

3.3.1 ClO, SO_2 and NO_2 in the Mid-Latitude Coastal Boundary Layer

Figure 3.2 shows the temporal variations of DSCDs for ClO, SO_2 and NO_2 measured at KGAWO during the measurement period. The mean DSCDs of ClO, SO_2 and NO_2 at a 3° elevation angle were 2.3 (±1.2) × 10^{14}, 8.0 (±6.3) × 10^{15} and

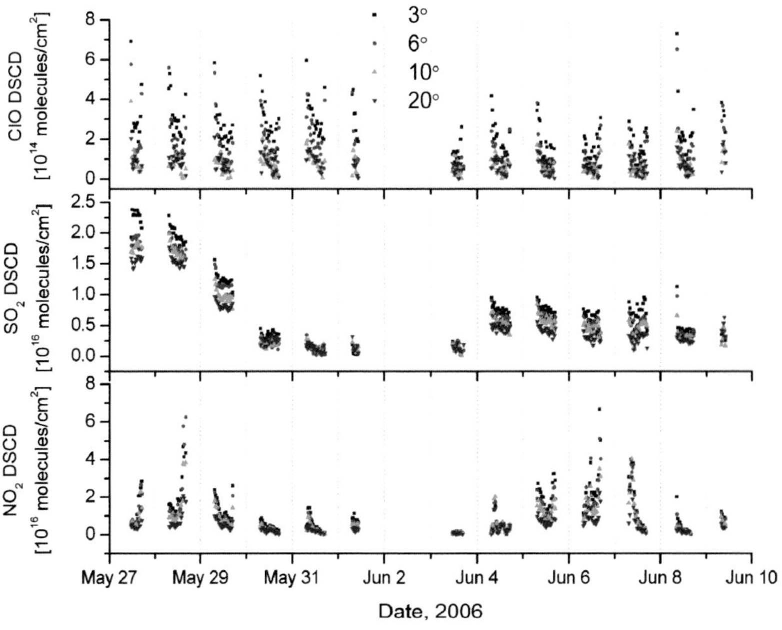

Fig. 3.2 ClO, SO_2 and NO_2 differential slant column densities (DSCD) at different elevation angles determined by the MAX-DOAS system. The date at the tic-mark denotes 0:00 h (local time) for the given day

1.2 (±1.1) × 10^{16} molecules/cm^2, with maximum values of 7.3 × 10^{14}, 2.4 × 10^{16} and 6.7 × 10^{16} molecules/cm^2, respectively.

The mean DSCDs of ClO, SO_2 and NO_2 observed during the measurement period can be converted to a mixing ratio. Mixing ratios can be calculated in the following manner on the assumption that the change of air density was negligible in the lowest 1 km of the atmosphere:

$$M = \frac{DSCD}{dAMF} \cdot \frac{1}{z} \tag{1}$$

where M is the mixing ratio, z is the height of the trace gas layer, and dAMF is a differential air-mass factor (AMF[α = 3°])–AMF[α = 90°]). The AMFs in this study were calculated using the Monte Carlo radiative transfer model described in von Friedeburg et al. (2002), which included multiple Rayleigh and Mie scattering, surface albedo, refraction, and full spherical geometry. Calculated dAMFs were 10.2 and 14.5 for 304 nm (for ClO and SO_2) and 412 nm (for NO_2) wavelengths, respectively, assuming a 30° solar zenith angle, 5% ground albedo, and a pure Rayleigh case for the troposphere. The mixing ratios converted from the DSCDs of ClO, SO_2 and NO_2 observed during the measurement period are shown in Fig. 3.3. The mean mixing ratios for ClO, SO_2 and NO_2 were 8.4 (±4.3), 296 (±233) and 305 (±284) pptv, respectively, if it is assumed that the trace gases were well-mixed within the 1-km height of the boundary layer.

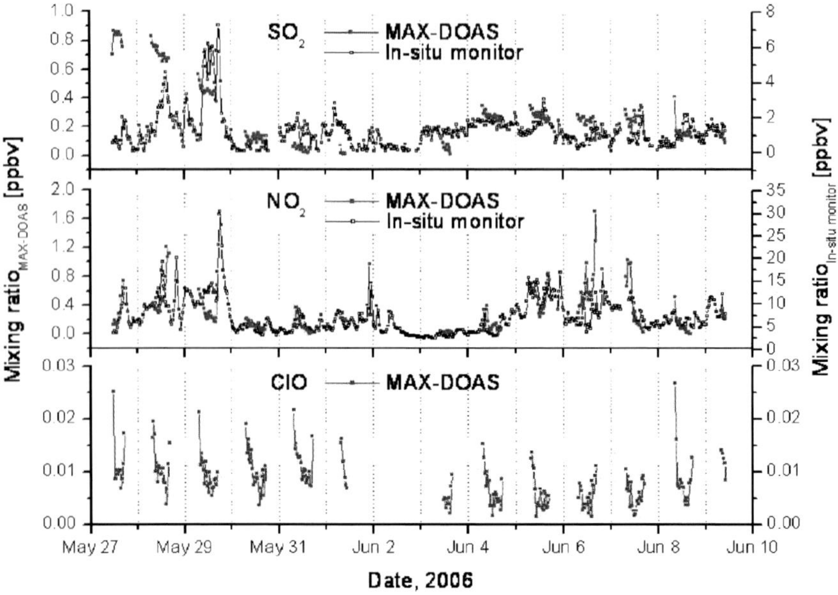

Fig. 3.3 Mixing ratios of SO_2 and NO_2 determined by the MAX-DOAS system and in situ monitors. ClO mixing ratios were determined by the MAX-DOAS system. The date at the tic-mark denotes 0:00 h (local time) for the given day

As shown in Figures 3.2 and 3.3, the diurnal variation of DSCD for ClO shows maxima in the morning and evening, and a minimum around noon. These changes have been predicted by a halogen chemistry model in the marine boundary layer (von Glasow et al., 2002). This feature can be explained by the difference in the photolysis spectra of O_3 and ClX, in which ClX is photolyzed by longer wavelengths of sunlight than O_3 (von Glasow et al., 2002). The major source of gas-phase chlorine in the coastal area is the release of species such as Cl_2 and BrCl from sea-salt aerosol, following the uptake of the gas phase and subsequent aqueous-phase reactions of hypohalous acids (HOX, where X = Br, Cl, I) (Vogt et al., 1996; Saiz-Lopez et al., 2004). The formation of $ClNO_2$ and X_2 can proceed by uptake of N_2O_5 formed at night by combination of NO_3 and NO_2, leading to an accumulation of these species before sunrise. They are photolyzed by sunlight to release Cl that subsequently participates in gas-phase catalytic reaction cycles (Fan and Jacob 1992; Vogt et al., 1996). This effect is more pronounced with higher levels of NO_x. The reactive halogen species in the troposphere are mainly removed by their reaction with proxy radicals, such as HO_2. The HO_2 mixing ratios are sufficiently high during the day to reduce the DSCD of ClO, resulting in the noon minimum. The relative lack of shorter wavelengths of sunlight at higher solar zenith angles leads to an increase in the DSCD of ClO, which produces the morning and evening peaks. Moreover, the high DSCD of ClO in the morning and evening could be the result of a higher relative humidity during those hours.

Table 3.2 SO_4^{2-}/SO_2 ratios determined for each PM1 sample collected for 24 h

	May 27	May 29	May 30	May 31	June 04	June 05	June 06	June 07	June 08
SO_4^{2-}/SO_2 ratio	6.05	1.36	5.98	0.82	1.37	0.76	1.77	1.41	2.74

NO_2 plays an important role in atmospheric chemistry by acting as a catalyst in the photochemical production of O_3. If halogens are present, NO_x mixing ratios are further reduced by reactions of NO_2 with ClO, yielding XNO_3 (Platt and Hönninger 2003). The initial high mixing ratios of NO_2 enhance chlorine activation, resulting in increases of ClO and an upward shift of O_3 destruction rates (von Glasow et al., 2002). The probabilities and rates of heterogeneous reactions for the release of reactive halogens increase with a higher concentration of acids (Fan and Jacob 1992).

The mixing ratios of NO_2 and SO_2 measured by the MAX-DOAS system and in situ monitors are plotted in Fig. 3.3. Relatively high NO_2 and SO_2 levels were observed during two periods: 27–29 May 2005, and 4–7 June 2005. The discrepancy in SO_2 and NO_2 trends between the MAX-DOAS system and in situ monitors during the first event period of 27–29 May 2005 could be due to the inhomogeneity of the air masses. The major difference between the MAX-DOAS system and in situ instruments is the fact that the passive MAX-DOAS system relies on the sun as its light source. Therefore, instrumental differences associated with measurement principles (i.e. line-integrating versus point measurement) need to be considered for an evaluation of MAX-DOAS performance (Lee et al., 2005a). The KGAWO site is occasionally affected by the long-range transport of pollutants from the Asian continent. The air mass transported from the Asian continent could be in the upper part of the boundary layer over the KGAWO site, and may have spread horizontally and towards the ground during the first event period. It is believed that an aged air mass had an impact during the first event period since the NO concentration during the first event period was lower than that recorded during the second period. This is supported by satellite images indicating the transport of an air mass from the Asian continent, and by high SO_4^{2-}/SO_2 ratios of PM1 samples collected by an URG PM1 sampler, as summarized in Table 3.2. In addition to the long-range transport of pollutants, local sources of pollution such as increased traffic around nearby resorts on weekends could have contributed to the high concentrations of NO_2 and SO_2 during the first measurement period, particularly on 29 May 2006. The increased concentrations of NO_2 and SO_2 observed during the second period on a weekend could be attributed to local sources.

3.3.2 ClO, SO_2 and NO_2 Measurement in the Plumes of a Fossil Fuel Power Plant

ClO, SO_2 and NO_2 were detected in each scan through the plume. The cross-sectional distribution of slant column densities was characterized by maximum SCDs in the viewing direction (elevation angle) toward the center of the plume at

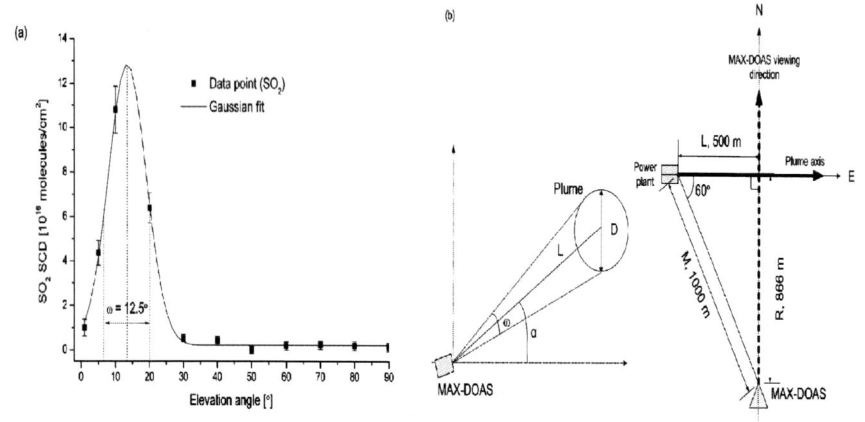

Fig. 3.4 (**a**) Cross-section of the fossil fuel power plant plume for SO_2, and (**b**) MAX-DOAS measurement geometry

any elevation angle, as shown in Fig. 3.4(a). SCDs measured during the measurement period were as high as 1.2×10^{16}, 1.6×10^{17} and 2.2×10^{17} molecules/ cm^2 in the center of the plume for ClO, SO_2 and NO_2, respectively.

A cross-section of the plume was estimated to determine mixing ratios of species measured from the plume (see Fig. 3.4[a]). After the plume cross-section (assumed to be circular) was fitted with a Gaussian distribution, the plume diameter, D, was determined from the width, ω, (full width at half maximum) as: $D \approx L \times \omega \times \cos(\varphi)$, where φ is the observation azimuth angle. The plume width, ω, was 13.2° for ClO, 12.5° for SO_2 and 15.2° for NO_2 (based on data collected on 11 January 2004 and $\varphi \approx 0°$ (see Fig. 3.4[b]), which correspond to plume diameters of ≈ 200 (±30), 190 (±30) and 230 (±31) m for ClO, SO_2 and NO_2, respectively. Based on the measured maximum SCD (ClO ≈ 8.3×10^{15}, SO_2 ≈ 1.8×10^{17} and NO_2 ≈ 1.3×10^{17} molecules/cm^2), the calculated mixing ratios (= SCD × cos(0°)/D) were ~15.5, ~354 and ~210 ppbv for ClO, SO_2 and NO_2, respectively. The SO_2 mixing ratio was above the standard established for air pollutants emitted from a power plant in Korea (150 ppbv for SO_2), while the mixing ratio of NO_2 was below the standard (250 ppbv for NO_2).

3.4 Conclusion

While MAX-DOAS is a new and emerging technique based on the scattered sunlight DOAS method, the ground-based MAX-DOAS system allows the determination of relatively precise levels of absorbers in the boundary layer using simple instrumentation. In this study, MAX-DOAS proved to be successful in applications that monitored trace gases in power plant plumes and ambient air. ClO, SO_2 and NO_2 have been directly observed in the mid-latitude coastal boundary layer by the MAX-DOAS technique. We demonstrated that the MAX-DOAS technique is capable of measuring trace species in the boundary layer with high sensitivity. The diurnal behavior of ClO was consistent with a source based on the heterogeneous processing of sea-salt aerosol. The ClO level was as high as ~27 pptv (with a mean of 8.4 pptv) during the measurement period, and is comparable to ClO levels of up to 15 pptv in the boundary layer over the Great Salt Lake (Stutz et al., 2002) and 30 pptv in the polar boundary layer (Tuckermann et al., 1997). Mean SO_2 and NO_2 levels observed during the measurement period were 296 (±233) and 305 (±284) ppbv, respectively. High SO_2 and NO_2 concentrations measured during the two event periods might have an effect on atmospheric halogen chemistry through reactive halogen release processes by the attack of strong acids on sea-salt aerosols, and through reducing processes by the reaction of ClO with NO_2 (Platt and Hönninger 2003). The ClO emitted from sea salt as a consequence of recycling processes has a long atmospheric lifetime of up to several days (Barrie and Platt 1997; Tuckermann et al., 1997; Platt and Hönninger 2003), and may therefore, play an important role in atmospheric chemistry. Further observations are required to quantify the seasonal effect of halogen species originating from sea-salt aerosol on tropospheric chemistry (in particular, ozone chemistry) of the mid-latitude region.

The ground-based MAX-DOAS system has also proven to be suitable for the measurement of ClO, SO_2 and NO_2 emissions from a power plant by scanning the plume from fixed positions. Mixing ratios calculated from cross-sectional distributions of ClO, SO_2 and NO_2 in fossil fuel power plant plumes were ~15.5, ~354 and ~210 ppbv, respectively. The emission strength can be determined if the wind speed is provided. The ground-based MAX-DOAS technique is considerably cheaper and simpler than other available remote sensing methods, and shows great potential for the routine, remote monitoring of industrial stacks. It is also suitable for studies of the transport, dispersion and chemistry of plumes.

Acknowledgements This work was supported by the Korea Science and Engineering Foundation through the Advanced Environmental Monitoring Research Center (ADEMRC) of the Gwangju Institute of Science and Technology (GIST).

References

Barrie L. and Platt, U. (1997), Artic tropospheric chemistry: An overview, *Tellus*, 49B, 450–454.

Bobrowski N., Hönninger G., Galle B., and Platt U. (2003), Detection of bromine monoxide in a volcanic plume, *Nature*, 423, 273–276.

Bogumil K., Orphal J., Homann T., Voigt S., Spietz P., Fleischmann O.C., Vogel A., Hartmann M., Bovensmann H., Frerik J., and Burrows J.P. (2003), Measurements of molecular absorption spectra with the SCIAMACHY pre-flight model: Instrument characterization and reference data for atmospheric remote sensing in the 230–2380 nm region, *J.Photoch. Photobiol. A*, 157, 167–184.

Chance K.V. and Spurr R.J.D. (1997), Ring effect studies: Rayleigh scattering, including molecular parameter for rotational Raman scattering, and the Fraunhofer spectrum, *Appl. Optics*, 36, 5224–5230.

Fan S.-M. and Jacob D.J. (1992), Surface ozone depletion in Artic spring sustained by bromine reaction on aerosols, *Nature*, 359, 522–524.
Farman J.C., Gardiner B.G., and Shaklin J.D. (1985), Large losses of total ozone in Antarctica reveal seasonal ClO_x/NO_x interaction, *Nature*, 315, 207–210.
Fish D.J. and Jones R.L. (1995), Rotational Raman scattering and the ring effect in zenith-sky spectra, *Geophys. Res. Lett.*, 22, 811–814.
Greenblatt G.D., Orlando J.J., Burkholder J.B., and Ravishankara A.R. (1990), Absorption measurements of oxygen between 330 and 1140 nm, *J. Geophys. Res.*, 95, 18, 577–582.
Hausmann M. and Platt U. (1994), Spectroscopic measurement of bromine oxide and ozone in the high Arctic during polar sunrise experiments 1992, *J. Geophys. Res.*, 99, 25, 399–413.
Hebestreit K., Stutz J., Rosen D., Matveiv V., Pelg M., Luria M., and Platt U. (1999), DOAS measurements of tropospheric bromine oxide in mid-latitudes, *Science*, 283, 55–57.
Hönninger G., von Friedeburg C., and Platt U. (2004), Multi axis differential optical absorption spectroscopy (MAX-DOAS), *Atmos. Chem. Phys.*, 4, 231–254.
Lee C., Choi Y.J., Jung J.S., Lee J.S., Kim Y.J., and Kim, K.H. (2005a), Measurement of atmospheric monoaromatic hydrocarbons using differential optical absorption spectroscopy: Comparison with on-line gas chromatography measurements in urban air, *Atmos. Environ.*, 39, 2225–2234.
Lee C., Kim Y.J., Tanimoto H., Bobrowski N., Platt U., Mori T., Yamamoto K., and Hong C.S. (2005b), High ClO and ozone depletion observed in the plume of Sakurajima volcano, *Geophys. Res. Lett.*, 32, DOI 10.1029/2005GL023785.
Leser H., Hönninger G., and Platt U. (2003), MAX-DOAS measurements of BrO and NO_2 in the marine boundary layer, *Geophys. Res. Lett.*, 30, DOI 10.1029/2002GL015811.
Marchand M., Bekki S., Lefèvere F., Hauchecorne A., Godin-Beckmann S., and Chipperfield M.P. (2004), Model simulations of the northern extravortex ozone column: Influence of past changes in chemical composition, *J. Geophys. Res.*, 109, DOI 10.1029/2003JD003634.
McElroy C.T., McLinden C.A., and McConnell J.C. (1999), Evidence for bromine monoxide in the free troposphere during the Arctic polar sunrise, *Nature*, 397, 338–341.
McElroy M.B., Salawitch R.J., Wofsy C.S., and Logan J.A. (1986), Reductions of Antarctic ozone due to synergistic interactions of chlorine and bromine, *Nature*, 321, 759–762.
Meller R. and Moortgat G.K. (2000), Temperature dependence of the absorption cross sections of formaldehyde between 223 and 323 K in the wavelength range 225–375 nm, *J. Geophys. Res.*, 105, 7089–7101.
Mozurkewich M. (1995), Mechanisms of the release of halogen atom sea-salt particles by free radical reactions, *J. Geophys. Res.*, 100, 14, 199–207.
Platt U. (1994), Differential optical absorption spectroscopy (DOAS). In M.W. Sigrist (Eds.), *Monitoring by Spectroscopic Techniques*, Wiley, New York, pp. 27–84.
Platt U. and Hönninger G. (2003), The role of halogen species in the trosposphere, *Chemosphere,* 52, 325–338.
Salawitch R.J. (2006), Atmospheric chemistry: Biogenic bromine, *Nature*, 439, 275–277.
Sander R. and Crutzen P.J. (1996), Model study indicating halogen activation and ozone destruction in polluted air masses transported to the sea, *J. Geophys. Res.*, 101, 9121–9138.
Schall C. and Heumann K. (1993), GC determination of volatile organoiodine and organobromine compounds in seawater and air samples, *Fresen. J. Anal. Chem.*, 346, 717–722.
Saiz-Lopez A. and Plane J.M.C. (2004), Novel iodine chemistry in the marine boundary layer, *Geophys. Res. Lett.*, 31, DOI 10.1029/2003GL019215.
Saiz-Lopez A., Plane J.M.C., and Shillito J.A. (2004), Bromine oxide in the mid-latitude marine boundary layer, *Geophys. Res. Lett.*, 31, DOI 10.1029/2003GL018956.
Simon F.G., Schneider W., Moortgat G.K., and Burrows J.P. (1990), A study of the ClO absorption cross-section between 240 and 310 nm and the kinetics of the self-reaction at 300 K, *J. Photoch. Photobiol. A*, 55, 1–23.
Solomon S. (1990), Progress towards a quantitative understanding of Antarctic ozone depletion, *Nature*, 347, 347–354.
Solomon S., Garcia F.S., Rowland F.S., and Wuebbles D.J. (1986), On the depletion of Antarctic ozone, *Nature*, 321, 755–758.

Stutz J. and Platt U. (1996), Numerical analysis and error estimation of differential optical absorption spectroscopy measurements least-squares methods, *Appl. Optics,* 35, 6041–6053.

Stutz J., Ackermann R., Fast J.D., and Barrie L. (2002), Atmospheric reactive chloride and bromine at the Great Salt Lake, Utah, *Geophys. Res. Lett.*, 29, DOI 10.1029/2002GL014812.

Tuckermann M., Ackermann R., Gölz C., Lorezen-Schmidt H., Senne T., Stutz J., Trost B., Unold W., and Platt U. (1997), DOAS-observation of halogen radical-catalysed arctic boundary layer ozone destruction during the ARCTOC-campaigns 1995 and 1996 in Ny-Ålesund, Spitshergen, *Tellus*, 49B, 533–555.

Vandaele A.C., Hermans C., Simon P.C., Carleer M., Colin R., Fally S., Mérienne M.-F., Jenouvrier A., and Coquart B. (1997), Measurements of the NO_2 absorption cross-section from 42000 cm^{-1} to 10000 cm^{-1} (238–1000 nm) at 220 K and 294 K, *J. Quant. Spectrosc. Radiative. Transf.*, 59, 171–184.

Van Roozendael M. and Fayt C. (2001), *WinDOAS 2.1 Software User Mannual.* (Uccle, IASB/BIRA).

Vogt R., Crutzen P.J., and Sander R. (1996), A mechanism for halogen release from sea-salt aerosol in the remote marine boundary layer, *Nature*, 383, 327–330.

Wagner T. and Platt U. (1998), Satellite mapping of enhanced BrO concentrations in the troposphere, *Nature*, 395, 486–490.

von Friedeburg C., Hönninger G., and Platt U. (2005), Multi-axis-DOAS measurements of NO_2 during the BAB II motorway emission campaign, *Atmos. Environ.*, 39, 977–985.

von Glasow R., Sander R., Bott A., and Crutzen P.J. (2002), Modelling halogen chemistry in the marine boundary layer 1. Cloud-free MBL, *J. Geophys. Res.*, 107(D17), DOI 10.1029/2001JD000942.

Wayne R.P., Poulet G., Biggs P., Burrows J.P., Cox R.A., Crutzen P.J., Hayman G.D., Jenkin M.E., Bras G.L., Moortgat G.K., Platt U., and Schindler R.N. (1995), Halogen oxides: Radicals, sources and reservoirs in the laboratory and in the atmosphere, *Atmos. Environ.*, 29, 2677–2881.

Wilmouth D.M., Hanisco T.F., Donahue N.M., and Anderson J.G. (1999), Fourier transfer ultraviolet spectroscopy of the A2P3/2 X2P3/2 transition of BrO, *J. Phys. Chem. A*, 103, 8935–8945.

Chapter 4
Laser Based Chemical Sensor Technology: Recent Advances and Applications

Frank K. Tittel, Yury A. Bakhirkin, Robert F. Curl, Anatoliy A. Kosterev, Matthew R. McCurdy, Stephen G. So, and Gerard Wysocki

Abstract There is an increasing need in many chemical sensing applications ranging from environmental science to industrial process control as well as medical diagnostics for fast, sensitive, and selective trace gas detection based on laser spectroscopy. The recent availability of continuous wave near infrared diode lasers-, mid-infrared quantum cascade and interband cascade distributed feedback (QC and IC DFB) lasers as mid-infrared spectroscopic sources addresses this need. A number of spectroscopic techniques have been demonstrated. For example, the authors have employed infrared DFB QC and IC lasers for the detection and quantification of trace gases and isotopic species in ambient air by means of direct absorption, cavity-enhanced, and photoacoustic spectroscopy. These spectroscopic techniques offer an alternative to non-spectroscopic techniques such as mass spectrometry (MS), gas chromatography (GC) and electrochemical sensors. The sensitivity and selectivity that can be achieved by both techniques (excluding electrochemical sensors) are similar, but the sensor response time, instrumentation size and cost of ownership for spectroscopic techniques can be advantageous as compared to MS-GC spectrometry.

Keywords: Laser absorption spectroscopy, cavity-enhanced and photoacoustic spectroscopy, near infrared diode lasers, mid infrared quantum cascade lasers, chemical sensing of trace gases

4.1 Introduction

Infrared laser absorption spectroscopy (LAS) is an extremely effective tool for the detection and quantification of molecular trace gases. The demonstrated detection sensitivity of LAS ranges from ppmv, ppbv to even sub-ppbv levels depending on the specific gas species and the detection method employed (Curl and Tittel 2002;

Rice University, Electrical and Computer Engineering Department, MS-366, 6100 Main St., Houston, TX 77005, USA

Y.J. Kim and U. Platt (eds.), *Advanced Environmental Monitoring*, 50–63.

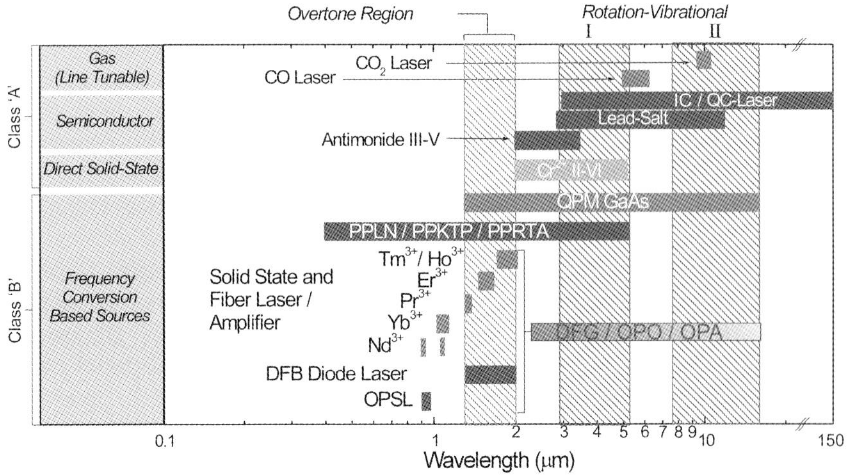

Fig. 4.1 Infrared laser sources and spectral coverage. PPLN, periodically poled lithium niobate; DFB, distributed feedback; OPSL, optically pumped solid state laser; DFG, difference frequency generation; OPO, optical parametrical oscillator; OPA, optical parametric amplifier; QPM, quasi phase matched; IC, interband cascade; QC, quantum cascade

Tittel et al. 2003). The spectral region of fundamental vibrational molecular absorption bands from 3 to 24 μm is the most suitable for high sensitivity trace gas detection. However, the usefulness of the laser spectroscopy in this region is limited by the availability of convenient tunable laser sources. Real-world applications require the laser source to be compact, efficient, reliable and operate at near room-temperatures. Existing options as shown in Fig. 4.1 include lead salt diode lasers, coherent sources based on difference frequency generation (DFG), optical parametric oscillators (OPOs), tunable solid state lasers, quantum and interband cascade lasers (QCLs and ICLs). Sensors based on lead salt diode lasers are typically large in size and require cryogenic cooling, since these lasers operate at temperatures <90 K. DFG sources (especially bulk and waveguide PPLN based) have been shown to be robust and compact (Yanagawa et al. 2005; Richter et al. 2006).

Recent advances in QCLs and ICLs fabricated by band structure engineering offer an attractive new source option for mid-infrared absorption spectroscopy with ultra-high resolution and sensitivity (Kosterev et al. 2002a, b) (See Fig. 4.1). The most technologically developed mid-infrared QC laser source to date is based on type-I intersubband transitions in InGaAs/InAlAs heterostructures (Capasso et al. 2000; Faist et al. 2002; Beck et al. 2002; Mann et al. 2003; Evans et al. 2004; Troccoli et al. 2004; Diehl et al. 2006). More recently ICLs based on type-II interband transition have been reported in the 3–5 μm region (Yang et al. 2002; Yang et al. 2004; Bradshaw et al. 2004; Mansour et al. 2006).

The vast majority of chemical substances have vibrational fundamental bands in the 3–24 μm region, and the absorption of light by rotational–vibrational transitions of these bands provides a nearly universal means for their sensitive and selective detection. Furthermore, near infrared spectroscopy from 1.3 to 3 μm can be used effectively in the quantification of numerous trace gas species. This application can use ultra-reliable, room temperature, single frequency distributed feedback (DFB) lasers that were primarily developed for optical telecommunications with output powers of tens of milliwatts. These lasers access molecular overtone or combination band transitions that are typically a factor of 30–300 weaker than the mid-infrared (mid-IR) fundamental transitions. However, in some cases this can be compensated by the judicious choice of the appropriate photoconductive or photovoltaic detector.

Pulsed and continuous wave (cw) DFB-QCLs allow the realization of compact, narrow linewidth sources combining single-frequency operation and substantially high powers (from tens to hundreds of milliwatts) at mid-IR wavelengths (4–24 μm) and temperatures that are attainable with thermoelectric cooling. The large wavelength coverage available with QCLs allows the detection, quantification and monitoring of numerous molecular trace gas species, especially those with resolved rotational–vibrational spectra. The high QCL output power permits the use of advanced detection techniques that significantly improve the detection sensitivity of trace gas species and decrease the complexity and size of the overall trace gas sensor architecture. This includes photoacoustic, laser absorption and cavity-enhanced spectroscopy. For example, in cavity ringdown spectroscopy (CRDS) (Berden et al. 2000) and integrated cavity output spectroscopy (ICOS), an effective absorption pathlength of hundreds of meters can be obtained in a compact device (Bakhirkin et al. 2004). In the following sections, a number of examples of laser-based analytical techniques based on absorption spectroscopy, laser photoacoustic spectroscopy and cavity-enhanced spectroscopy will be presented.

4.2 Chemical Sensing Based on Tunable Thermoelectrically Cooled CW Quantum Cascade Lasers

The development of laser spectroscopic techniques strongly relies on increasing the availability of new tunable laser sources. For applications in the mid-IR molecular fingerprint region, QCLs have proved to be convenient and reliable light sources for the spectroscopic detection of trace gases (Capasso et al. 2000). Spectroscopic applications usually require single-frequency operation. This is achieved by introducing a DFB structure into the QCL active region. Although DFB QCLs show high performance and reliability characteristics, the range of wavelength tuning of the emitted laser radiation is limited by the wavelength tuning range of the DFB structures. Typically the maximum thermal tuning range of DFB-QCLs is ~2 cm^{-1} achieved by laser current injection control or 10 cm^{-1} by varying the temperature of the QCL chip. The development of bound-to-continuum QC lasers (Faist et al. 2001) has addressed the issue of wide frequency tunability. Bound-to-continuum QC lasers have an intrinsically broader gain profile, because the lower state of the laser transition is a relatively broad continuum. A luminescence spectrum of

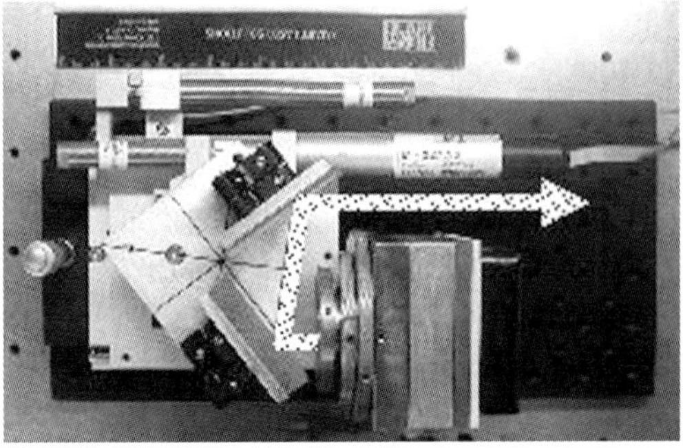

Fig. 4.2 Compact tunable external grating cavity quantum cascade laser

297 cm^{-1} FWHM (full width at half maximum) at room temperature was observed for $\lambda \approx 10\,\mu m$ QCL devices employing bound-to-continuum transitions (Maulini et al. 2004). To take advantage of the broadband gain of such QCLs, an external cavity (EC) configuration can be used for wavelength selection (Wysocki et al. 2005; Peng et al. 2003). Recently, even broader gain profiles with FWHM of ~350 cm^{-1} were achieved by using a heterogeneous quantum cascade structure based on two bound-to-continuum designs emitting at 8.4 and 9.6 µm (Maulini et al. 2006). The development of a QC laser spectrometer for high resolution spectroscopic applications and multi species trace gas detection in the mid-IR through the design and implementation of a novel EC-QCL architecture is described by Wysocki et al. 2005. The instrument depicted in Fig. 4.2 employs a piezo-activated cavity mode tracking system for mode-hop free operation. The mode-tracking system provides independent control of the EC length, diffraction grating angle and laser current. The system performance and spectroscopic application capability was demonstrated by studying nitric oxide (NO) absorption features at ~1945 cm^{-1} with a gain medium operating at ~5.2 µm. The EC-QCL exhibited a coarse tuning range of ~35 cm^{-1} and a continuous mode-hop free fine tuning range of 1.2 cm^{-1} as shown in Fig. 4.3. Wide wavelength tunability and a narrow laser linewidth of <30 MHz, which allowed resolving spectral features separated by less than 0.006 cm^{-1} (see inset of Fig. 4.3) makes such an EC-QCL an excellent light source suitable for high resolution spectroscopic applications and multiple species trace gas detection. The flexibility of this arrangement makes it possible to use it with other QC lasers at other wavelengths without changing the EC configuration.

4.3 Trace Gas Detection Based on Laser Photoacoustic Spectroscopy

Photoacoustic spectroscopy (PAS), based on the photoacoustic effect, in which acoustic waves result from the absorption of laser radiation by a selected target compound in a specially designed cell, is an effective method for sensitive trace

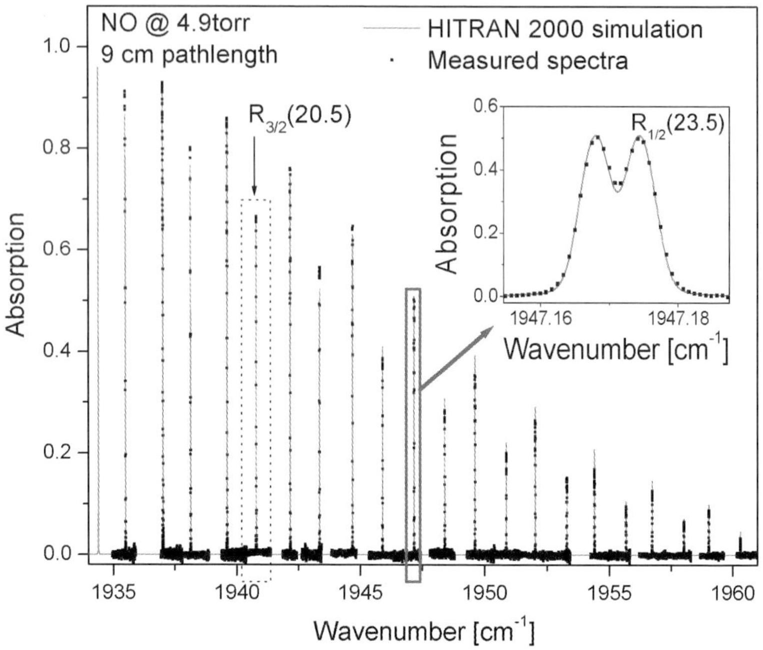

Fig. 4.3 Nitric oxide absorption spectra measured at different EC-QCL grating angles. The narrow laser linewidth allows a spectral selectivity of $<0.006\,cm^{-1}$

gas detection. In contrast to other infrared absorption techniques, PAS is an indirect technique in which the effect on the absorbing medium and not the direct light absorption is detected. Light absorption results in a transient temperature effect, which then translates into kinetic energy or pressure variations in the absorbing medium that can be detected with a sensitive microphone. PAS is ideally a background-free technique, since the signal is generated only by the absorbing gas. However, background signals can originate from nonselective absorption of the gas cell windows (coherent noise) and external acoustic (incoherent) noise. PAS signals are proportional to the pump laser intensity and therefore, PAS is most effective with high power laser excitation. A sensitivity of 8 ppmv was demonstrated with only 2 mW of modulated diode laser power in the CH_4 overtone region (Liang et al. 2000; Gomes et al. 2006). The implementation of high power cw DFB-QCL excitation in the fundamental absorption region leads to considerably improved trace gas detection sensitivity.

A recently introduced novel approach to photoacoustic detection of trace gases utilizing a quartz tuning fork (QTF) as a sharply resonant acoustic transducer was first reported in 2002 (Kosterev et al. 2002c; Kosterev et al. 2005a). The basic idea of quartz-enhanced photoacoustic spectroscopy (QEPAS) is to invert the common PAS approach and accumulate the acoustic energy not in a gas-filled cell but in a sharply resonant acoustic transducer. A natural candidate for such an transducer is crystalline quartz, because it is a low-loss piezoelectric material. A variety of packaged quartz crystals for use in timing applications is commercially available. Readily available low-frequency quartz elements are quartz tuning forks (QTF) intended for use in electronic clocks as frequency standards. These QTFs resonate at 32 768 (2^{15}) Hz in vacuum. A typical QEPAS absorption detection module (ADM) consisting of a QTF equipped with an accoustic micro-resonator is shown in Fig. 4.4. Only the symmetric vibration of a QTF (i.e. when the two QTF prongs bend in opposite directions) is piezo-electrically active. The laser beam is focused between the prongs of the QTF and its wavelength is modulated at $f_m = f_0/2$ frequency, where f_0 is the QTF resonant frequency. A lock-in amplifier is used to demodulate the QTF response at f_0. Spectral data can be acquired if the laser wavelength is scanned. To increase the effective interaction length between the radiation-induced sound and the QTF, an acoustic gas-filled micro-resonator can be added similarly as in the traditional PAS approach. Acoustically, a QTF is a quadrupole, which results in excellent environmental noise immunity. Sound waves from distant acoustic sources tend to move the QTF prongs in the same direction, thus resulting in no photoacoustic response. Advantages of QEPAS compared to conventional resonant photoacoustic spectroscopy include QEPAS sensor immunity to environmental acoustic noise, a simple absorption detection module design, no spectrally selective element is required, applicable over a wide range of pressures, including atmospheric pressure and its capability to analyze small gas samples, down to 1 mm^3 in

Fig. 4.4 QEPAS absorption detection module with micro-resonator

volume. QEPAS has already been demonstrated in trace gas measurements of NH_3 (Kosterev et al. 2004a), CO_2 (Weidmann et al. 2004; Wysocki et al. 2006), N_2O (Kosterev et al. 2005b), HCN (Kosterev et al. 2006), CO in propylene (Kosterev et al. 2004b) and CH_2O (Horstjann et al. 2004; Angelmahr et al. 2006).The measured normalized noise equivalent absorption coefficient for H_2O is 1.9×10^{-9} cm^{-1} $W/Hz^{-1/2}$ in the overtone region at 7306.75 cm^{-1} is the best among the tested trace gas species to date using QEPAS and is indicative of fast vibrational–translational relaxation of the initially excited states. An experimental study of the long-term stability of a QEPAS-based NH_3 sensor (Kosterev et al. 2005a) showed that the sensor exhibits very low drift, which allowed data averaging over >3 h of continuous concentration measurements.

Formaldehyde (H_2CO) is widely used in the manufacture of building materials and numerous household products. Outgassing of formaldehyde from these materials may lead to elevated indoor levels, particularly for poorly ventilated structures. It is also an important intermediate species in the oxidation of hydrocarbons in both combustion systems and in the troposphere. Thus, H_2CO may be present in substantial concentrations in both indoor and outdoor air samples. Tropospheric H_2CO concentration measurements provide a means of validating photochemical model predictions that play a key role in our understanding of tropospheric ozone formation chemistry (Wert et al. 2003).

Formaldehyde is a pungent-smelling, colorless gas that causes a variety of effects (including watery and/or burning eyes, nausea, and difficulty in breathing) in some humans exposed to H_2CO levels of only 100 ppbv. Known to cause cancer in animals, it is also a suspected human carcinogen. Formaldehyde is detectable by scent in humans in concentrations of 0.07–1.2 ppmv in air. The US Occupational Safety and Health Administration (OSHA) has established permissible exposure limit (PEL) of 0.75 ppmv averaged for an 8 h work day and short-term exposure limit (STEL) of 2 ppmv averaged over 15 min (Occupational Safety and Health Standards, 2006). At levels ten times the PEL, respiratory protection is required. Other organizations have taken a more aggressive approach to H_2CO management. For example, the threshold limit value (TLV) for H_2CO established by the American Conference of Governmental Industrial Hygienists is 0.3 ppmv as a “ceiling limit” not to be exceeded at any time, and has classified it as a “suspected human carcinogen”. A committee of the National Academy of Sciences, working on behalf of NASA, has extensively reviewed the toxicity of H_2CO and has established spacecraft maximum allowable concentrations for various times of exposure (Crossgrove 1994–2000). Their recommended upper limits range from 0.4 ppmv for short-term (1 h) exposure down to 0.04 ppmv for 7–180 day space missions.

A novel continuous-wave mid-IR DFB ICL was utilized to detect and quantify H_2CO using quartz-enhanced photoacoustic spectroscopy (Horstjann et al. 2004). The sensor architecture is depicted in Fig. 4.5. Unlike the Fabry–Perot gain chip, which is used in the widely tunable EC-QCL configuration described in Section 4.2 it is possible to use an ICL or QCL chip with an embedded

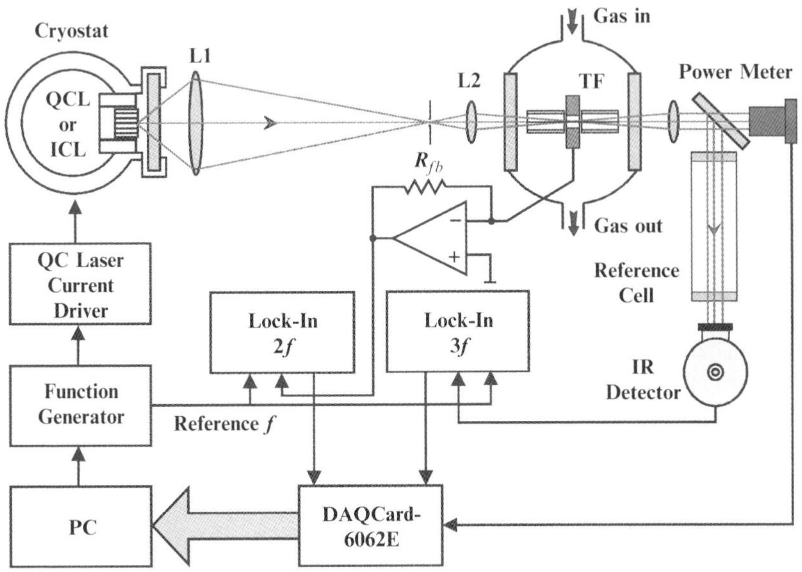

Fig. 4.5 Schematic diagram of a QCL- or ICL-based QEPAS sensor platform

DFB structure that provides a single-frequency output. This allows the ICL or QCL to be matched to a specific target analyte with a limited tuning range of ~1 cm^{-1} with current control and a few cm^{-1} with temperature control. The ICL was operated at liquid-nitrogen temperatures and provided a single-mode output power of up to 12 mW at 3.53 μm (2832.5 cm^{-1}). The noise equivalent (1 σ) detection sensitivity of the sensor was measured to be 1.1×10^{-8} cm^{-1} $W/Hz^{-1/2}$ for H_2CO in ambient air with 75% relative humidity, which corresponds to a detection limit of 0.20 ppmv for a 1 s sensor time constant and 4.6 mW laser power delivered to the QEPAS absorption detection module.

Recently, Wojcik et al., 2006 demonstrated the performance of a novel infrared photoacoustic laser absorbance sensor for gas-phase species using an amplitude-modulated (AM) QCL and a quartz tuning fork microphone. A photoacoustic signal is generated by focusing 5.3 mW of a Fabry–Perot QC laser operating at 8.41 μm between the tines of a quartz tuning fork which served as a transducer for the transient acoustic pressure wave. The sensitivity of this sensor was calibrated using the infrared absorber Freon-134 a by performing a simultaneous absorption measurement using a 31 cm absorption cell. The power and bandwidth normalized noise equivalent absorption sensitivity (NEAS) of this sensor was determined to be 2.0×10^{-8} cm^{-1} $W/Hz^{-1/2}$, which translates into noise equivalent concentration of 40 ppbv for the available QCL laser power and a data acquisition time of 1 s.

4.4 Trace Gas Sensors using a High-Finesse Optical Cavity

Sensitive laser absorption spectroscopy often requires a long effective optical pathlength of the probing laser beam in media that are to be analyzed. Traditionally, this requirement is satisfied using an optical multipass cell. Such an approach can be difficult to implement in certain field applications, requiring compact gas sensor configurations. For example, a typical commercial 100 m long multipass cell has a volume of 3.5 L. An alternative way to obtain a long optical path is to make the light bounce along the same path between two parallel ultralow-loss dielectric mirrors. An effective optical pathlength of several kilometers can be obtained in a very small volume. The light leaking out of such an optical cavity can be used to characterize the absorption of the intracavity medium. Presently a variety of techniques exists to perform high sensitivity absorption spectroscopy in a high-finesse optical cavity (Berden et al. 2000). Two methods are cavity ringdown spectroscopy (CRDS) (Kosterev et al. 2001) and integrated cavity output spectroscopy (ICOS) (Bakhirkin et al. 2004).

A gas analyzer based on a continuous-wave mid-infrared quantum cascade laser operating at ~5.2 μm and off-axis integrated cavity output spectroscopy (OA-ICOS) has been developed to measure NO concentrations in human breath (Bakhirkin et al. 2004). A compact sample cell, 5.3 cm in length and with a volume of ~80 cm^3, which is suitable for online and off-line measurements during a single breath cycle, was designed and tested. A noise-equivalent (signal-to-noise ratio of 1) sensitivity of 10 ppbv of NO was achieved. The combination of ICOS with wavelength modulation resulted in a 2 ppbv noise-equivalent sensitivity. The total data acquisition and averaging time was 15 s in both cases. In 2005, a pulsed, non-cryogenic cavity-enhanced spectrometer that operates at 5.2 μm was developed (Silva et al. 2005).

Subsequently, a nitric oxide sensor based on a thermoelectrically cooled, cw DFB QCL laser operating at 5.45 μm (1835 cm^{-1}) and off-axis ICOS combined with a wavelength-modulation technique was developed to determine NO concentrations at the sub-ppbv levels that are essential for a number of applications, such as medical diagnostics (specifically in detecting NO in exhaled human breath) and environmental monitoring (Bakhirkin et al. 2006; McCurdy et al. 2006). Exhaled nitric oxide (eNO) is an important biomarker in many respiratory diseases (Namjou et al. 2006). Exhaled NO levels have been extensively studied in asthma. These measurements may be clinically useful in other chronic respiratory conditions, such as chronic obstructive pulmonary disease, particularly if the NO contributions are partitioned into alveolar and conducting airway regions. eNO levels are generally in the range of 4–15 ppbv in healthy human subjects and 10–160 ppbv in subjects with untreated asthma when breath is collected at the standard 3 L min^{-1}, in accordance with the American Thoracic Society recommendations.

The sensor as shown in Fig. 4.6 employs a 50 cm-long high-finesse optical cavity that provides an effective path length of 700 m. A noise equivalent minimum detection limit of 0.7 ppbv with a 1 s observation time was achieved (Bakhirkin et al. 2006). A wavelength modulated signal for a calibrated NO concentration of 23.7 ppbv was fitted using a general linear fit procedure (Kosterev et al. 2002d) (see Fig. 4.7).

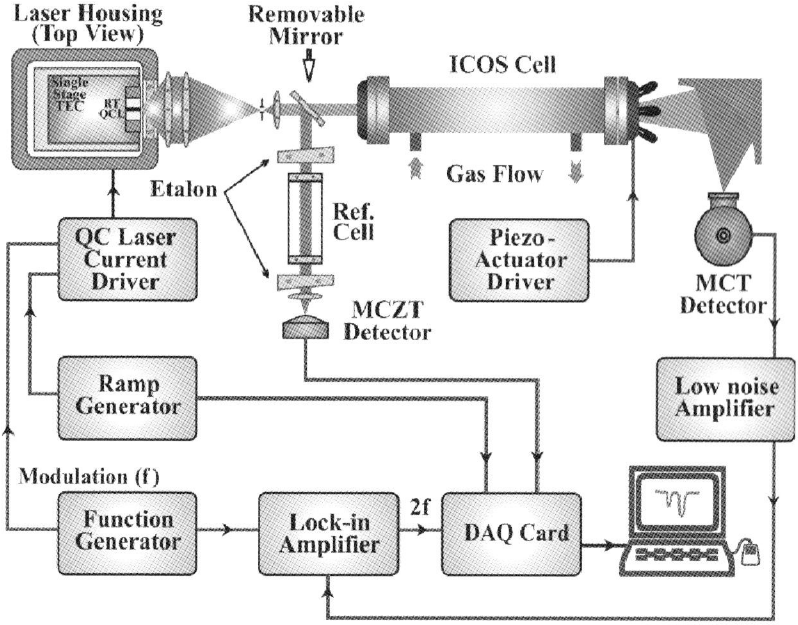

Fig. 4.6 Schematic diagram of a CW-TEC-DFB QC laser-based nitric oxide off-axis integrated cavity output spectrometer

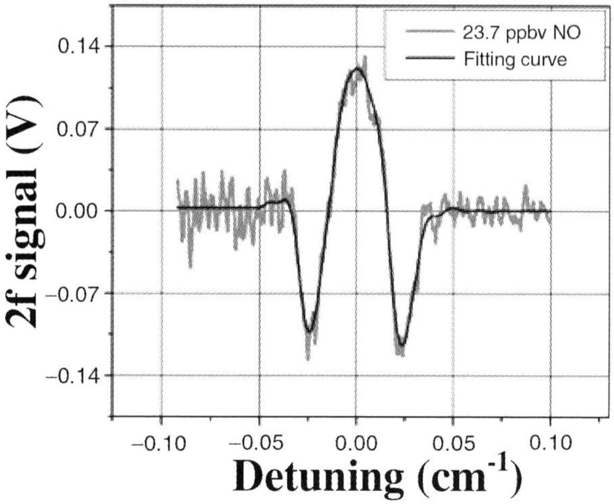

Fig. 4.7 2f OA-ICOS based NO absorption spectrum

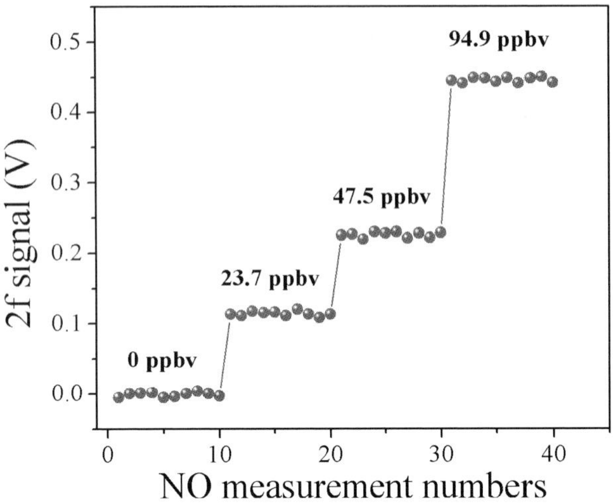

Fig. 4.8 Calibration of an OA-ICOS NO sensor

A 1 σ deviation of the amplitude obtained as a fit result corresponds to 0.7 ppbv. Step concentration measurements of NO are depicted in Fig. 4.8. Each measurement was repeated ten times and the amplitude of the wavelength-modulated spectroscopic signal was retrieved and plotted in Fig. 4.8. More recently (Nelson et al. 2006), a detection sensitivity of 0.03 ppbv was achieved with 30 s averaging time with a path length of 210 m, corresponding to an absorption coefficient of 1.5×10^{-10} cm^{-1}.

Detection of formaldehyde using off-axis ICOS with a 12 mW ICL was recently demonstrated (Miller et al. 2006). A 3.53 µm continuous-wave, mid-infrared, distributed feedback ICL was used to quantify H_2CO in gas mixtures containing ≈1–25 ppmv of H_2CO. Analysis of the spectral measurements indicates that a H_2CO concentration of 150 ppbv would produce a spectrum with a signal to noise ratio of 3 for a data acquisition time of 3 s. This is a relevant sensitivity level for formaldehyde monitoring of indoor air, occupational settings, and on board spacecraft in long duration missions in particular as the detection sensitivity improves with the square root of the data acquisition time.

4.5 Summary

Compact, sensitive, and selective gas sensors based on near infrared and mid-infrared semiconductor lasers have been demonstrated to be effective in numerous real-world applications. These applications include such diverse fields as environmental monitoring (e.g. CO, CO_2, CH_4 and H_2CO are important carbon gases in global warming and ozone depletion studies), industrial emission measurements (e.g. fence line perimeter monitoring in the petrochemical industry, combustion sites, waste incinerators, down gas well monitoring, gas pipeline and compressor station safety), urban (e.g. automobile traffic, power generation) and rural emissions (e.g. horticultural greenhouses, fruit storage and rice agro-ecosystems), chemical analysis and process control for manufacturing processes (e.g. semiconductor, pharmaceutical, food), detection of medically important molecules (e.g. NO, CO, CO_2, NH_3, C_2H_6 and CS_2), toxic gases, drugs, and explosives relevant to law enforcement and public safety, and spacecraft habitat air-quality and planetary atmospheric science (e.g. such planetary gases as H_2O, CH_4, CO, CO_2 and C_2H_2).

Acknowledgements We are grateful to Drs. C. Gmachl, F. Capasso, J. Faist, R. Maulini and R. Yang for their invaluable scientific support. Financial support of the research performed by the Rice group was provided by the National Aeronautics and Space Administration via awards from the Johnson Space Center, Houston, Texas and Jet Propulsion Laboratory, Pasadena, California, the Pacific Northwest National Laboratory, Richland, Washington, the National Science Foundation and the Welch Foundation, the Office of Naval Research via a sub-award of Texas A & M University and the Texas Advanced Technology Program.

References

Angelmahr M., Miklos A., and Hess P., (2006), Photoacoustic spectroscopy of formaldehyde with tunable laser radiation at the parts per billion level, *Appl. Phys. B*, 85, 285–288.

Bakhirkin Y.A., Kosterev A.A., Roller C., Curl R.F., and Tittel F.K., (2004), Mid-infrared quantumcascade laser based off-axis integrated cavity output spectroscopy for biogenic no detection, *Appl. Opt.*, 43, 2257–2266.

Bakhirkin Y.A., Kosterev A.A., Curl R. Tittel F.K., Yarekha D.A., Hvozdara L., Giovannini M., and Faist J., (2006), Sub-ppbv nitric oxide concentration measurements using cw room-temperature quantum cascade laser based integrated cavity spectroscopy, *Appl. Phys. B*, 82, 149–154.

Beck M., Hofstetter D., Aellen T., Faist J., Oesterle U., Ilegems M., Gini E., Melchior H. (2002), Continuous wave operation of a mid-infrared semiconductor laser at room temperature, *Science*, 295, 301–305.

Berden G., Peeters R., and Meijer G., Cavity ring-down spectroscopy, (2000), Experimental schemes and applications, *Int. Rev. Phys. Chem.*, 19, 565–607.

Bradshaw J.L., Breznay N.P., Bruno J.D., Gomes J.M., Pham J.T. Towner F.J., Wortman D.E., Tober R.L., Monroy C.J., and Olver K.A. (2004), Recent progress in the development of type II interband cascade lasers, *Physica E*, 20, 479–485.

Capasso F., Gmachl C., Paiella R., Tredicucci A., Hutchinson A.L., Sivco D.L., Baillargeon J.N., and Cho A. Y. (2000), New frontiers in quantum cascade lasers and applications, *IEEE Sel. Top. Quantum Electron.*, 6, 931–947.

Crossgrove R.E., Spacecraft maximum allowable concentrations for selected airborne contaminants, *Natl. Acad. Sci.*, 1–4, 1994–2000.

Curl R.F. and Tittel F.K. (2002), Tunable infrared laser spectroscopy, *Ann. Rep. Prog. Chem. Sect.*, 98, 219–272.

Diehl L., Bour D., Corzine S., Zhu J., Hoefler G., Loncar M., Troccoli M., and Capasso F. (2006), High-power quantum cascade lasers grown by low-pressure metal organic vapor-phase epitaxy operating in continuous wave above 400 K, *Appl. Phys Lett.*, 88, 201115.

Evans A., Yu J.S., Slivken S., and Razeghi M. (2004), Continuous-wave operation of l ~ 4.8 mm quantum-cascade lasers at room temperature, *Appl. Phys. Lett.*, 85, 2166–2168.

Faist J., Hofstetter D., Beck M., Aellen T., Rochat M., and Blaser S. (2002), Bound-to-continuum and two-phonon resonance quantum-cascade lasers for high duty cycle, high temperature operation, *IEEE. J. Quantum Electron.*, 38, 533–546.

Faist J., Beck M., Aellen T., and Gini E. (2001), Quantum cascade laser based on bound-to-continuum transition, *Appl. Phys. Lett.*, 78, 147–149.

Gomes M., Da Silva, Miklos A., Falkenroth A., and Hess P. (2006), Photoacoustic measurement of N_2O concentrations in ambient air with a pulsed optical parametric oscillator, *Appl. Phys. B*, 82, 329–336.

Horstjann M., Bakhirkin Y.A., Kosterev A.A., Curl R.F., and Tittel F. K. (2004), Formaldehyde sensor using interband cascade laser based quartz-enhanced photoacoustic spectroscopy, *Appl. Phys. B*, 79, 799–803.

Kosterev A.A., Curl R.F., Tittel F.K., Kochler R., Gmachl C., Capasso F., Sivco D.L., Cho A.Y., Wehe S., and Allen M. (2002b), Thermoelectrically cooled quantum cascade laser based sensor for the continuous monitoring of ambient atmospheric carbon monoxide, *Appl. Opt.*, 41, 1169–1173.

Kosterev A.A. and Tittel F.K. (2002a), Chemical sensors based on quantum cascade lasers, *IEEE J. Quantum Electron.* 38, 582–591.

Kosterev A.A. and Tittel F.K. (2004a), Ammonia detection by use of quartz-enhanced photoacoustic spectroscopy with a near-IR telecommunication diode laser. *Appl. Opt.*, 43, 6213–6217.

Kosterev A.A., Tittel F.K., Serebryakov D., Malinovsky A., and Morozov A. (2005a), Applications of quartz tuning fork in spectroscopic gas sensing, *Rev. Scient. Ins.*, 76, 043105.

Kosterev A.A., Mosely T.S., and Tittel F.K. (2006), Impact of humidity on quartz enhanced photoacoustic spectroscopy based detection of HCN, *Appl. Phys. B*, 85, 295–300.

Kosterev A.A., Bakhirkin Y.A., and Tittel F.K. (2005b), Ultrasensitive gas detection by quartz-enhanced photoacoustic spectroscopy in the fundamental molecular absorption bands region, *Appl. Phys. B*, 80, 133–138.

Kosterev A.A., Bakhirkin Y.A., Curl R.F., and Tittel F.K. (2002c), Quartz-enhanced photoacoustic spectroscopy, *Opt. Lett.*, 27, 1902–1904.

Kosterev A.A., Malinovsky A.L., Tittel F.K., Gmachl C., Capasso F., Sivco D.L., Baillargeon J. N., Hutchinson A.L., and Cho A.Y. (2001), Cavity ring-down spectroscopy of NO with a single frequency quantum cascade laser, *Appl. Opt.*, 40, 5522–5529.

Kosterev A.A., Bakhirkin Y.A., Tittel F.K., Blaser S., Bonetti Y., and Hvozdara L. (2004b), Photoacoustic phase shift as a chemically selective spectroscopic parameter, *Appl. Phys. B (Rapid Commun.)*, 78, 673–676.

Kosterev A.A., Curl R.F., Tittel F.K., Koehler R., Gmachl C., Capasso F, Sivco D.L., Cho A.Y. (2002d), Transportable automated ammonia sensor based on a pulsed thermoelectrically cooled quantum-cascade distributed feedback laser *Appl. Opt.* 41, 573–578.

Liang G.C., Hon-Huei Liu Kung A.H., Mohacsi A., Miklos A., and Hess P. (2000), Photoacoustic trace detection of methane using compact solid-state lasers, *J. Phys. Chem. A*, 104, 10179–10183.

Mann C.H., Yang Q.K., Fuchs F., Bronner W., Kiefer R., Koehler K., Schneider H., Kormann R., Fischer H., Gensty T., and Elsaesser W. (2003), Quantum cascade lasers for the mid-infrared spectral range: Devices and applications, B. *Adv. Solid State Phys.* Kramer (Ed.), Springer Verlag, Berlin, Heidelberg, 43, 351–368.

Mansour K., Qiu Y., Hill C.J., Soibel A., and Yang R.Q. (2006), Mid-Infrared interband cascade lasers at thermoelectric cooler temperatures, *Electron. Lett.*, 42, 1034–1036.

Maulini R., Mohan A., Giovannini M., Faist J., and Gini E. (2006), External cavity quantum-cascade laser tunable from 8.2 to 10.4 μm using a gain element with a heterogeneous cascade, *Appl. Phys. Lett.*, 88, 201113–201116.

Maulini R., Beck M., Faist J., and Gini E. (2004), Broadband tuning of external cavity bound-to-continuum quantum-cascade lasers, *Appl. Phys. Lett.*, 84, 1659–1661.

McCurdy M., Bakhirkin Y.A., and Tittel F.K. (2006), Quantum cascade laser-based integrated cavity output spectroscopy of exhaled nitric oxide, *Appl. Phys. B*, 85, 445–452.

Miller J.H., Bakhirkin Y.A., Ajtai T., Tittel F.K., Hill C.J., and Yang R.Q., (2006), Detection of formaldehyde using off-axis integrated cavity output spectroscopy with an interband cascade laser, *Appl. Phys. B*, 85, 391–396.

Namjou K., Roller C.B., Reichl T.E., Jeffers J.D., McMillen G.L., McCann P.J., and Camp M.A. (2006), Determination of exhaled nitric oxide distributions in a diverse sample population using tunable diode laser absorption spectroscopy, *Appl. Phys. B*, 85, 427–435.

Nelson D.D., McManus J.B., Herndon S.C., Shorter J., Zahniser M.S., Blaser S., Hvozdara L., Mueller A., Giovannini M., and Faist J. (2006), Characterization of a near-room temperature, continuous-wave quantum cascade laser for long-term, unattended monitoring of nitric oxide in the atmosphere, *Opt. Lett.*, 31, 2012–2014.

Occupational Safety and Health Standards, in *Standards-29 CFR*. 2006 .

Peng C., Luo G., and Le H.Q. (2003), Broadband, continous, and fine-tune properties of external-cavity thermoelectric-stabilized mid-infrared quantum-cascade lasers, *Appl. Opt.*, 42, 4877–4882.

Richter D. and Weibring P. (2006), Ultra high precision mid-IR spectrometer: design and analysis of an optical fiber pumped difference frequency generation source, *Appl. Phys. B*, 82, 479–486.

Silva M.L., Sonnenfroh D.M., Rosen D.I., Allen M.G., and O'Keefe A. (2005), Integrated cavity output spectroscopy measurements of nitric oxide levels in breath with a pulsed room–temperature quantum cascade laser, *Appl. Phys. B*, 81, 705–710.

Tittel F. K., Richter D., and Fried A. (2003), Mid-Infrared Laser Applications in Spectroscopy. Solid State Mid-Infrared Laser Sources, I.T Sorokina, and K.L. Vodopyanov (Eds.), *Springer Verlag*, *Topics Appl. Phys.*, 89, 445–510.

Troccoli M., Bour D., Corzine S., Hofler G., Tandon A., Mars D., Smith D.J., Diehl L., and Capasso F. (2004), Low-threshold continuous-wave operation of quantum-cascade lasers grown by metalorganic vapor phase epitaxy, *Appl. Phys. Lett.*, 85, 5842–5844.

Weidmann D., Kosterev A.A., Tittel F.K., Ryan N., and McDonald D. (2004), Application of widely electrically tunable diode laser to chemical gas sensing with quartz-enhanced photoacoustic spectroscopy, *Opt. Lett.*, 29, 1837–1839.

Wert B.P., et al., (2003), Signatures of terminal alkene oxidation in airborne formaldehyde measurements during TexAQS 2000, *J. Geophys. Res.*, (*Atmospheres*), 108 (D3), 4104.

Wojcik M.D., Phillips M.C., Cannon B.D., and Taubman M.S. (2006), Gas phase photoacoustic sensor at 8.41 mm using quartz tuning forks and amplitude modulated quantum cascade lasers, *Appl. Phys. B*, 85, 307–313.

Wysocki G., Kosterev A.A., and Tittel F.K. (2006), Influence of molecular relaxation dynamics on quartz-enhanced photoacoustic detection of CO_2 at $\lambda = 2\,\mu m$, *Appl. Phys. B*, 85, 301–306.

Wysocki G., Curl R.F., Tittel F.K., Maulini R., Bulliard J.M., and Faist J. (2005), Widely tunable mode-hop free external cavity quantum cascade laser for high resolution spectroscopic applications, *Appl. Phys. B*, 81, 769–777.

Yanagawa T., Kanbara H., Tadanaga O., Asobe M., Suzuki H., and Yumoto J. (2005), Broadband difference frequency generation around phase-match singularity, *Appl. Phys. Lett.*, 86, 161106.

Yang R.Q., Hill C.J., Yang B.H., Wong C.M., Muller R.E., and Echternach P.M. (2004), Continuous-wave operation of distributed feedback interband cascade lasers, *Appl. Phys. Lett.*, 84, 3699–3701.

Yang R.Q., Bradshaw J.L., Bruno J.D., Pham J.T., and Wortman D.E. (2002), Mid-infrared type II interband cascade lasers, *IEEE. J. Quant. Electron.*, 38, 547–558.

Chapter 5
Atmospheric Monitoring With Chemical Ionisation Reaction Time-of-Flight Mass Spectrometry (CIR-TOF-MS) and Future Developments: Hadamard Transform Mass Spectrometry

Kevin P. Wyche, Christopher Whyte, Robert S. Blake, Rebecca L. Cordell, Kerry A. Willis, Andrew M. Ellis, and Paul S. Monks

Abstract Chemical ionisation reaction mass spectrometry (CIR-MS) is a more general version of proton transfer reaction mass spectrometry (PTR-MS) in which alternative chemical ionisation schemes are possible. This concept has been realised in a new instrument based on time-of-flight mass spectrometry (TOF-MS) and has been applied to the measurement of a range of trace atmospheric volatile organic compounds (VOCs) and oxygenated volatile organic compounds (OVOCs) (Blake et al., 2004 and Wyche et al., 2005). Initial results have demonstrated the instrument to be capable of recording the entire mass spectrum in "real time" (ca. 1 min) with sensitivities in the order of 0.1 counts ppbV^{-1} s^{-1} in each unit mass channel. This article constitutes a brief overview of the CIR-TOF-MS instrument and several of its applications. A short account is also given of the "next generation" instrument which is under development. This new instrument will combine rapid beam modulation with Hadamard transformation of the detector output and should improve the detection sensitivity by more than an order of magnitude over the current CIR-TOF-MS instrument.

Keywords: Proton transfer reaction mass spectrometry (PTR-MS), chemical ionisation reaction (CIR-MS) mass spectrometry, volatile organic compound, aerosol, Hadamard transform

5.1 Introduction

Proton transfer reaction mass spectrometry (PTR-MS) is now an established technique for identifying and quantifying trace volatile organic compounds (VOCs) in air. It utilises the hydronium ion (H_3O^+) as the primary chemical ionisation (CI) reagent, and achieves quantification by allowing reaction to take place in a drift

Department of Chemistry, University of Leicester, Leicester, UK

Y.J. Kim and U. Platt (eds.), *Advanced Environmental Monitoring*, 64–76.

tube with a fixed reaction time. This technique shows great potential in a variety of applications, including the real-time monitoring of atmospheric species and pollution episodes (Hewitt et al., 2003), medical diagnostics via breath analysis (Lirk et al., 2004) and forensic investigations.

The proton transfer reaction will proceed only if the proton affinity of the analyte is greater than that of H_2O. This excludes the bulk components of air, such as nitrogen and oxygen, but includes the majority of organic compounds (with the main exception being the short chain alkanes). The ionisation process tends to be quite 'soft' which helps to minimise ion fragmentation and therefore assist compound identification in complex mixtures. In contrast to its most well-established competitor technique, gas chromatography/mass spectrometry (GC/MS), PTR-MS continuously collects data "online" on timescales of about 1 s.

Standard PTR-MS instruments employ quadrupole mass spectrometers (Hansel et al., 1995), which suffer from several drawbacks such as relatively low mass resolution ($m/\Delta m = 100$) and limited mass range (generally poor transmission above 1000 Da). However, the most notable weakness of quadrupole mass spectrometers is that they act as mass filters, and therefore are unable to capture the entire mass spectrum in a given instant. In an attempt to circumvent these problems, we have developed a PTR-MS system based on time-of-flight mass spectrometry (TOF-MS), which benefits from a mass resolution ($m/\Delta m$) in excess of 1000, a potentially limitless mass range, and most significantly the ability to observe all mass channels simultaneously (Blake et al., 2003). However, we have also shown that the system can function with chemical ionisation reagents other than proton donating species, such as NO^+ and O_2^+ (Blake et al., 2006). This more general technique is called chemical ionisation reaction time-of-flight mass spectrometry (CIR-TOF-MS).

The new University of Leicester CIR-TOF-MS has been applied to the monitoring of atmospheric volatile organics in the contemporary urban boundary layer and has recently taken part in several key environment chamber studies. These studies are briefly described here, along with an indication of new developmental work in progress, focusing particularly on a Hadamard transform version of the CIR-TOF-MS instrument.

5.2 Experimental

The University of Leicester CIR-TOF-MS instrument, which is shown schematically in Fig. 5.1 has been described in detail elsewhere (Blake et al., 2004). In brief, the ion source/drift tube assembly (constructed in house) consists of a series of seven stainless steel electrodes separated by insulating spacers. The drift region is operated at a relatively high pressure (~8–10 mbar) in order to optimise the ion yield. Hence, under standard operating drift tube voltages, an E/N ratio (where E = electric field strength and N = gas number density) of around 150 Td is achieved (1 Townsend [Td] = 10^{-17} V cm^2). In contrast to most standard PTR-MS systems, we employ a radioactive ion source (^{241}Am which emits α particles with energy of

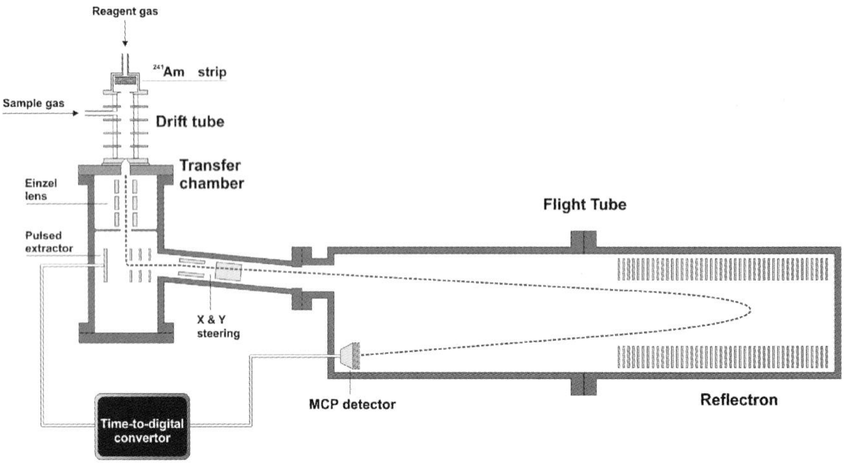

Fig. 5.1 Schematic diagram of the CIR-TOF-MS instrument

the order 5 MeV) to generate the CI reagent. The benefits of using a radioactive ion source include the lack of any substantial settling time after a change in source operating conditions, such as a change in the identity of CI reagent gas. The time-of-flight mass spectrometer, supplied by Kore technology (Ely, UK), is of the orthogonal variety with a duty cycle of approximately 3% for a mass scan of 0–100 Da. The instrument has a mass resolution in excess of 1000.

In order to limit water contamination, all alternative reagent gases employed in place of water vapour are first passed through a 1 m long (1/8″ outer diameter) stainless steel coil held at −77°C, in a bath of dry ice and acetone, before injection into the ion source at 52 sccm. The alternative reagents included nitric oxide in N_2 (600 ppmV, BOC special gases), oxygen (BOC special gases, N5.0 grade) and ammonia (BOC special gases, MS grade). Mass scans are typically conducted with an accumulation period of 60 seconds.

VOC quantification can be achieved in the CIR-TOF-MS via experimentally derived compound sensitivities, obtained from standard gas calibration mixtures, or theoretically from knowledge of the drift cell kinetics. Instrument sensitivity is defined as the number of ion counts acquired by the instrument per 1 ppbV of sample, following normalisation of the entire mass scan to 10^6 primary CI reagent ion counts per second (units: normalised counts per second (ncps) ppbV^{-1}). Theoretical quantification can be achieved using equation (E1) below, which is derived by applying simple kinetic arguments for the proton-transfer reaction (R1), assuming pseudo–first-order conditions are obeyed (i.e. $i(H_3O^+)_0 > i(MH^+)$):

$$i(MH^+) \approx i(H_3O^+)_0[M]kt \quad (E1)$$

$$H_3O^+ + M \rightarrow MH^+ + H_2O \quad (R1)$$

The concentration of analyte M can be deduced by measuring the ratio $i(MH^+)/i(H_3O^+)_0$, where: $i(MH^+)$ is the normalised protonated analyte ion signal (ncps), $i(H_3O^+)_0$ is the normalised hydronium signal (10^6 counts per second), [M] is the analyte concentration, *k* is the proton transfer rate constant (molecules cm^3 $molecule^{-1}$ s^{-1}) and *t* is the ion-molecule reaction time in the drift tube (Hansel et al., 1995; Blake et al., 2004). Equation (E1) can be generalised to CI reagents other than H_3O^+.

5.3 Example Results and Discussion

5.3.1 Chemical Ionisation Reaction-Mass Spectrometry (CIR-MS)

Figure 5.2 displays the raw CIR-MS mass spectrum, recorded in the absence of sample, when NO^+ was employed as the primary reagent ion as an alternative to hydronium. The data presented in Fig. 5.2 clearly demonstrates that the CIR-TOF-MS instrument is capable of generating a clean source of NO^+. Similarly clean spectra have been observed for other CI reagents, although a small amounts of H_3O^+ is still observable at m/z = 19 in this particular example (Wyche et al., 2005 and Blake et al., 2006).

As an example of where the use of NO^+ might be useful, consider the mass spectra of acetone and propanal generated by proton transfer from H_3O^+. The resulting spectra, shown in the left panel of Fig. 5.3 are virtually identical, both dominated by the $[MH]^+$ peak at the same *m/z*. This illustrates one of the problems of proton transfer, namely its frequent inability to distinguish between isobaric species.

However, when NO^+ is employed as the CI reagent the mass spectra for these two test compounds change markedly. Acetone is observed to form an ion/molecule association complex (*m/z* = 88) as its dominant product, whereas its aldehyde counterpart did not. In contrast, propanal tends to undergo fragmentation on charge transfer from NO^+, with the resultant dominant peaks occurring at *m/z* = 29 (HCO^+, $C_2H_5^+$) and 27 ($C_2H_3^+$). There is also a small but detectable $[M\text{-}H]^+$ peak for propanal which is missing from the spectrum of acetone. These examples are illustrative of more general findings, also observed in SIFT-MS studies of aldehydes and ketones (Spanel et al. 1997), which can be summarised : (1) saturated ketones generally react with NO^+ to form ion/molecule association complexes, with a number of fragment ions observed in lighter mass channels whose abundance increases with increasing

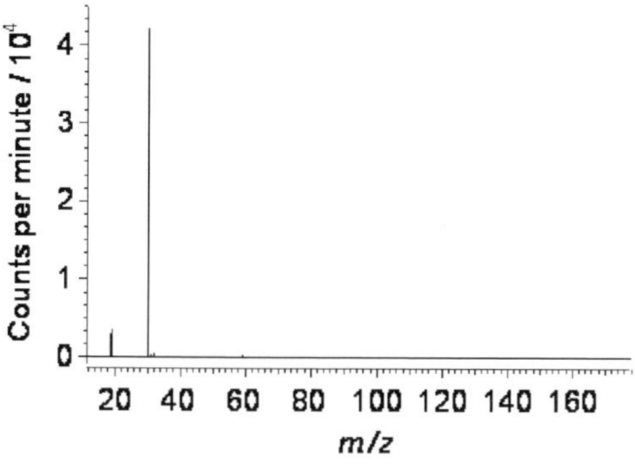

Fig. 5.2 Raw mass spectrum taken using NO^+ as the primary reagent ion in the absence of an analyte

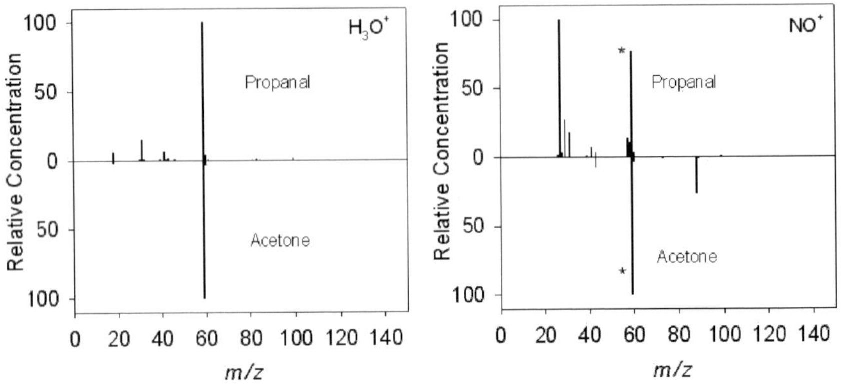

Fig. 5.3 Mass spectra for acetone and propanal generated with either H_3O^+ (left panel) or NO^+ ions (right panel) as the CI reagent. Asterisked peaks are present as a consequence of residual water contamination, i.e. ion generation by proton transfer from H_3O^+. It should be noted that the mass spectra above have been subjected to a background subtraction which removes the contribution from the CI reagent ion

chain length; (2) aldehydes generally fragment more extensively in their reactions with NO^+, typically producing formyl, alkyl and acylium ions.

The significance of the data in Fig. 5.3 lies in the benefits accrued by switching from H_3O^+ to NO^+. This switch provides distinct 'fingerprints' in the spectrum derived from NO^+ which makes it possible, at least in principle, to distinguish the two isobaric compounds. This effect has been shown to be quite general and it is even possible in some cases to discriminate between structural isomers by this means (Wyche et al., 2005).

5.3.2 Atmospheric Monitoring

During January 2005, the CIR-TOF-MS instrument was employed in the ACCENT-sponsored OVOC measurement intercomparison held at the SAPHIR atmospheric simulation chamber (Jülich, Germany) (for details regarding the chamber itself, see Karl et al., 2004). The intercomparison campaign comprised five separate experiments during which a series of 14 different atmospherically significant OVOCs were monitored at three distinct concentration levels, under various humidities and ozone content.

Figure 5.4 displays the real-time data obtained from the chamber measurements. This focuses on acetaldehyde, and shows both CIR-TOF-MS results (circles) alongside the concentrations (blue line) expected on the basis of the amount of acetaldehyde injected into the SAPHIR chamber. The CIR-TOF-MS measurements were carried out using H_3O^+ as the CI reagent. Data was accumulated for approximately 3 h at the initial concentration (region A in Fig. 5.4) before being diluted in two successive steps (regions B and C). Inspection of Fig. 5.4 reveals the high quality of the CIR-TOF-MS measurements, with all three measuring periods (A, B and C) and both dilution phases clearly captured, with the instrument tracking the expected concentrations throughout. Based on 1 min measurements, the accuracy, precision and signal to noise ratio in this instance were as good as 12.5%, 15.4% and 8.1% respectively, for region A (where the acetaldehyde concentration is in the region of 10 ppbV). Full details regarding the performance and validation of the CIR-TOF-MS technique during the SAPHIR OVOC intercomparison can be found in Wyche et al., 2006.

Combining the benefits of fast, accurate and precise on-line data acquisition with the multichannel capability of time-of-flight mass spectrometry, allows the construction of a highly detailed picture of the sample matrix and its evolution with time. The CIR-TOF-MS technique allows even the most complex and transient events to be monitored in high detail. Fig. 5.5 demonstrates this by showing real-time data recorded in all mass channels during one of the OVOC intercomparison experiments.

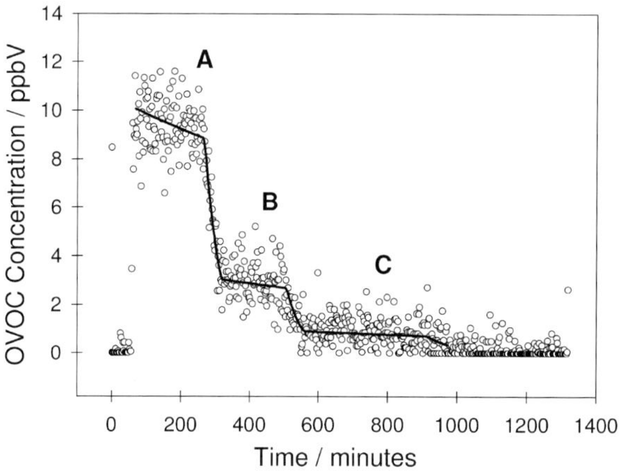

Fig. 5.4 Real-time CIR-TOF-MS measurements for acetaldehyde (circle points) along with estimated concentrations (blue line) derived from the known amount of acetaldehyde injected into the SAPHIR chamber

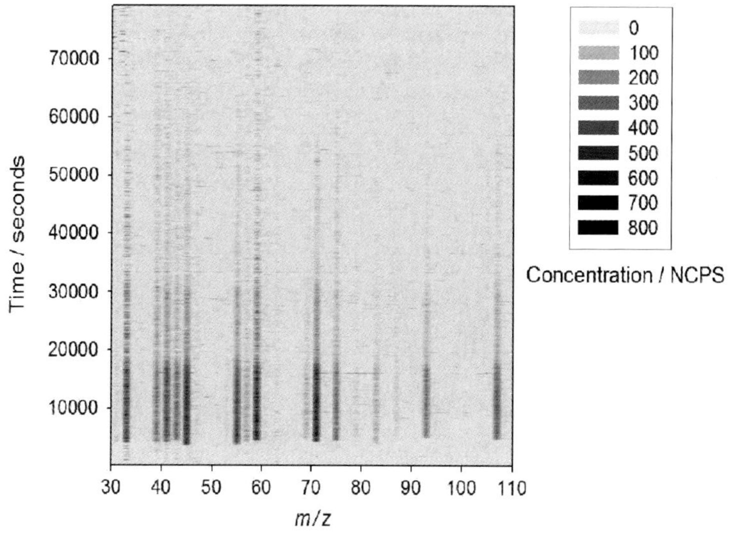

Fig. 5.5 Multi channel time-of-flight mass spectrum recorded in real time over an entire experiment during the OVOC intercomparison campaign

In order to demonstrate the quality of CIR-TOF-MS data acquisition and to further examine the sensitivity of the instrument to trace compounds during real-time measurement, example VOC data recorded during urban air sampling are presented in Fig. 5.6. Air from outside of the university was extracted into the CIR-TOF-MS instrument via a Teflon sampling tube over a period lasting one week. Leicester is a medium-sized U.K. city (population ~300,000 people) and the sample site is located approximately one mile from the city centre, where substantial quantities of anthropogenic VOCs and OVOCs are expected. Figure 5.6 shows the temporal profile recorded over a week period for two common urban pollutants, acetone and 1-propanol.

The two traces in Fig. 5.6 display the typical signature of the urban atmosphere, with strong peaks around early–mid morning during the city 'rush hour' where commuter traffic results in elevated levels of anthropogenic VOCs and OVOCs.

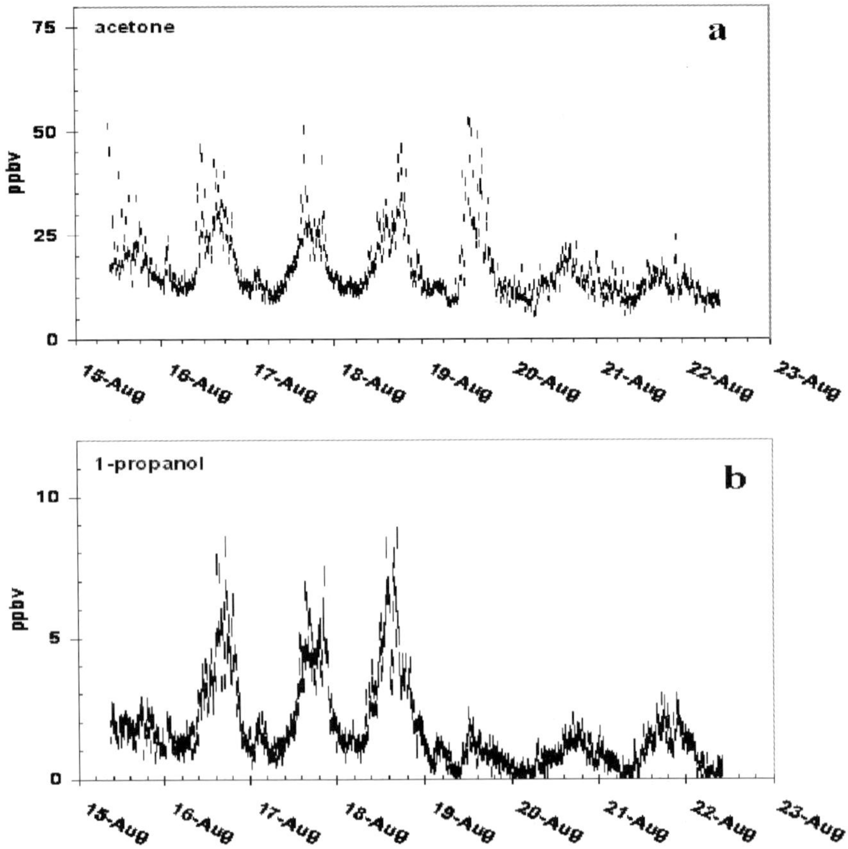

Fig. 5.6 Urban atmosphere measurements for **a** acetone and **b** 1-propanol over a 1 week period

5.3.3. *Secondary Organic Aerosol Formation*

Section co-authors: **Axel Metzger, Josef Dommen, Jonathan Duplissy and Urs Baltensperger** (Laboratory for Atmospheric Chemistry, Paul Scherrer Institute, Ch.).

Secondary organic aerosol (SOA) is a major contributor to the total mass of particulate matter in the modern urban troposphere (Baltensperger 2005). Classical SOA formation and growth schemes involve both homogeneous nucleation and heterogeneous condensation of the heavy, low volatility oxidation products of certain precursor VOCs. Additionally, the incorporation of lower molecular weight gas phase compounds into the aerosol has also received renewed attention recently in light of findings by Jang and Kamens (2001)and Kalberer et al., (2004), who have provided evidence for the formation and existence of oligomer species in the aerosol.

Recently the CIR-TOF-MS was employed in a series of environment chamber experiments at the Paul Scherrer Institut (Zurich), investigating the gas phase evolution during SOA formation. SOA formation was initiated by the photooxidation of either isoprene, α-pinene or 1,3,5- trimethylbenzene (see Paulsen et al. 2005 for chamber details).

Figure 5.7 presents the temporal evolution of several key gas phase compounds measured by the CIR-TOF-MS during the simulation chamber photooxidation of 1,3,5-trimethylbenzene. The VOC profiles presented in Figure 5.7 include the precursor, 1,3,5-trimethylbenzene (circles), the oxidation products 3,5-dimethyl-3(2H)-furanone, 3,5-dimethyl-5(2H)-furanone, 2 methyl-4-oxo-2-pentenal and methyl maleic anhydride (nominally isobaric species, the sum of which is monitored in mass channel 113- squares), along with signal traces for ions with $m/z = 214$ (triangles) and 232 (diamonds).

The compounds monitored with protonated masses 214 and 232 constitute the largest gas phase compounds detected to date during this type of SOA study. High

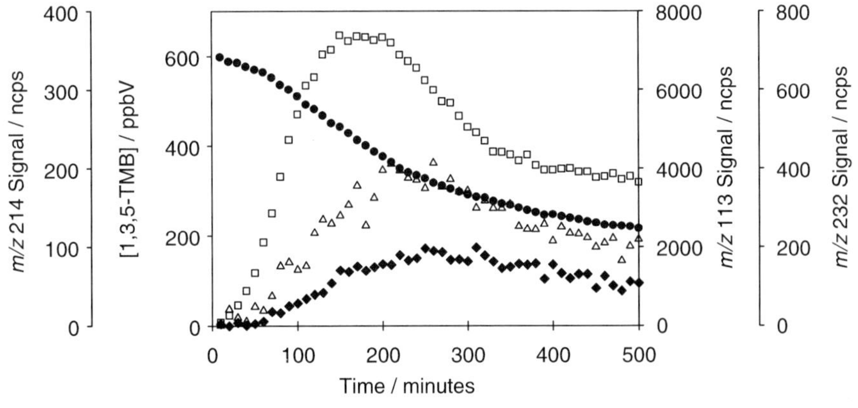

Fig. 5.7 Evolution of various key gas phase compounds during the photooxidation of 1,3,5-trimethylbenzene. See text for further details

resolution data from the CIR-TOF-MS suggest that the compound with mass 232 is 1,3,5-trimethyl-4-nitrooxy-6,7-dioxa-bicyclo[3.2.1]oct-2-en-8-ol (designated species TM135BPNO3 in MCM V3.1), a first-generation oxidation product of the precursor. The mass 214 compound has not yet been identified.

Nucleation begins approximately 133 min after the chamber lights were switched on and roughly 100 min after the first appearance of the oxidation products of the precursor.

5.4 Hadamard Transform CIR-TOF-MS

5.4.1 Introduction

While CIR-TOF-MS offers the advantages of quantitative multi-channel, real-time analysis, a higher detection sensitivity would be welcome. The chemical ionisation reaction ion source is a continuous ion source, whereas the time-of-flight mass spectrometer operates in pulsed mode; i.e. when a packet of ions is injected into the flight tube there is an inherent dead time, dictated by the time it takes for the slowest (heaviest) ions in the packet to reach the detector, during which no new ions can be injected into the flight tube. This corresponds to a low duty cycle, which for our CIR-TOF-MS instrument is estimated to be <5%. This means that >95% of all ions produced by the ion source are not analysed. Increasing the duty cycle should clearly provide a corresponding increase in instrument sensitivity. An improvement in sensitivity could allow for the monitoring of rapid fluxes and the accurate detection of VOC/OVOC species at high pptV mixing ratios over timescales of the order seconds. We are seeking to achieve this by exploiting Hadamard transform TOF-MS (Brock et al., 2000).

5.4.2 Methodology

The duty cycle of a time-of-flight instrument can be improved through the application of a continuous ion beam multiplexing technique. Using this method, instead of analysing a single discrete ion packet, multiple ion packets are passed into the TOF-MS for analysis according to a pre-determined pseudo random binary sequence (PRBS). The PRBS has n elements (derived from $n = 2^m - 1$, where m is an integer) with near equal quantities of '1's (total of $[n + 1]/2$) and '0's (total $[n-1]/2$). Under these operating conditions a '1' controls an ion packet detection event and a'0' controls the removal of the ion beam by some means. The net effect is that the detector records a superposition of multiple, overlapping time-of-flight spectra, but by deconvolution based on knowing the PRB sequence, the actual time-of-flight mass spectrum can be evaluated. This post detection processing is based

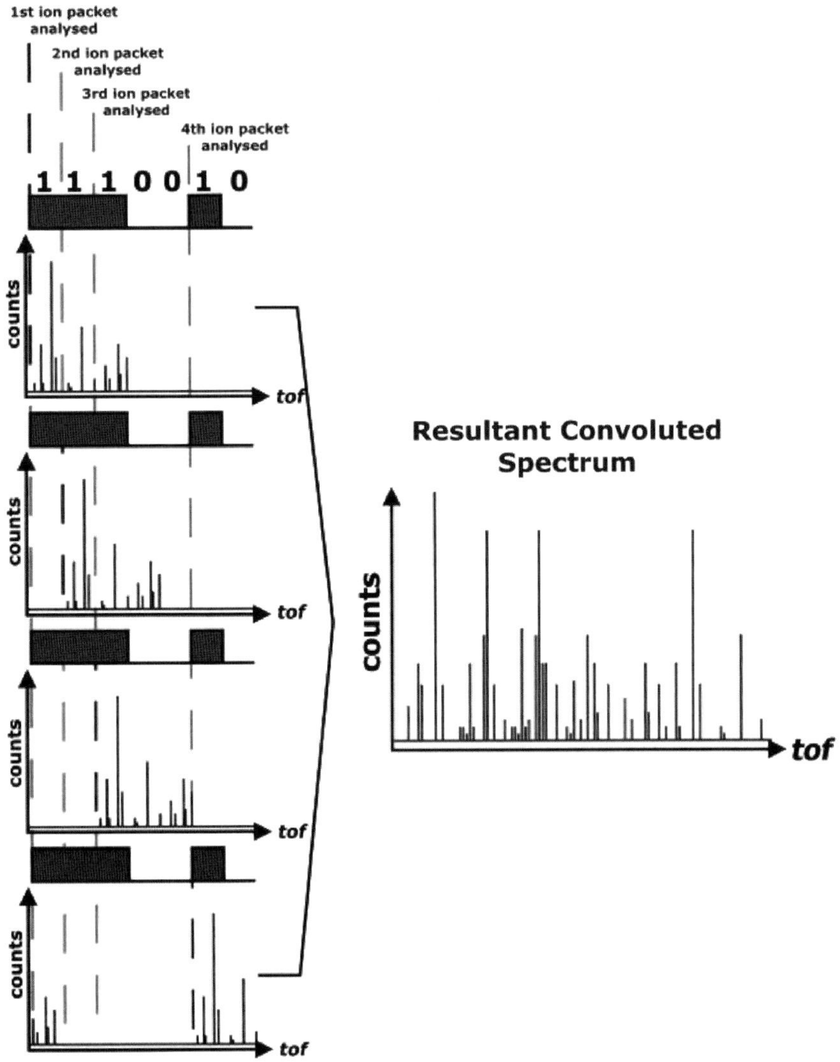

Fig. 5.8 Methodology for the analysis of overlapping, sequential time-of-flight spectra. The individual spectra on the left, delayed in time according to the PRB sequence at the top of the figure, are summed to give the resultant multiplexed ion signal shown on the right

on Hadamard transformation. With this scheme, because of the approximately equal numbers of '1's and '0's the duty cycle increases to 50%. The essence of this procedure is summarised in Fig. 5.8.

Mathematically, the multiplexing technique is equivalent to a matrix multiplication in which a square convolution matrix (whose elements are dictated by the chosen PRBS) multiplies the one-dimensional time-of-flight ion intensity data array to produce the convoluted spectrum. To retrieve the underlying time-of-flight spectrum, one therefore needs to perform the reverse, which requires multiplying the detector output array by the inverse of the convolution matrix.

In practice the ion beam modulation is achieved via a Bradbury–Nielson gate (BNG), which in our system consists of a 20 mm aperture around which two sets of interleaved wire electrodes are wound. When the electrodes are biased to opposite potentials, an electric field is produced in order to diverge the ion beam along two separate paths, whose trajectories are such that when they pass into the mass analyser neither will strike the detector (see Fig. 5.9(a)), effectively switching off detection. Correspondingly when no voltage is applied to the BNG the ion beam is transmitted along a straight path for detection as shown in Fig. 5.9(b). The application of the BNG electric field is controlled by the PRBS.

With a potential duty cycle of 50%, the Hadamard TOF-MS variant should yield compound detection sensitivities that are more than an order of magnitude greater than those of the original CIR-TOF-MS instrument. In addition, as a consequence of the transformation process the signal-to-noise ratio is expected to also improve, by a factor of $n/2$. This CIR-HT-TOF-MS instrument is currently under development in our laboratory.

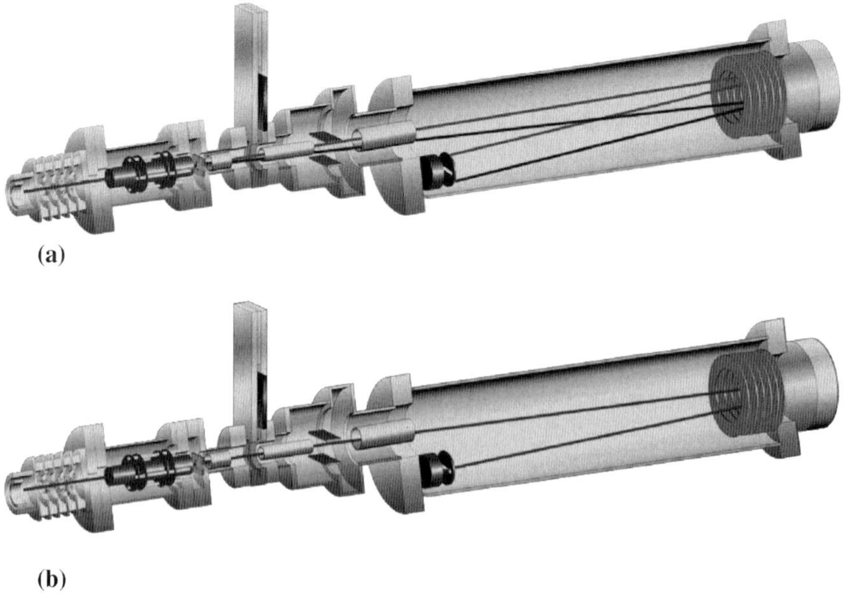

Fig. 5.9 A schematic representation of the instrument using a BNG to diverge the continuous ion beam along two separate paths that do not strike the active part of the detector **(a)** and to transmit the ions into the TOF-MS analyser **(b)** for detection

5.5 Conclusions

The work described in this paper demonstrates the successful development of the first CIR-TOF-MS instrument for the fast monitoring of complex mixtures of VOCs and OVOCs. The technique is highly versatile and can provide ppbV detection sensitivities on one-minute time scales. The potential benefits of employing the TOF-MS are considerable, with the sample matrix being probed in more depth and detail than has been possible previously.

Work has also been taking place on the 'next generation' instrument, which employs ion beam multiplexing in a Hadamard transform instrument. Encoding the ion beam in this manner in the TOF-MS system should further improve the instrument sensitivity by an order of magnitude.

References

Baltensperger U., Dommen J., Paulsen D., Alfarra M., Coe R., Fisseha R., Gascho A., Gysel M., Nyeki S., Sax M., Steinbacher M., Prevot A., Sjogren S., and Weingartner E. (2005), Secondary organic aerosols from anthropogenic and biogenic precursors, *Faraday Discuss.*, 130, 265–278.

Blake R.S., Whyte C., Hughes C.O., Ellis A.M., and Monks P.S. (2004), Demonstration of proton-transfer reaction time-of-flight mass spectrometry for real-time analysis of trace volatile organic compounds, *Anal. Chem.*, 76(13), 3841–3845.

Blake R.S., Wyche K.P., Ellis A.M., and Monks P.S. (2006), Chemical ionization reaction time-of-flight mass spectrometry: Multi-reagent analysis for determination of trace gas composition, *Int. J. Mass Spectrom.*, 254(1–2), 85–93.

Brock A., Rodriguez N., and Zare R.N. (2000), Characterization of a Hadamard transform time-of-flight mass spectrometer, *Rev. Sci. Instrum.*, 71(3), 1306–1318.

Hansel A., Jordan A., Holzinger R., Prazeller P., Vogel W., and Lindinger W. (1995), Proton-transfer reaction mass-spectrometry–online trace gas-analysis at the ppb level, *Int. J. Mass Spectrom.*, 150, 609–619.

Hewitt C.N., Hayward S., and Tani A. (2003), The application of proton transfer reaction-mass spectrometry (PTR-MS) to the monitoring and analysis of volatile organic compounds in the atmosphere, *J. Environ. Monitor.*, 5(1), 1–7.

Jang M. and Kamens R.M. (2001), Atmospheric secondary aerosol formation by heterogeneous reactions of aldehydes in the presence of a sulfuric acid aerosol catalyst, *Environ. Sci. Technol.*, 35, 4758–4766.

Kalberer M., Sax M., Steinbacher M., Dommen J., Prevot A., Fisseha R., Weingartner E., Frankevich V., Zenobi R., and Baltensperger. (2004), Identification of polymers as major components of atmospheric organic aerosols, *Science*, 303, 1659–1662.

Karl M., Brauers T., Dorn H.P., Holland F., Komenda M., Poppe D., Rohrer F., Rupp L., Schaub A., and Wahner A. (2004), Kinetic study of the OH-isoprene and O_3-isoprene reaction in the atmosphere simulation chamber, SAPHIR, *Geophys. Res. Lett.*, 31(5).

Lirk P., Bodrogi F., and Rieder J. (2004), Medical applications of proton transfer reaction-mass spectrometry: Ambient air monitoring and breath analysis, *Int. J. Mass Spectrom.*, 239(2–3), 221–226.

Paulsen D., Dommen J., Kalberer M., Prevot A.S.H., Richter R., Sax M., Steinbacher M., Weingartner E., and Baltensperger U. (2005), Secondary organic aerosol formation by irradiation of 1,3,5-trimethylbenzene-NOx-H2O in a new reaction chamber for atmospheric chemistry and physics, *Environ. Sci. Technol.*, 39(8), 2668–2678.

Smith D., Diskin A.M., Ji Y. F., and Spanel P. (2001), Concurrent use of $H3O^+$, NO^+, and O-2($^+$) precursor ions for the detection and quantification of diverse trace gases in the presence of air and breath by selected ion-flow tube mass spectrometry, *Int. J. Mass Spectrom.*, 209(1), 81–97.

Spanel P., Ji Y.F., and Smith D. (1997), SIFT studies of the reactions of $H3O^+$, NO^+ and O-2($^+$) with a series of aldehydes and ketones, *Int. J. Mass Spectrom.*, 165, 25–37.

Wyche K.P., Blake R.S., Willis K.A., Monks P.S. and Ellis A.M. (2005), Differentiation of isobaric compounds using chemical ionization reaction mass spectrometry, *Rapid Commun. Mass Spectrom.*, 19(22), 3356–3362.

Wyche K.P., Blake R.S., Ellis A.M., Monks P.S., Koppman R., Brauers T., and Apel E. (2006), Technical Note: Performance of chemical ionization reaction time-of-flight mass spectrometry (CIR-TOF-MS) for the measurement of atmospherically significant oxygenated volatile organic compounds, *Atmos. Chem. Phys, 7,* 609–620.

Chapter 6
Continuous Monitoring and the Source Identification of Carbon Dioxide at Three Sites in Northeast Asia During 2004–2005

Fenji Jin[1], Sungki Jung[1], Jooll Kim[1], K.-R. Kim[1], T. Chen[2], Donghao Li[2], Y.-A. Piao[2], Y.-Y. Fang[2], Q.-F. Yin[3], and Donkoo Lee[4]

Abstract We conducted continuous monitoring and the source identification of carbon dioxide at Gosan, Seoul (Korea) and Yanbian during 2004–2005. The data reported are in situ continuous 1-year measurements of atmospheric CO_2 from the Gosan, Seoul and Yanbian stations. One-minute averages of near-surface atmospheric CO_2 concentration were obtained using a measurement system based on non-dispersive infrared (NDIR) analysis using the NOAA/ESRL (National Oceanic & Atmospheric Administration/Earth System Research Laboratory) standard with high precision monitoring data. The background CO_2 concentration of the complete measurement data was determined using the Advanced Global Atmospheric Gases Experiment (AGAGE) statistical pollution identification procedure for removing pollution episode data. The background characteristics at Gosan are discussed in detail. The background concentration of CO_2 showed quite evident diurnal and seasonal variation. The diurnal variation shows a maximum in the nighttime and a minimum in the daytime, and the seasonal cycle shows a maximum in spring and a minimum in summer. Background data at Seoul and Yanbian also show a similar trend. In addition, we applied a hybrid receptor model driven by three-dimensional synoptic meteorology from the HYSPLIT4 (HYbrid Single-Particle Lagrangian Integrated Trajectory) model to determine CO_2 relative emission strength contributions from the Northeast Asia region as observed from Gosan. Modeling results from Seoul and Yanbian are also presented—they are important in creating a full potential source region map of the Northeast Asia region, as observations from Gosan are limited to the wind patterns crossing the station. Results indicate that there appears to be a large potential source region in the northeastern and eastern parts of China.

Keywords: Carbon dioxide, continuous monitoring, emission strength, long-range transport, trajectory

[1] *School of Earth and Environmental Science, Seoul National University, Seoul, Korea*

[2] *Yanbian University, Yanji, Jilin, China*

[3] *Huaiyin Teacher's College, Huaian, Jiangsu, China*

[4] *College of Agriculture and Life Sciences, Seoul National University, Seoul, Korea*

Y.J. Kim and U. Platt (eds.), *Advanced Environmental Monitoring*, 77–89.

6.1 Introduction

The role of greenhouse gases derived human activities in the radiative forcing of climate change has been well documented in the scientific reports of the Intergovernmental Panel on Climate Change (IPCC 2001). Carbon dioxide (CO_2) is one of the representative greenhouse gases. Before the Industrial Era, which began around 1750, the atmospheric carbon dioxide concentration had remained constant at 280 ± 10 ppmv for several thousand years. It has risen continuously since then, reaching 380 ppmv in 2004. The continuous rise in levels of atmospheric CO_2 is caused by anthropogenic emissions of CO_2. In particular, three-quarters of these emissions are due to fossil fuel burning (IPCC 2001).

Continuous observation of CO_2 mixing ratios in the atmosphere is very important, providing a basis for studies of the global carbon cycle and CO_2-induced climatic change. For this reason atmospheric CO_2 has been monitored at many sites worldwide for many years. Northeast Asia is of special interest because the economic growth of this region during the last 50 years could have significant implications in terms of the release of industrial pollution into the environment, including anthropogenic release of CO_2 into the atmosphere. Although measurement data obtained through frequent flask sampling can show seasonal variations, it is difficult to determine detailed variations such as diurnal variation and CO_2 concentrations affected by air mass transport on an hourly timescale. With the goal of establishing a background reference site for future research, the Gosan station at Jeju Island, Korea was created in 2003.

Deriving the background concentration from the continuous atmospheric measurement of a specific species is an important objective for which many different methods have been developed, such as the weighted method (Gras 2001; Zhou 2003), the European Monitoring and Evaluation Program (EMEP) daily wind direction sector allocations, the Nuclear accident model (NAME), the Lagrangian dispersion model defined air mass origins, and the AGAGE statistical pollution identification procedure (O'Doherty et al. 2001). For this study we have utilized the AGAGE statistical method of background calculation (O'Doherty et al. 2001).

An important application of measurement data is the modeling of the emission source and the magnitude of emissions from the surrounding region. Air-mass back trajectories have often been used in combination with observational data to identify potential source areas of air pollutants and determine their respective contribution at the receptor sites. Many statistical methods exist for this purpose, including residence time analysis, and hybrid receptor models such as PSCF (Potential source contribution function), CWT (concentration weighted trajectory), and RTWC (residence time weighted concentration) (Ashbaugh 1983; S. Reimann 2004). Here, we apply the RTWC model of back trajectory calculated in consecutive time steps, 24 times a day and CO_2 concentration in the same steps. Following the example of Gosan station, we have gradually expanded our *station* to other sites, namely Yanbian (China) and Seoul (Korea) to extend the potential source region identified.

The objectives of this study are source identification of the carbon dioxide in the Northeast Asia region and model calculations of relative emission strengths in these regions through in-situ continuous monitoring measurement at Gosan, Seoul and Yanbian. These results may have significant implications, as accurate CO_2-emission data is becoming important with efforts to reduce global atmospheric pollution such as the Kyoto protocol.

6.2 Methodology

6.2.1 Site Description

Figure 6.1 shows a map of the Northeast Asia region with three of the measuring stations used in this study. A detailed description follows.

The Gosan station is located near the southwestern tip of the Jeju Island, south of the Korean peninsula (126°11′00″ E, 33°11′70″ N; altitude: 70 m above mean sea level). Because of its remote and relatively unpolluted location, Gosan station is considered to be an ideal location for measuring the atmospheric composition of air masses considered to be representative of the background concentrations in Northeast Asia (Carmichael et al. 1997; Chen et al. 1997). For this reason, Gosan station has been included in numerous cooperative research projects, like ACE-Asia and Atmospheric Brown Cloud, as well as being incorporated into research networks such as the CSIRO-LOFLO network and the Advanced Global Atmospheric Gases Experiment (Prinn et al. 2000) networks. We have conducted high-frequency continuous monitoring at Gosan, Jeju, since 2003.

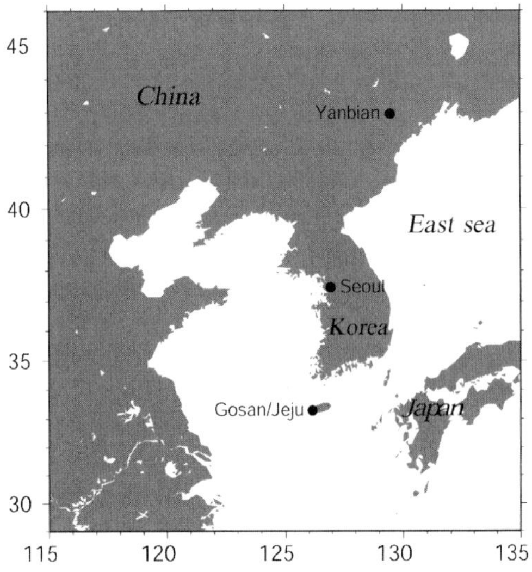

Fig. 6.1 Atmospheric monitoring sites showing Gosan/Jeju island station, Seoul station and Yanbian station, respectively

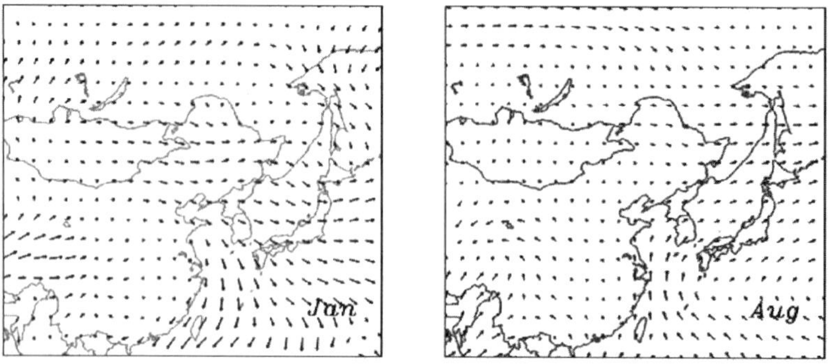

Fig. 6.2 Monthly wind for the Northeast Asia region from NCEP (2005)

Although too close to local pollution sources to be valuable as background monitoring stations, Seoul and Yanbian stations could still be used to observe regional pollution and thus be utilized for broadening the potential source region study area. Seoul station (37°27′N, 126°57′E) is located in Seoul National University, and is considered representative of atmospheric CO_2 mixing ratios in a highly industrialized city. Measurements from Seoul will prove to be valuable in determining the baseline of the Korean peninsula as well as research into pollution sources in Korea and China. Continuous measurements have been conducted there since 2005. Yanbian station (42°32′ N, 129°18′ E) is located at Yanbian University, China. Yanbian city is a semi-developed city with a population of just over two million. The relatively less developed nature of its surroundings could be ideal for studying the air mass outflow from the Asian continent, with an emphasis on pollution from China. Continuous measurements of atmospheric CO_2 have been conducted there since 2004.

Analysis of the monthly wind provided by the US National Centers for Environmental Prediction (NCEP) global data assimilation system (GDAS) for the Northeast Asia region shows that the prevailing winds in the region are northwesterly in winter and southwesterly in summer (Fig. 6.2).

6.2.2 Analytical Method

Atmospheric CO_2 concentration is measured using a Licor 6262 non-dispersive infrared (NDIR) analyzer system (Fig. 6.3). The system is composed of four parts, inlet part, controller part, calibration part and detector part. Ambient air is supplied into the system via 10 mm o.d. Dekoron tubing 40 m up to the intake tower. Air is drawn in via a vacuum pump through pressure release valves set at 6 psi to remove excess air and allow high continuous flow through the mainline and passed through a 7 µm in-line filter to remove particles and a Nafion-dryer to remove moisture.

NOAA/ESRL (National Oceanic & Atmospheric Administration/Earth System Research Laboratory) reference standards were used as the calibration reference, with

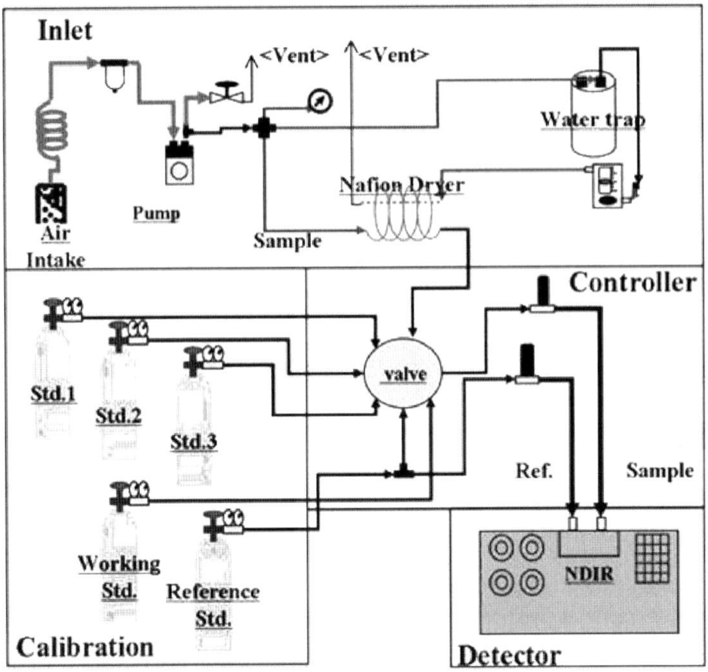

Fig. 6.3 Continuous CO_2 monitoring system diagram

3 standards per station in the range of 300–400 ppmv. The sample precision through an NDIR analyzer was below ±0.1 ppmv. The NDIR system was installed at each of the stations; at Gosan in April 2004, at Yanbian in January 2005 and at Seoul in March 2005. Air samples were analyzed at a frequency of two to three times per minute.

6.2.3 Background Determination

In previous studies different methods were used to sort the observations by air mass origins, in an attempt to separate regional and/or local pollution events from the background measurement (Gras 2001; Zhou 2003). Because air masses from various sources are influencing the in-situ site, the approach of selecting a certain wind sector as the background condition requires careful measurement and analysis of the wind patterns of the monitoring location, and is not easy to implement in all locations. So in the present study the AGAGE statistical pollution identification procedure was selected to determine the background concentration (O'Doherty et al. 2001).

AGAGE statistical pollution identification procedure determines the pollution events of a given day by examining the trends for 60 days before and after it.

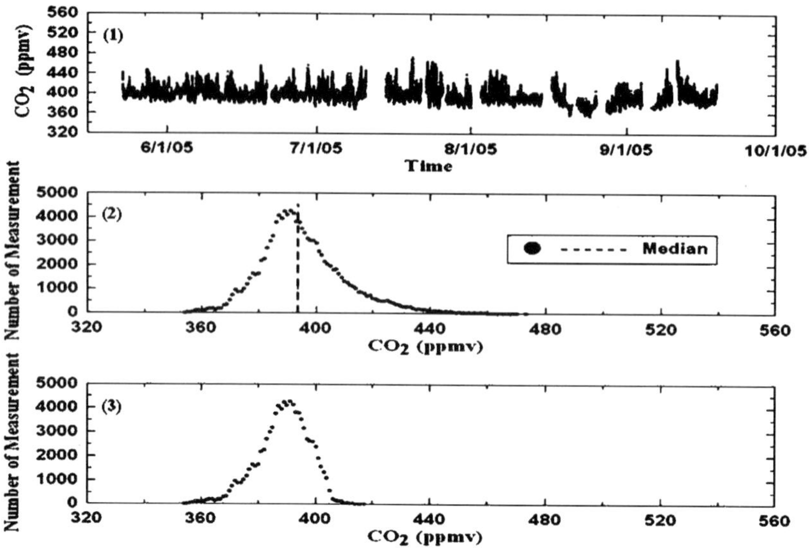

Fig. 6.4 Pollution identification for Seoul, measurements, (**1**) plot of 121-day window around selected measurement, (**2**) median (hashed line) values of all sample points, (**3**) remaining baseline data with Gaussian distribution

A Gaussian distribution is assumed for the background during this 121-day period, and events that deviate from the median of the distribution by more than a certain factor (typically 2–3 σ) are labeled as pollution events. Details of the method are explained in the Appendix of O'Doherty's paper (O'Doherty et al. 2001).

Figure 6.4 depicts the validity of the AGAGE statistical pollution identification process. Figure 6.4(1) shows the time series with identified pollution events of CO_2 from May 18 and September 18, 2005 at Seoul. Figure 6.4(2) shows that the histogram indicates that the majority of CO_2 data during this period is centered about 390 ppmv and the long tail feature of high CO_2 concentration values are less frequent. After applying the AGAGE statistical pollution identification procedure, the remaining data (background data) show a structure similar to a Gaussian distribution (Fig. 6.4(3)).

6.2.4 Hybrid Receptor Model

From April 2004 to March 2005, three-day back trajectories for every hour were calculated by the HYbrid Single-Particle Lagrangian Integrated Trajectory (HYSPLIT) model (Draxler and Rolph 2003) with 6-hourly archived meteorological data provided from the US National Centers for Environmental

Prediction (NCEP) global data assimilation system (GDAS), known as the final run (FNL) data.

In order to investigate potential source regions of CO_2 we combined these trajectories with measured concentrations at the station. For this we have used the method of Seibert et al. (1994) which computes the mean concentration for each grid cell after superimposing a grid to the domain of trajectory computations using the following formula:

$$\overline{C_{ij}} = \frac{1}{\sum_{l=1}^{M} \tau_{ijl}} \sum_{l=1}^{M} (c_l) \tau_{ijl}$$,

where $\overline{C_{ij}}$ is a relative measure of potential source region strength, i and j are the indices of the horizontal grid, l is the index of the trajectory, M is the total number of trajectories, c_l is the concentration (minus the background concentration) measured during arrival of trajectory l, and τ_{ijl} is the residence time of the trajectory l spent over grid cell i, j. This concentration field method was used for several compounds to investigate their source (Stohl 1996; Charron et al., 2000; Ferrarese et al., 2002; Reimann et al., 2005). The domain of the calculated trajectories was superimposed with a 0.5° × 0.5° grid. For the calculation of residence time, we used the method of Poirot et al. (1986) with adjustments applied for geometry.

A high value of $\overline{C_{ij}}$ means that, on average, air parcels passing over cell (i,j) result in high concentrations at the receptor site. But because measured concentrations are distributed equally to all grid cells passed by the appropriate trajectory, the approach used is susceptible to underestimation of the spatial gradients of the true emission field (Stohl 1996). In order to eliminate low confidence level areas, a point filter was applied to the model results. This increases the confidence level of the results but also reduces the area studied.

6.3 Results and Discussion

6.3.1 Data

Plots of the continuous monitoring data for each of the stations are shown in Fig. 6.5.

The gray points represent the pollution flagged data using the statistical pollution detection algorithm described in section 6.2.3, and the black points are the remaining background data. The frequency and amplitude of the pollution events at Yanbian and Seoul are noticeably larger, due to the urban characteristics of their locations. A more detailed discussion of the background data follows, and the pollution data will be discussed in the next section.

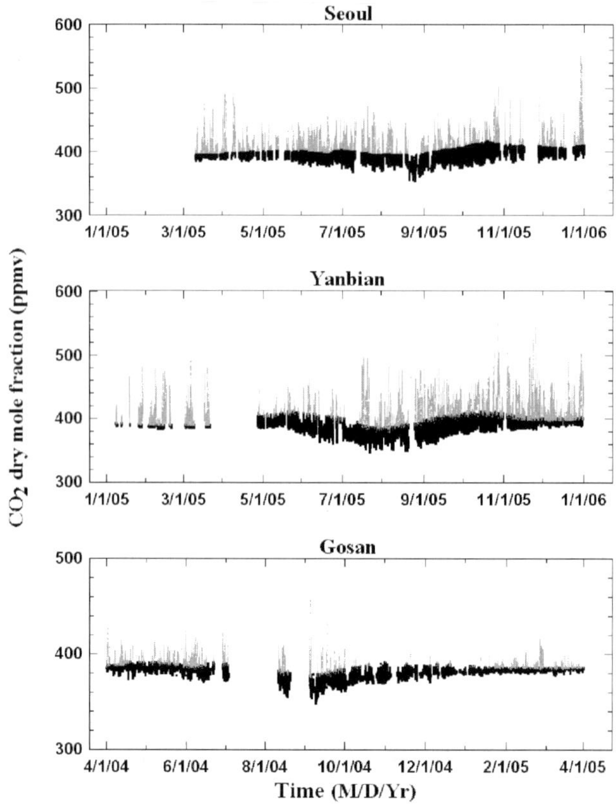

Fig. 6.5 Continuous monitoring (1 min) CO_2 data measured at each of the Northeast Asia Atmospheric Monitoring Network sites. Time shown is the local time

6.3.2 *Atmospheric CO_2 Background Characteristics*

Applying the AGAGE statistical pollution identification procedure we derived the background CO_2 concentrations in Gosan, Seoul and Yanbian.

6.3.2.1 Diurnal Variation

Background concentrations of CO_2 at Seoul, Yanbian and Gosan show diurnal variation (Fig. 6.6), higher values in the nighttime and lower values in the daytime. The diurnal variation is relatively small in winter and large in summer, due to the effect of photosynthesis by plants being stronger in daytime and summer than in nighttime and winter.

All of the stations show higher concentrations in the nighttime and lower concentrations in the daytime. The high-peak at Gosan appears at approximately 6 h

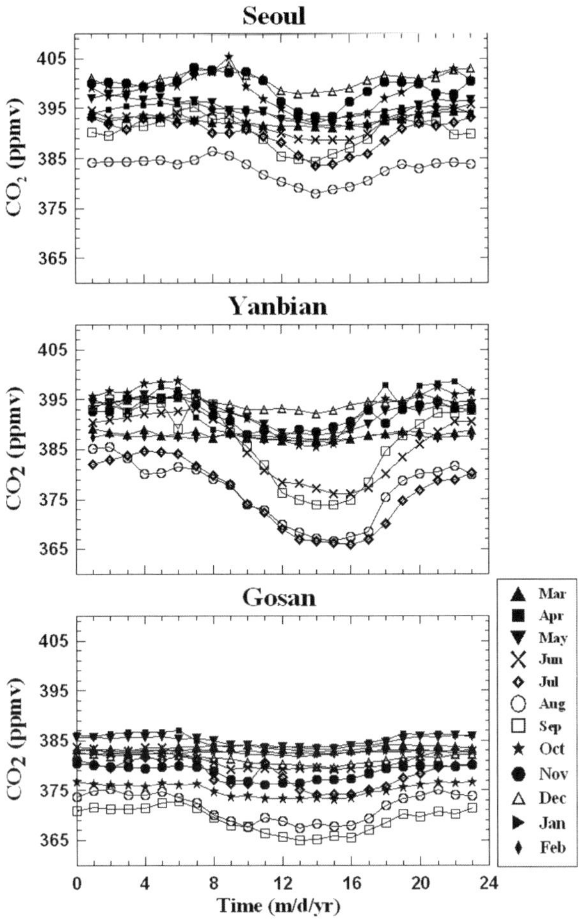

Fig. 6.6 Averaged CO_2 diurnal variation of background data at Seoul, Yanbian and Gosan in the measurement period. Hour in the shows the local time in each site

and the low peak around 13–15 h. The high-peak for Yanbian is at around 6–8 h, with the low-peak at 14–16 h. Seoul shows a high-peak at 7–9 h and a low-peak at 13–15 h. All of the times are in local time. One reason for this trend is the respiration in nighttime and photosynthesis in daytime by plants. Another reason is the change in the boundary layer, which changes with the earth's surface temperature. This change will enhance the low concentrations in the afternoon and the high concentrations in the early morning.

The amplitude of the diurnal variation is largest in summer, due to increased influence from plants. The smallest diurnal variation can be seen in Gosan, with Yanbian showing the largest variations. This can be attributed to the relatively low influence of terrestrial plants in Gosan, while Yanbian station is more influenced by terrestrial plants due to the surrounding forests.

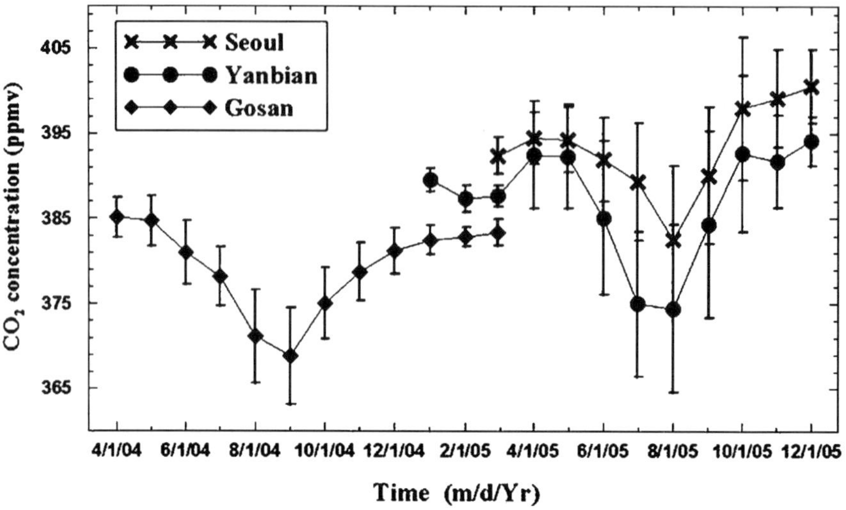

Fig. 6.7 CO_2 averaged seasonal cycle of background data at Seoul, Yanbian and Gosan with standard deviation indicated as error bars in the measurement period

6.3.2.2 Seasonal Variation

The averaged CO_2 seasonal variation for the background data is shown in Fig. 6.7.

There was an obvious seasonal cycle, with a maximum occurring in April and a minimum in September at Gosan. The measurements from Gosan show a low peak at September, but not having a large enough number of measurements in August may have caused this phenomenon. Future measurements will verify this anomaly. CO_2 mixing ratios declined rapidly during the period May–September, and climbed fleetly during the period September–December. The averaged CO_2 seasonal amplitude was up to 15 ppmv in Gosan.

Figure 6.7 also shows seasonal cycles in Yanbian and Seoul, with maxima occurring in April and December respectively, with a minimum occurring in August at both sites. The averaged CO_2 seasonal amplitude was up to 18 ppmv in Seoul and 20 ppmv in Yanbian.

The seasonal variation of Seoul and Yanbian reflects the periodicity of terrestrial vegetation growth in the middle of NH.

6.4 Potential Source Region and Relative Emission Strength

The following figure shows the potential pollution source region and its relative emission strength derived from the hybrid receptor model (Fig. 6.8). Deduced from our model, the plain regions in northern China could be a potent source region for CO_2 in Gosan. The maximum relative emission strength is about 16 ppmv above background concentration.

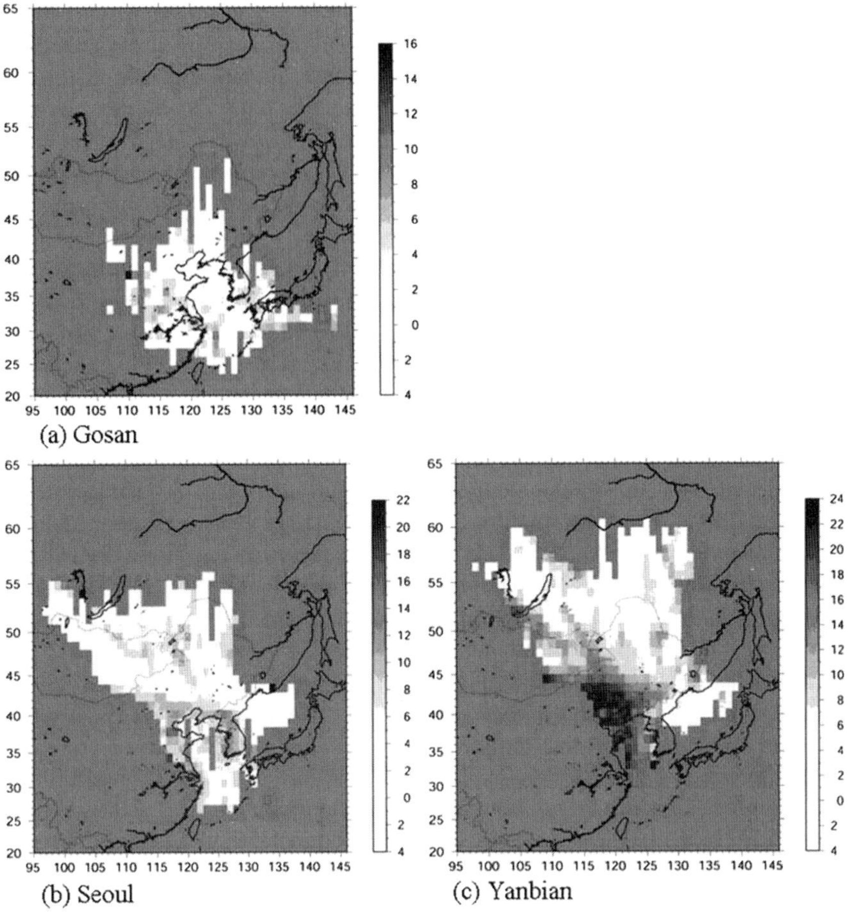

Fig. 6.8 CO_2 average concentration above background (ppmv), red color means high relative emission strength, yellow color means low relative emission strength. Units show concentration above background

During the sampling period, Seoul was mostly affected by sources located in eastern China and inside the Korean peninsula. The maximum relative emission strength is about 22 ppmv above background concentration. Yanbian was mostly affected by sources located in eastern China and inside the Korean peninsula. The maximum relative emission strength is about 22 ppmv above background concentration. Yanbian was mostly affected by sources located in eastern China. The maximum relative emission strength is about 24 ppmv above background concentration. The relative strength in Yanbian appears to be stronger than in Seoul.

The relative strength in Gosan is appears to be weakest, and the values in Yanbian to be the strongest.

In this study, the region around Shanghai, one of the most densely industrialized regions in Northeast Asia shows relatively low relative emission strength because the prevailing winds over the site at Gosan are rarely flowing from there. Though the hybrid receptor model has proven to be effective in identifying pollution sources, it has some limitations. Some of the modeling results show an unexpectedly large source pollution coming from the oceans. A possible explanation is that the trajectories with high concentration pass over the oceans. Further refinement of the model will help to explain or solve this problem.

6.5 Conclusions

We report the results of continuous atmospheric CO_2 measurements in the period from April 2004 to March 2005 at Gosan station, and in 2005 at Yanbian and Seoul stations using high-quality sampling data, by using the high-frequency high-precision NDIR CO_2 analysis system. Although measurement data through frequent flask sampling can show seasonal variations, it is difficult to know detailed variations such as diurnal variation and CO_2 concentrations affected by air mass transport on an hourly timescale. Pollution episode data and background data were separated using the AGAGE statistical method.

Atmospheric CO_2 background concentrations measured at Gosan, Seoul and Yanbian in recent years show typical diurnal and seasonal variations, higher values in the nighttime and lower values in the daytime, higher values in winter and lower values in summer. This is due to the terrestrial biosphere in NH and to photosynthesis by plants, and to the dilution of the boundary layer, etc.

Potential source regions in Northeast Asia were observed for anthropogenic atmospheric CO_2 by applying trajectory statistics. The results show the possibility of CO_2 potential sources in the plain regions in northern China contributing to Gosan, while contributions from northern and eastern China seem to be detected at Seoul and Yanbian. Seoul and Yanbian stations are important in creating a full potential source region map of the Northeast Asia region, as observations from Gosan are limited by the air masses transported to Gosan.

Further refinement of the modeling method will improve the accuracy of the potential source regions and relative emission strengths. With these improvements, and as atmospheric CO_2 data is accumulated over longer periods, we will be able to discuss long-term trends in the potential source regions and relative emission strengths in the Northeast Asia region.

Acknowledgement This study was supported in part by the Korea Science and Engineering Foundation, the project title of Cooperative Research for Restoration of Degraded Ecosystems in Northeast Asian Regions. We would like to thank the help of Mi-young Ko and the Gosan Weather Station staff in management of the CO_2 continuous system at Gosan station.

References

Ashbaugh L. (1983), A statistical trajectory technique for determining air pollution source regions, *J. Air pollut. Control. Assoc.*, 33(11), 1096–1098.

Carmichael G.R., Hong M.-S., Ueda H., Chen L.-L., Murano K., Park J.K., Lee H., Kim Y., Kang C., and Shim S. (1997), Aerosol composition at Cheju Island, Korea, *J. Geophys. Res.*, 102, D5,6047–6062.

Charron A., Plaisance H., Sauvage S., Coddeville P., Galloo J.C., and Guillermo R. (2000), A study of the source-receptor relationships influencing the acidity of precipitation collected at a rural site in France, *Atmospheric Environ.*, 34, 3665–3674.

Chen L.-L., Carmichael G.R., Hong M.-S., Ueda H., Shim S., Song C.H., Kim Y.P., Arimoto R., Savoie D., Murano K., Park J.K., Lee H.-G., and Kang C. (1997), Influence of continental outflow events on the aerosol composition at Cheju Island, South Korea, *J .Geophys. Res.*, 102, D23, 28551–28574.

IPCC. (2001), Climate Change, 2001. The scientific basis. Intergovernmental Panel on Climate Change. Cambridge University Press, UK.

Draxler R.R. and Rolph G.D. (2005), HYSPLIT (HYbrid Single-Particle Lagrangian Integrated Trajectory) Model access via NOAA ARL READY Web site (http://www.arl.noaa.gov/ready/hysplit4.html). NOAA Air Resources Laboratory, Silver Spring, MD.

Ferrarese S., Longhetto A., Cassardo C., Apadula F., Bertoni D., Giraud C., and Gotti A. (2002), A study of seasonal and yearly modulation of carbon dioxide sources and sinks, with a particular attention to the Boreal Atlantic Ocean, *Atmospheric Environ.*, 36, 5517–5526.

Gras J.L. (2001), Aerosol black carbon at Cape Grim, by light absorption, *Baseline atmospheric program (Australia),* 1997–1998, 20–26.

O'Doherty S., Simmonds P.G., Cunnold D.M., Wang H.J., Sturrock G.A., Fraser P.J., Ryall D., Derwnet R.G., Weiss R.F., Salamech P., Miller B.R., and Prinn R.G. (2001), In situ chloroform measurements at Advanced Global Atmospheric Gases Experiment atmospheric research stations from 1994 to 1998, *J. Geophys. Res.,* 106, D17, 20, 429–20.

Poirot R.L. and Wishinski P.R. (1986), Visibility, sulfate and air mass history associated with the summertime aerosol in northern Vermont, *Atmospheric Environ.*, 20, 1457–1469.

Prinn R.G., Weiss R.F., Fraser P.J., Simmonds P.G., Cunnold D.M., Alyea F.N., O'Doherty S., Salameh P., Miller B.R., Huang J., Wang R.H.J., Hartley D.E., Harth C., Steele L.P., Sturrock G.A., Midgley P.M., and McCulloch A. (2000), A history of chemically and radiatively important gases in air deduced from ALE/GAGE/AGAGE, *J. Geophys. Res.,* 105, D14, 17,751–17.

Reimann S., Schaub D., Stemmler K., Folini D., Hill M., Hofer P., and Buchmann B. (2004), Halogenated greenhouse gases at the Swiss High Alpine Site of Jungfraujoch (3580 m asl). Continuous measurements and their use for regional European source allocation, *J. Geophys. Res.,* 109, D05307, doi: 10.1029/2003JD003923.

Seibert P., Kromp-Kolb H., Baltensperger U., Jost T., Schwikowski M., Kasper A., and Puxbaum H. (1994), Trajectory analysis of aerosol measurements at high alpine sites, in *Transformation of Pollutants in the Troposphere*, edited by P. M. Borrell et al., pp. 689–693, SPB Acad., Hague, Netherlands.

Stohl A. (1996), Trajectory statistics-A new method to establish source-receptor relationships of air pollutants and its application to the transport of particulate sulfate in Europe, *Atmospheric Environ.*, 30, 579–587.

Zhou L.X., Tang J., Wen Y.P., Li J.L., Yan P., and Zhang X.C. (2003), The impact of local winds and long-range transport on the continuous carbon dioxide record at Mount Waliguan, China, *Tellus,* 55B, 145–158.

Chapter 7
Aircraft Measurements of Long-Range Trans-Boundary Air Pollutants over Yellow Sea

Sung-Nam Oh[1], Jun-Seok Cha[2], Dong-Won Lee[2], and Jin-Su Choi[2]

Abstract Airborne gaseous and particulate matter was measured above downwind ocean areas from China to the western region of the Korean peninsula for 10 days in the spring and autumn of 2005.

The main objectives of this study were to investigate the spatial distribution of pollution in the ocean atmosphere between Korea and China, and to improve our understanding of acidic deposition in the Korean region in relation to processes affecting the transport of long-range trans-boundary air pollutants from China. The scientific payload on-board an ultra-light aircraft included measurements of the concentrations of SO_2, NO_x, O_3 and particulates. Meteorological profiles (air temperature, winds and humidity) were recorded simultaneously at a ground site.

A six-manned aircraft (PA-31-350 type) served the flight scenarios of eight altitude and two azimuth levels for measuring gas fluxes and particulate depositions in the first intensive observation period (IOP) during April 15–25 and a second period during October 15–25, 2005.

The overall mean concentrations of air pollutants in the atmospheric boundary layer were in the range 2.58–6.63 ppbv for SO_2, 3.74–4.24 ppbv for NO_x and 48.8–54.28 ppbv for O_3 in both observation periods. Measurements at different altitudes revealed that the pollutants were normally observed in high concentrations in the atmospheric boundary layer. However, the longitudinal measurements of SO_2 concentration showed higher values in areas over the Yellow Sea, though values decreased when approaching the Korean peninsula. Enhanced mass concentrations of SO_2 were observed for altitudinal measurements when a strong westerly air stream occurred in the low level boundary layer from China to Korea.

Aerosol number concentrations varied significantly in the range 32–4,640 ea/cm^3 for aircraft measurements during both periods. The differences in NO_x level between aircraft and surface measurements play an important role in the chemical form and size of particulate matter.

[1] *Meteorological Research Institute (METRI), Korea Meteorological Administration (KMA), 460-18 Shindaebang-dong, Dongjak-gu, Seoul 156-720, Korea*

[2] *Global Environment Research Center, National Institute of Environment Research, Environmental Research Complex, Gyeongseo-dong, Seo-gu, Inchon 404-708, Korea*

Y.J. Kim and U. Platt (eds.), *Advanced Environmental Monitoring*, 90–106.

Keywords: Airborne gaseous, aircraft measurements, long-range trans-boundary, aerosol sulfate, air pollutants of SO_2, NO_x, O_3

7.1 Introduction

The pacific coast of East Asia is a region in which human activities impose a heavy burden in the form of atmospheric air pollutants such as sulfur compounds and aerosols, which exhibit an average flow duration from a day to a week in the low layer of the troposphere (Kim et al. 2001). The emissions of SO_2 and NO_x have been decreasing or have at least remained steady since 1980 in Western Europe and North America; however, they are still growing in the Asian region, especially in northeast countries that are under the shadow of the westerly belt of China. Although the rate of SO_2 emission in China has been generally decreasing, the total amount of SO_2 emitted is still large (Hatakeyama et al. 2004; Streets et al. 2001). Industrialization in China is one of the main causes of acidic pollution and aerosol.

The spatial distribution of ocean areas between Korea and China is one of the important parameters that must be taken into account for an assessment of the environmental impact of pollution in this region. As a consequence, aircraft measurements have been incorporated into pollution studies in order to investigate the long-range transport and distribution of atmospheric pollutants, as demonstrated by Prospero (1999), Jaffe et al. (1999), Clarke et al. (2001) and Nunnermacker et al. (2004). Aircraft measurements of air pollution have also been carried out by Kim et al. (1997) and Han et al. (2006) over ocean areas around the Korean peninsula. However, these measurements were not sufficient for a complete analysis of the long-range transport of air pollution or a thorough estimation of the spatial and temporal distributions of pollution over these ocean areas.

Aerosol sulfate contributes to cloud formation as major cloud condensation nuclei, and to atmospheric greenhouse effects through atmospheric radiative forcing. The transport of atmospheric particulates in Northeast Asia is closely related to Asian dust, namely yellow sand dust, which was thoroughly investigated during the ACE-Asia campaign in the spring of 2001 (Arimoto et al. 2004).

The study of the long-range transport of air pollutants has been carried out by several aircraft observation campaigns over the seas between Korea and continental China by various countries of Northeast Asia since 1995 (Hatakeyama et al. 1998; Kim et al. 2001).

However, in order to precisely analyze the source–receptor relationship of pollutants in this region and to clarify the effects of pollutants on the regional atmospheric environment and global climate, it is very important to analyze the dynamics of atmospheric chemical constituents emitted by natural and anthropogenic sources in both receptor and source regions. A study of the source–receptor relationship of air pollution was organized under a Korean initiative in 1996 in the form of an expert working group for the investigation of the long-range trans-boundary air pollutants in Northeast Asia (LTP) with representatives from Korea, China and Japan, and a number of collaborative experiments have since been performed.

The objectives of our aircraft measurements are to study the spatial distribution of pollution in the ocean atmosphere between Korea and China, and to improve our understanding of acidic deposition in the Korean area in relation to processes affecting long-range trans-boundary transport of air pollutants from China. The measurements are used to identify the sources and characterize the dispersion fluxes of atmospheric pollutants over the ocean domain from China to west Korea. Experiments were carried out during April and October, 2005.

7.2 Measurements, Instrumentation and Meteorology

7.2.1 Flight Strategy

A twin propeller type Chieftain aircraft (US PIPER Co., see Fig. 7.1) was flown over the downwind area from China above the Yellow Sea between China and Korea and inland of Korea for comparisons with ground measurements of air pollution and meteorological data. The dimensions of the aircraft were 10.55 × 3.96 × 12.40 m, and it had a capacity of 350 HP and a normal cruising speed of 370 km/h. The sampling nozzle was designed to minimize the loss of sampling compounds and was attached to the bottom of the aircraft. The mass flow controller (MFC, Tylon) was supplemented to the sampling system to keep the flow rate constant following changes in altitude.

Four flight strategies, as shown in Fig. 7.2, were adapted with a change in altitude (from 300 to 2,300 m) in round trips along the longitudinal and latitudinal flight tracks from the starting point of Kimpo Airport in Seoul for each measurement period. Experiments were mainly carried out in the afternoon during successive days during April 15–21 and October 15–22, 2005.

Fig. 7.1 The aircraft (left) and sampling nozzle (right) installed on the bottom of the aircraft (aircraft size: 10.55 m (L) × 3.96 m (H) × 12.40 m (W), cruising time: 6.9 h, landing weight: 700 LB (317.5 kg), cruising distance: 1.97 km, limit altitude: 3,000 m, speed: 150 knots (277.8 km/h))

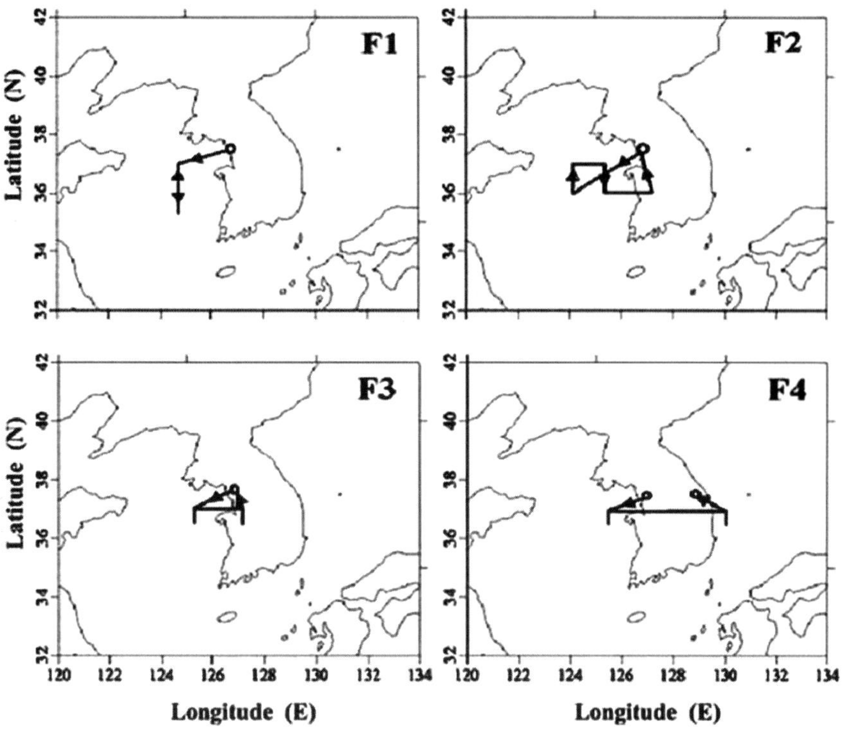

Fig. 7.2 Aircraft flight tracks for the measurement periods

7.2.2 Instruments

Ambient air for gas analysis was introduced into the cabin of the aircraft through a 3/8-in. Teflon tube connected to an air inlet at the bottom of the aircraft.

SO_2 was measured continuously with a UV pulse fluorescence TECO model Thermo-43C (Trace Level). NO_x concentration was measured with a chemiluminescence NO_x analyzer (TECO model 43S), which was modified to measure total oxides of nitrogen. Ozone was monitored with an absorption ozone analyzer (Model 49C, TECO) with 4-s switching of the light path for its dual-cell system using a UV photometric method (Table 7.1).

Aerosol number concentrations in five size bins ranging from 0.3 to 10 μm were continuously monitored with an optical particle counter (Rion, KC-01C). Aerosol mass concentrations were monitored with PM10 at six ground stations on the Korean peninsula for comparison with aircraft measurements.

Gas materials and aerosol states were measured every 10 s and the mean value of the data was averaged over a 1-min interval. The collected data was statistically analyzed with respect to hourly and daily values, though aerosol number concentration was analyzed over 2-min intervals. The geographical position of the aircraft was determined by a GPS β pilot system.

Table 7.1 Description of relevant instruments equipped on the aircraft

Gases	Instruments	Range	Response time	Precision
SO_2	UV fluorescence/TEI	<100 ppb	80 s (10-s average)	0.2 ppb
NO_x	Chemiluminescence/TEI	<100 ppb	60 s (10-s average)	0.05 ppb
O_3	UV photometric/TEI	<100 ppb	20 s (10-s average)	1 ppb
Particle Number	Aerodynamic particle counter/TSI-3320	<1,000 cm^3	1.4 min	
Position & Altitude	GPS/Garmin, GPS-II	Position (°) & altitude (m)		
Atmospheric Vertical Structure	Radiosonde/USA-AIR3	30 hpa	1 s	

SO_2 and NO_x were also routinely calibrated with the span gases. Detection limits for SO_2 and NO_x were 0.2 ppb in the case of the 10-s average and 0.05 ppb in the case of the 2-min average. The ozone monitor with a detection limit of 1 ppb was also calibrated before and after each flight. All calibrations were performed through the complete Teflon inlet line. Analyzed gas and geographical location data were stored in a portable computer. General descriptions for relevant aircraft instrumentation are shown in Table 7.1 Typical flight routes consisted of vertical profiles up to 3 km and horizontal trans-sections along the longitude over the Yellow Sea.

Aerosol samplings were also carried out at the Tea-An ground station, which is located on the western tip of Korea (36°44′N, 126°08′E, 20 m above the mean sea level).

7.2.3 Atmospheric Vertical Structures and Meteorology

On April 15, the observation area around Tea-An was covered by a high-pressure system with westerly and southwesterly winds. On April 18, the observation area was dominated by two high-pressure systems centered around North China and the South-East Sea of Japan. It was cloudy and wind blew mainly from the west and southwest with a strong wind speed of 10 m/s at a 300-m flight level. On October 15, the Korean region had a comparable day of strong wind from a low-pressure system in the northwest region of China.

The depth of the atmospheric boundary mixing layer was identified using the vertical profiles of potential temperature and the water vapor mixing ratio, which were measured by radiosonde at the Tae-An ground station at a time corresponding with the aircraft flight, as shown in Fig. 7.3a, b. The observation station of KMA at Seosan reported a northwesterly wind with a speed between 2 and 4 m/s at the ground station. The afternoon soundings suggested that the atmospheric mixing layer in the region reached a height between 1,000 and 2,000 m.

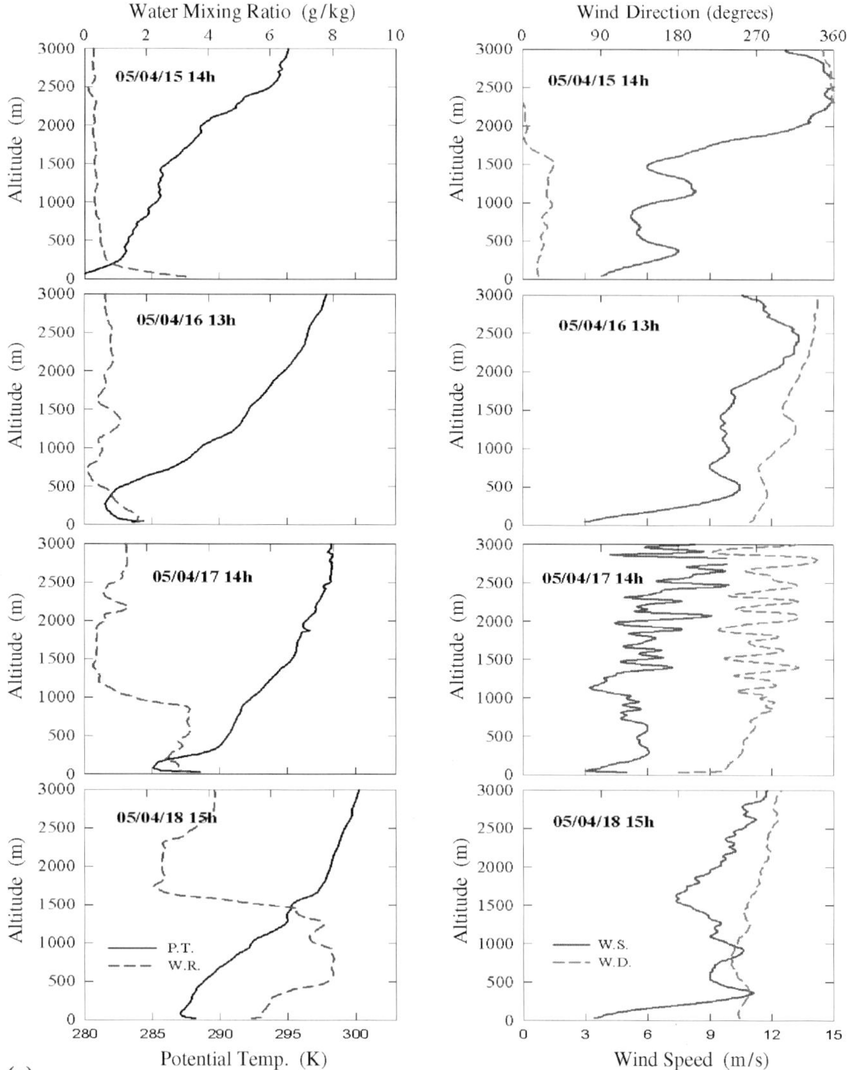

Fig. 7.3 **a** Atmospheric vertical profiles using 5-s sensitivity observations of potential temperature (P.T.), water mixing ratio (W.R.), wind speed (W.S.) and direction (W.D.) at Padori Tae-An on April 15–18, 2005

(continued)

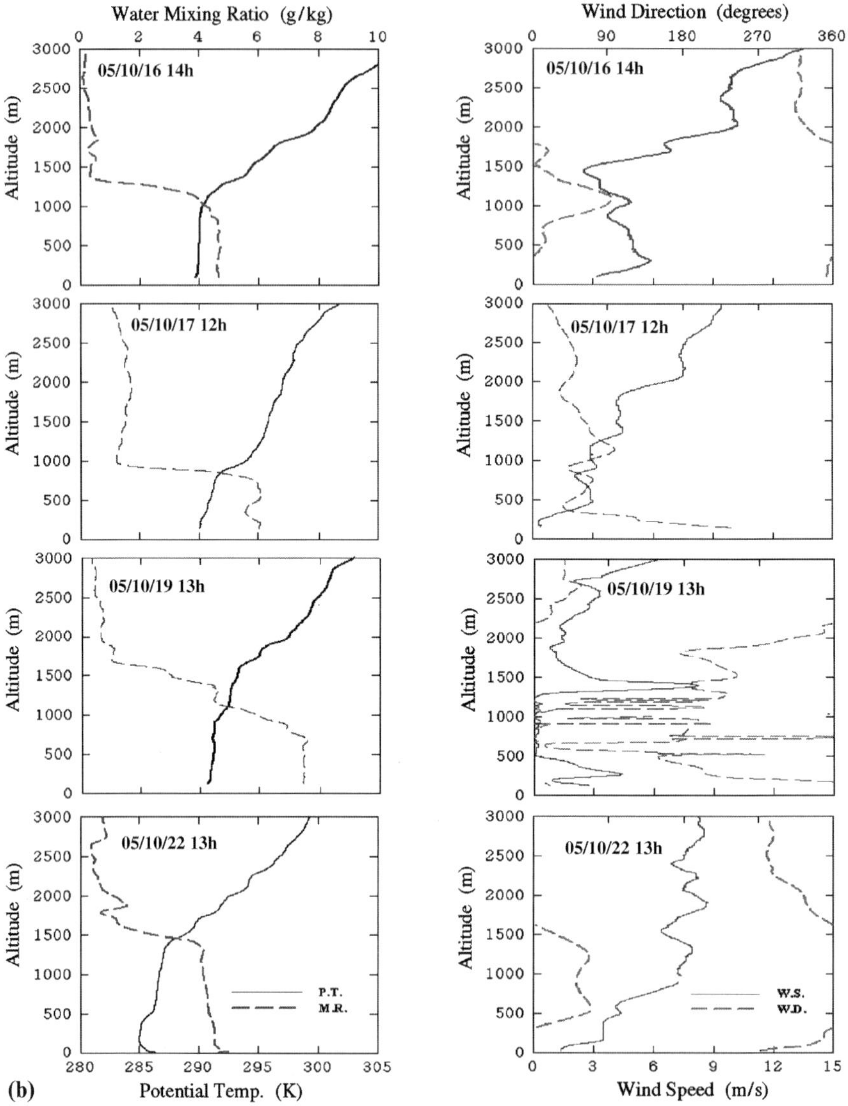

Fig. 7.3 b (continued) Atmospheric vertical profiles using a 5-s sensitivity observations of potential temperature (P.T.), water mixing ratio (W.R.), wind speed (W.S.) and direction (W.D.) at Padori Tae-An on October 16–19, 2005

7.3 Results and Discussion

7.3.1 Gaseous Species

Statistical analyses of aircraft measurements are shown in Table 7.2, which include the mean and standard deviation of gas concentrations during the measurement periods. Mean gas concentrations (ppbv) of SO_2, NO_x and O_3 were respectively 2.58, 4.24 and 54 in April and 6.63, 3.74 and 48.8 in October, 2005.

Since the observation area was affected by a high-pressure system on flight days, we expected that the highly polluted air mass emitted in this area would be confined to lower altitudes. The concentrations of the three gases at various altitudes were only measured on April 18 and October 17. The results showed that the highest concentrations were measured at an altitude of around 500 m on April 18 and 800 m on October 17, as shown in Figs. 7.4 and 7.5. The maximum concentration at these altitudes was higher than 20 ppb for SO_2 and higher than 10 ppb for NO_x during both flight days. The results seem to indicate that the high concentrations at low altitude were a result of the transport of air pollution emission from China, while the high concentration at 500 m was the result of strong wind from the west. The concentrations of O_3 were greater than 80 ppb (and up to 110 ppb on April 18) at these altitudes on both observation days. The lower concentrations of these species at higher altitudes were related to the highest water mixing ratio. Lower concentrations of SO_2 and NO_x were measured at an altitude of 1,000 m. The enhanced SO_2 concentration occurred within the low-level westerlies in association with the anti-cyclonic flow over southern China and the cyclonic circulation over Manchuria.

Table 7.2 Daily mean concentrations (ppbv) of SO_2, NO_x and O_3 during the measurement periods

		April (2005)						October (2005)					
		15 (F1)	16 (F2)	17 (F4)	18 (F3)	21 (F1)	Total	15 (F1)	16 (F2)	17 (F3)	19 (F4)	22 (F1)	Total
SO_2	Mean	0.48	1.83	1.46	7.76	1.39	2.58	17.34	3.47	9.31	2.16	0.82	6.63
	S.D.	0.40	1.86	0.66	6.75	1.07		10.52	1.53	7.84	1.93	0.37	
	Max.	1.20	9.25	3.63	21.47	4.98		35.94	6.73	30.38	9.09	1.81	
	Min.	0.01	0.02	0.49	0.26	0.13		2.17	1.15	0.47	0.11	0.06	
NO_x	Mean	2.11	2.77	2.84	9.14	4.36	4.24	6.71	3.30	3.37	4.00	1.14	3.74
	S.D.	0.19	0.50	0.61	4.56	1.92		2.24	1.30	1.84	3.66	0.61	
	Max.	2.58	3.61	4.49	15.02	9.25		13.44	7.41	6.70	18.67	3.71	
	Min.	1.82	1.58	2.09	2.71	2.59		2.17	1.53	1.00	0.52	0.43	
O_3	Mean	44	55	64	67	42	54	57	48	61	44	35	49
	S.D.	3	5	5	26	4		11	4	19	5	3	
	Max.	49	63	72	118	52		69	56	91	64	44	
	Min.	39	47	52	36	36		38	42	34	30	28	

F1, F2, F3 and F4 represent the different flight schedules, as shown in Fig. 7.2

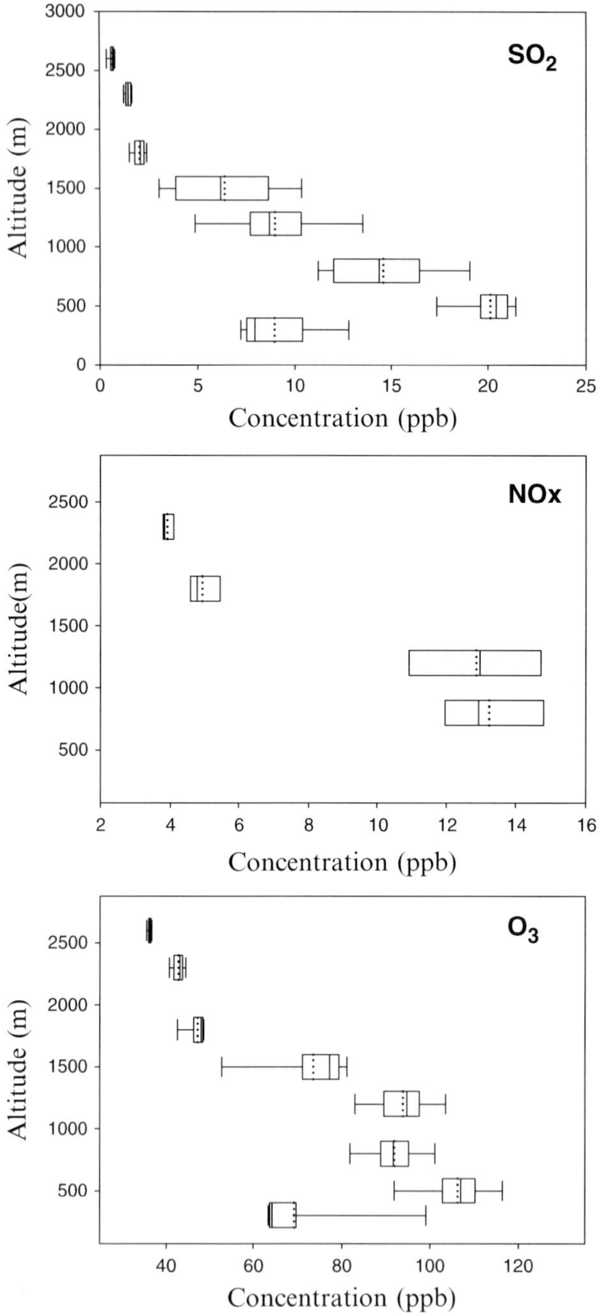

Fig. 7.4 Altitudinal distributions of the mass concentrations for SO_2, NO_x and O_3 on April 18, 2005

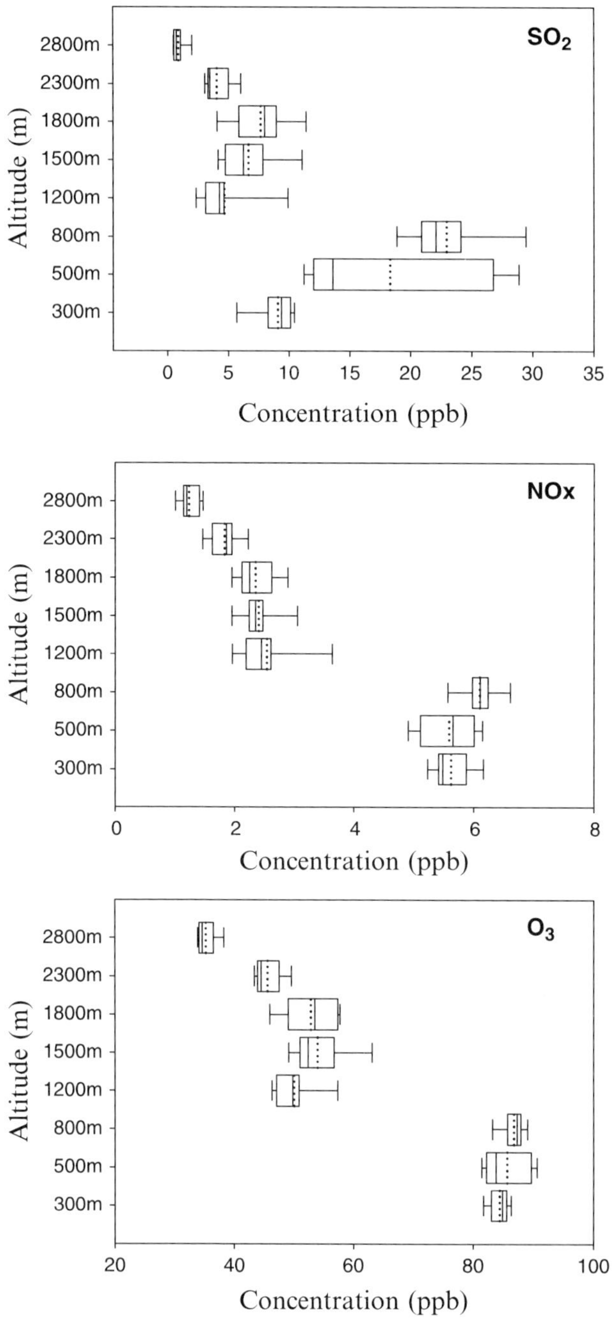

Fig. 7.5 Altitudinal distributions of the mass concentrations for SO_2, NO_x and O_3 on October 17, 2005

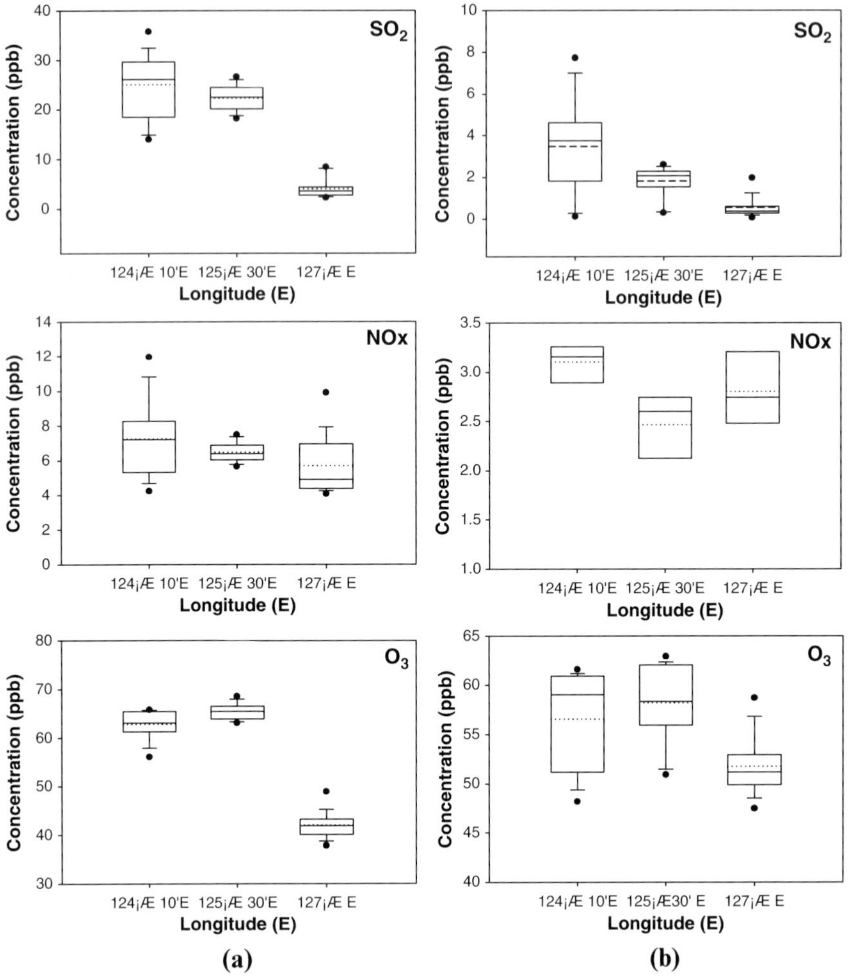

Fig. 7.6 Longitudinal distributions of mass concentrations for SO_2, NO_x and O_3 on April 16 **(a)** and October 15 **(b)**, 2005

The longitudinal distribution of 1-min average SO_2 and O_3 concentrations decreased with the progress of the flight towards the Korean peninsula, as shown in Fig. 7.6. The NO_x concentration did not show any trend with flight progress during the observation periods.

For source analysis to classify the pathway of the air mass, a divided northeast region was applied to the backward trajectories of air pollution and revealed a mostly northwest flow on the surface level, as shown in Fig. 7.7. The high concentration of SO_2 was recorded when most of the air flow passed through region III in April and region II in October (Fig. 7.8). The above pattern of SO_2 concentration was almost comparable to that of the urban atmosphere. After this episode of high-SO_2 plume transport, SO_2 and aerosol number concentrations decreased

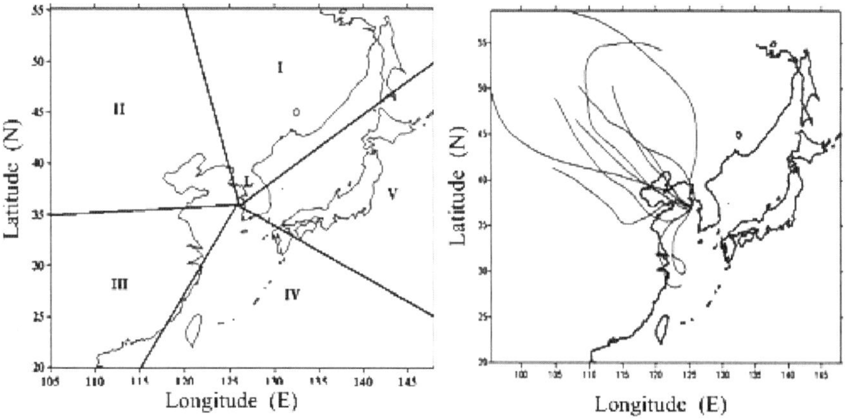

Fig. 7.7 Divided air-pollution regions in Northeast Asia **(a)** and HYSPRIT backward trajectory **(b)** for the measurement period

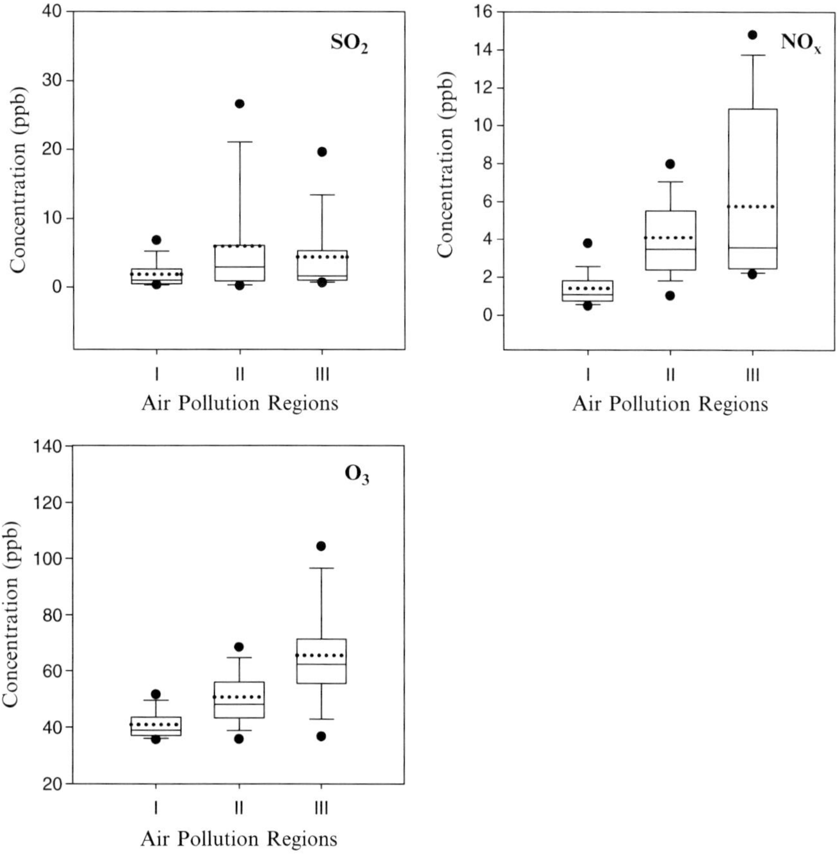

Fig. 7.8 Box-plots of air pollutant concentrations in relation to regional distributions (see Fig. 7.6a) for 2005

markedly with a northerly and northwesterly flow in association with the passage of a cold front.

Park (1998) showed that the Shandong peninsula had a high density of SO_2 emission, with an annual emission amount per 10,000 km^2 in 1996 of more than 0.1 Mt. This region is located just 300–500 km west of the Taean region in Korea. Therefore, favorable meteorological conditions could lead to the transport of sulfur compounds from the Shandong region to the west coast of Korea within a day.

7.3.2 *Aerosol and Ozone*

Ozone concentration decreased markedly with altitude from 52–63 ppb to 32–41 ppb on April 18, but aerosol number concentration showed a significant increasing trend.

The concentrations of fine particles of the atmosphere over inland and ocean areas measured at the Taean background site showed no variation of altitudinal concentrations during the period April 15–21 but wide variation during October 15–22, as shown in Fig. 7.9. The overall mean mass concentration was 13.0 µg/m^3.

However, the amount of ammonia gas over open oceans should decrease due to deposition. Thus, gaseous nitric acid produced during transport over the oceans should deposit onto large particles. Nitrate is known to exist in a coarse mode in aerosols in relatively clean air (Jordan et al. 2003; Shimohara et al. 2001; Song and Carmichael 2001) because gaseous nitric acid quickly deposits on pre-existing large particles. Only fine liquid particles of sulfuric acid, also formed during the transport, can absorb ammonia gas and undergo neutralization (Hatakeyama et al. 2004). These facts help to explain the situation observed over the western Pacific region close to the Asian continent.

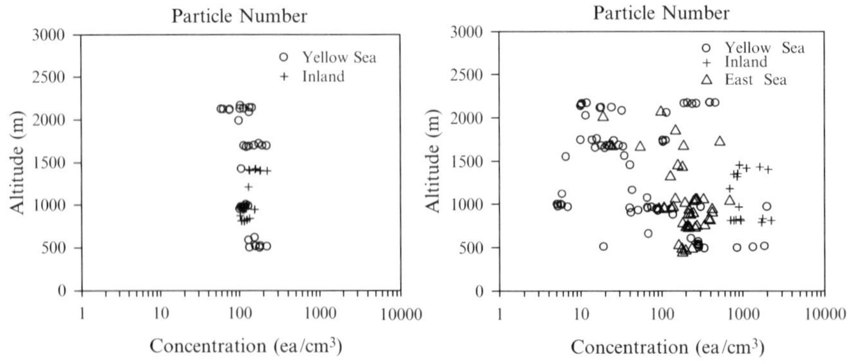

Fig. 7.9 Vertical distributions of particle number over the Yellow Sea and East Sea obtained by aircraft and inland of Korea at Tea-An on April 21(a) and October 21(b), 2005

7.3.3 Flux of Air Pollutants

Studies on the flux of air pollutant transport in the atmospheric boundary layer using aircraft measurements has been introduced by Lelieved et al. (1989). The flux estimation of air pollutant transport in the region of the Korean peninsula has been conducted by Kim et al. (1997) and Han et al. (2006) using aircraft data gathered from 1997 to 2004. Their results indicated that the influx of SO_2 was approximately five times higher than the outflux when considering yearly flux variation, and that there was a decreasing long-term trend since 1998. NO_x outflux averaged 0.095 ton/km/h and was three times higher than the SO_2 outflux. The influx and outflux of O_3 showed an even distribution on a yearly basis, except for 2002 (influx of 5.45 ton/km/h).

It is important to understand the degree to which the net flux of air pollution transport varies in the region of the Korean peninsula, including the west ocean area as influenced by influx from China and outflux to the Korea region.

In this study, the flux amount (F) at the aircraft flight level was calculated using the following similarity relationship:

$$F = 10^{-3} CHLU \sin Q \tag{1}$$

where F is the transport flux (ton/h/km), C the concentration (ppbv) of air pollutants, H the height of the atmospheric boundary layer (m), L the horizontal flight distant (km), U the mean wind speed (km/h) in the boundary layer, and Q the angle between the aircraft flight and mean wind directions.

Table 7.3 Influx and outflux calculations of air pollution transport in the boundary layer over the west Korean ocean region during the aircraft measurement period

Sampling time	Longitude	Concentration (ppb)			H (m)	L (km)	U	F (ton/h/km)		
		SO_2	NO_x	O_3				SO_2	NO_x	O_3
'05.04.15	124°30′	1.18	2.3	46	500	111	−2.14	−0.01	−0.35	−0.38
'05.04.16	124°10′	3.5	3.1	56	800	111	8.48	0.24	0.15	2.96
'05.04.16	125°30′	1.8	2.5	58	800	111	8.48	0.13	0.12	3.03
'05.04.16	127°	0.6	2.8	52	800	111	8.48	0.04	0.14	2.72
'05.04.17	124°10′	1.2	2.5	63	800	111	5.08	0.05	0.07	1.98
'05.04.17	125°30′	1.2	2.5	62	800	111	5.08	0.05	0.07	1.94
'05.04.17	127°	2.3	3.9	67	800	111	5.08	0.1	0.11	2.1
'05.04.18	124°30′	12	12.6	88	1,500	111	8.45	1.57	1.14	8.62
Average							in	0.31	0.26	3.33
							out	0.01	0.35	0.38
'05.10.16	124°10′	4.8	3	50	1,200	111	−1.51	−0.09	−0.04	−0.7
'05.10.16	125°30′	2.7	2.1	45	1,200	111	−1.51	−0.05	−0.03	−0.64
'05.10.16	127°	2.5	5.4	45	1,200	111	−1.51	−0.05	−0.07	−0.64
'05.10.17	124°30′	17.7	5.8	85.7	1,000	111	−1.6	−0.29	−0.06	−1.057
'05.10.19	125°	1.7	1.8	44.8	1,500	111	0.4	0.01	0.008	0.208
Average							in	0.01	0.008	0.208
							out	0.12	0.05	0.76

The results indicate that influxes were larger than outfluxes in April but smaller in October, as shown in Table 7.3 The larger outflux than influx of October suggests emission from the Korean region. The outflux of NO_x was on average 0.35 and 0.05 ton/km/h in April and October, respectively. The average influx of O_3 was 3.33 ton/km/h and 0.21 ton/km/h during the measurement periods in April and October, respectively. The influx of the three gases into the Korean peninsula was extremely large on April 18, when the wind in the boundary layer blew strongly from a northwesterly direction.

7.4 Conclusions

Observations of the long-range transport of the trace gases $SO_{2,}$ NO_x and O_3 in the atmospheric boundary layer were conducted over the ocean area "transport corridor" between the Korean peninsula and China using measurements obtained by aircraft during April 15–22 and October 15–25, 2005. The objectives of gathering measurements by aircraft were to study the spatial distribution of pollution in the ocean atmosphere between Korea and China, and to improve our understanding of acidic deposition in the Korean area in relation to processes affecting the transport of long-range trans-boundary air pollutants from China.

Aircraft measurements of gas concentrations (ppbv) for SO_2, NO_x and O_3 were 2.58, 4.24 and 54 in April and 6.63, 3.74 and 48.8 in October, respectively. Changes in concentration of the three gases with respect to altitude were only investigated on April 18 and October 17, 2005. The highest concentrations of the three gases were recorded around an altitude of 500 m on April 18 and at 800 m on October 17, 2005. The highest concentrations of SO_2 at these altitudes were greater than 20 ppb, while the concentration of NO_x was higher than 10 ppb for both flight days. The results suggest that the high concentrations at low altitude were the result of the transport of air pollution emission from China, while the high concentration at 500 m was due to strong wind from the west. The concentrations of O_3 were greater than 80 ppb (and up to 110 ppb on April 18) at these altitudes on both observation days. The lower concentrations of these species at higher altitudes were related to the highest water mixing ratio. Lower concentrations of SO_2 and NO_x were measured at an altitude of 1,000 m.

Ozone concentration decreased markedly with altitude from 52–63 ppb to 32–41 ppb on April 18, whereas aerosol number concentration showed a significant increasing trend.

The concentrations of fine particle in the atmosphere over inland and ocean areas measured at the Taean background site showed no variation with respect to altitude during the period April 15–21, but exhibited wide variation during October 15–22, as shown in Fig. 7.9. The overall mean mass concentration was 13.0 $\mu g/m^3$. Flux analysis indicated that influxes of air pollution were larger than outfluxes in April but smaller in October. The larger outflux than influx suggests emission from the Korean region. The influx of the three gases into the Korean peninsula was extremely large in spring, when the wind in the boundary layer blew strongly from a northwesterly direction. Correlations between O_3 and NO_x showed interesting changes that were related to the distance from the emission source. O_3 and NO_x showed a negative correlation, which indicates that the pollutants in the air masses are very clean. During the process of transport, most of the acidic sulfates and acidic gases were mixed with regional air pollutants such as chlorides and nitrates existing in the metropolitan Seoul area.

The pathway of the air mass based on backward trajectory analysis indicated that air pollution transport was affected predominantly by a northwesterly flow at ground level. The high concentration of SO_2 was measured when the most of the air flow passed through region III of central China in April and region II of southern China in October. Flux analysis revealed that the influx was greater than the outflux in April but smaller in October for the west ocean of Korea.

Acknowledgements The research described in this paper was conducted with financial support from the Long Range Trans-boundary Air-Pollution project of the Korea Ministry of Environment, Korea in 2005.

References

Arimoto R., Zhanf X., Huebert B.J., Kang C.H., Savoie D., Propspero J.M., Sage S., Schloesslin C.A., Khaing H., and Oh S.N. (2004), Chemical composition of atmospheric aerosols from Zhenbeitai, People's Republic of China, and Gosan, South Korea, during ACE-Asia, *J. Geophys. Res.*, 109,(D19S04).

Clarke A.D., Collins W.G., Rasch P.J., Kapustin V.N., Moore K., Howell S., and Fuelberg H.E. (2001), Dust and pollution transport on global scales: Aerosol measurements and model predictions, *J. Geophys. Res.*, 106, 32555–32569.

Han J.S., Ahn J.Y., Hong Y.D., Konh B.J., Lee S. J., and Sun W.Y. 2006, The vertical distribution patterns of long range transported SO_2 in Korea peninsula, *Journal of Korean Society for Atmospheric Environment.*, 22(1), 99–106.

Hatakeyama S., (1998), *Data of IGAC/APARE/PEACAMPOT Aircraft and Ground-based Observations*:91 –95 Collective Volume [CD-ROM], Center for Global Environment Research, National Institute for Environmental Studies, Tsukuba, Japan.

Hatakeyama S., Takami A., Sakamaki F., Mukai H., Sugimoto N., and Shimizu A. (2004), Aerial measurement of air pollutants and aerosols during 20–22 March 2001 over the East China Sea, *J. Geophys. Res.*, 109, D13304, DOI 10.1029/2003JD004271.

IGAC. (1998), from http://saga.pmel.noaa.gov/aceasia/prospecus 122198.html.

Jeffe D.A., Anderson T., Covert D., Kotchenruther R., Trost B., Danielson J., Simpson W., Berntsen T., Karlsdottir S., Blake D., Harries J., Carmichael G., and Uno I. (1999), Transport of Asian air pollutant to North America, *Geophys. Res. Lett.*, 26, 26711–26714.

Jordan C.E., Dibb J.E., Anderson B.E., and Fuelberg H.F. (2003), Uptake of nitrate and sulfate on dust aerosols during TRACE-P, *J. Geophys. Res.*, 108(D20), 8817, DOI 10.1029/2002JD003101.

Kim B.G., Cha J.S., Han J.S., Park I.S., Kim J.S., Na J.G., Choi D.L., Ahn J.Y., and Kang C.G. (1997), Aircraft measurement of SO_2, NO_x over Yellow Sea Area, *Journal of Korea Air Pollution Research Association*., 13(5), 361–369.

Kim B.G., Han J.S., and Park S.U. (2001), Transport of SO_2 and aerosol over the Yellow sea, *Atmos. Environ.*, 35, 727–737.

Lelieved J., Janson F.W., and Dop H.V. (1989), Assessment of pollutant fluxes across the frontiers of the Federal Republic of Germany on the basis of aircraft measurement, *Atmos. Environ.*, 23, 939–951.

Nunnermacker L.J., Weinstein-Lloyd J., Kleinman L., Daum P.H., Lee Y.N., Springston S.R., Klotz P., Newman L., Neuroth G., and Hyde P. (2004), Ground-based and aircraft measurements of trace gases in Phoenix, Arizona (1998), *Atmos. Environ.*, 38, 4941–4956.
Prospero J.M. (1999), Long-term measurements of the transport of African mineral dust to the southeastern Unites States: Implications for regional air quality, *J. Geophys. Res.*, 104, 15917–15927.
Shimohara T., Oishi O., Utsunomiya A., Mukai H., Hatakeyama S., Eun-Suk J., Uno I., and Murano K. (2001), Characteristics of atmospheric air pollutants at two sites in northern Kyushu, Japan—Chemical form, and chemical reaction, *Atmos. Environ.*, 35, 667–681.
Song C.H. and Carmichael G. (2001), A model study of the evolution processes of dust and sea salt particles during long range transport, *J. Geophys. Res.*, 106, 18131–18154.
Streets D., Tsai N.Y., Akimoto H., and Oka K. (2001), Trends in emissions of acidifying species in Asia, 1985–1997, *Water Air Soil Poll.*, 130, 187–192.

Chapter 8
Optical Remote Sensing for Characterizing the Spatial Distribution of Stack Emissions

Michel Grutter[1], Roberto Basaldud[1], Edgar Flores[1], and Roland Harig[2]

Abstract In this contribution, optical methods based on passive FTIR (Fourier Transform Infrared) and DOAS (Differential Optical Absorption Spectroscopy) techniques have been used to characterize the dispersion of gas emissions from industrial sources. Portable, zenith-looking, passive-DOAS instruments measured the horizontal distribution of an SO_2 plume from a power plant in a coastal town of Mexico. The column density of this gas was measured while making traversals across the plume with a car and a boat downwind from the emission source. The cross sections measured at different distances from the source are used to characterize the horizontal dispersion and to estimate emission fluxes. In addition, a Scanning Infrared Gas Imaging System (SIGIS) was used to acquire passive IR spectra at 4 cm^{-1} resolution in a two-dimensional array, from which a false-color image is produced representing the degree of correlation of a specific gaseous pollutant. The 24-h, real-time animations of the SO_2 plume help us to understand dispersion phenomena in various atmospheric conditions. The wealth of information retrieved from these optical remote sensors provides an alterative method for evaluating the results from plume dispersion models.

Keywords: Industrial emissions, optical remote sensing, passive DOAS, passive FTIR, plume dispersion

8.1 Introduction

The dispersion of atmospheric pollutants from stacks is often characterized by mathematical models which predict the ground-level concentrations from meteorological, topographical and emission data. Tall stacks do not eliminate the

[1] *Centro de Ciencias de la Atmósfera, Universidad Nacional Autónoma de México, 05410 México D.F. México*

[2] *Institut für Messtechnik, Technische Universität Hamburg-Harburg, 21079 Hamburg, Germany*

Y.J. Kim and U. Platt (eds.), *Advanced Environmental Monitoring*, 107–118.

pollution to the atmosphere, but they do aid in reducing ground-level concentrations and their potentially harmful or damaging effects (Schnelle and Partha 2000). The tall stack, however, may not always provide the best solution when the atmospheric conditions do not favor long range dispersion. The most widely used models for regulatory purposes are the deterministic Gaussian plume models. The performance of such models is commonly evaluated by experimentally determining the concentration of a pollutant at a number of receptor points and comparing the measurements with the model output. This task can be expensive, time consuming and the results may not represent the general performance of the model over the total area of interest.

In this contribution, we demonstrate how measurements, based on optical remote sensing, are used to characterize the spatial distribution of stack emissions. The plume's structure is determined by two methods based on the radiative absorption and emission of the polluting gases. The first technique measures the amount of absorption of solar UV by molecules like SO_2 and NO_2. The radiation is spectroscopically analyzed so that the column concentration of these gases can be monitored while moving along a path perpendicular to the propagation of the plume. This measurement is a cross-section of the **horizontal distribution** of the plume at a specific distance from the source. The second technique uses a spectroscopic analysis of the natural thermal radiation from the plume and surroundings. These thermal emissions are analyzed with a passive infrared sensor and used to identify the characteristic emission/absorption properties of specific constituents in the plume. A two-dimensional image of the **vertical distribution** of the plume is constructed by scanning the area around the emission source.

The two measurement techniques were implemented in Manzanillo, a coastal town with 138,000 inhabitants (INEGI 2005), located on the Pacific coast of Mexico, in the State of Colima (19.03 N, 104.19 W). A large, oil-fired power plant, with a maximum capacity of producing 1,900 MW of electricity, is located approximately 3 km south of the town center. The oil used in this plant has a high content of sulphur (3–4%) such that the plume produced during combustion is high in SO_2. In the following sections, the measurement techniques will be described and selected results will be presented.

8.2 Passive DOAS

The Differential Optical Absorption Spectrometer (DOAS) is a widely used technique for the continuous measurement of atmospheric gases (Platt 1994; Platt et al. 1979). In its configuration for active sensing, where the light traveling along an open path is provided by a synthetic radiation source, low detection limits for the ambient concentrations of various gases (O_3, SO_2, HCHO, NO_2, several hydrocarbons, aromatic compounds, etc.) can be achieved. This technique is based on the spectral analysis of the differential absorption by molecules in the ultraviolet and visible part of the spectrum. The broader extinction of UV light due to other

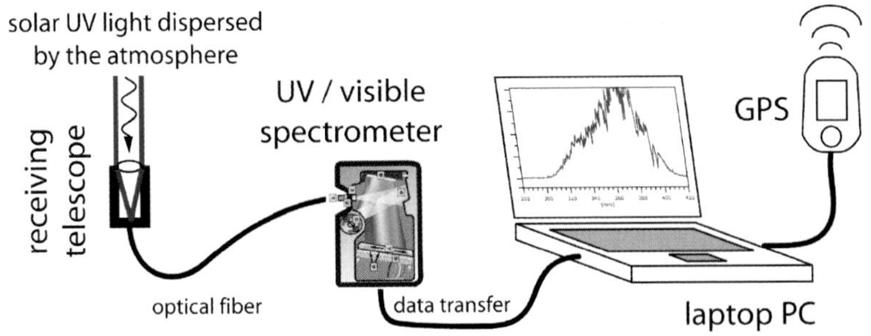

Fig. 8.1 Schematic diagram of the portable passive DOAS instrument used for measuring column gas concentrations on a moving platform

processes such as dispersion by fine particles is cancelled when processed and thus not taken into account.

In the passive sensing configuration, as is in the case when dispersed light from the sun (i.e. from a blue sky) is used as the radiation source, the spectral analysis is based on the differential absorption of a particular gas or group of gases. The total amount of molecules in an atmospheric column is determined between the altitude where the radiation is dispersed and the position of the observer. These techniques have been used to identify and quantify atmospheric gases both in industrial (Lohberger et al. 2004) and volcanic (Bobrowski et al. 2003; Lee et al. 2005) plumes.

The scheme used for making passive DOAS measurements in this study is presented in Fig. 8.1. The dispersed solar UV light is collected with a narrow field-of-view (<20 mrad) telescope that was built in-house. This consists of a concave lens (f = 100 mm), a bandpass optical filter (240–400 nm) and a 200-μm diameter optical fiber. The light is analyzed with a hand-held spectrometer (Ocean Optics, model S2000), at a resolution of 0.44 nm between 280 and 420 nm. This devise uses a UV holographic grating, a 2048 element CCD detector and has no moving parts; hence, it is appropriate for this type of application. The spectra are recorded using an interface based on LabView that couples each acquisition with a longitude-latitude fix from a GPS receiver. User defined parameters along with dark and background spectra are entered prior to each measurement along a trajectory.

The spectra are evaluated following these general steps:

- Reference spectra of the target gases are generated for each spectrometer through the convolution of a high-resolution reference spectrum from the literature or a spectral library. For this purpose the instrumental line shape (ILS), determined by measuring the emission line from a low-pressure mercury lamp, is used to produce a reference spectrum with the true resolution of the spectrometer.
- A dark spectrum is subtracted from the measured spectrum (blue trace in Fig. 8.2) to correct the baseline offset.

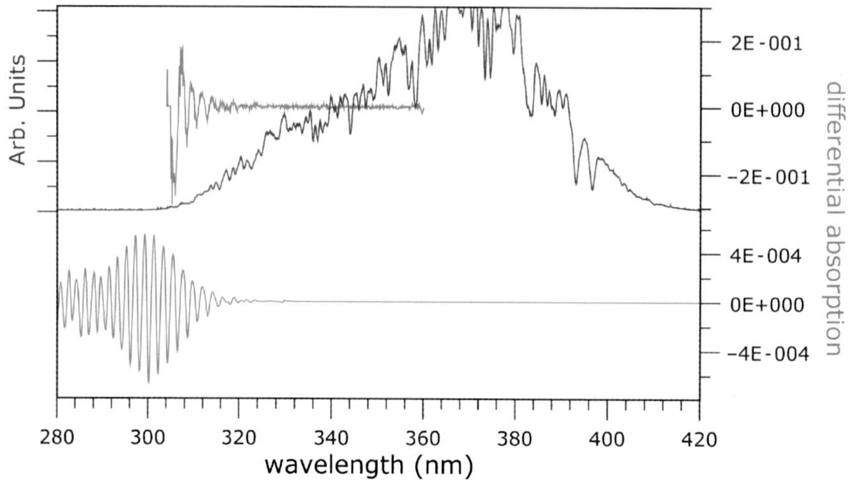

Fig. 8.2 Measured spectrum (blue trace) from solar dispersed light in the direction of the zenith and across a plume. The orange traces are differential absorptions of SO_2 from the measurement (upper) and from the reference (lower) spectra

- A differential absorption spectrum (upper orange trace in Fig. 8.2) is created from the background spectrum measured outside the gas plume and by applying a high-pass filter.
- The resulting spectrum is fitted to the reference spectrum (lower trace) using the software DOASIS (Kraus 2003).
- SO_2 is analyzed by fitting the reference (Vandaele et al. 1994) in the region 306–317 nm and taking into account the interference of O_3 absorptions. NO_2 (Harder et al. 1997) is evaluated in the 350–390 nm region.

Figure 8.3 presents the results from traversals of a plume over the ocean made from a small boat that sailed across the Manzanillo Bay. Fig. 8.3a is a map of the region with the red dot marking the power plant location. The trajectory of the boat is drawn on the map as a thick black line starting close to the location marked by the green star and moving in the direction marked by the arrows. The positions of the plume crossings are shown on the boat trajectory in false colors whose scale is given in the same figure. The wind rose in Fig. 8.3b shows that winds were predominately from the southeast, coinciding with the positions where the plume was detected.

Figure 8.3c is a plot of the SO_2 column concentration along the trajectory of the boat. The first large peak in this plot represents the first pass below the plume at 2.5 km (in a straight line) downwind of the source. The SO_2 profile reveals that the width of the plume at this distance was 1,250 m, which can be interpreted as its horizontal dispersion profile. The second and forth passes, approximately 5 km from the emission source, have widths of 1,500 m. The third pass and furthest from the source (10 km) had an approximate width of 2,800 m.

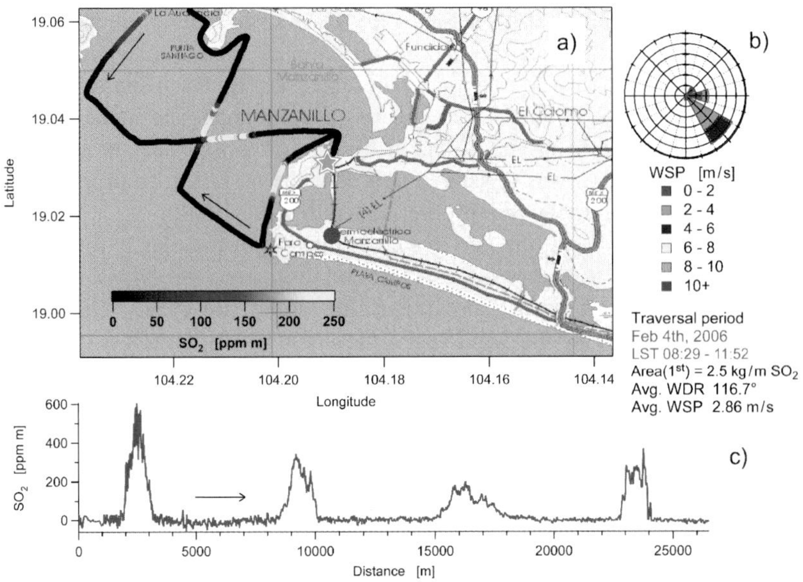

Fig. 8.3 The passive-DOAS measurements shown here characterize the spatial distribution of the SO_2 in a plume from a power plant (red dot) in Manzanillo, Mexico. In this figure are **a**. the map of the region with the boat trajectory, **b**. the wind chart and **c**. the column SO_2 concentration measured along this path

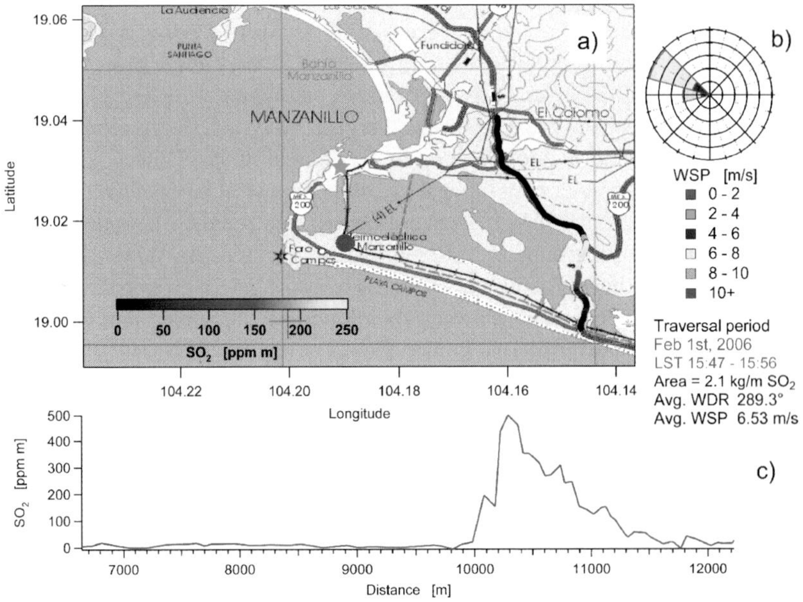

Fig. 8.4 Results from a passive-DOAS measurement done from the land along a roadway east from the power plant (red dot) in Manzanillo, Mexico. Frame **a**. is a map with the boat trajectory, **b**. is the wind chart and **c**. is the column SO_2 measured along this path

Traversals of the plume with the passive DOAS were more often done on land while driving a car along a highway. One of the many measurements done during the one-week period of the campaign is presented in Fig. 8.4. Here the wind was coming from the WNW as often occurred during the afternoon. The plume was carried across the lagoon and measured by the instrument, as can be seen in Fig. 8.4a, c. The peak concentration was registered at a distance of 7.7 km from the emission source and, according to the SO_2 profile, its width was ~1,500 m.

In addition to providing the horizontal distribution of the plume, these profiles were used to estimate the emission fluxes of SO_2 and NO_2. This is done by conversion to units in kg/m^2 and integrating under the curve. The resulting values are multiplied by the wind speed measured at a 15 m tower close to the emission source.

8.3 Passive FTIR

8.3.1 Radiative Transfer Model

Passive remote sensing of gas clouds is based on the analysis of infrared radiation absorbed and emitted by the molecules in the clouds. The propagation of radiation through the atmosphere is described by radiative transfer theory. One method for modeling the radiative transfer is the division of the atmosphere along the optical path into layers which are assumed to be homogeneous with regard to all physical and chemical properties. Each of these layers absorbs a fraction of the radiation entering the layer but also emits radiation. Both of these processes depend on the properties of the layer, such as composition, temperature and pressure.

In order to model the radiation from the sky, many layers are required due to the different temperatures and pressures at different altitudes. However, to describe the basic spectral characteristics of a gas cloud in the lowest layer of the atmosphere, as measured by a ground-based passive infrared spectrometer, a model with only two layers and a background is sufficient in most cases. As shown in Fig. 8.5, radiation from the background, e.g., a surface, propagates through the gas cloud (Layer 2) and the atmosphere between the cloud and the spectrometer (Layer 1). Layers 1 and 2 are considered homogeneous with regard to all physical and

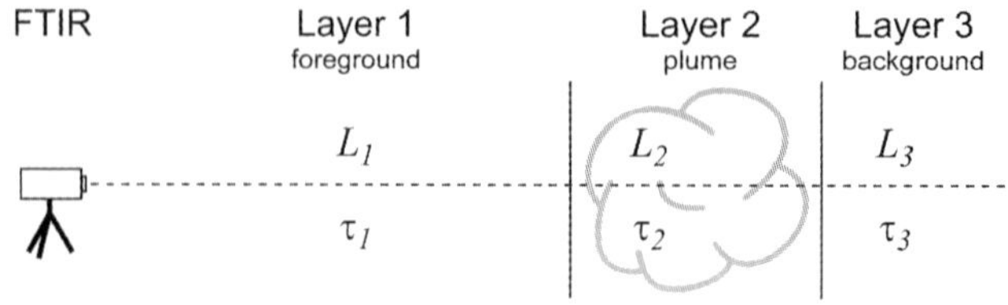

Fig. 8.5 Simple radiative transfer model for the identification of gas pollutants from passive infrared spectra

chemical properties within each layer. The radiation containing the signatures of the atmosphere, the gas cloud, and the background radiation is measured by the spectrometer.

In this model, the spectral radiance at the entrance aperture of the spectrometer L_1 can be described by

$$L_1 = (1-\tau_1)B_1 + \tau_1\left[(1-\tau_2)B_2 + \tau_2 L_3\right], \quad (1)$$

where τ_i is the transmittance of layer i, B_i is the spectral radiance of a blackbody at the temperature T_i of layer i, and L_3 is the radiance that enters the layer of the cloud from the background. All quantities in Eq. 1 are frequency dependent. If the background of the field of view is a surface, the radiation entering the cloud contains radiation emitted by the surface and reflected radiation, i.e., ambient radiation and radiation from the sky. The contribution of scattering is in this case neglected.

If the temperatures of the Layers 1 and 2 are equal (B_1=B_2), Eq. 1 can be simplified to:

$$L_1 = B_1 + \tau_1\tau_2\left(L_3 - B_1\right). \quad (2)$$

The radiance difference $\Delta L = L_1 - L_3$ is given by

$$\Delta L = (1-\tau_1\tau_2)\Delta L_{13}, \quad (3)$$

where $\Delta L_{13} = B_1 - L_3 = B_2 - L_3$. In this work the term "radiance" is used as a simplifying synonym for the correct term "spectral radiance." Thus, the analysis of the spectrum allows detection, identification, and quantification of the species contained in the gas cloud.

8.3.2 Method for the Detection of SO_2

In the present work, the passive FTIR technique is used to examine the plume shape from a power plant based on the identification of SO_2 under the influence of different meteorological conditions. The detection method is based on the approximation of a measured spectrum by reference spectra, which has been described elsewhere (Harig and Matz 2001; Harig et al. 2002). First, the spectrum of the brightness temperature $T_{br}(\sigma)$ is calculated. The detection is performed in three steps. In the first step, the mean brightness temperature is subtracted and the signatures of the target compound (SO_2) and atmospheric species (H_2O, O_3) are fitted to the resulting spectrum. Moreover, the baseline is approximated by a least-squares fit. In the next step, the contributions of all fitted signatures (i.e. atmospheric species and baseline) except the signature of the target compound (SO_2 in this case) are subtracted from the measured spectrum. In the final step, the coefficient of correlation between the corrected spectrum, i.e., the result of the subtraction, and a reference spectrum in a specific spectral range, is calculated. The calculation is performed for three different column densities of the target compound. The maximum value for this coefficient is used as an indicator for the presence of the gas, as shown by the lighter colors in Fig. 8.7.

Reference spectra with different column densities are calculated by convolution of high-resolution transmittance spectra (e.g., calculated with absorption coefficients computed using HITRAN/FASCODE (Rothman et al. 2003; Smith et al. 1978) with a real instrument line shape function (Harig 2004)).

8.3.3 Scanning Imaging Remote Sensing System

SIGIS (Scanning Infrared Gas Imaging System) is an imaging remote sensing system based on the combination of an interferometer with a single detector element and a scanning mirror. The system is comprised of the interferometer OPAG 22 (Bruker Daltonics, Leipzig, Germany), a telescope, an azimuth-elevation-scanning mirror actuated by stepper motors, a data processing and control system with a digital signal processor (FTIR DSP), an image processing system (Video DSP), a video camera and a PC for control and display of the results (Fig. 8.6).

The maximum optical path difference of the interferometer configuration used in this work is 1.8 cm, resulting in a spectral resolution of approximately 0.5 cm^{-1}. For the visualization of gas clouds, however, a spectral resolution of 4 cm^{-1} is used. The choice of resolution has been evaluated as a good trade-off between the goals of a low detection limit, sufficient selectivity and a short measurement time. The signal-to-noise ratio improves with decreasing spectral resolution for a constant

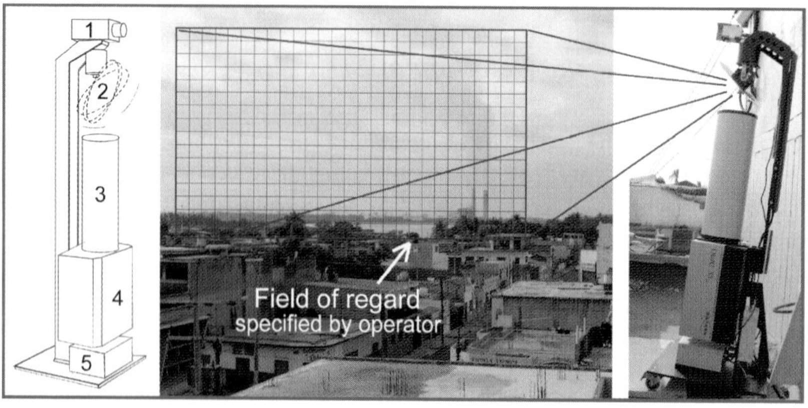

Fig. 8.6 Schematic diagram (left) and photograph (right) of the Scanning Infrared Gas Imaging System (SIGIS), with (**1**) video camera, (**2**) azimuth-elevation-scanning mirror, (**3**) telescope, (**4**) spectrometer and (**5**) DSP system. The field of regard is illustrated in the center

measurement time (Harig 2004). On the other hand, higher resolution yields higher selectivity. At $4\,cm^{-1}$, six two-sided interferograms per second can be measured.

For the visualization of pollutant clouds, the scanning mirror is sequentially positioned to an array of points within the field of regard. The size and the direction of the field of regard and the spatial resolution (i.e. the angle between adjacent fields of view) are variable. Each interferogram measured by the interferometer is recorded by the FTIR DSP system. The Fourier transformation is performed by the DSP, the spectrum is transferred to the PC and analyzed using the identification method described in section 8.3.2. The results are visualized by overlaying the false color images to a video image. For each target compound in the spectral library, various false color images visualizing the results of the detection, identification and quantification algorithms are produced. The SO_2 molecule was found to be a good tracer for the visualization of the emission plume due to its high abundance in this particular case, although a great variety of gases can be used as identifiers in other applications. Simultaneous with the analysis and visualization of one interferogram by the FTIR DSP and the PC, the scanning mirror is set to the following position to record the next interferogram, and so on.

8.3.4 Measurements

The instrument was placed on the roof of a 3-story building in downtown Manzanillo. The location, that was 2.5 km north of the power plant, is marked by a green star in Figs. 8.3a and 8.4a. Various images of the exhaust gas plume of the plant were recorded. The images in Fig. 8.7 are taken from an 8-h sequence, which captured the SO_2 distribution around the source during distinctly different atmospheric conditions. The coefficient of correlation between the measured and processed spectra and a reference spectrum of SO_2 is shown as a false color representation (as described in section 8.3.2) and superimposed on the video image. A complete scan was made every 3 min from which the evolution of the plume can be followed in detail. The eight images shown in Fig. 8.7 demonstrate the large variability in the spatial distribution of SO_2 during an eight and a half hour period.

It is evident from this particular example that the westerly sea breeze pushes the power plant's pollution towards the land throughout the afternoon with some deviations as seen in the 15:17 frame. At around 20:00 hours, the wind direction changes and the land breeze begins to dominate. Particularly interesting is that when the plume has already changed its direction towards the sea as can be seen in the 21:06 frame, an air mass which had been collecting pollution over the land within the last several hours passes the observation window from left to right (shown as the red area at the left in the 21:06 frame). This can be more clearly seen in a video animation which was created for the interpretation. The easterly land breeze then predominates throughout the night as shown in the last frame of Fig. 8.7.

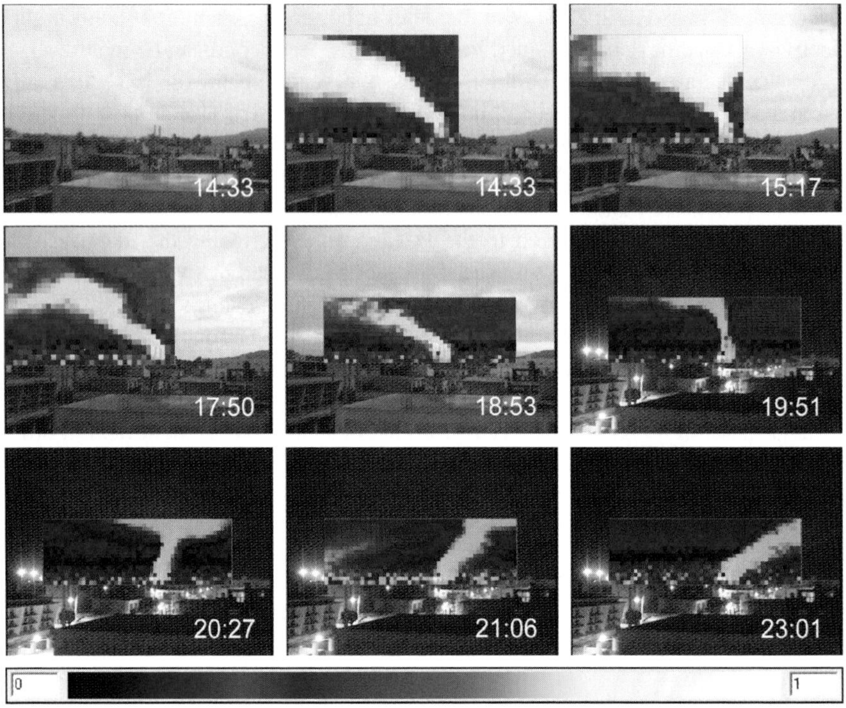

Fig. 8.7 Sequence of video and SIGIS images showing the SO_2 distribution on Feb 2, 2006 in Manzanillo, Mexico. The local time is given on the bottom right

8.4 Conclusions

The results presented here show how the spatial distribution of the gaseous pollutants from an industrial emission source can be measured with optical remote sensors. The passive DOAS instrument was successful in obtaining horizontal profiles of the plume by making transects perpendicular to the plume's axis and at different distances from the source. From these measurements, that were performed both from sea and land, indirect estimates of the emission fluxes can be obtained. Similarly, the temporal evolution of the horizontal and vertical plume structure was continuously measured with a passive FTIR instrument, which is capable of detecting pollutants from distances greater than 2.5 km from the source. The Scanning Infrared Gas Imaging System can continuously register two-dimensional images of gaseous species which allows the monitoring and animation of fundamental plume propagation properties (degree of dispersion, direction, speed, etc.) not only during the day but also quite impressively during the night.

This contribution describes two state-of-the-art methodologies which can contribute to the field of advanced environmental monitoring with their ability to detect and visualize clouds of potentially toxic pollutants from a distance. Additionally, the measurement of the spatial distribution of emission source with these techniques can be used as an alternative and convenient way to evaluate the performance of plume dispersion models commonly used to diagnose these situations. Finally, given the potential effects which the emissions of large quantities of pollutants have on global climate and the health of humans, plants and the ecosystems, one cannot underestimate the need to reduce the emissions by exploiting newer and better technologies.

Acknowledgements This work has been funded through project CONACYT-CFE-2004-C01-44. The authors would like to thank Stefan Kraus and the University of Heidelberg for making the software DOASIS available, the M. in. Sc. student Andrés Hernández and the personnel at the Environmental Protection Department from CFE in Manzanillo for their valuable help during the field campaign and the mechanical workshop (M.A. Meneses and A. Rodriguez) at CCA-UNAM for helping build part of the instruments.

References

Bobrowski N., Hönninger G., Galle B., and Platt U. (2003), Detection of bromine monoxide in a volcanic plume, *Nature*, 423, 273–276.

Harder J.W., Brault J.W., Johnston P.V., and Mount G.H. (1997), Temperature dependent NO_2 cross sections at high spectral resolutions, *J. Geophys. Res.*, 102(D3), 3861–3880.

Harig R. (2004), Passive remote sensing of pollutant clouds by FTIR spectrometry: Signal-to-noise ratio as a function of spectral resolution, *Appl. Opt.*, 43(23), 4603–4610.

Harig R. and Matz G. (2001), Toxic cloud imaging by infrared spectroscopy: A scanning FTIR system for identification and visualization, *Field Anal. Chem. Technol.*, 5(1–2), 75–90.

Harig R., Matz G., and Rusch P. (2002), Scanning Infrared Remote Sensing System for Identification, Visualization, and Quantification of Airborne Pollutants, *Proc. SPIE*, 4575, 83–94.

INEGI. Conteo de población y vivienda 2005, Instituto Nacional de Estadística Geográfica e Informática, 2005.

Kraus S. *DOAS Intelligent System*, University of Heidelberg in cooperation with Hoffman Messtechnik GmbH, 2003.

Lee C., Kim Y.J., Tanimoto H., Bobrowski N., Platt U., Mori T., Yamamoto K., and Hong C.S. (2005), High ClO and ozone depletion observed in the plume of Sakurajima volcano, Japan, *Geophys. Res. Lett.*, 32, L21809.

Lohberger F., Hönninger G., and Platt U. (2004), Ground-based imaging differential optical absorption spectroscopy of atmospheric gases, *Appl. Opt.*, 43, 4711–4717.

Platt U. (1994), Differential optical absorption spectroscopy (DOAS). (In M.W. Sigrist (Ed), *Air monitoring by spectroscopy techniques* (pp. 27–83), Wiley Interscience, New York).

Platt U., Perner D., and Pätz H.W. (1979), Simultaneous measurement of atmospheric CH_2O, O_3 and NO_2 by differential optical absorption, *J. Geophys. Res.*, 84, 6329–6335.

Rothman, L.S., Barbe A., Benner D.C., Brown L.R., Camy-Peyret C., Carleer M.R., Chance K., Clerbaux C., Dana V., Devi V.M., Fayt A., Flaud J.-M., Gamache R.R., Goldman A., Jacquemart D., Jucks K.W., Lafferty W.J., Mandin J.-Y., Massie S.T., Nemtchinov V., Newnham D.A., Perrin A., Rinsland C.P., Schroeder J., Smith K.M., Smith M.A.H., Tang K., Toth R.A., Vander Auwera J., Varanasi P., and Yoshino K. (2003), The HITRAN molecular spectroscopic database: edition of 2000 including updates through 2001, *J. Quant. Spectrosc. Radiat. Transfer,* 82, 5–44.

Schnelle K.B. and Partha R.D., *Atmospheric Dispersion Modeling Compliance Guide,* (McGraw-Hill, New York, 2000).

Smith H.J.P., Dude D.J., Gardner M.E., Clough S.A., Kneizys F.X., and Rothman L.S. FASCODE- Fast Atmospheric Signature Code (Spectral Transmittance and Radiance), pp. Report AFGL-TR-78-0081, Air Force Geophysics Laboratory Technical, Hanscom AFB, MA., 1978.

Vandaele A.C., Simon P.C., Guilmot J.M., Carleer M., and Colin R. (1994), SO_2 absorption cross section measurement in the UV using a Fourier transform spectrometer, *J. Geophys. Res.*, 99, 25599.

Section 2
Atmospheric Environmental Monitoring

Chapter 9
Mass Transport of Background Asian Dust Revealed by Balloon-Borne Measurement: Dust Particles Transported during Calm Periods by Westerly from Taklamakan Desert

Y. Iwasaka[1], J.M. Li[2], G.-Y. Shi[3], Y.S. Kim[1,a], A. Matsuki[1,b], D. Trochkine[1,c], M. Yamada[1], D. Zhang[4], Z. Shen[5], and C.S. Hong[1]

Abstract The dust storm which is caused by low pressure activities in China and Mongolia has been investigated by many investigators, but very thin dust clouds, which can be frequently detected in every season (we call it background Asian dust here) by lidar in Japan, Korea, and China but not by satellite, have attracted very few investigators since detection of the cloud is not easy. It, however, has been suggested that the background Asian dust also plays an important role in the biogeochemical cycle of dust in east Asia and west Pacific regions through long range transport of dust particles by westerly winds, and information of outflow rate of background dust particles over the dust source areas is strongly desired since previous investigations were made mostly in the down wind regions (Iwasaka et al. 1988; Matsuki et al. 2002; Trochkine et al. 2002). According to the balloon-borne measurements made under the calm weather condition in 2001–2004 at Dunhuang (40°00′N, 94°30′E), China, mass flux of background Asian dust due to westerly wind was about 50 ton/km^2/day over the Taklamakan desert (about 4 to 6 km altitudes) and

[1] *Institute of Nature and Environmental Technology, Kanazawa University, Kanazawa, Japan*

[2] *Graduate School of Environmental Studies, Nagoya University, Nagoya, Japan*

[3] *Institute of Atmospheric Physics, Chinese Academy of Science, Beijing, China*

[4] *Faculty of Environmental and Symbiotic Sciences, Prefectural University of Kumamoto, Kumamoto, Japan*

[5] *Cold and Arid Regions Environmental and Engineering Research Institute, Chinese Academy of Science, Lanzhou, China*

[a] *Now: Institute of Environmental and Industrial Medicine, Hanyang University, Seoul, Korea*

[b] *Now: Laboratorire de Meteorologie Physique, Universite Blaise Pascal, Aubie re CEDEX, France*

[c] *Now: Institute for Water and Environmental Problems, Siberian Branch of Russian Academy of Science, Barnaul, Russia*

Y.J. Kim and U. Platt (eds.), *Advanced Environmental Monitoring*, 121–135.

total mass of mineral dust transported by westerly from the Taklamakan desert to downwind will be about 1.4 × 10^7 ton/year. From those values it is suggested that background Asian dust transported from the Taklamakan desert is very important and more investigations are desired to clarify the effect of the background Asian dust to environment and climate in east Asia and west Pacific regions.

Keywords: Background Asian dust, background KOSA, balloon-borne measurement, mass flux of Asian dust particle

9.1 Introduction

Asian dust layers with the peak height of about 4–6 km and layer thickness of about 0.5–2 km were frequently detected in Japan, Korea, and China by lidar (e.g., Iwasaka et al. 1988; Kwon et al. 1997; Sakai et al. 2000; Murayama et al. 2001). Those Asian dust events have been called background KOSA (or weak KOSA) since meteorological observatories give no report of dust and few typical dust storms are identified in arid and/or semi-arid regions of China and Mongolia. Additionally it is hardly possible to detect such types of dust events through satellite imaging.

Recently aircraft-borne measurements were made to understand the nature of background KOSA (Mori et al. 1999; Trochkine et al. 2002; Matsuki et al. 2002, 2003) since it is hardly possible, from lidar measurements, to know the mixing state of KOSA particles and other types of aerosols such as sea salt, sulfate particles and others, and consequently we cannot estimate mass concentration and/or mass flux of background KOSA.

It has been believed for a long time that the effect of the Asian continental air becomes extremely small in Japan, Korea, and east coast of China in summer since the global air-circulation pattern and land surface vegetation largely change in summer in the Asian-Pacific region. However, this is speculative since there have been few observations of long-range transport of atmospheric constituents in the free troposphere owing to technical difficulty. Matsuki et al. (2003) suggested that the effect of Pacific high is very strong below about 4 km over Japan islands but not above about 4-km altitude even in summer, and that major particles were mineral dust particles in coarse mode above about 4 km even in summer since westerly winds actively transports atmospheric constituents including KOSA particles in the free troposphere.

Iwasaka et al. (2004) suggested, on the basis of lidar measurements made at Dunhuang, China, the Taklamakan desert as a possible pool which can flow out Asian background dust in every season. Both mountain-valley winds and westerly above about 5 km were suggested as important system transporting background Asian dust from the Taklamakan desert to down wind (Iwasaka et al. 2003b and 2004). However it is hardly possible, as mentioned above, to estimate quantitatively outflow rate or mass flux of dust particles from their lidar measurements.

In 2004, balloon-borne measurements have been made, following the balloon and lidar measurements in 2001–2003, at Dunhuang (40°00′ N, 94°30′ E), China to understand the mass flux of dust particles supplied in to the atmosphere from the Taklamakan desert under calm weather conditions. The observations showed that lots of dust particles diffused up to the free troposphere (~5 km) and were transported out by westerly over the Taklamakan desert (here we called it Asian background dust and distinguish it from the Asian dust caused by cyclone in the Asian continent). Here we showed how to measure and to estimate the mass of background Asian dust and discussed possible importance of contribution of background dust in mass budget in Asia-Pacific regions.

9.2 Mass Flux of Dust Due to Westerly over Taklamakan Desert

9.2.1 Balloon-Borne OPC to Measure Aerosol Size and Concentration

Figure 9.1 shows the balloon train (configuration of instruments) to measure aerosol size and number concentration with balloon-borne optical particle counter (OPC). The detailed specification of the balloon-borne OPC was described already (e.g., Hayashi et al. 1998) and we shortly described the outline of it here. We used the forward scattering effect of particles to measure particle size and number concentration, and the OPC contained semi-conductor laser as light source and photodiode as detector of scattering light from the aerosols. The

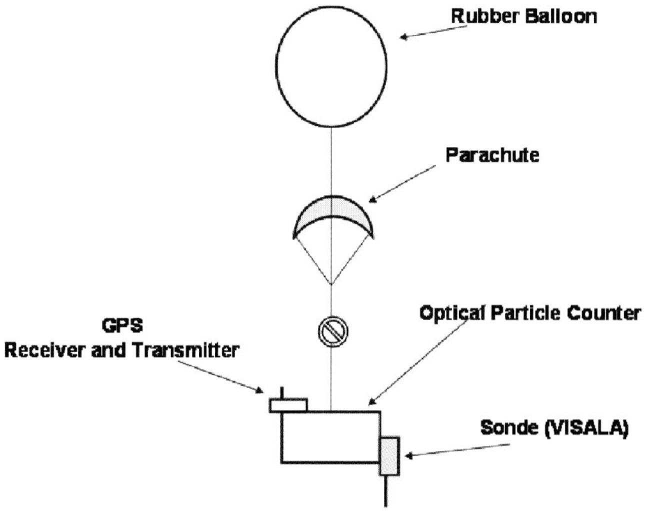

Fig. 9.1 Balloon train to measure aerosol number concentration and size by balloon-borne optical particle counter

output signals from the detector were transferred by radio wave with wavelength of 400 MHz. Particle sizing was made at particle diameters of 0.3, 0.5, 0.8, 1.2 and 3.6 μm, and modified those to diameters of 0.3, 0.5, 0.8, 1.2, 2.0, 3.0, 5.0 and 7.0 μm in measurements on October 24, 2004 to obtain more detailed number-size distributions.

9.2.2 Observation of Wind Speed and Direction

The receiving antenna of GPS (geographical positioning system) signal and transmitter which transfers the signal to the balloon launching site was mounted on the balloon to monitor the balloon position during the flight of the balloon. Wind speeds and directions were estimated from analysis of the balloon trajectories.

9.2.3 Atmospheric Temperature and Humidity

Atmospheric temperature and humidity were monitored by meteorological radio sonde of VISALA Co. Ltd. and those signals also were transferred by radio with wavelength of 400 MHz. The humidity sensor, according to VISALA Go. Ltd., can not work properly under the atmospheric temperatures lower than 213 K (−60°C), and therefore the values of relative humidity shown in Fig. 9.2 have some uncertainties above 13 km.

9.2.4 Measurements Under Calm Weather Condition

The balloon-borne measurements were made under the calm weather conditions; wind speeds lower than 2 m/s near the ground and cloudiness lower than 1/8, and therefore the observed results can be recognized as ones showing background levels. The local meteorological observatory gave no reports of dust storm. Figure. 9.2 shows aerosol number and size distributions, atmospheric temperatures, and wind speeds over Dunhuang (40°00′ N, 94°30′ E), China on January 11, 2002; August 27, 2002; February 24, 2003; March 24, 2003 and October 24, 2004.

The particle concentrations were high near the ground and gradually decreased according to increase of height, but many peaks of aerosol layer were identified and corresponded well to temperature inversion and humidity changes. Roughly speaking the coarse particles with their radius larger than 1.2 μm seem to be well mixed from near the boundary to about 5 km and westerly wind dominated above about 5 km. Those features are found also in the measurements made on other days (August 17,

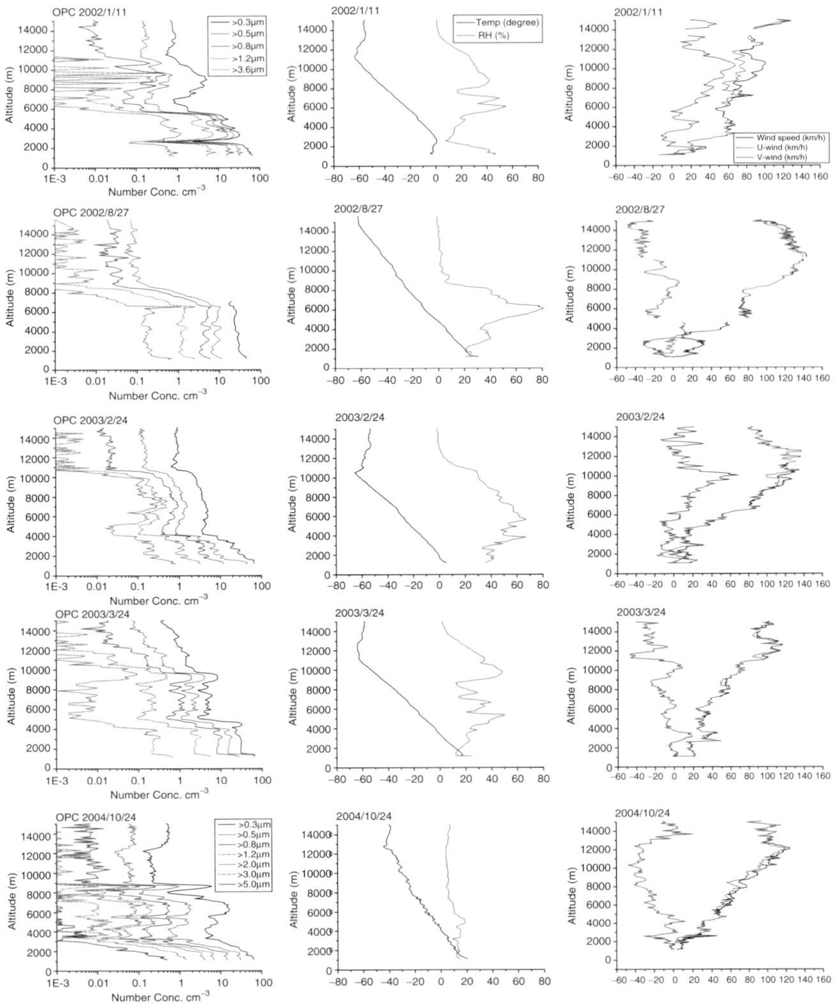

Fig. 9.2 Vertical profile of number-size distribution (cm^{-3}), Atmospheric temperature (°C), Relative Humidity (%), Wind speed (km/h), U-wind speed (km/h) and V-wind speed (km/h) on January 11, 2002, August 27, 2002, February 24, 2003, March 24, 2003 and October 24, 2004. U-wind represent the longitudinal components of wind, positive values in U-wind indicate westerly flows. V-wind represent the latitudinal components of wind, positive values in V-wind indicate northerly flows

2001; October 17, 2001; April 30, 2002; and March 22, 2004, those measurements were made without GPS signal detection and are not shown here and available to see in Kim et al. 2003 and Iwasaka et al. 2003b)

The wind, here, is divided to two components: U-wind and V-wind. U-wind and V-wind refer to the longitudinal component (positive values indicate westerly) and latitudinal component (positive values indicate northerly flows) of winds, respectively.

9.3 Mass Flux of Background Asian Dust Transported by Westerly over Taklamakan Desert

Iwasaka et al. (2003a) and Yamada et al. (2005), on the basis of chemical element analysis and observation of morphology of the particles collected with the balloon-borne impactor during the campaign of balloon-borne measurements made at Dunhuang, China corresponding to the balloon-borne OPC measurements, showed that dust particles dominated in the coarse mode size in the free troposphere over the Taklamakan desert (for example; 86% in August 29, 2002 and 99% in March 24, 2003). Therefore we, according to the analytical procedure presented by Trochkine et al. (2002) and Matsuki et al. (2002), estimated mass flux of dust due to dominating westerly in the free troposphere over the Taklamakan desert from the particle size and number concentrations, and winds observed with the OPC sonde assuming that coarse mode particles are mostly composed of dust particles (Fig. 9.3).

We first estimate volume concentration with equation (1) on the assumption that the shape of coarse mode particles (certainly dust particles) was oval having long axis a and short axis b (Okada et al. 2001) in order to confirm quantitatively that coarse mode particles were dominant in volume (and mass) concentration of particulate matter. V_j, the volume concentration of aerosol (cm^3/cm^3) in altitude layer j,

$$V_j = (4/3)\pi \sum{}_i a_i b_i^2 n(r_i)j \tag{1}$$

where, r_i is the geometric mean radius of the ith size-bin, $n(r_i)_j$ is the number concentration of particles within in layer j, and j is the layer number. The values of a, b, and r have the following relations (Okada et al. 2001),

$$a_i \times b_i = r_i, \tag{2}$$

and

$$a : b = 1.4 : 1 \tag{3}$$

As shown in Table 9.1, the total volume percentage of large particles (diameter > 1.2 μm) ranged from 88.2% to 94.6% in the height region of 2–10 km, and it is strongly suggested that mass of the particulate matter is largely dependent on coarse mode particle mass and possible to neglect contribution of the fine particles to total mass of particulate matter.

Multiplying the volume concentration at j-layer by the density of mineral dust, we can obtain the mass concentration of aerosol particle (g/cm^3) at the j-layer, C_j,

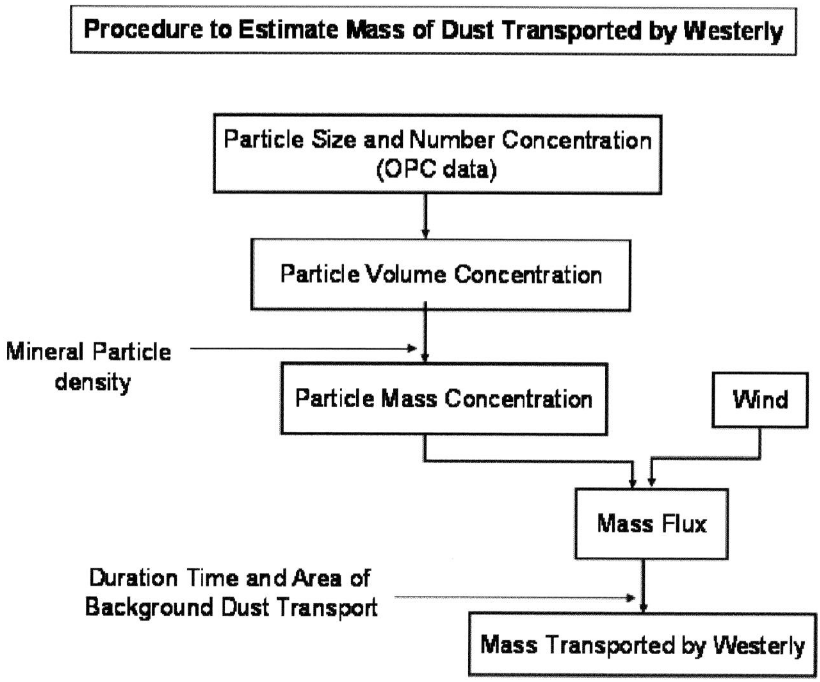

Fig. 9.3 Mass of background Asian dust transported by westerly wind is evaluated on the basis of measurements of aerosol size and number concentrations. Mineral particle density (or external mixing ratio of dust particles) is deduced from analysis of particulate matter collected in the free troposphere

Table 9.1 Volume percentage of large particles

	Date				
Size range	Jan 11, 2002	Aug 27, 2002	Feb 24, 2003	Mar 24, 3003	Oct 24, 2004
Diameter > 0.8 μm	98.5%	93.3%	97.3%	95.5%	96.0%
Diameter > 1.2 μm	94.6%	88.2%	94.5%	86.9%	92.2%

$$M_j = V_j \times \rho, \tag{4}$$

where ρ is the density of mineral and 2.6 g/cm^3 was assumed (Ishizaka and Ono 1982). From the wind speeds and directions at *j*-layer, we can estimate mass flux over the Taklamakan desert area from the following relation,

$$F_j = M_j \times Wind_j, \tag{5}$$

As described before, we focus on mass of dust particles transported long-range by westerly. Equation (3) was modified to equation (4) to obtain the mass flux due to the westerly, F_j westerly,

$$F_j \text{ westerly} = M_j \times [\text{westerly component of Wind}_j] \tag{6}$$

9.4 Results and Discussion

It is suggested, from intensive observations of dust particles from the Taklamakan desert, that the geomorphological feature of the Talim basin and local wind systems on the Taklamakan desert were important factors causing background Asian dust (Iwasaka et al. 2003b, 2004). Therefore it is of interest to discuss mass of the Asian background dust transported by westerly from the Taklamakan desert to downwind.

9.4.1 Mass Concentration Obtained from the Balloon-Borne Measurements in the Free Troposphere

The mass concentrations of mineral dust estimated on the basis of equations (1) and (2) are listed in Table 9.2. High values were frequently observed in the ground atmosphere (1–2 km), and interestingly relatively high concentrations (50 μg/m^3 or larger than it) were frequently detected above about 2 km. Most of the observations showed that mass concentration rapidly decreased above about 5 km, and strongly

Table 9.2 Mass concentration in the height region where westerly wind dominated (μg/m^3)

	Date					
Height	Jan 11, 2002	Aug 27, 2002	Feb 24, 2003	Mar 24, 2003	Oct 24, 2004	Average
4–5	169.8	59.8	10.4	24.2	3.1	53.5
5–6	71.4	60.0	28.7	3.5	3.6	33.4
6–7	1.0	80.9	41.0	4.9	3.8	26.3
7–8	0.8	10.0	71.0	4.8	0.8	17.5
8–9	3.8	*	32.4	4.4	0.8	10.4
9–10	7.4	*	29.3	20.4	0.0	14.3
10–11	0.3	*	3.6	6.3	0.0	2.6
11–12	0.4	*	0.1	0.9	0.0	0.4
12–13	0.1	*	0.1	0.1	0.0	0.1
13–14	0.1	*	0.1	0.1	0.1	0.1
14–15	0.1	*	0.1	0.1	0.1	0.1

* The values were not estimated since measurements of number concentration of the particles with $D > 0.3\,\mu m$ were not good

confirmed suggestion that dust particles were actively removed by westerly winds whose speeds increased above about 5 km (Iwasaka et al. 2003b, 2004).

Iwasaka et al. (1988), from lidar measurements at Japan, estimated total mass loading of weak KOSA centered around the altitude of 300 K potential temperature (1.5–4.5 km) as 1.9–25 μg/m^3 assuming that most of the particles were composed of dust particles and size distribution of particulate matter at those heights was the same with the measurements made near the ground. Therefore this value contained some uncertainties. Trochkine et al. (2002), from aircraft borne measurements over Japan islands, showed dust particle load of 2.3–2.6 μg/m^3 in the altitudes of 3–5 km and suggested importance of long-range transport of dust particles in the free troposphere during calm weather periods.

Taking into consideration that background dust concentrations possibly decrease during long-range transport from the source area to down wind through gravitational deposition and others, the concentration of background KOSA observed over Japan islands is expected to be lower than the values obtained over the Taklamakan desert.

9.4.2 Horizontal Mass Flux Estimated From the Balloon-Borne Measurements

Figure 9.4 shows the vertical profile of horizontal mass fluxes by westerly winds estimated from the values in Fig 9.2 with equations (1), (2) and (3). Most of the profiles suggested active transport of dust in the region where westerly dominated, but fluxes were found to be extremely low on October 24, 2004 compared with other values possibly due to appearance of very stable atmosphere. The lapse rate of temperature is obviously less than the dry adiabatic lapse rate just above the

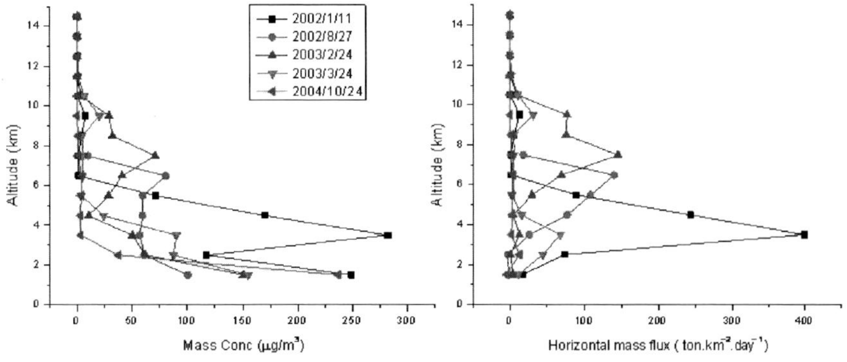

Fig. 9.4 Vertical profile of estimated mass concentration (μg/m^3) and horizontal mass flux (ton/km^2/day) on January 11, 2002; April 30, 2002; August 27, 2002; February 24, 2003; March 24, 2003 and October 24, 2004

Table 9.3 Average, maximum and minimum horizontal mass fluxes of dust (Unit: ton/km^2/day)

	Height (km)										
Flux	4–5	5–6	6–7	7–8	8–9	9–10	10–11	11–12	12–13	13–14	14–15
Average flux	68.7	46.8	44.0	34.3	22.0	30.4	5.3	0.8	0.3	0.2	0.1
Maximum	242.8	108.1	140.2	145.3	75.8	77.1	10.7	2.1	0.4	0.3	0.1
Minimum	3.0	2.5	1.4	1.2	1.3	0.1	0.1	0.1	0.1	0.1	0.1

ground, and the relative humidity was quite low (<25%). Under the very stable atmospheric conditions, vertical mixing of dust particles becomes very low and supply of dust particles to the regions above about 5 km will be depressed.

The vertical distribution of minimum, average and maximum horizontal mass fluxes of dust are summarized in Table 9.3, showing that the average horizontal mass flux was about in the range of 44.0–68.7 ton/km^2/day in the region of 4–7 km where westerly winds are strong and transport actively dust particles diffusing up from the ground surface of the desert.

It can be strongly suggested, from large horizontal mass flux in the free troposphere shown in Fig. 9.4 and Table 9.3, that mineral dust are effectively transported long-range to downwind by westerly wind in the troposphere, especially on the surface of about 300 K potential temperature, over the Taklamakan desert.

Iwasaka et al. (1988) and Matsuki et al. (2002) stressed the important contribution of background KOSA to biogeochemical cycle of metals and atmospheric constituents in east Asia–west Pacific region since their mass flux in the free troposphere is not negligible comparing with the values of severe KOSA. Matsuki et al. (2002), on the basis of aircraft borne measurements during the days without dust storm episode, estimated the horizontal fluxes due to westerly wind to be in the range of 0.2–7 ton/km^2/day in 2–6 km over Nagoya, Japan. As described before, various deposition and dilution processes are expected during long-range transport of dust, and therefore the difference between the values of both regions possibly suggested those removal and dilution processes.

Zhao et al. (2003), from their model simulations of dust transports in spring of 2001, estimated the horizontal mass fluxes in spring to be in the range of 0.06–2.9 ton/km^2/day over the Sea of Japan and in the range of 0.06–0.35 ton/km^2/day over the North Pacific. It is reported that nearly 20 sandstorm events happened in 2001 spring, at China, and their results can be recognized as examples showing effect of severe KOSA events.

In Table 9.4, the horizontal mass flux estimated here is compared with those obtained over the Sea of Japan, Japan and North Pacific Ocean. In this study, average fluxes during five observations were in the range of 30–68 ton/km^{-2}/day at heights of 4–10 km (Table 9.3). The comparison is made on the basis of very limited measurements and model calculations but possible important contribution of background Asian dust is strongly suggested.

Table 9.4 Horizontal mass fluxes of dust particles: Comparison with other investigations

References	Present study	Zhao et al. 2003	Matsuki et al. 2002	Zhao et al. 2003
Site	Dunhuang, China	Japan Sea (40°N, 130°E)	Nagoya, Japan	North Pacific Ocean (42°N, 140°W)
Condition	Calm weather	Spring	Spring (Calm weather)	Spring
Method	Balloon-borne measurement	NARCM (Northern aerosol regional climate model)	Aircraft observation	NARCM (Northern aerosol regional climate model)
Altitude (km)	4–10	1–10	2–6	1–10
Flux (ton/ km^2/day)	30–68 (Average value)	0.06–2.9 (5–250 μg/ m^2/s)	0.2–7	0.06–0.35 (5–30 μg/ m^2/s)

9.4.3 Importance of Mass of Dust Particles Transported Under Calm Weather Conditions

As described before, horizontal mass flux of dust particles by westerly winds is very large in the free troposphere and it is expected that lots of dust particles are transported by westerly to downwind. The mass transported by westerly winds is given by following relation (Fig. 9.5),

$$\text{Mass} = \text{F (horizontal flux)} \times \text{cross section of the region where westerly is dominant} \times \text{Duration time} \quad (7)$$

where mass flux by westerly winds was already defined by equation (6) and duration time means the time when background dust is transported by westerly winds in calm weather over the Taklamakan desert.

Mass of the dust transported by westerly winds as background dust is roughly estimated to be about 1.4×10^7 ton/year assuming that flux is 5.78×10^2 tons/km^2/day averaging values of 4–5 and 5–6 km in Table 9.3, cross section 800 km^2 considering approximately area of 2 km (height) × 400 km (length in south–north direction), and duration time of 300 days a year.

This value, if westerly wind transports most of the particles which diffuse up to about 5 km from the ground, can be considered as mass of particles which are supplied into the atmosphere and transported long-range as background dust from the Taklamakan desert. The emission of background dust per unit surface area is estimated to be about 4.4×10 ton/km^2/year assuming about 3.2×10^5 km^2 of the surface of Taklamakan desert (emission of background Asian dust). Iwasaka

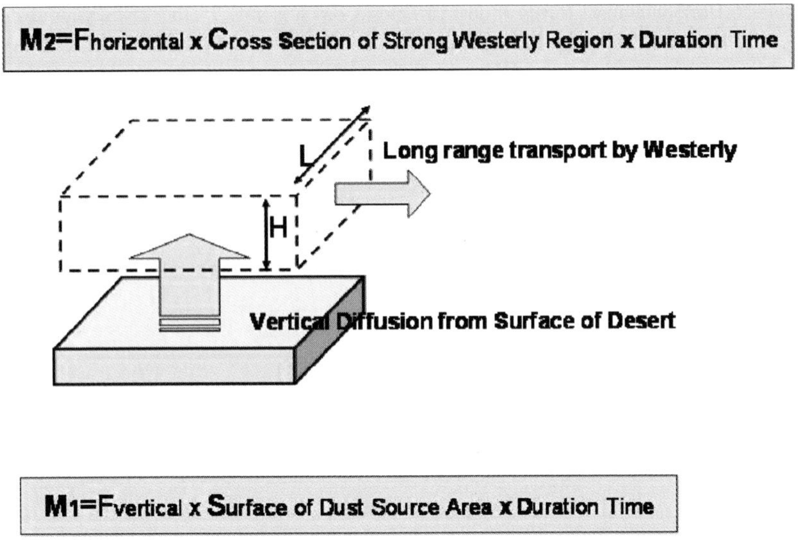

Fig. 9.5 Relation of horizontal mass flux (F horizontal) and mass of background Asian dust flowing out from the Taklamakan desert by westerly wind. Mass of dust particles transported by westerly wind (M2) is approximately equal to mass of dust diffusing into the atmosphere (M1) under the calm weather condition over the Taklamakan desert (see Text)

et al. (1983), on the basis of lidar measurements in Japan, estimated mass load of single severe KOSA, 1.63×10^6 ton, and it is easily suggested that the mass of the flowing out dust is not of negligible levels. Arao and Ishizaka (1986), on the basis of solar radiation measurements at Japan, estimated mass load of Asian dust to be in $4.1–5.3 \times 10^6$ton/year in 30°–40°N region. It is impossible to compare directly the present values and their estimations, but possible to suggest at least important potential of background Asian dust. Xuan et al. (2004) estimated the dust emission of PM10 from the source area in east Asia was 1.04×10^7 ton/year, and their estimation and the present one show reasonable correspondence considering that Xuan et al. (2004) treated all areas as possible dust source regions and only the particles with size larger than 10 µm. Zhao et al. (2003) estimated emission of 21.5 tons/km^2 in spring from source area of Mongolia and China on the basis of numerical modeling. Comparing the present estimations to values given by Zhao et al. (2003), the load of background dust from the Taklamakan desert is a little larger. One possible reason is duration time of background dust from the Taklamakan desert. Background Asian dust can be supplied into the atmosphere in not only spring but also other seasons, and total contribution will become effective. Most of arid and semi-arid regions largely change their surface, especially vegetation, according to seasons, but the Taklamakan desert seems to have potential as source of dust always. Another reason is that we treat only the surface of the Taklamakan desert to estimate emission per unit surface, but their simulation covered very wide area and value of emission was normalized by the surface which included not only strong source areas but also the area having extremely low emission strength. If they treat the mass of emission on only arid area, the value will largely increase. Therefore it is impossible to compare directly the present observations to them, but we can suggest at least that the contribution of background dust from the Taklamakan desert is very large from view point mass balance of dust in east Asia and west Pacific region.

9.5 Summary and Conclusion

On the basis of balloon-borne measurements made at Dunhuang, China during calm weather periods, we discussed mass of background Asian dust transported by westerly from the Taklamakan desert to down wind, and the following results are suggested:

1. The average mass concentrations of background Asian dust were about within the range of 2.6–53.5 $\mu g/m^3$ in the free troposphere over the Taklamakan desert.
2. The average horizontal mass fluxes of background Asian dust were changing within the range of 5.3–68.7 $ton/km^2/day$ in the free troposphere, and mass of dust transported by westerly was about 1.4×10^7 ton/year.
3. It is suggested that the westerly wind largely contributes to long range transport of background Asian dust particles in the free troposphere, and this confirmed previous researches suggesting importance of westerly long-range transport of Asian dust not only in dust storm events but also in calm weather conditions in east Asia to west pacific regions (Duce et al. 1980; Arimoto et al. 1985; Iwasaka et al. 1983, 2004; Uematsu et al. 1983; Gao et al. 1992; Xiao et al. 1997; Uno et al. 2001; Matsuki et al. 2002, 2003; Zhao et al. 2003).

These suggest the possible important contribution of background Asian dust to biogeochemical cycles of not only severe Asian dust but also background Asian dust in east Asia and west Pacific region. The measurements made here, however, are on the basis of only the limited number of balloon-borne measurements which the GPS sensor were mounted on, and much more measurements are desired in the future to obtain better understanding of background Asian dust and its environmental effects. Difference of mass concentration and of mass flux between in dust source regions and in down wind is possibly due to deposition of dust, diffusion in meridian direction and others. It is also necessary to assess the difference and to discuss what processes make the difference in order to understand environmental effects of the fall of background Asian dust.

Acknowledgement This study was supported by the Japan Ministry of Education, Culture, Sports, Science and Technology (Grant-in-Aid for Specially Promoted Research, 10144104), and the Japan Society for the Promotion of Science (Inter-Research Centers Cooperative Program, Stratospheric Physics and Chemistry Based on Balloon-borne Measurements of Atmospheric

Ozone, Aerosols, and Others). Staff members of Dunhuang City Meteorological Bureau gave us kind and helpful technical support during the balloon-borne measurements.

References

Arao K. and Ishizaka Y. (1986). Volume and mass of yellow sand dust in the air over Japan as estimated from atmospheric turbidity. *J. Meteorol. Soc. Japan.*, 64, 79–94.

Arimoto R., Duce R.A., Ray B.J., and Unni C.K. (1985), Atmospheric trace elements at Enewetak Atoll: 2. Transport to the ocean by wet and dry deposition, *J. Geophys. Res.*, 90, 2391–2408.

Duce R.A., Unni C.K., Ray B.J., Prospero J.M., and Merrill J.T. (1980), Long-range atmospheric transport of soil dust from Asia to tropical North Pacific: Temporal variability, *Science*, 209, 1522–1524.

Gao Y., Arimoto R., Zhou M.Y., Merrill J.T., and Duce R.A. (1992), Relationships between the dust concentrations over Eastern Asia and the remote North Pacific, *J. Geophys. Res.*, 97, 9867–9872.

Hayashi M., Iwasaka Y., Watanabe M., Shibata T., Fujiwara M., Adachi H., Sakai T., Nagatani M., Gernandt H., Neuber R., and Tsuchiya M. (1998), Size and number concentration of liquid PSCs: Balloon-borne measurements at Ny-Alesund, Norway in winter of 1994/95, *J. Meteorol. Soc. Japan*, 76, 549–560.

Ishizaka Y. and Ono A. (1982), Mass size distribution of the principal minerals of yellow sand dust in the air over Japan, *Idojaras*, 86, 249–253.

Iwasaka Y., Minoura H., and Nagaya K. (1983), The transport of spatial scale of Asian dust-storm clouds: A case study of the dust-storm event of April 1979, *Tellus*, 35B, 189–196.

Iwasaka Y., Yamato M., Imasu R., and Ono A. (1988), Transport of Asian dust(KOSA) particles: Importance of weak KOSA events on the geochemical cycle of soil particles, *Tellus*, 40B, 494–503.

Iwasaka Y., Shi G.-Y., Yamada M., Matsuki A., Trochkine D., Kim Y.S., Zhang D., Nagatani T., Shibata T., Nagatani M., Nakata H., Shen Z., Li G., and Chen B. (2003a), Importance of dust particles in the free troposphere over the Taklamakan Desert: Electron microscopic experiments of particles collected with a balloon borne particle impactor at Dunhuang, China, *J. Geophys. Res.*, 108, No. D23, DOI 10.1029/2002JD003270.

Iwasaka Y., Shibata T., Nagatani T., Shi G.-Y., Kim Y.S., Matsuki A., Trochkine D., Zhang D., Yamada M., Nagatani M., Nakata H., Shen Z., Li G., Chen B., and Kawahira K. (2003b), Large depolarization ratio of free tropospheric aerosols over the Taklamakan Desert revealed by lidar measurements: Possible diffusion and transport of dust particles, *J. Geophys. Res.*, 108, No. D23, DOI 10.1029/2002JD003267.

Iwasaka Y., Shi G.-Y., Kim Y.S., Matsuki A., Trochkine D., Zhang D., Yamada M., Nagatani T., Nagatani M., Shen Z., Shibata T., Nakata H. (2004), Pool of dust particles over the Asian continent: Balloon-borne optical particle counter and ground-based lidar measurements at Dunhuang, China, *Environ. Monit. Assess.*, 92, 5–24.

Kim Y.S., Iwasaka Y., Shi G.Y., Nagatani T., Shibata T., Trochkine D., Matsuki A., Yamada M., Chen B., Zhang D., Nagatani M., Nakata H. (2004), Dust particles in the free atmosphere over desert areas on the Asian continent: Measurements from summer 2001 to summer 2002 with balloon-borne optical particle counter and lidar, Dunhuang, China, *J. Geophys. Res.*, 109, D19S26, DOI 10.1029/2002JD003269.

Kwon S.-A., Y. Iwasaka T. Shibata T. Sakai (1997), Vertical distribution of atmospheric particles and water vapor densities in the free troposphere: Lidar measurement in spring and summer in Nagoya, Japan, *Atmos. Environ.*, 31, 1459–1467.

Matsuki A., Iwasaka Y., Trochkine D., Zhang D., Osada K. and Sakai T. (2002). Horizontal mass flux of mineral dust over East Asia in spring: Aircraft-borne measurements over Japan, *J. Arid Land Stud.*, 11, 337–345.

Matsuki A., Iwasaka Y., Osaka K., Matsunaga K., Kido M., Inomata Y. Trochkine D., Nishita C., Nezuka T., Sakai T., Zhang D., and Kwon S.A. (2003), Seasonal dependence of the long-range transport and vertical distribution of free tropospheric aerosols over eastAsia: On the basis of aircraft and lidar measurements and Isentropic trajectory analysis, *J. Geophys. Res.,* 108, No. D23, DOI 10.1029/2002JD003266108.

Mori I., Iwasaka Y., Matsunaga K., Hayashi M., and Nishikawa M. (1999), Chemical characteristics of free tropospheric aerosols over Japan sea coast: Aircraft-borne measurements, *Atmos. Environ.*, 33, 601–609.

Murayama T., Sugimoto M., Uno I., Kinoshita K., Aoki K., Hagiwara N., Liu Z., Matsui I., Sakai T., Arar K., Sohn B.J., Won J.G., Yoon S.C., Li T., Zhou J., Hu H., Abo M., Iokibe K., Koga R., and Iwasaka Y. (2001), Ground-based network observation of Asia dust events of April 1998 in East Asia, *J. Geophys. Res.,* 106, 18345–18359.

Okada K., Heintzenberg J., Kai K., and Qin Y. (2001), Shape of atmospheric mineral particles collected in three Chinese arid-regions, *Geophys. Res. Lett.*, 26, 3123–3126.

Trochkine D., Iwasaka Y., Matsuki A., Zhang D., and Osada K. (2002), Aircraft borne measurements of morphology, chemical elements, and number-size distributions of particles in the free troposphere in spring over Japan: Estimation of particle mass concentration, *J. Arid Land Stud.*, 11, 327–335.

Uematsu M., Duce R.A., Prospero J.M., Chen L., Merrill J.T., and Mcdonald R.L. (1983), Transport of mineral aerosol from Asia over the North Pacific Ocean, *J. Geophys. Res.*, 88, 5343–5352.

Uno I., Amanom H., Emori S., Kinoshita K., Matsui I., and Sugimoto N. (2001), Trans-Pacific yellow sand transport observed in April 1998, *J. Geophys. Res.*, 106, 18331–18344.

Xiao H., Carmichael G.R., Durchenwald J., Thornton D., and Bandy A. (1997) Long-range transport of Sox and dust in East Asia during the PEX B experiment, *J. Geophys. Res.*, 102, 28589–28612.

Yamada M., Iwasaka Y., Matsuki A., Trochkine D., Kim Y. S., Zhang D., Nagatani T., Shi G.-Y., Nagatani M., Nakata H., Shen Z., Chen B., and Li G. (2005), Feature of dust particles in the spring free troposphere over Dunhuang in northwestern China: Electron microscopic experiments on individual particles collected with a balloon-borne impactor, *Water Air Soil Pollut.*, Focus/5, 231–250.

Zhao T.L., Gong S.L., Zhang X.Y., and McKendry I.G. (2003), Modeled size-segregated wet and dry deposition budgets of soil dust aerosol during ACE-Asia 2001: Implications for trans-Pacific transport, *J. Geophys. Res.*, 108, 8665–8673.

Chapter 10
Identifying Atmospheric Aerosols with Polarization Lidar

Kenneth Sassen

Abstract A variety of types of aerosol particles, both natural and human-made, are commonly suspended in the atmosphere. Different aerosol types have characteristic shapes, but basically fall into two categories: spherical and irregular. Haze and forest fire smoke particles are examples of the former, and desert dust and biogenic debris (e.g., pollen) of the latter. It is shown here that the capability of polarization lidar systems to sense the exact shape of particles makes it a powerful tool to remotely identify many types of aerosols. This is particularly important in the study of how aerosols may affect the properties of clouds.

Keywords: Aerosol backscattering, aerosol shape, polarization lidar

10.1 Introduction

Aerosols suspended in the atmosphere have a variety of impacts ranging from human health to climate change. Layers of aerosols directly affect the radiative balance of the Earth–atmosphere system through scattering and absorption, thus increasing the local solar albedo (potentially cooling the surface and heating the atmosphere), and indirectly by modifying cloud particle phase and size distribution. These indirect aerosol effects on climate are highly uncertain (IPCC 2001). As examples, indirect aerosol effects include changes in the cloud condensation nuclei concentration or type, which affects water cloud albedo and the likelihood of precipitation development, while the ability for some particles, like mineral dusts, to serve as ice nuclei that affect cirrus cloud formation and the phase of supercooled clouds. Because the source (i.e., chemical composition) of aerosols determines to a large degree their ability to interact with light and affect clouds, it is important to be able to remotely determine aerosol type even at considerable distances from their source.

Geophysical Institute, University of Alaska Fairbanks, 903 Koyukuk Drive, Fairbanks, Alaska 99775 USA 907-474-7845, 907-474-7290 (FAX)

Y.J. Kim and U. Platt (eds.), *Advanced Environmental Monitoring*, 136–142.

Fortunately, the polarization lidar technique (Sassen 2000, 2005) has the unique ability to discriminate between spherical and nonspherical particles, and thus determine unambiguously the thermodynamic phase of clouds as well as identify the type of atmospheric aerosols because of its sensitivity to exact particle shape. Spherical aerosols (e.g., haze and aqueous smoke particles) produce no change in the polarization state of backscattered light, whereas nonspherical particles can generate considerable depolarization depending on the exact particle shape and, to some extent, on particle size relative to the incident wavelength (Mishchenko and Sassen 1998). For example, Asian dust storm aerosols have a highly irregular shape (Okada et al. 2001) and typically generate linear depolarization ratios (δ, the ratio of the returned laser powers in the planes of polarization orthogonal and parallel to that transmitted) of about 0.2–0.3 (Murayama et al. 2001; Sassen 2002), which is similar to some cirrus ice crystal clouds. If the particle dimensions are smaller than the incident wavelength, reduced depolarization will be measured for a particular nonspherical particle. However, those aerosols most likely to affect cloud properties have dimensions similar to (visible and near-infrared) lidar wavelengths, indicating that multiple-wavelength depolarization measurements are especially useful and could provide particle size estimates (Sassen et al. 2001). Such data, especially when combined with Raman (Wandinger 2005) or high spectral resolution lidar (Eloranta 2005) data, are very promising for characterizing the type, composition, and size of aerosols.

10.2 The AFARS Dataset

Current research at the Arctic Facility for Atmospheric Remote Sensing (AFARS) involves three polarization lidars (at 0.532, 0.694, 1.06, and 1.574 μm wavelengths) to study clouds, aerosols, and their interactions, as well as a 94-GHz polarimetric Doppler radar and various visible and infrared radiometers. In support of Aqua and Terra satellite overpasses over AFARS (64.86° latitude and −147.84° longitude), regular remote sensing observations involving mainly the "turnkey" cloud polarization lidar (CPL) are being obtained. The CPL is based on a high power (1.5 J), 0.1 Hz, ruby (0.694 μm) laser transmitter, and a two-channel receiver using a 25-cm diameter telescope (Sassen et al. 2001). The current ~3.0-y AFARS CPL dataset (as well as earlier midlatitude data from Salt Lake City, Utah) can be viewed at http://corona.gi.alaska.edu/AFARS/. Below we provide typical CPL data from various aerosol types sampled at AFARS.

10.3 AFARS Polarization Lidar Aerosol Studies

Time-averaged (approximately 10–30 min) vertical CPL profiles of linear depolarization ratios and relative returned laser power for three distinct types of boundary-layer aerosols over Fairbanks are given in Fig. 10.1. In each case the

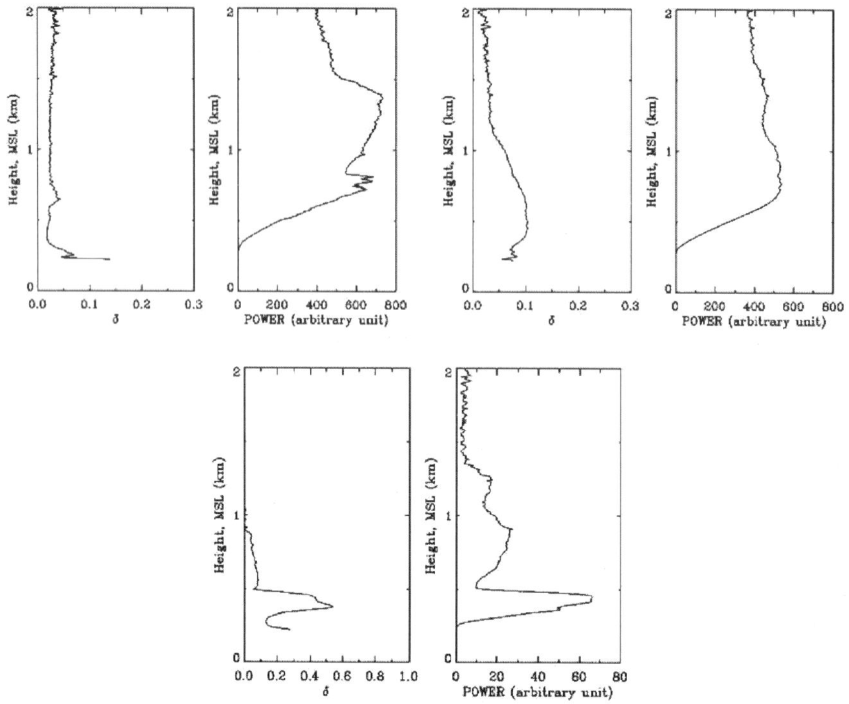

Fig. 10.1 Vertical profiles of CPL linear depolarization ratio δ and range-normalized relative returned laser power for three distinct types of Arctic aerosols, including Arctic haze (sampled on 3/31/04), tree pollen (5/14/04), and urban ice fog (12/24/04)

effect of the incomplete transmitter/receiver overlap is evident by the near-zero returned powers below about 300-m MSL (or ~100 m above the ground level). Also note that the δ values are what are referred to as the "total" depolarization ratios, because they are calculated from the sum of molecular and aerosol backscattering. Pure molecular backscattering yields δ of ~0.03 at the ruby wavelength.

At the upper left are profiles from an episode of Arctic haze, which, as in this case, often produces reduced horizontal visibility in winter and spring at high latitudes. The particles are aqueous droplets of ammonium sulfate solutions, derived photochemically from midlatitude air pollution. Because the particles are therefore spherical, near-zero depolarization is measured. Note that the peak returned power in the haze at ~1.4 km corresponds to the minimum δ value, slightly lower than the pure molecular value. (These droplets are too small and dispersed to generate any depolarization increase through the multiple scattering process.) Slight depolarization increases at lower levels are probably attributable to local urban pollution sources.

Next in Fig. 10.1 is shown the laser depolarizing effects of airborne biogenic debris, which are a common occurrence in the spring and summer at Fairbanks, and probably at many other locations as well. At the upper right are data from an occurrence

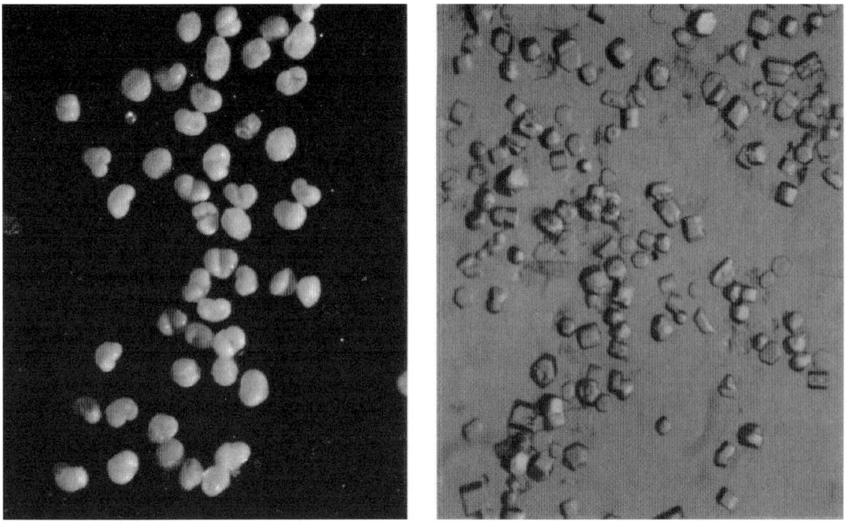

Fig. 10.2 Scanning electron microscope image of a) tree pollen (80x) at left collected during a solar corona display at Fairbanks, and at right, b) ice fog crystals (140x) showing a mixture of hexagonal and other habits (courtesy of Walter Tape)

of birch tree pollen grains. The presence of these relatively large (~20 μm) pollen during boreal forest green-out is indicated by the vividly colored solar corona. Although these particles are near-spherical in shape (see Fig. 10.2a), they are sufficiently nonspherical to generate $\delta \approx 0.1$. Cottonwood-type tree and firewood seeds, which are much larger than pollen grains and are often visible, swirling in the air, appear to generate a similar amount of depolarization.

Finally, shown at the bottom are profiles obtained from an ice fog generated at frigid temperatures (~ –40°C and colder) from urban water vapor sources, particularly local power plants. The returned power peak at ~ 0.4 km height represents the ice fog layer, whereas the weaker signals aloft are probably from a mixture of Arctic haze and dispersed ice crystals. The relatively minute ice crystals usually display the typical hexagonal symmetry of naturally formed ice crystals, although unusual crystal forms are often observed in ice fogs (see Fig. 10.2b). These ice crystals generate δ up to 0.6, which is significantly higher than those values in cirrus clouds at similar temperatures (Sassen and Benson 2001).

We now provide height versus time CPL displays for additional types of aerosols studied at AFARS. Fig. 10.3a shows data obtained during an active forest fire season where a number of regional fires surrounding Fairbanks were burning. The result was a number of elevated smoke layers up to a height of ~8.0 km, as well as a dense smoke-filled boundary layer below ~4.5 km. Note the gravity waves present in the smoke layers. In contrast to the δ of ~0.3 in the broken cirrus cloud layer between 9.0 and 11.0 km height, the fresh smoke layers generate near-zero depolarization. This indicates that the aerosol was dominated by spherical aqueous droplets containing organic solutions liberated during the

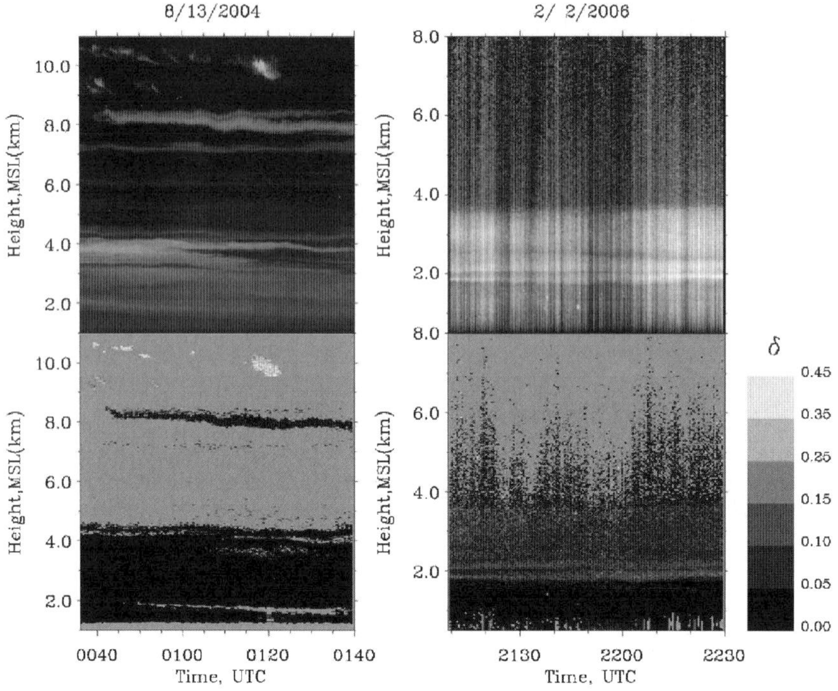

Fig. 10.3 Two examples of CPL height versus time displays of returned laser power (top, based on a logarithmic gray scale, where white is the strongest signal) and linear depolarization ratio (note δ scale at bottom right) for, (a) a smoky period at Fairbanks, and (b) a volcanic ash plume from the Augustine volcano south of Fairbanks

combustion process. In smoke layers that have aged and dried out, higher depolarization values are observed, presumably due to the crystallized remnants of evaporated droplets.

The final CPL aerosol study example is given in Fig. 10.3b, which shows a volcanic eruption cloud during late January 2006 from the Alaskan Augustine volcano. This volcano is ~850 km to the south of AFARS with a summit height of 1.26 km MSL, and weather conditions were briefly favorable in early February to transport the volcanic debris to our area. Volcanic aerosol transport models and satellite imagery confirm that the eruption cloud passed over AFARS at this time in the ~2.0 to 4.0 km height interval. (Note that the irregular returned power display at top was caused by variable ice fog plumes present below the beam cross-over point.) Although a fresh volcanic plume was previously observed by a Raman lidar (Pappalardo et al. 2004), these data are apparently the first laser depolarization measurements of a volcanic eruption plume in the troposphere. (However, Hayashida et al. (1984) appear to have sampled a lower stratospheric ash layer during the 2–5 month period following the El Chichón volcanic eruption in 1982, which yielded 0.10–0.15 δ values that are quite similar to those measured here.)

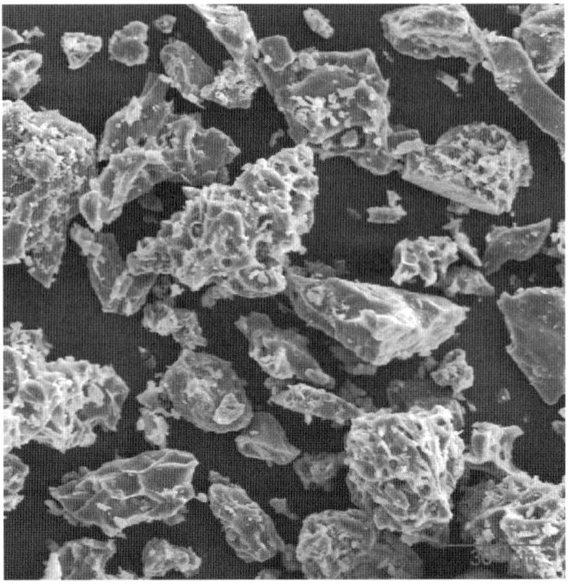

Fig. 10.4 Scanning electron microscope image (note that the sizes of three particles in the lower right corner are about 30 microns) of ash fallout from the late-January 2006 eruption of the Augustine volcano (courtesy of K. Dean of the Alaska Volcano Observatory). Because of the ~24-h transport time predicted by trajectory analysis to reach AFARS, the larger silica glass and feldspar crystals would have fallen out. Even the smallest particles have a highly irregular shape, however.

This amount of depolarization leaves little doubt that the aerosol is nonspherical, as is confirmed by the surface volcanic ash fallout sample shown in Fig. 10.4, which was collected 120 km from the volcano.

10.4 Conclusion and Outlook

We have shown several examples of ruby (0.694 μm wavelength) polarization lidar observations of various aerosols sampled in the Arctic region at AFARS in Alaska, including unique findings from an ash cloud from a nearby volcanic eruption. It is also interesting that boreal plants and trees produce floating pollen and seeds that generate noticeable amounts of laser depolarization. The sensitivity of the laser backscatter depolarization technique to exact particle shape means that there is a strong potential for identifying aerosol type (i.e., shape, composition, and relative size), especially when collected at two or more wavelengths and combined with quantitative lidar methods, which can separate out molecular scattering and determine backscatter-to-extinction ratios. We look forward at AFARS to a program involving depolarization measurements at four lidar wavelengths combined with new nitrogen Raman data to more fully evaluate the potential of researching aerosols with lidar. Particularly promising for aerosol research is the 1.574-μm wavelength eye-safe scanning lidar, which has been designed to collect complete Stokes parameter, circular depolarization, and differential polarization data.

Acknowledgements This research is being supported by NASA grant # NNG04GF35G and NSF-MASINT (Measures and Signatures Intelligence) IRS 92-60000147.

References

Eloranta E.E. (2005), High spectral resolution lidar. In C. Weitkamp (Ed.), *Lidar*, (pp. 143–163). New York: Springer Press.

Hayashida S., Kobayashi A., and Iswasaka Y. (1984), Lidar measurements of stratospheric aerosol content and depolarization ratios after the eruption of El Chichón volcano: Measurements at Nogoya, Japan. *Geof. Int.*, 23, 277–288.

IPCC (2001), *Climate change science: An analysis of some key questions. National Academy of Science*, retrieved from http://books.nap.edu/books/0309075742/html/.

Mishchenko M.I., and Sassen K. (1998), Depolarization of lidar returns by small ice crystals: An application to contrails, *Geophys. Res. Lett.*, 25, 309–312.

Murayama T., et al. (2001), Ground-based network observation of Asian dust events of April 1998 in east Asia, *J. Geophys. Res.*, 106, 18,345–18,360.

Okada O., Heintzenberg J., Kai K., and Qin Y. (2001), Shape of atmospheric mineral particles collected in three Chinese arid-regions, *Geophys. Res. Lett.*, 28, 3123–3126.

Papapalardo G., et al. (2004), Raman lidar observations of aerosol emitted during the 2002 Etna eruption, *Geophys. Res. Lett.*, 31, L05120, DOI 10.1029/2003GL019073.

Sassen K. (2000), Lidar backscatter depolarization technique for cloud and aerosol research. In M. L. Mishchenko et al. (Eds.), *Light Scattering by Nonspherical Particles* (pp. 393–416). New York: Academic Press.

Sassen K. (2002), Indirect climate forcing over the western US from Asian dust storms, *Geophys. Res. Lett.*, DOI 10.1029/2001GL014051.

Sassen K., (2005), Polarization lidar. In C. Weitkamp (Ed.), *Lidar* (pp.19–42). New York: Springer Press.

Sassen K., and Benson S. (2001), A midlatitude cirrus cloud climatology from the Facility for Atmospheric Remote Sensing: II. Microphysical properties derived from lidar depolarization, *J. Atmos. Sci.*, 58, 2103–2112.

Sassen K., Comstock J.M., Wang Z., and Mace G.G. (2001), Cloud and aerosol research capabilities at FARS: The facility for atmospheric remote sensing, *Bull. Am. Meteor. Soc.*, 82, 1119–1138.

Wandinger U. (2005), Raman lidar. In C. Weitkamp (Ed.), *Lidar* (pp.241–271). New York: Springer Press.

Chapter 11
A Novel Method to Quantify Fugitive Dust Emissions Using Optical Remote Sensing

Ravi M. Varma[1,*], Ram A. Hashmonay[1], Ke Du[2], Mark J. Rood[2], Byung J. Kim[3], and Michael R. Kemme[3]

Abstract This paper describes a new method for retrieving path-averaged mass concentrations from multi-spectral light extinction measured by optical remote sensing (ORS) instruments. The light extinction measurements as a function of wavelength were used in conjunction with an iterative inverse-Mie algorithm to retrieve path-averaged particulate matter (PM) mass distribution. Conventional mass concentration measurements in a controlled release experiment were used to calibrate the ORS method. A backscattering micro pulse lidar (MPL) was used to obtain the horizontal extent of the plume along MPL's line of sight. This method was used to measure concentrations and mass emission rates of PM with diameters ≤10 μm (PM_{10}) and PM with diameters ≤2.5 μm ($PM_{2.5}$) that were caused by dust from an artillery back blast event at a location in a desert region of the southwestern United States of America.

Keywords: Fugitive dust, emission estimation, optical remote sensing, particulate matter, PM_{10}, $PM_{2.5}$, Mie theory, FTIR, transmissometer, micro pulse lidar

11.1 Introduction

Particulate matter (PM) emissions from fugitive sources are a major concern because of their contribution to degradation of air quality. Several studies in the past have shown that PM with diameters ≤10 μm (PM_{10}) and PM with diameters

[1] *ARCADIS, 4915 Prospectus Drive Suite F, Durham, NC 27713, USA*

[2] *Department of Civil & Environmental Engineering, University of Illinois at Urbana-Champaign, 205 N. Mathews Ave., Urbana, IL 61801, USA*

[3] *U.S. Army ERDC—CERL, 2902 Farber Drive, Champaign, IL 61822 USA*

**Current affiliation: Department of Physics, National University of Ireland, University College Cork, Cork, Ireland; Tel: +353-21-4903294, Fax: +353-21- 4276949*

Y.J. Kim and U. Platt (eds.), *Advanced Environmental Monitoring*, 143–154.

≤2.5 μm ($PM_{2.5}$) have adverse effects on human health in the areas surrounding these sources. This paper presents a method for quantifying fugitive dust emissions using optical remote sensing (ORS) techniques. The method makes use of path-integrated multi-spectral light extinction measurements in a vertical plane by ORS instruments downwind of a fugitive PM source. The light extinction measurements are used to retrieve path-averaged $PM_{2.5}$ and PM_{10} mass concentrations using inversion of the Mie extinction efficiency matrix for a range of size parameters. This retrieval needs to be calibrated against a standard mass concentration measurement for the specific dust of interest. This novel method was applied for dust plumes generated by artillery back blast in a desert area of the southwestern United States of America (USA).

The ORS instruments used for mass concentration retrieval in this study include two monostatic active open path-Fourier transform infrared spectrometers (OP-FTIRs) and two open path-visible laser transmissometers (OP-LTs). In addition to the above ORS instruments, a backscattering micro pulse lidar (MPL, Sigma Space Corporation, Maryland, USA) was also used to obtain the horizontal extent of the PM plume along the MPL's line of sight. In this study, the mass concentration retrieval part of the ORS method was calibrated using concurrent measurements by DUSTTRAK (DT) aerosol monitors (Model 8520, TSI, Inc., Minnesota, USA) in the corresponding particle size ranges that were traced back to particulate mass filter calibrations. Once the mass concentrations were retrieved from each set of ORS instruments in a vertical plane of measurement, the emission rates across the plane were computed by fitting a bivariate Gaussian function to the data (Hashmonay et al. 2001). The ORS method was applied for PM mass emission quantification due to an artillery back blast.

11.2 Method

The ORS method to retrieve PM mass concentrations relies on multi-spectral path-integrated light extinction measurements. Light extinction (absorption + scattering) as a function of wavelength can be computed by Mie theory if the optical properties and the size distribution of the PM are known. The optical properties are expressed in terms of the complex refractive index, m, and are defined as (Kerker 1969):

$$m = n - in\kappa \tag{1}$$

where the real part of the refractive index, n, is defined as the ratio between the wavelength of light in free space, $\lambda_{0,}$ and the wavelength in matter, λ. The imaginary part of the complex refractive index, $n\kappa$, is the absorption factor, and the absorption index (κ) is related to the absorption coefficient of the matter. If the PM is non-absorbing, then the light extinction is due to scattering alone and the refractive index will have a non-zero real part and a zero imaginary part. In such a scenario, the real part of the refractive index is nearly constant for a wide range of non-absorbing wavelengths.

The extinction efficiency is a function of the particle diameter, d, and λ and m. When a broad spectral band of incident light is detected, the extinction efficiency for each combination of particle sizes and wavelengths need to be calculated. Our method for calculating the Mie extinction efficiencies follows the recurrence procedure in Wickramasinghe (1973). These extinction efficiency values are organized in a matrix, $Q_{e\,ij}$, where each row (i) is for a different λ and each column (j) is for a different particle size range. The matrix, $Q_{e\,ij}$, is calculated by using a predetermined constant refractive index for the relevant spectral range (which is corrected by a calibration procedure) and for particle diameters up to 20 μm. The extinction coefficient, σ_e, is expressed in cm^{-1} as a function of wavelength and for a given size distribution as (Hashmonay and Yost 1999):

$$\sigma_{e_i} = \frac{\pi}{4} \sum_j Q_{e_{ij}} N_j d_j^2 \tag{2}$$

where N_j is the particle number density at the jth particle size class in cm^{-3} and d_j is the mean particle diameter of the jth particle size range in centimeter. The value of N_j is the unknown that needs to be calculated by the inversion method. The ORS measurements provide the extinction spectra (σ_{ei} in Eq. 2) that is an input to the iterative inversion algorithm. The kernel matrix for the inversion procedure is $Q_{e\,ij}$, which includes the extinction efficiency values for a range of wavelengths and particle sizes. We use a multiplicative relaxation algorithm for the inversion procedure (Chahine 1970), which was originally developed for radiative transfer applications. There is no need to pre-define a particle size distribution and the algorithm iteratively converges to the unique best fit regardless of the first guess. Multiplying the resulting particle size distribution by its corresponding size bins' mean diameter cubed and particle density provides the mass distribution.

To illustrate the inversion, we use a typical ORS dust extinction spectrum collected using one OP-FTIR and one OP-LT with collinear optical paths. This experiment investigated the optical extinction properties of dust samples collected from an artillery back blast site at a desert location in the southwestern USA. A centrifugal blower released the dried and sifted dust approximately normal to the beam paths in a controlled manner. During this simulation, 1-s data were acquired using the OP-LT (at 0.67 μm) and 4-s data were acquired using the OP-FTIR (2–13.5 μm spectral range) at 4 cm^{-1} resolution. The baseline of absorbance (optical depth) spectra was flat at zero when no dust was encountered by the light except for the absorption bands for H_2O vapor and CO_2 (spectral bands around 2.5 μm, 4.2 μm, and from 5.5 to 7 μm). These absorption bands in the infrared wavelengths show up in the spectra with or without dust encountered in the ORS beam path. The spectrum (open diamonds) shown in Fig. 11.1 is an example with dust encountering the light beam from the OP-FTIR and OP-LT. The OP-LT extinction at 0.67 μm is shown as the far left data point in the spectrum, and is averaged over the time period of the OP-FTIR data acquisition. The broad absorption feature of the back blast dust at around 10 μm is much stronger when compared to other dust samples studied by the authors. The sharp singularity-like feature between 11.2 and 11.6 μm is similar to other previously sampled dusts. These two unique features together enable us to identify the

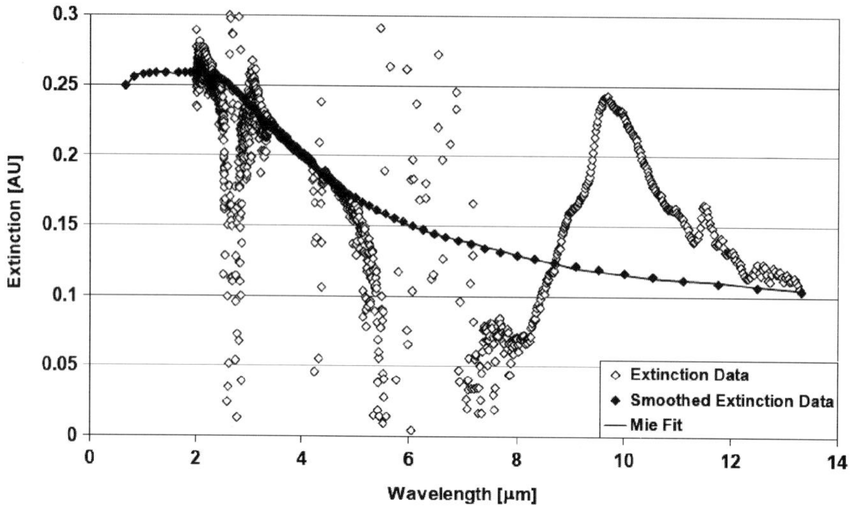

Fig. 11.1 Multi-spectral light extinction spectra by ORS instruments (OP-FTIR and OP-LT)

suspended dust that is generated by the artillery back blast during the field campaign in the southwestern USA.

Seven wavelengths were chosen from the OP-FTIR spectrum that avoid unique dust features and other gaseous absorption peaks so that the light extinction values at each wavelength are due to scattering alone. The wavelengths used to obtain the extinction values for the smoothed extinction distribution are 2.4, 3.5, 3.8, 4.1, 4.4, 4.9, and 13.2 μm (in addition to the OP-LT wavelength of 0.67 μm). A triangle-based cubic interpolation was performed between those wavelengths to generate 90 interpolated data points that were used to compute the extinction efficiency (kernel) matrix ($Q_{e_{ij}}$). The interpolated data set enhances the over-determination of the inversion problem and ensures a unique solution. The resulting interpolated extinction values define the baseline offset (when dust is encountered by light) that excludes the extinction caused by H_2O vapor and CO_2 and avoids other gaseous and PM absorption features. These interpolated extinction values are shown in the Fig. 11.1 as solid diamonds. Also, we used 54 bins that describe particle diameters ranging from 0.25 to 20 μm on a log scale. The first 43 bins that correspond to PM_{10} are of interest for this study. For illustration, we assumed an initial complex refractive index (1.6, 0) which is representative of airborne dust from the desert (Grams et al. 1974) for a range of non-absorbing wavelengths. To achieve proper apportionment of PM_{10} and $PM_{2.5}$ in the retrieved mass distributions, a correction was made to this value as explained in the calibration section below. After the matrix was computed, we inverted it to get number concentrations for all the 54 diameter bins (mid-point of each diameter range in the log scale). The retrieved size (mass) distribution up to 10 μm, which is of importance to characterize PM_{10}, is shown in Fig. 11.2. Mie calculations were then completed in the forward direction to see if the retrieved extinction spectrum from the derived size distribution fit well with the input extinction data (solid line in Fig. 11.1). $PM_{2.5}$ concentration can

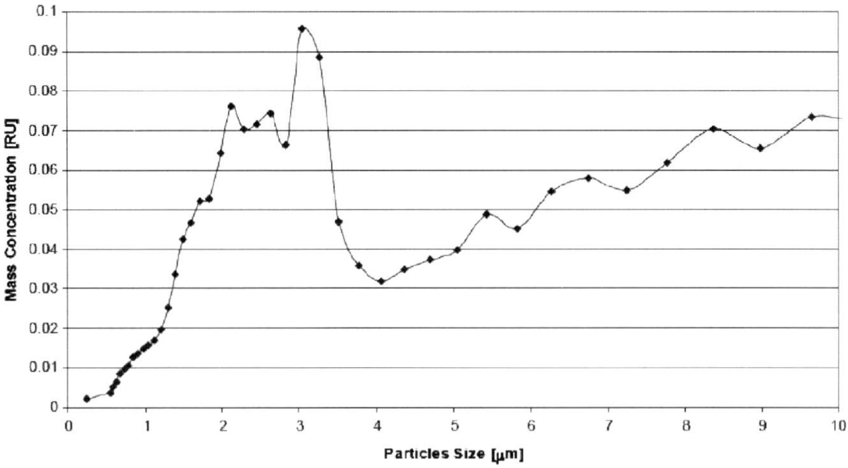

Fig. 11.2 Mass distribution retrieved from multi-spectral ORS measurement. RU = relative mass units

then be calculated by adding the mass concentrations in all particle size bins up to 2.5 μm, and PM_{10} concentration can be calculated by adding the mass concentrations in all particle size bins up to 10 μm.

11.3 Calibration of ORS Method

Calibration of the ORS method was done with a two-step approach by comparing $PM_{2.5}$ and PM_{10} mass concentration values retrieved by the ORS method to $PM_{2.5}$ and PM_{10} measured mass concentration values that were obtained by a pair of calibrated DT aerosol monitors. The DT monitors measure particle mass concentration of PM_{10} or $PM_{2.5}$ when using the corresponding inlet attachment. The first step in the ORS calibration is to retrieve the right apportionment of $PM_{2.5}$ and PM_{10} in the dust plume and the second step is to get the right mass concentration value for both $PM_{2.5}$ and PM_{10}. An experiment was performed with a controlled release of dust from the artillery back blast site in a tent with a 3-m beam path. The tent was used to enclose the optical path of the ORS to avoid influence from ambient wind, and to ensure as uniform a dust plume as possible throughout its beam path. The monostatic OP-FTIR and the OP-LT were placed close to each other so that the combined optical beams formed the ORS optical beam path for the experiment. The dust plume was generated and introduced into the ORS beam path using a blower to entrain the dust and a fan to disperse the dust uniformly in the beam path. Two DT aerosol monitors were placed inside the tent mid-way along the ORS beam path, with one DT using a $PM_{2.5}$ inlet attachment and the other using a PM_{10} inlet attachment. The ORS and DT measurements were temporally synchronized and each data point for comparison is an average of 10 s. The $PM_{2.5}$ and PM_{10} mass concentrations were retrieved from ORS extinction measurements, as explained in the previous section. The $PM_{2.5}$ and PM_{10} data collected by the DT aerosol samplers were averaged for the same period as the ORS data. The DT samplers were previously calibrated against filter-based mass concentration measurements at the artillery facility (and prior to this experiment) for $PM_{2.5}$ and PM_{10}.

The ORS-retrieved PM_{10} data were plotted against the corrected PM_{10} from the DT samplers with the PM_{10} inlet attachment, and the ORS retrieved $PM_{2.5}$ data were plotted against the corrected $PM_{2.5}$ from the corresponding DT sampler. The slopes of both scatter plots were different, which means that the ORS method needs to be corrected to obtain the right apportionment of $PM_{2.5}$ and PM_{10} in the dust plume. The goal was to complete the ORS mass concentration retrieval algorithm to obtain the correct PM_{10} and $PM_{2.5}$ apportionment in the dust plume (as seen by the DT aerosol samplers), while using only one calibration factor (named apportionment factor). This goal was achieved by iteratively changing the apportionment factor that multiplies the initially assumed refractive index of 1.6, re-computing $Q_{e_{ij}}$, and then inverting the matrix to retrieve the mass distribution. This process was repeated until the slopes of both the PM_{10} and $PM_{2.5}$ scatter plots were nearly the same. Factorizing the real part of the complex refractive index adjusted the apportionment to be traceable to the mass filters' apportionment. A factor of 0.86 (equivalent to a refractive index of 1.37) provided the most accurate apportionment for about 70 dust plume measurements during the calibration experiments, and was used in the inversion method for the entire dataset that was collected from the artillery back blast site. The evolution of retrieved mass distribution for one of the extinction spectra measured at the artillery site is shown in Fig. 11.3 for three refractive index values. It is apparent

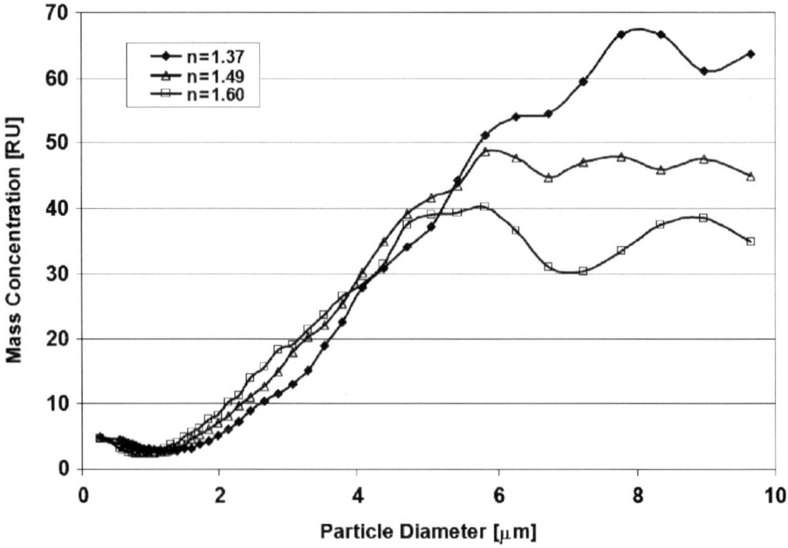

Fig. 11.3 Dependence of mass distribution on refractive index. n = refractive index, RU = relative mass units

that changing the refractive index values in the ORS method calibration effectively changes the retrieved apportionment of $PM_{2.5}$ and PM_{10} in the dust plume.

Once the right apportionment is achieved, then a common multiplicative factor is used to get the correct PM_{10} and $PM_{2.5}$ mass concentrations. This multiplicative factor (named mass factor) is to make the slope of the ORS mass concentrations scatter plot against the DT aerosol sampler derived mass concentrations to 1. The above two calibration factors (apportionment and mass) will be different for different kinds of dust samples and the calibration procedure needs to be performed on a case to case basis. The calibration scatter plots for the artillery back blast dust are as shown for PM_{10} and $PM_{2.5}$ in Fig. 11.4a and b, respectively. The R^2 values from the comparison are 0.78 for PM_{10} and 0.72 for $PM_{2.5}$. These excellent correlations between the open path and point measurements were obtained despite the fact that the pair of DT samplers and ORS instruments had imperfect synchronization of the temporal and spatial measurements. After the calibration, the data from the artillery back blast field campaign were corrected for the ORS mass concentration retrieval.

The first step of the calibration procedure to obtain the apportionment factor reduces the uncertainty from assuming optical properties of the PM material (i.e., real part of refractive index). Both steps of the calibration rely on accurate and concurrent $PM_{2.5}$ and PM_{10} measurements using a standard method (e.g., calibrated DT aerosol monitor).

11.4 Application of ORS Method for PM Emission Rate Estimation

The ORS method to retrieve aerosol mass concentrations was applied to measure $PM_{2.5}$ and PM_{10} mass emission rates from artillery back blast at a desert location in the southwestern USA during October 27, 2005. Tests were carried out on improved gun placements (i.e., artillery were located on surfaces that had been modified to enhance ease of firing and potentially mitigate dust emissions). The extinction spectra of soil dust and canon smoke were separated from each other as a result of field tests that occurred at the artillery site when it rained on one day of the test. The rain moistened the soil and caused the artillery back blast plume to contain only soot associated with the discharge of the propellant, but no dust. From our past experience with black carbon extinction spectra (graphite and black smoke), and from this day of measurements we confirmed that the soot particles had a flat, constant extinction value across visible and infrared wavelengths (due to absorption). During our calibration experiments, with the dust collected from an artillery site, we observed a constant ratio between the peak value at 10 μm (0.24 in Fig. 11.1) and the minimum value at 8 μm (0.07 in Fig. 11.1). This ratio for pure dust from this site was 3.4 (= 0.24/0.07). Therefore, the contribution from soot particles in the mass emission rate calculations were separated from that of the dust particles by subtracting a constant value of extinction throughout the spectrum (calculated by

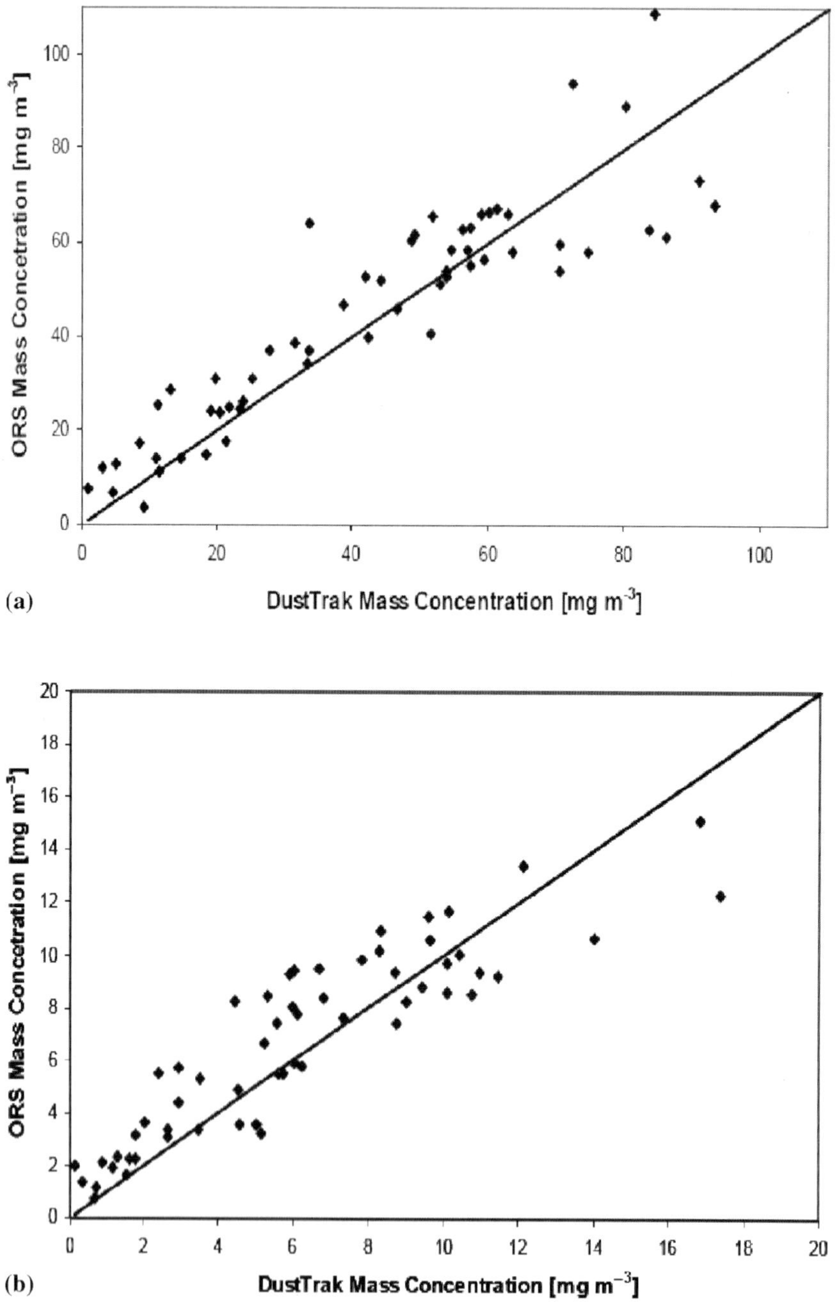

Fig. 11.4 **a** PM_{10} calibration curve ($y = 1.00x$, $R^2 = 0.78$). **b** $PM_{2.5}$ calibration curve ($y = 1.00x$, $R^2 = 0.72$)

solving a simple algebraic equation), thereby maintaining the known ratio between the values at the maximum to the minimum of the dust absorption feature.

An example of the application of the ORS method to quantify PM mass emission is explained below. A full presentation and discussion of the data are beyond the scope of this paper. We used two pairs of OP-FTIR/OP-LT ORS instruments to measure mass concentrations of PM across the plumes that were generated by artillery back blast events (Fig. 11.5). Both pairs of ORS instruments were located next to each other and their beams crossed at about 30 m downwind of the artillery gun location. The first ORS beam path was close and parallel to the ground, and directed toward a retroreflector that was placed at the bottom of a tower (scissor lift) and 100 m away from the location of the first pair of OP-FTIR/OP-LT (hereafter, ORS lower beam path). The second ORS beam path was elevated in such a way that the OP-FTIR/OP-LT pair was looking at a retroreflector placed 15 m above the ground on the same tower as the retroreflector that was at the ground level (hereafter, ORS upper beam path). The dust plume generated from the back blast was brought to the vertical measurement plane of the ORS by a predominantly southwesterly wind. The MPL was another ORS instrument used in this field campaign (Du et al. 2006), which was placed along the same path of the other ORS systems, but was located behind their location by 410 m. Backscatter data from the MPL were used to determine the horizontal profile of plumes along the ORS line of sight (shown inside the box in Fig. 11.5).

Intermittent artillery firing took place between 11:35 AM and 3:30 PM on October 27, 2005. Both OP-FTIRs (the lower one looking at the ground-level retroreflector and the elevated one looking at the elevated retroreflector) collected roughly 10-s spectra at 0.5 cm^{-1} resolution, and were averaged over the duration of

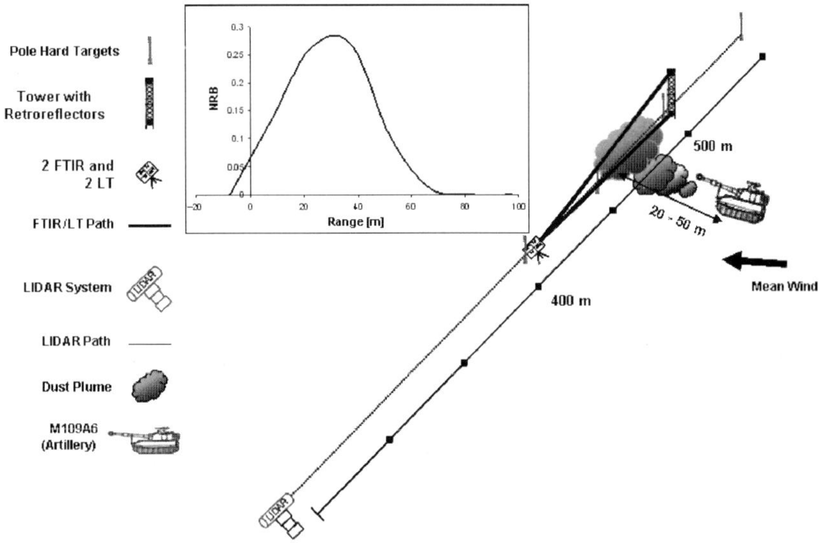

Fig. 11.5 Setup on October 27, 2005, at an artillery facility in southwestern USA. MPL backscattered intensity profile is shown in the box along ORS lower beam path

each plume event (corresponding to each shot). Data collected from each OP-LT were averaged for the same time-interval as the corresponding OP-FTIR. The combined spectra for each plume event were used in conjunction with an inversion algorithm to retrieve $PM_{2.5}$ and PM_{10} mass along each beam path. The following calculations were made for one plume from an artillery firing event, identified as number 172 during the field campaign. The baseline extinction values, averaged for this plume event from the ORS lower beam path and the ORS upper beam path, are shown in Table 11.1. These points are representative of the baseline features of the ORS optical depth spectrum by excluding the gaseous as well as PM absorption features, as explained in the ORS method for the mass concentration retrieval procedure discussed before. We used the calibrated ORS method to compute $PM_{2.5}$ and PM_{10} mass concentrations along the ORS lower and upper beam paths. From the retrieved mass concentrations, $PM_{2.5}$ and PM_{10} were calculated for both the ORS lower and upper beam paths (Table 11.2). The differences in path-integrated mass concentrations between the lower and upper beam paths provide vertical dilution information of PM mass.

The vertical extent of each plume can be calculated from the plume dilution information obtained from the two pairs of OP-FTIR/OP-LT beam paths. Following a previous study (Hashmonay et al. 2001) we fitted a bivariate Gaussian function to the ORS data. This method, which is also called the radial plume mapping method, is being used for gas pollutant emission rate quantification, and is also applicable for quantifying PM mass emission rate. For estimating the emission rate by Hashmonay

Table 11.1 Extinction values used from ORS spectrum for mass retrieval

Wavelength (μm)	Lower ORS beam path	Upper ORS beam path
0.67	0.194	0.100
2.4	0.235	0.078
3.5	0.254	0.067
3.8	0.236	0.062
4.1	0.221	0.061
4.4	0.213	0.058
4.9	0.201	0.055
13.2	0.130	0.040

Table 11.2 Path-averaged PM mass in ORS beam paths

	$PM_{2.5}$ mass concentration (mg m^{-3})	PM_{10} mass concentration (mg m^{-3})
Upper ORS beam path	0.36	1.2
Lower ORS beam path	0.42	4.1

et al. (2001), the parameters of the assumed Gaussian function were iteratively calculated from the path-integrated OP-FTIR gas absorption measurements from five retroreflectors in the vertical measurement plane used in that study. These parameters are the normalization coefficient, peak location on the ground along the OP-FTIR line of sight, and the horizontal and vertical standard deviations of the Gaussian plume profile. In this study, we used the vertical dilution information from mass concentrations to obtain the vertical standard deviation of the plume profile. Instead of several retroreflectors placed on the ground (as in the case of Hashmonay et al. 2001), we used the MPL to obtain horizontal peak location as well as the horizontal standard deviation for the Gaussian plume function. The MPL was operated in the field, aligned collinear to the ORS lower beam path and across the plume, to get the plume dimension along its line of sight. The MPL profile that was averaged for this plume event is shown inside the box in Fig. 11.5. The *x*-axis of this plot is along the ORS lower beam path. The origin of this plot is fixed at the OP-FTIR/OP-LT pair location (not the MPL). From this plume profile along the ORS lower beam path, we assessed the peak location at 35 m from origin with a standard deviation of 19 m. We fixed the vertical plume peak location on the ground.

Once the mass-equivalent plume profile was obtained, we integrated the mass concentration over the entire plane of measurement (both for $PM_{2.5}$ and PM_{10} separately). These plane-integrated mass concentrations (in units of g m^{-1}), in conjunction with the wind vector normal to the plane (in m s^{-1}) provided the mass emission rate of dust generated from the artillery back blast (in g s^{-1}). Fig. 11.6 shows the reconstructed mass-equivalent plume profile in the vertical plane of measurement for this back blast event of roughly 20-s duration. The upper ORS beam path and the tower with retroreflectors are also shown, for reference, as dark vertical bars at a crosswind distance of 100 m. An average wind speed of 4.5 m s^{-1} and a wind direction of 17° normal to the plane of measurement were measured for the duration of the entire event. This wind vector and the reconstructed PM mass concentration profile (for both $PM_{2.5}$ and PM_{10}) integrated over the plane of measurements were

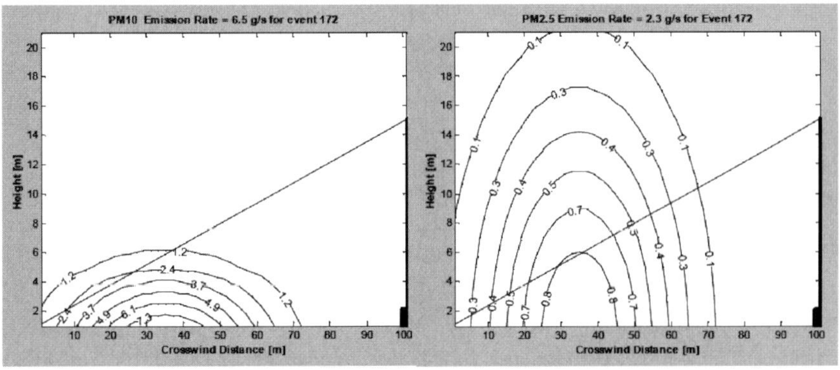

Fig. 11.6 Reconstructed PM_{10} and $PM_{2.5}$ mass-equivalent plume profiles. The contour values are in mg m^{-3}

used to compute the emission rate of PM mass across the plane of measurements. The PM_{10} mass emission rate for this event was calculated as 6.5 g s^{-1} and the $PM_{2.5}$ mass emission rate was calculated as 2.3 g s^{-1}. The total amount of PM_{10} and $PM_{2.5}$ generated for the duration of this plume was 130 and 46 g, respectively.

11.5 Summary

A new and innovative optical remote sensing (ORS) method was described here to obtain particulate matter (PM) concentrations and mass emission rates that are generated by fugitive dust events. Multi-spectral path-integrated measurements were used as part of the ORS system. This method was successfully calibrated by determining calibration factors using concurrent conventional mass concentration measurements in a closed chamber with controlled dust plume releases. Field measurements were carried out at a southwestern USA desert location where dust plumes were generated from the shock of artillery back blast. We developed a simple method for eliminating the error in dust mass emission rate estimations introduced by soot particles from the cannon by maintaining the ratio of maximum to minimum extinction values around a prominent dust absorption feature in the spectrum, as estimated during calibration experiments. The horizontal extent of the plumes along the ORS line of sight were obtained by backscatter measurements from a micro pulse lidar. PM with diameters $\leq 10\,\mu m$ (PM_{10}) and PM with diameters $\leq 2.5\,\mu m$ ($PM_{2.5}$) mass emission rates and concentrations were obtained by making multiple ORS measurements on a vertical plane downwind of the PM source in conjunction with the wind vector.

Acknowledgements The authors like to thank Dr. Jack Gillies and Dr. Hampden Kuhns, Desert Research Institute (DRI) for providing the DUSTTRAK calibration data against DRI filter-based measurements made during our combined field measurements. This research was sponsored by the Strategic Environmental Research and Development Program grant number CP-1400.

References

Chahine M.T. (1970), Inverse problems in radiative transfer: Determination of atmospheric parameters, *J. Atmos. Sci.*, 17, 960–967.

Du K., Rood M., Kim B., Kemme M., Hashmonay R., and Varma R. (2006, June), Optical Remote Sensing of Dust Plumes Using Micropulse Lidar, (Paper presented at the 99th Annual Meeting of the Air & Waste Management Association, New Orleans, Louisiana, 315, 5 pp).

Grams G.W, Blifford Jr., I.H., Gillette D.A., and Russel P.B. (1974), Complex refractive index of airborne soil particles. *J. Appl. Meteor.*, 13, 459–471.

Hashmonay R.A. and Yost M.G. (1999), On the application of OP-FTIR spectroscopy to measure aerosols: Observations of water droplets. *Environ. Sci. Technol.*, 33(7), 1141–1144.

Hashmonay R.A., Natschke D.F., Wagoner K., Harris D.B., Thompson E.L., and Yost M.G. (2001), Field evaluation of a method for estimating gaseous fluxes from area sources using open-path Fourier transform infrared. *Environ. Sci. Technol.*, 35, 2309–2313.

Kerker M. (1969), *The Scattering of Light*, (New York: Academic Press).

Wickramasinghe N.C. (1973), *Light Scattering Functions for Small Particles with Applications in Astronomy*, (London: Adam Hilger)

Chapter 12
Raman Lidar for Monitoring of Aerosol Pollution in the Free Troposphere

Detlef Müller, Ina Mattis, Albert Ansmann, Ulla Wandinger, and Dietrich Althausen

Abstract Geometrical, optical, and microphysical properties of free-tropospheric pollution have been determined with multiwavelength Raman lidar at Leipzig, Germany. Long-term observations carried out at fixed times (three times per week) since 1997 show advection of different aerosol types such as anthropogenic pollution from North America, forest-fire smoke from North America and Siberia, pollution from polar areas, and Saharan dust. Up to 45% off all regular observations indicate free-tropospheric pollution. On average, 20–25% of columnar optical depth was contributed by these layers. In extreme cases, the fraction of optical depth was considerably higher. At times pollution was found around 10–12 km height. Geometrical depth of the layers in many cases exceeded 1 km. Mean Ångström exponents of the layers varied from as low as 0.7 for Saharan dust to as high as 1.7 for anthropogenic pollution from North America. Individual measurements show significantly lower, respectively higher values. Lidar ratios in general were larger at 355 nm than at 532 nm. One remarkable exception is aged forest-fire smoke for which we find a reversed spectral dependence. Results for the Leipzig lidar site may be contrasted to results on European pollution outflow observed with Raman lidar at the southwest coast of Portugal. We also find strong differences with respect to South and Southeast Asian pollution observed during several field campaigns in the Indian Ocean.

Keywords: Free troposphere, inversion, multiwavelength lidar, particle properties, pollution, Raman lidar, transport

12.1 Introduction

Optical and microphysical properties of particles in the free troposphere are only poorly understood (Jacob et al. 1999; Collins et al. 2000; McKendry et al. 2001; Collins et al. 2002; Creilson et al. 2003; Prather et al. 2003). Properties of boundary-layer aerosols

---

Leibniz Institute for Tropospheric Research, Permoserstraße 15, 04318 Leipzig, Germany

Y.J. Kim and U. Platt (eds.), *Advanced Environmental Monitoring*, 155–166.

that originate from local and regional emissions of particles and gases usually are very different from free-tropospheric particles. In climate models, particles are mainly present in the boundary layer over the continents and are deposited after 2 to 4 days or within a range of less than 2000 km within the source region (Rodhe 1999). In contrast, free-tropospheric particles are often advected over large distances from other continents. Transport times may easily exceed 1 week. The direct radiative effect of free-tropospheric particles may be of minor importance. However, such particles may have considerable impact on, e.g., cloud formation processes, which thus renders them highly important with respect to the aerosol indirect effect.

The main difficulty of documenting and characterizing particles in the free troposphere is the limitation of observational techniques. Satellite passive sensors and Sun photometers cannot separate between particles in the boundary layer and pollution in the free troposphere. The retrieved parameters rather describe a mixture of particle types. Airborne observations are limited to short periods during field campaigns and/or restriction on access to areas of interest.

Aerosol conditions have been routinely monitored with the stationary multi-wavelength Raman lidar of the Leibniz Institute for Tropospheric Research since 1997. Measurements were carried out three times per week (Monday noon and after sunset, Thursday after sunset) in the framework of the German aerosol research lidar network (AFS–Deutsches Lidarnetzwerk) from 1997 to 2000 (Bösenberg et al. 2001), and within the European Aerosol Research Lidar Network (EARLINET) since 2000 (Bösenberg et al. 2003). The data provide information on the frequency of occurrence of free-tropospheric particle layers, their height distribution, as well as optical and microphysical properties of such particle layers.

In this contribution we give a brief overview on some properties of these free-tropospheric layers. Section 12.2 describes the instrument and the methodology used for data analysis. Section 12.3 presents a measurement example and some statistical information. Section 12.4 gives a brief summary.

12.2 Methodology

The Raman lidar is described in detail by Mattis et al. (2004). A Nd:YAG laser is used for generating laser pulses at 355, 532, and 1064 nm wavelength. The pulse repetition rate is 30 Hz. The laser beam is emitted on one optical axis and expanded 15-fold before it is transmitted into the atmosphere. A Cassegrain telescope is used for collecting the backscattered light. The diameter of the main mirror is 1 m. The signals are separated with beam splitters and interference filters according to wavelength. Backscatter signals are collected at the three emitted wavelengths, at 387 and 607 nm resulting from Raman scattering from nitrogen (355 and 532 nm primary wavelength), and at 408 nm resulting from Raman scattering from water vapor (355 nm primary wavelength). In addition, the system detects the component of light cross-polarized to the plane of polarization of the outgoing beam at 532 nm.

A detailed description of data analysis may be found in Ansmann and Müller (2005). Briefly, profiles of the particle volume extinction coefficients are derived at 355 and 532 nm wavelength with the use of the nitrogen vibrational Raman signals detected at 387 and 607 nm (Ansmann et al. 1990). Errors usually are in the range of 10%–30%. Particle backscatter coefficients at 355, 532, and 1064 nm are calculated with the Raman method (Ansmann et al. 1992). The Raman signal at 607 nm is also used as reference signal for the 1064 nm signal. Uncertainties can be kept to 5%–15%.

The particle extinction-to-backscatter (lidar) ratio is determined at 355 and 532 nm wavelength. This quantity is sensitive to particle type, as it contains information on particle size and particle light absorption. The uncertainty may vary between 20% and 60%. The Ångström exponent (Ångström 1964) describes the slope of the spectrum of the extinction coefficient. The parameter is a qualitative measure of particle size. In our case, this parameter was determined for the wavelength pair at 355/532 nm. The uncertainty varies between 20% and 60%. The depolarization ratio of the particles is calculated from the total elastic-backscatter signal at 532 nm and the component cross-polarized to the state of polarization of the emitted light beam (Cairo et al. 1999). This parameter allows us to identify depolarizing mineral-dust particles. Uncertainties in general can be kept to below 10%.

One important fact to consider is the incomplete overlap between outgoing laser beam and receiver field of view of the detector telescope. Accordingly it is not possible to derive trustworthy information on particle extinction and particle lidar ratio below approximately 1000 m height. In contrast, particle backscatter coefficients determined with the Raman method can be derived to close to the ground (minimum height 60–120 m), because ratios of signal profiles are taken, which cancels the geometrical overlap effect. A description of this technique is given by Wandinger and Ansmann (2002).

Microphysical particle properties, e.g., particle size distribution, effective radius, and complex refractive index are determined with an inversion algorithm that is used for routine analysis of the optical particle properties. The algorithm is described in detail by Müller et al. (1999a,b; 2001) and Veselovskii et al. (2002). The algorithm requires particle backscatter coefficients measured at three wavelengths (355, 532, and 1064 nm), and extinction coefficients measured at two wavelengths (355 and 532 nm) as input information. In general, we select individual height layers of the optical profiles, average the optical data for these height layers, and then carry out data inversion. Usually, effective radius can be derived with an accuracy of 30%. In that respect, one has to consider the limited range of measurement wavelengths. The fraction of particles in the coarse mode of the particle size distribution under circumstances cannot be completely retrieved. Veselovskii et al. (2005) show that according to that limitation it may not be possible to derive particle effective radius larger than approximately 2 μm.

The real part of the complex refractive index is derived to an absolute accuracy of 0.05–0.1. The imaginary part is derived to its correct order of magnitude if it is $< 0.01i$. If the imaginary part is $> 0.01i$ the accuracy is on the order of ±50%. The complex refractive index is derived as wavelength-independent quantity.

Single-scattering albedo is calculated from the derived microphysical particle properties with a Mie-scattering code (Bohren and Hufmann 1983). That parameter is defined as the ratio of particle scattering to particle extinction (scattering + absorption). It is one of the key input parameters in studies on climate forcing by aerosols.

12.3 Results

12.3.1 Measurement Example

Figure 12.1 shows the example of forest-fire smoke observed in July 2004. Particles were detected to approximately 8 km height. A characteristic feature of pollution plumes that are transported across intercontinental distances from North America to Central Europe are filament-like structures. Additional examples may be found in Mattis et al. (2003) and Müller et al. (2005).

Figure 12.1 also shows the profiles of the particle Ångström exponent and particle lidar ratios. The Ångström exponent varies around 0.5–1 in the central part of the lofted layer. The particle lidar ratio is around 55–65 sr at 532 nm, and 30–40 sr at 355 nm. The numbers for the Ångström exponent are quite characteristic for aged forest-fire smoke. In general, we find values <1.5 for such plumes (Müller et al. 2005). The lidar ratio at 532 nm characteristically is larger than the one at 355 nm. In contrast, anthropogenic pollution from North America generally shows lidar ratios that are larger at 355 nm than at 532 nm. This spectral feature is mainly caused by the large particle size of forest-fire smoke compared to the much smaller particles of anthropogenic pollution. Despite the large size of the particles, see following paragraph, the depolarization ratio of around 2% indicates that the particles in the center of the smoke plume do not deviate much from spherical shape.

Figure 12.2 shows particle effective radius and single-scattering albedo. Effective radius varies between 0.2 and 0.25 μm. Single-scattering albedo varies between 0.85 and 0.9 at 532 nm wavelength. We do not find strong differences between single-scattering albedos of aged forest-fire smoke and light-absorption characteristics of anthropogenic pollution from North America.

Travel time of the relatively fresh smoke plume was on the order of 5 days. For comparison, we detected forest-fire smoke transported from Siberia to Central Europe in 2003 (Damoah et al. 2004; Mattis et al. 2003; Müller et al. 2005). The plumes traveled more than 10 days from their source region before they were detected over Leipzig. We found particle effective radii as large as 0.4 μm and Ångström exponents as low as 0 in those well-aged smoke plumes.

12.3.2 Statistical Information

The long-term observations reveal a variety of aerosol types over the Leipzig lidar site. These aerosol types were identified on the basis of their optical and microphysical particle properties. In addition, backward trajectory analysis with HYSPLIT (Draxler and Hess, 2003) and chemical-tracer modelling with FLEXPART (Stohl et al. 1998; Stohl and Seibert 1998) was used to identify source regions of

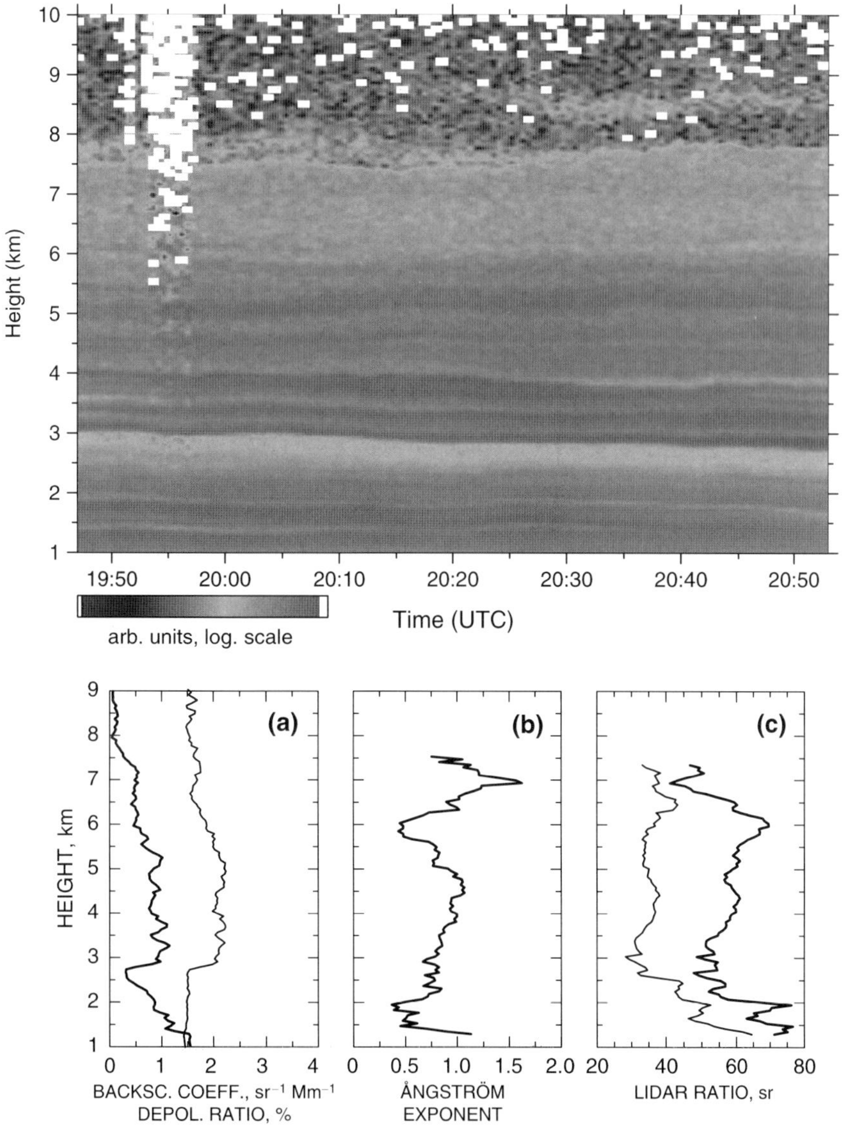

Fig. 12.1 (top) Range-corrected backscatter signal at 1064-nm wavelength of forest-fire smoke observed on 22 July 2004. (bottom) Profiles of **(a)** the particle backscatter coefficient (thick line) and the depolarization ratio (thin line) at 532nm wavelength, **(b)** the Ångström exponent for the wavelength range from 355–532nm, and **(c)** the particle lidar ratio at 355 nm (thin line) and 532 nm (thick line)

the observed free-tropospheric plumes. Fig. 12.3 shows the frequency distribution of events of aerosol transport from North America, the polar region north of 70° N, and the Sahara to Leipzig.

In summary, we used 640 measurements for that statistical analysis which encompasses the time from 1997 to 2004. We find a rather characteristic temporal

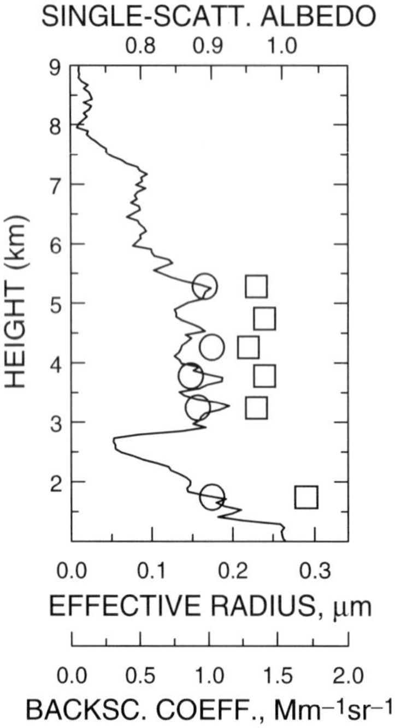

Fig. 12.2 Particle effective radius (squares) and single-scattering albedo (circles). Also shown is the profile of the particle backscatter coefficient at 532 nm

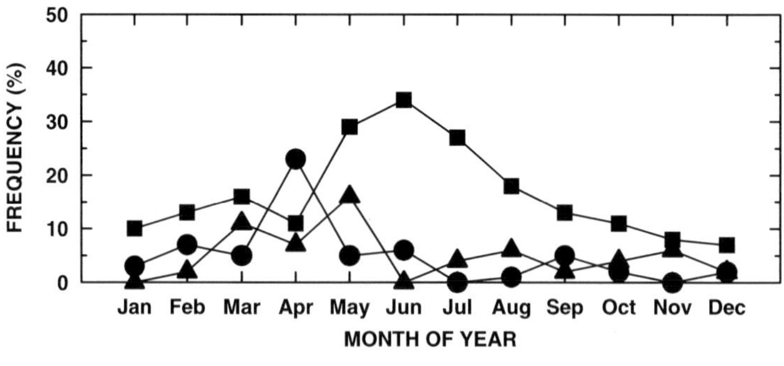

Fig. 12.3 Relative frequency of occurrence of (squares) pollution from North America, (bullets) pollution from areas north of 70° N, and (triangles) Saharan dust

distribution of advection of pollution from North America with a clear maximum in late spring/early summer. At that time, the forest-fire season reaches its maximum intensity in North America. However, there also is a constant advection of anthropogenic pollution from the heavily industrialized areas at the east coast of the United States and Canada. Accordingly we also find pollution advection during winter time. A more detailed analysis is needed in order to separate the measurement cases according to forest-fire smoke and anthropogenic pollution.

With respect to aerosols transported from the polar region we assume that at least part of these plumes contained so-called Arctic Haze (Heintzenberg et al. 2003; Müller et al. 2004; Shaw 1995). That aerosol type consists of aged anthropogenic pollution which is transported from industrialized regions of North America and Europe to polar areas. There the aerosols remain within the polar vortex until they are ejected back to North America and Europe, which mainly occurs during spring time of each year.

Saharan dust is observed approximately six times per year over Leipzig, which is much less than what is usually observed over South Europe. For instance, regular Raman lidar observations at the EARLINET station at Napels (Italy), showed Saharan dust in 61 of 270 observations (Pisani 2006). In general, dust may be observed over Leipzig in almost any month of the year with a higher probability in spring time.

At the present state of data analysis, we find particles in the free troposphere in up to 45% of all regular observations carried out at the Leipzig Raman lidar site. It should be noted that in some cases residual aerosol layers from the planetary boundary layer of the preceding day may be responsible for some of the aerosol load. For that reason, we currently analyze radiosonde data of temperature and relative humidity to identify these critical measurement cases. Furthermore, chemical tracer modeling with FLEXPART will be used for an improved source identification of the lofted layers.

The long-term Raman lidar observations show that on average 20% of column optical depth is contributed by particles in the free troposphere (Mattis et al. 2004). The mean bottom height at which these layers occur is at or above 2 km in 50% of all cases. The top height of the lofted layers is at or above 3 km height in 50% of all investigated cases. The mean geometrical depth of the layers is 1 km or higher in 50% of all cases. In that respect we did not yet separate between measurements in which one single pollution layer was present or an accumulation of several thin layers (see Fig. 12.1), i.e., in such cases we counted the bottom of the lowest layer as minimum and the top of the upper most layer as maximum height of the lofted layers.

In some extreme cases, we find particles up to 12 km height. These cases are linked to forest-fire smoke transported from Canada to Central Europe. According to Fromm et al. (2000, 2005) so-called "pyro-cumulonimbus" events, which are cases of violent convection resulting from extreme heat in large-scale forest fires may carry smoke particles into the lower stratosphere.

Data on free-tropospheric particles have been collected with the mobile Raman lidar (Althausen et al. 2000) of the institute during several field campaigns since 1997. We observed European anthropogenic pollution, which was lofted above the marine boundary layer, in the southwest of Portugal during the Second Aerosol Characterization Experiment (ACE 2) in June/July 1997 (Ansmann et al. 2001; Ansmann et al. 2002; Müller et al. 2002), and pollution outflow from South and Southeast Asia during the Indian Ocean Experiment (INDOEX) in February/March 1999 and March 2000 (Franke et al. 2003; Müller et al. 2003). The lofted particle plumes typically appeared in heights between 1 and 4 km above sea level. The particle layers were clearly separated from the marine boundary layer. Two more

Table 12.1 Characteristic values of particle Ångström exponent (å) for the wavelength range from 355 to 532 nm and the lidar ratio (S_{532}) at 532 nm. Also shown is the spectral dependence of the lidar ratios measured at 355 nm and 532 nm. The last column shows the ratio of optical depth in the free troposphere (τ_{ft}) to optical depth of the atmospheric column (τ_{to}). The numbers in the last three rows of the last column describe the ratio of optical depth in the planetary boundary layer to optical depth in the atmospheric column.

Aerosol type	Experiment	å	S_{532} (sr)	S_{355}/S_{532}	τ_{ft}/τ_{to}
	Free troposphere				
Anthropogenic pollution	ACE 2	1.4	35–55	>1	0.63
Anthropogenic pollution from North America	EARLINET	1.7	30–50	>1	0.22
Forest-fire smoke from North America	EARLINET	1.0	40–65	<1	0.3
Saharan dust	EARLINET	0.5	50–70	>1	0.5
North Indian pollution	INDOEX	1.2	50–80	>1	0.53
South Indian pollution	INDOEX	0.9	30–50	>1	0.53
Southeast Asia pollution	INDOEX	1.5	30–70	>1	0.31
Clean marine (October 1999)	INDOEX	0.1	<35	~1	0.33
	Planetary boundary layer				
Anthropogenic pollution in the European PBL	EARLINET	1.4	40–65	>1	0.22
Clean marine	ACE 2	0.3	20–25	~1	0.37
Clean marine (mean conditions)	INDOEX	–	20–30	~1	0.23

campaigns were carried out in the framework of INDOEX in July and October to document clean marine conditions. However, even in summer we observed free-tropospheric particle layers which were advected from Africa and Saudi Arabia across the North Indian Ocean. The field sites were at Sagres (37.01° N, 8.95° W) during ACE 2 and in the Maldives (4.1° N, 73.3° E) during INDOEX, respectively.

Table 12.1 compares some characteristic optical features of lofted plumes observed over Central Europe and at the field sites. For comparison reason we also add some results on particle properties in the continental boundary layer over Central Europe (EARLINET) and the marine boundary layer (ACE2 and INDOEX). Details may be found in the literature cited above. For instance, lowest Ångström exponents are found for the clean marine boundary layer during ACE 2, and the clean marine conditions that prevailed in the free troposphere in October 1999 during INDOEX. Anthropogenic pollution from North America and Europe in general is characterized by higher Ångström exponents than pollution that consists of a mixture of anthropogenic pollution and biomass-burning observed during INDOEX. With respect to the INDOEX area, we found different transport channels of the free-tropospheric plumes, i.e., pollution was carried from North India, South India, and Southeast Asia to the field site (Franke et al. 2003). Comparably small Ångström exponents are found for forest-fire smoke from North America and for dust from the Sahara.

We find lidar ratios at 532 nm ranging from <35 sr for clean marine air to as high as 80 sr for strongly-light absorbing pollution from North India. On average, the

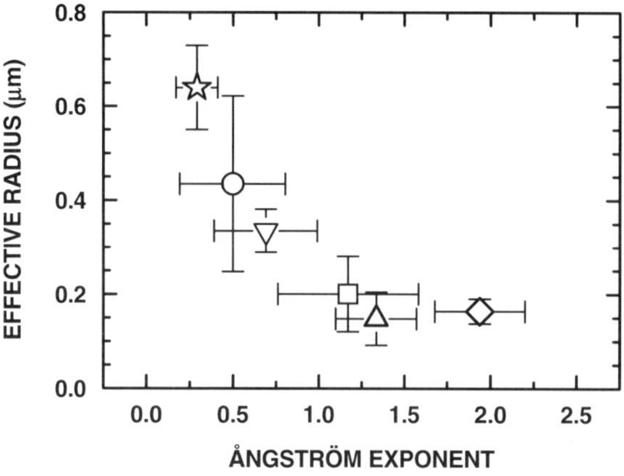

Fig. 12.4 Particle effective radius and Ångström exponent of (diamond) anthropogenic pollution from North America, (triangle, upward) anthropogenic pollution outflow observed during ACE 2, (square) anthropogenic pollution/biomass-burning aerosols observed during INDOEX, (triangle, downward) forest-fire smoke from North America, (circle) marine conditions in October 1999 during INDOEX, and (star) clean marine conditions during ACE 2. Ångström exponents differ in part from respective numbers given in table 12.1. In this figure we only show results of measurement cases that were used for the retrievel of microphysical particle parameters.

lidar ratio at 355 nm is larger than the lidar ratio at 532 nm. The exception is forest-fire smoke for which we find a reversed spectral dependence.

Figure 12.4 finally shows an example of the derived microphysical particle parameters. Details may be found in Müller et al. (2002, 2003, 2005). Particle effective radius has been determined for cases of anthropogenic pollution and forest-fire smoke transported from North America to Central Europe and European pollution outflow observed during ACE 2. Mixtures of anthropogenic pollution and biomass-burning aerosols were observed during INDOEX. We find comparably small particles for anthropogenic pollution. Particles are larger for mixtures of anthropogenic pollution and biomass-burning smoke. The largest pollution particles are observed for forest-fire smoke. Sea-salt particles in the marine environment are at the top end of particle size. As with respect to single-scattering albedo, we usually observed low light absorption of the particles. Mean values in general were >0.9 at 532 nm wavelength. However, South Asian pollution often was much more light-absorbing. For instance, single-scattering albedo was as low as 0.8 for pollution advected from North India to the Indian Ocean (Müller et al. 2003).

12.4 Summary

Multiwavelength Raman lidar observations carried out at a Central European site since 1997 as well as field campaigns carried out in Europe and South Asia provide us with a comprehensive view on natural and human-made pollution in the free troposphere. We observed different aerosol types that vary considerably in mean particle size, e.g., anthropogenic pollution vs. forest-fire smoke from North America, and particle single-scattering albedo, e.g., anthropogenic European pollution vs. South Asian pollution. The results show the merit of multiwavelength Raman lidar sounding which provides us with a vertically resolved view on aerosol conditions.

Acknowledgement The observations at the Leipzig EARLINET site presented in this work were funded by the European Commission under grant EVR1-CT1999-40003. EARLINET is currently funded by the European Commission under grant RICA-025991. We thank A. Stohl (Norsk institutt for luftforskning (NILU), Kjeller, Norway) for providing the results from FLEXPART simulations. The trajectory model HYSPLIT is available at http://www.arl.noaa.gov/ready/hysplit4.html.

References

Althausen D., Müller D., Ansmann A., Wandinger U., Hube H., Clauder E., and Zörner S. (2000), Scanning 6-wavelength 11-channel aerosol lidar, *J. Atmos. Ocean. Technol.*, 17, 1469–1482.

Ångström A. (1964), The parameters of atmospheric turbidity, *Tellus*, 16, 64–75.

Ansmann A. and Müller D. (2005), Lidar and atmospheric aerosol particles. In C. Weitkamp (Ed.), *Lidar. Range-Resolved Optical Remote Sensing of the Atmosphere*, Springer, New York, pp. 105–141.

Ansmann A., Riebesell M., and Weitkamp C. (1990), Measurements of atmospheric aerosol extinction profiles with a Raman lidar, *Opt. Letts.*, 15, 746–748.

Ansmann A., Wandinger U., Riebesell M., Weitkamp C., and Michaelis W. (1992), Independent measurement of extinction and backscatter profiles in cirrus clouds by using a combined Raman elastic-backscatter lidar, *Appl. Opt.*, 31, 7113–7131.

Ansmann A., Wagner F., Althausen D., Müller D., Herber A., and Wandinger U. (2001), European pollution outbreaks during ACE 2: Lofted aerosol plumes observed with Raman lidar at the Portuguese coast, *J. Geophys. Res.*, 106, 20,725–20,733.

Ansmann A., Wagner F., Müller D., Althausen D., Herber A., von Hoyningen-Huene W., and Wandinger U. (2002), European pollution outbreaks during ACE 2: Optical particle properties inferred from multiwavelength lidar and star/Sun photometry, *J. Geophys. Res.*, 107, EID 4259, DOI 10.1029/2001JD001109.

Bohren C.F. and Huffman D.R. (1983), *Absorption and Scattering of Light by Small Particles*, Wiley, New York, NY.

Bösenberg J. et al. (2001), The German aerosol lidar network: Methodology, data, analysis. Report No. 317. Max Planck Institute for Meteorology, Hamburg, Germany.

Bösenberg J. et al. (2003), EARLINET: A European aerosol research lidar network to establish an aerosol climatology. Report No. 348. Max Planck Institute for Meteorology, Hamburg, Germany.

Cairo F., Di Donfrancesco G., Adriani A., Lucio P., and Federico F. (1999), Comparison of various linear depolarization parameters measured by lidar. *Appl. Opt.*, 38, 4425–4432.

Collins D.R., Johnsson H.H., Seinfeld J.H., Flagan R.C., Gassó S., Hegg D.A., Russell P.B., Schmid B., Livingston J.M., Öström E.K., Noone J., Russell L.M., and Putaud J.P. (2000), In situ aerosol-size distributions and clear-column radiative closure during ACE 2. *Tellus Ser. B*, 52, 498–525.

Collins W.J., Rasch P.J., Eaton B.E., Fillmore D.W., Kiehl J.T., Beck C.T., and Zender C.S. (2002), Simulation of aerosol distributions and radiative forcing for INDOEX: Regional climate impacts, *J. Geophys. Res.*, 107, EID 8028, DOI 10.1029/2000JD000032.

Creilson J.K., Fishman J., and Wozniak A.E. (2003), Intercontinental transport of tropospheric ozone: A study of its seasonal variability across the North Atlantic utilizing tropospheric ozone residuals and its relationship to the North Atlantic oscillation, *Atmos. Chem. Phys.*, 3, 2053–2066.

Damoah R., Spichtinger N., Forster C., James P., Mattis I., Wandinger U., Beirle S., Wagner T., and Stohl A. (2004), Around the world in 17 days—hemispheric-scale transport of forest fire smoke from Russia in May 2003, *Atmos. Chem. Phys.*, 4, 1311–1321.

Draxler R.R., and Hess G.D. An overview of the HYSPLIT_4 modelling system for trajectories, dispersion, and deposition, Austral. Meteorol. Magazine, 47, 295–308, 1998.

Franke K., Ansmann A., Müller D., Althausen D., Venkataraman C., Shekar Reddy M., Wagner F., and Scheele R. (2003), Optical properties of the Indo-Asian haze layer over the tropical Indian Ocean, *J. Geophys. Res.*, 108, EID 4059, DOI 10.1029/2002JD002473.

Fromm M., Alfred J., Hoppel K., Hornstein J., Bevilacqua R., Shettle E., Servranckx R., Li Z., and Stocks B. (2000), Observations of boreal forest fire smoke in the stratosphere by POAM III, SAGE II, and lidar in 1998, *Geophys. Res. Letts.*, 27, 1407–1410.

Fromm M., Bevilacqua R., Servranckx R., Rosen J., Thayer J.P., Herman J., and Larko D. (2005), Pyro-cumulonimbus injection of smoke to the stratosphere: Observations and impact of a super blowup in northwestern Canada on 3–4 August 1998, *J. Geophys. Res.*, 110, EID 08205, DOI 10.1029/2004JD005350.

Heintzenberg J., Tuch T., Wehner B., Wiedensohler A., Wex H., Ansmann A., Mattis I., Müller D., Wendisch M., Eckhardt S., and Stohl A. (2003), Arctic haze over Central Europe, *Tellus Ser. B*, 55, 796–807.

Jacob D.J., Logan J.A., and Murti P.P. (1999), Effect of rising Asian emissions on surface ozone in the United States, *Geophys. Res. Letts.*, 26, 2175–2178.

Mattis I., Ansmann A., Wandinger U., and Müller D. (2003), Unexpectedly high aerosol load in the free troposphere over central Europe in spring/summer 2003, *Geophys. Res. Letts.*, 30, EID 2178, DOI 10.1029/2003GL018442.

Mattis I., Ansmann A., Müller D., Wandinger U., and Althausen D. (2004), Multiyear aerosol observations with dual-wavelength Raman lidar in the framework of EARLINET, J. Geophys. Res., 109, EID D13203, DOI 10.1029/2004JD004600.

McKendry I.G., Hacker J.P., Stull R., Sakiyama S., Mignacca D., and Reid K. (2001), Long-range transport of Asian dust to the Lower Fraser Valley, British Columbia, Canada, *J. Geophys. Res.*, 106, 18,361–18,370.

Müller D., Ansmann A., Wagner F., and Althausen D. (2002), European pollution outbreaks during ACE 2: Microphysical particle properties and single-scattering albedo inferred from multiwavelength lidar observations, *J. Geophys. Res.*, 107, EID 4248, DOI 10.1029/2001JD001110.

Müller D., Franke K., Ansmann A., and Althausen D. (2003), Indo-Asian pollution during INDOEX: Microphysical particle properties and single-scattering albedo inferred from multi-wavelength lidar observations, *J. Geophys. Res.*, 108, EID 4600, DOI 10.1029/2003JD003538.

Müller D., Mattis I., Wehner B., Althausen D., Wandinger U., Ansmann A., and Dubovik O. (2004), Comprehensive characterization of Arctic haze from combined observations with Raman lidar and sun photometer, *J. Geophys. Res.*, 109, EID D13206, DOI 10.1029/2003JD004200.

Müller D., Mattis I., Wandinger U., Ansmann A., Althausen D., and Stohl A. (2005), Raman lidar observations of aged Siberian and Canadian forest-fire smoke in the free troposphere over Germany in 2003: Microphysical particle characterization, *J. Geophys. Res.*, 110, EID D17201, DOI 10.1029/2004JD005756.

Müller D., Wandinger U., and Ansmann A. (1999a). Microphysical particle parameters from extinction and backscatter lidar data by inversion with regularization: Theory, *Appl. Opt.*, 38, 2346–2357.

Müller D., Wandinger U., and Ansmann A. (1999b). Microphysical particle parameters from extinction and backscatter lidar data by inversion with regularization: Simulation, *Appl. Opt.*, 38, 2358–2368.

Müller D., Wandinger U., Althausen D., and Fiebig M. (2001), Comprehensive particle characterization from three-wavelength Raman-lidar observations, *Appl. Opt.*, 40, 4863–4869.

Pisani G. (2006), Optical characterization of tropospheric aerosol in the urban area of Napels. Dissertation. University of Napels, Italy.

Prather M., Gauss M., Berntsen T., Isaksen I., Sundet J., Bey I., Brasseur G., Dentener F., Derwent R., Stevenson D., Grenfell L., Hauglustaine D., Horowitz L., Jacob D., Mickley L., Lawrence M., von Kuhlmann R., Muller J.-F., Pitari G., Rogers H., Johnson M., Pyle J., Law K., Van Weele M., and Wild O. (2003), Fresh air in the 21st century?. *Geophys. Res. Letts.*, 30, EID 1100, DOI 10.1029/2002GL016285.

Rodhe H. (1999), Human impact on the atmospheric sulfur balance, *Tellus Ser. B*, 51, 110–122.

Shaw G.E. (1995), The Arctic haze phenomenon, *Bull. Am. Meteorol. Soc.*, 76, 2403–2413.

Stohl A. and Seibert P. (1998), Accuracy of trajectories as determined from the conservation of meteorological tracers, *Q. J. R. Meteorol. Soc.*, 125, 1465–1484.

Stohl A., Hittenberger M., and Wotawa G. (1998), Validation of the Lagrangian particle dispersion model FLEXPART against large scale tracer experiment data, *Atmos. Environ.*, 32, 4245–4264.

Veselovskii I., Kolgotin A., Griaznov V., Müller D., Wandinger U., and Whiteman D.N. (2002), Inversion with regularization for the retrieval of tropospheric aerosol parameters from multiwavelength lidar sounding, *Appl. Opt.*, 41, 3685–3699.

Veselovskii I., Kolgotin A., Müller D., and Whiteman D.N. (2005), Information content of multiwavelength lidar data with respect to microphysical particle properties derived from eigenvalue analysis, *Appl. Opt.*, 44, 5292–5303.

Wandinger U. and Ansmann A. (2002), Experimental determination of the lidar overlap profile with Raman lidar, *Appl. Opt.*, 41, 511–514.

Chapter 13
An Innovative Approach to Optical Measurement of Atmospheric Aerosols—Determination of the Size and the Complex Refractive Index of Single Aerosol Particles

Wladyslaw W. Szymanski[1], Artur Golczewski[1], Attila Nagy[2], Peter Gál[2], and Aladar Czitrovszky[2]

Abstract As a result of the intrinsic nature of elastic light scattering aerosol particles, the non-monotonic size dependence of the scattered light intensity influences the function of most single optical particle counters and spectrometers. In order to tackle the problem of the varying response of single particle spectrometers when refractive indices of aerosol change, we developed a system utilizing two laser illumination sources with different wavelengths (533 nm and 685 nm) and four detectors collecting the forward and backward scattered light from both illuminating beams. The new method aims to determine the size and refractive index of particles typically occurring in the atmosphere. We successfully tested this new method numerically for its capability to simultaneously determine particle size in the range from 0.1 to 10 μm, the real part of the refractive index spanning from 1.1 to 2, and the imaginary part of refractive index between 0 and 1. The first experimental results obtained with the prototype of the spectrometer verify the capability of the technique for accurate size measurement and real-time differentiation between non-absorbing and absorbing aerosol particles.

Keywords: Atmospheric aerosols, optical measurement, particle size, refractive index

13.1 Introduction

Aerosol particles affect our environment at the local, regional, and global level. They are now recognized as a significant environmental problem both in terms of health and the transfer of radiation through the atmosphere, thus also affecting the Earth's thermal budget. Atmospheric aerosols vary widely in space and time and

[1] *Faculty of Physics, University of Vienna, Boltzmanngasse 5, A-1090 Vienna, Austria*

[2] *Research Institute for Solid State Physics and Optics, Dept. of Laser Applications, Hungarian Academy of Science, H-1525 Budapest, P.O. Box 49, Hungary*

Y.J. Kim and U. Platt (eds.), *Advanced Environmental Monitoring*, 167–178.

thus continuous monitoring of their physical and optical properties is crucial. Single optical particle spectrometry based on elastic light scattering provides a convenient means of real-time measurement of atmospheric particles above about 100 nm. However, as a result of the intrinsic nature of elastic light scattering by aerosol particles, the non-monotonic size dependence of the scattered light intensity and its strong variation with the refractive index of particles influence the function of most single optical particle counters and spectrometers (Szymanski and Liu 1986; Hanusch and Jaenicke 1993; Czitrovszky and Jani 1993; McMurry et al. 1996; Kerker 1997; Renliang 2000; Chang et al. 2003). Therefore, the major disadvantage of this widely used type of instrumentation is the very high likelihood of variability in the size measurement with changing optical properties of the particles in question. Therefore, through using this instrumentation, particle sizing information substantially different from the actual values may be obtained, resulting as a consequence of an erroneous assessment of particle-related environmental impact.

13.2 Status of the Optical Particle Spectrometry

Optical particle counters and spectrometers measure light scattered from a single particle as it passes through an illuminated sensing volume of the instrument. The illumination source is typically monochromatic, either gas laser or semiconductor laser. In some cases, instruments with an incandescent light source are still being used, providing the advantage of smoother response characteristics (Cooke and Kerker 1975; Liu et al. 1985). The scattered light intensity is utilized as a measure of the particle size. In an ideal instrument, a monotonic relationship between the particle size and the scattered light intensity independent of the particle material would allow a unique particle sizing. However, in reality this is usually not the case (Barnard and Harrison 1988; Liu and Daum 2000). The instruments' response, given by the amount of light scattered into a given solid angle vs. particle size, is essentially a non-monotonic function of particle size. It depends strongly on the angular range into which the scattered light is collected and in general it also varies with the refractive index of the particulate material. The response can be calculated by means of the Mie theory of light scattering (Bohren and Hufman 1983).

Figure 13.1 shows the oscillatory character of response curves calculated by means of the Mie theory. The horizontal line indicates all particles with different sizes and refractive indices scattering exactly the same amount of light, hence yielding exactly the same response. It means that these particles are indistinguishable by such a spectrometer due to its multi-valued response. An integration of those intensity functions over an angular range, which corresponds to scattered light collected in a given optical particle spectrometer, results to a certain degree in flattening of oscillations, yielding sometimes a nearly single-valued response curves vs. particle diameter (Barnard and Harrison 1988; Szymanski et al. 2000). Such behavior can be achieved by assuming particles with a specific refractive

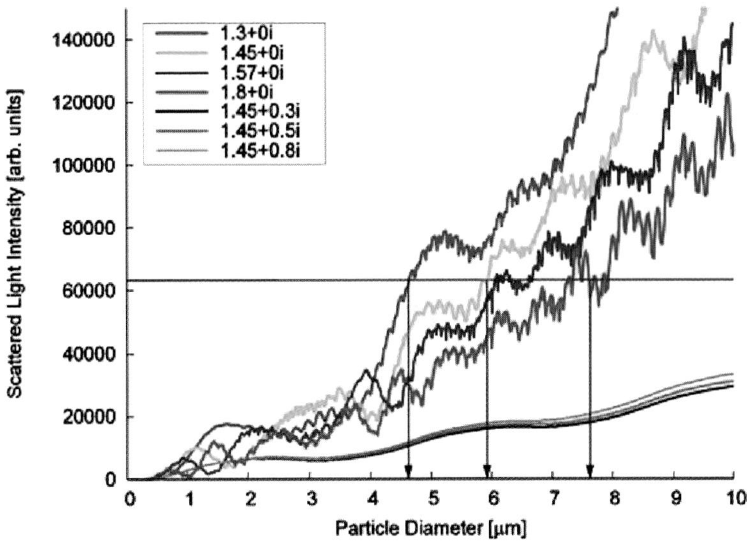

Fig. 13.1 Response curves for different complex refractive indices of particles in the size range 0.1–10 μm. The scattered light is collected in the solid angle enclosing $\theta = 10°–30°$ with respect to the laser beam propagation direction (0°)

index. In practice, however, the environmental particles and hence their complex refractive indices are frequently unknown, which may result in incorrect particle sizing with errors up to an order of magnitude as can be easily deducted from data shown in Fig. 13.1.

Although these effects were addressed in previous studies (Cooke and Kerker 1975; Hinds and Kraske 1986; Liu et al. 1983; Szymanski and Liu 1986; Hering and McMurry 1991; Pinnick et al. 2000), it should be pointed out that the deficiencies of the single optical particle spectrometry technique have not so far been solved in a satisfactory manner. Regardless of the actual optical design, in a conventional optical particle spectrometer (OPS) light scattered from a single particle is collected in one given angular range. This scattered light is detected and converted into a voltage pulse. Its magnitude relates to particle size but varies basically in the same way as does the scattered light shown in Fig. 13.1. It means that data inversion in such a case is basically an "ill-posed" problem, because the measurable quantity depends in a nonlinear way on three parameters—size, and the real and imaginary parts of the refractive index.

To quantify the possible sizing errors which may occur when measuring environmental particles by means of conventional optical particle spectrometers, sizing performance with scattered light collection in the angular range from 10 to 30° was simulated using a stochastic method. Assuming that the OPS in question was calibrated with non-absorbing particles having a complex refractive index of $m = 1.45 + 0i$, size measurement simulations were carried out in the size range

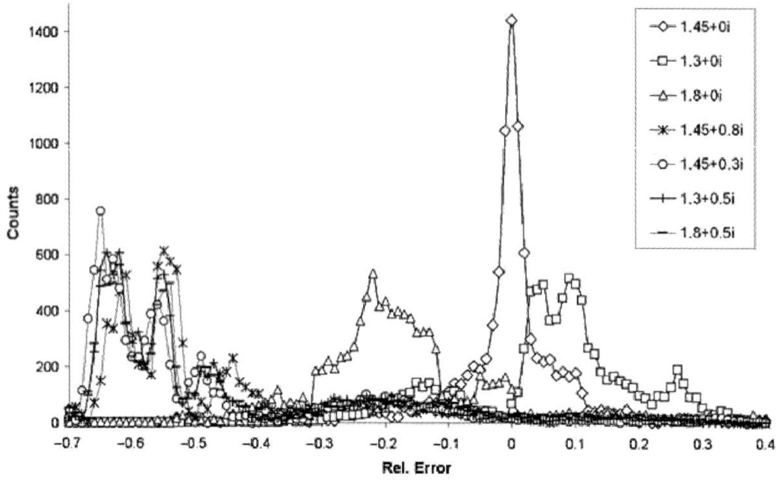

Fig. 13.2 Results of sizing simulations in particle range from 0.1 to 10 μm with an optical particle spectrometer collecting scattered light in the angular range from 10° to 30°

from 0.1 to 10 μm for particles having different complex refractive indices indicated in Fig. 13.2.

The quality of sizing, given here by the ratio of the difference between the numerically retrieved and actual size to the actual particle size, is called here the relative error as indicated in Fig. 13.2. Results of these numerically simulated measurements for randomly selected particle sizes and complex refractive indices are shown in Fig. 13.2. It is evident that depending on the refractive indices, either under-sizing or over-sizing of measured particles takes place except for particles with the same refractive index as calibration particles. It can be seen, that for all absorbing particles a systematic under-sizing occurs, which is nearly independent of the absorption of a particle given by the imaginary part of the refractive index.

An interesting approach to the particle sizing problem based on multi-angle differential light scattering was explored previously (Wyatt et al. 1988). Later some progress towards more complete optical particle analysis was achieved; however, a real breakthrough was not accomplished (Dick et al. 1994, 1996).

13.3 The Dual Wavelength Optical Particle Spectrometer

In order to overcome the problem of the varying responses of single optical particle spectrometers when refractive indices of aerosol change, we developed a system utilizing two illumination laser sources with different wavelengths of 533 nm and

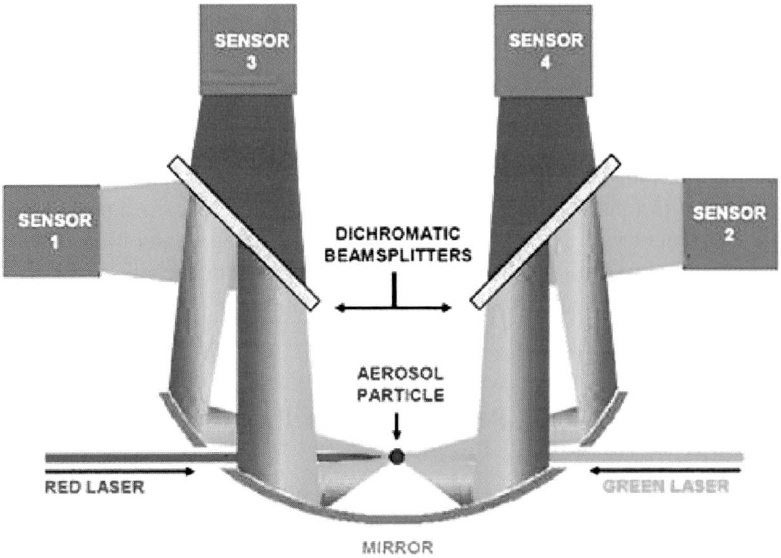

Fig. 13.3 The principle of the DWOPS. Utilization of two lasers and an appropriate scattering geometry ensures four independent signals from each aerosol particle. The aerosol flow is perpendicular to the plane of the diagram

685 nm. Four sensors collect forwards and backwards scattered light from each detected particle. We named this device the dual wavelength optical particle spectrometer (DWOPS). The concept of this technique was introduced earlier (Szymanski et al. 2002). Here we show the results of computational examination of its measuring performance, as well as the first experimental verification of the prototype. The schematic diagram of the DWOPS is shown in Fig. 13.3.

An aerosol particle passing the instrument's sensing volume scatters simultaneously both green and red light, which then is redirected by a parabolic mirror towards dichromatic beam splitters, which transmit the red and reflect the green light. In this way, separated monochromatic scattered light information is then registered by four sensors. This method delivers four independent scattered light fluxes, yielding four pulses from each detected aerosol particle.

Three parameters of a particle—its size, and the real and imaginary parts of the refractive index allow to exactly determine the amount of light scattered into a given angular range and hence the resulting pulse height. Providing that the particles in question are spherical, the described instrument directly yields the information about particle size and its complex refractive index based on the signal evaluation procedure which will be discussed here. The assumption of the sphericity of measured particles need not always be the case; however, experimental evidence shows that a substantial portion of atmospheric aerosols, especially in the submicrometer size range, are spherical, or nearly spherical (McMurry et al. 1996).

Probably the most interesting feature of this new instrumental approach is the detail that for spherical particles with different refractive indices, the DWOPS does not require any recalibration vs. some particle standards, as might be needed with currently existing single optical particle spectrometers. Common optical spectrometers are typically calibrated by means of polystyrene latex (PSL) particles with a refractive index $m = 1.59–0i$. Accurate size measurements of particles with refractive indices different from the PSL refractive index would require a new calibration of the instrument with apt particles (Liu et al. 1983; Hering and McMurry 1991). In the case of DWOPS, each single particle measurement results in information about the particle size and its complex refractive index.

13.4 Modeling of the Performance of the New Spectrometer

The strength of this method is its principal simplicity, as it is shown in Fig. 13.3. The measuring system of the DWOPS contains two colinear, monochromatic, unpolarized, focused laser light sources with different wavelengths ($\lambda 1$=685 nm and $\lambda 2$=532 nm) illuminating aerosol particles from opposite sides which are passing through the sensing volume. Light scattered from each single particle is then collected over angular ranges (10°–30° and 150°–170°) forward and backward with respect to the incoming beam of each light source. This yields a quadruple set of independent scattered light pulses from each detected single particle, which is in contrast to commonly operating optical particle spectrometers which provide a single pulse from each measured particle.

The information contained in these four scattered light pulses allows us, using our evaluation method described below, to determine for each measured particle the size and complex refractive index. For modeling of the DWOPS performance, a Monte–Carlo method was used. Each randomly generated particle is described by its size and complex refractive index $m = \text{Re}(m) + \text{Im}(m)$. Data related to such a particle is converted into four digitized responses (RF, RB, GF, GB). This set of four numbers can be superimposed with noise simulating an actual measuring situation. This quadruple of values is then compared with a precomputed table calculated using the Mie theory of light scattering (Brown 1996; Bohren and Hufman 1983). The table consists of four columns, in which each row was calculated for a particle with given size and a real and an imaginary part of the refractive index. The values in the precalculated table correspond to DWOPS responses based on the scattering geometry of each sensor. Fig. 13.4 shows the main steps of the simulation procedure.

Tables with various sizes were examined in order to optimize the evaluation performance from the point of view of system accuracy and computing time. Obviously, with the increasing number of rows (i.e. decreasing step size) in the table the numerical resolution of size and refractive index would increase. However, due to the high complexity of the morphology of light scattering curves vs. particle size and vs. complex refractive index, the numerical increase of resolution might

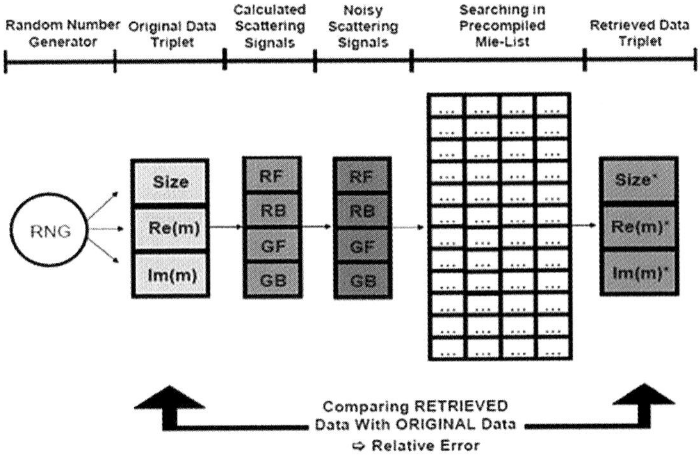

Fig. 13.4 Schematic presentation of the simulation procedure. Scattered light pulses are defined here as follows: RF = red laser, forward scattering; RB = red laser, backward scattering; GF = green laser, forward scattering; GB = green laser, backward scattering. The "*" sign indicates the retrieved values, usually different from the input values

yield single-valued results for a given quadruple of measured data (Nagy et al. 2005). In this contribution, the numerically investigated range of parameters covers the size of particles from 0.1 μm to 10 μm. The real part of examined refractive indices is between 1.1 and 2.0 and the imaginary part between 0 and 1, which corresponds to properties of most environmental aerosols. Randomly generated particles with a certain, known size and refractive index are used to obtain light scattering response values based on the DWOPS geometry. These values are then employed to recover the original particle data based on best fitting them to precalculated ideal responses.

The Monte–Carlo procedure was used to build up frequency diagrams of relative errors between the "measured" and ideal value for each particle. In the above specified size and complex refractive index space random numbers with uniform distribution were generated, yielding a particle ensemble with certain sizes and refractive indices. In this way, particles were numerically obtained and used for the exploration of capabilities of the DWOPS. The dimension of the evaluation table has a considerable impact on the accuracy and the speed of the data recovery. An evaluation table with 50 equidistant logarithmic steps in the above given size range from 0.1 to 10 μm was used in this study. The number of spacing intervals in the real part of the refractive index range was 19 with linear spacing (0.05 step size), and the number of sections in the imaginary part of the refractive index range was 21 with linear spacing (0.05 step size). Results of modelling the DWOPS performance using the Monte–Carlo procedure are shown in Fig. 13.5.

Depending on the investigated particle size range, an optical system of a spectrometer may sometimes provide multi-valued responses. For that reason the

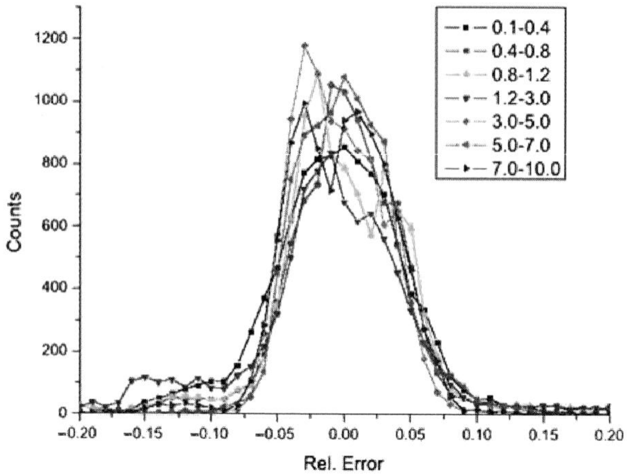

Fig. 13.5 Sizing accuracy for different particle size ranges (legend shows size ranges in μm). The noise level superimposed on the signal amounts to 5%

investigated particle size range interval was divided into subintervals covering following size ranges: 0.1–0.4 μm, 0.4–0.8 μm, 0.8–1.2 μm, 1.2–3 μm, 3–5 μm, 5–7 μm and 7–10 μm. It is evident that the instrument's performance is similar in all sizing intervals. Nearly bell-shaped distribution of relative error vs. counts proves the sizing feasibility of the optical system. Certain secondary structures depending on the size range can be observed; however, a pronounced maximum is unmistakable, yielding for the examined particles a unique sizing result with the relative sizing error of about ±5% (FWHM). With the increasing signal noise levels the width of these distributions increase, however, as long as the noise levels do not exceed about 15%, a unique signal evaluation can be obtained.

Similarly to the modeling of sizing errors we have investigated relative errors for the determination of particle refractive indices. The relative error for the assessment of the real part of the refractive index is within about ±10% of the actual value (FWHM). The accuracy of the determination of the imaginary part of the refractive index is somewhat lower. Here the relative error spreads within about ±15% (FWHM). These findings lead to the conclusion that the evaluation table with the above specified dimensions is suitable for accurate measurements within the targeted size range.

13.5 Measuring Performance of the New Spectrometer

The decisive proof of the feasibility of any prototype of a new measuring system is the experimental verification of the performance. For this study, well-defined laboratory aerosols with known sizes and complex refractive indices were used and are summarized in Table 13.1. The experimental arrangement is schematically shown in Fig. 13.6.

Table 13.1 Refractive indices of particle materials used in this study

Material	Complex refractive index
Polystyrene latex (PSL)	1.59 + 0.0*i*
Di-ethyl-hexyl-sebaccate (DEHS)	1.45 + 0.0*i*
Black ink (Staedtler Mars Ink No. 745)	1.70 + 0.32*i* (Hogan 1985)

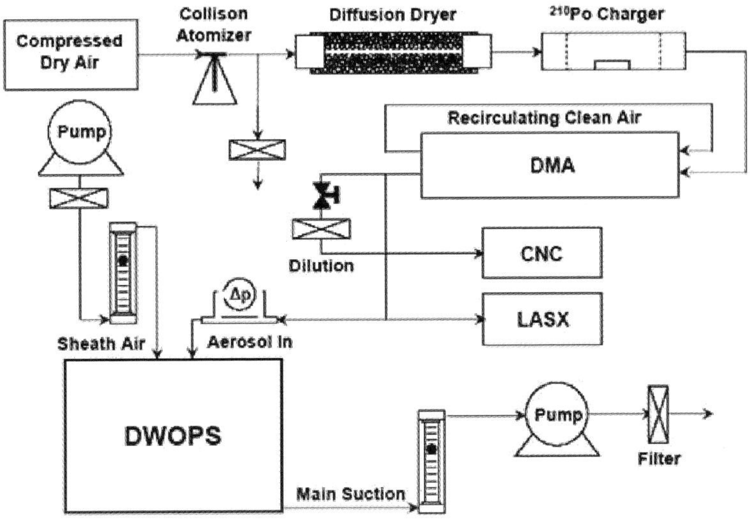

Fig. 13.6 Experimental arrangement for experimental performance verification of the DWOPS

All particles were generated from a solution or suspension using pneumatic atomization. PSL particles were obtained from diluted aqueous suspensions of certified particle size standards (Duke Scientific Corp.). DEHS particles were generated from an isopropyl alcohol-DEHS 0.5% solution. The absorbing, carbon-like particles were generated from a black ink 50% diluted in ultrapure distilled water.

The atomized (collison atomizer) and dried (diffusion dryer) particles were brought to charge equilibrium in a bipolar ion atmosphere obtained with a weak radioactive source (^{210}Po charger). Following the charging process, particles enter the differential mobility analyzer (DMA), where they move on trajectories determined by the clean airflow and the electric field resulting from the applied voltage (Liu and Pui 1975). The relationship between the voltage V and monodisperse particle size (D_{DMA}) extracted from the DMA is given by: D_{DMA} = (constant) $ne\mathrm{C}(D_{\mathrm{DMA}})V$.

$\mathrm{C}(D_{\mathrm{DMA}})$ is the particle slip correction, e is the elementary charge, n is the number of charges carried by a particle. An optimization of operating conditions of such a system yields test aerosols where the number of particles carrying $n > 1$ elementary charges is negligible.

Thus the obtained DEHS and black ink particles have exactly known sizes (D_{DMA}) with the width of the distribution of the order of 5% or better.

The atomization process of PSL suspensions results in test aerosols containing not only the required calibration particles with defined size but also those residual precursor particles. They are much smaller in size, typically below 100 nm, but due to their high numbers they may cause measurable pulses influencing the determination of the instrument's performance. For that reason these PSL particles were also size classified by means of the DMA. This process removes the residual particles yielding nearly perfectly monodispersed challenge aerosol of exactly known particle size. Condensational nuclei counter (CNC, TSI Inc.) and laser aerosol spectrometer (LASX, Particle Measuring Systems) were used to control the number, concentration and size of test aerosols generated with the atomizer-dryer-differential mobility analyzer system and to provide reproducible experimental conditions.

Analogously to the numerical experiments described above, actual particles with known properties were generated and used to study the capability of the DWOPS to optically determine the size of the particles in question and their complex refractive index. Results of these experiments are summarized in Table 13.2 and show clearly the exceptional sizing capability of the DWOPS along with the possibility to distinguish between absorbing and non-absorbing particles. The agreement between the nominal complex refractive indices given in Table 13.1 and refractive indices obtained from measurements with DWOPS is very credible.

Such measurements are not possible with optical particle spectrometers currently available. It is also evident that there is no need to recalibrate the instrument when measuring particles with various refractive indices, as is usual with currently existing single optical particle spectrometers. Providing that the particles in question are spherical, the described instrument directly yields information about particle size and its complex refractive index based on the signal evaluation procedure present. This experimental data, although limited by the current optical arrangement toward smaller particles by the laser power and toward larger particles by the DMA capability, verify the postulated performance of the DWOPS system (Szymanski et al. 2002). Measurements performed earlier with conventional spectrometers using aerosol particles with refractive indices corresponding to those used in this study showed that erroneous particle sizing up to 100% due to changes in complex refractive index may occur (Garvey and Pinnick 1983; Liu et al. 1983; Szymanski and Liu 1986).

Table 13.2 Values for particle size and its complex refractive index *m* obtained with the DWOPS

Material	D_{DMA} [μm]	D_{DWOPS} [μm]	Re (*m*)	Im (*m*)
DEHS	0.589	0.588 ± 0.01	1.55 ± 0.20	0.04 ± 0.04
DEHS	0.633	0.643 ± 0.04	1.45 ± 0.09	0.02 ± 0.03
DEHS	0.730	0.731 ± 0.09	1.46 ± 0.02	0 ± 0.04
DEHS	0.846	0.834 ± 0.03	1.46 ± 0.02	0 ± 0.04
PSL	1.000	0.949 ± 0.01	1.60 ± 0.02	0 ± 0.04
Black ink	0.750	0.728 ± 0.09	1.62 ± 0.08	0.34 ± 0.1
Black ink	0.845	0.811 ± 0.08	1.61 ± 0.20	0.32 ± 0.2

13.6 Summary

Single optical particle spectrometry, although very suitable for measurement of environmental particles in the size range from about 100 nm up to tens of micrometers, appears to have been stagnating in recent years. Modifications with regard to the source of illumination, decreasing size of the instrument, or comfort of handling were commenced; however, a factual innovation in terms of novel design yielding information beyond particle size distribution seems to be missing. The DWOPS system described here is a new approach to intensity-based optical particle spectrometry. The instrument's performance was numerically investigated in the size range from 0.1 to 10 μm and for a range of complex refractive indices occurring in environmental aerosols. First experimental results, although within a limited size range, verify the potential of this technique. It shows clear advantages over the traditional single optical particle measuring method. The problem of the variability of calibration curves with changing refractive indices and the issue of a non-monotonic response of the instrument with regard to the particle size—the well-known limitation of conventional systems—does not arise for the DWOPS, because the combined response of all four sensors and the described signal evaluation method determine directly the size, along with the complex refractive index of each detected particle. Numerical modeling and experimental data are very promising; however, further experimental studies with an optimized optical system, allowing measurements over the entire size range, also with nonhomogeneous spherical, as well as nonspherical particles, are planned and will be reported later. Making reasonable assumptions regarding the range of particle sizes and the range of the real and imaginary parts of refractive indices, it was numerically explored and experimentally verified that a unique measurement of particle size along with an unambiguous determination of the complex refractive indices of measured particles in real time is possible with the DWOPS system. This approach may pave the way for new applications for single optical particle measurement in atmospheric science and environmental monitoring.

Acknowledgements The support of this work was provided by Austrian Science Foundation (FWF), Proj. P15619 to W.W.S., by National Research and Development Program, Proj. NRDP3:3/005/2001, and by the ÖAD-WTZ, Proj. Nr. A13/2003.

References

Barnard J.C. and Harrison L.C. (1988), Monotonic responses from monochromatic optical particle counters, *Appl. Opt.*, 27,584–592.

Bohren C.F. and Hufman D.R. (1983), *Absorption and Scattering of Light by Small Particles.*, Wiley Interscience Publication, New York.

Chang H., Okuyama K., and Szymanski W.W. (2003), Experimental evaluation of optical properties of porous silica/carbon composite particles, *Aerosol Sci. Technol.*, 37, 735–751.

Cooke D.D. and Kerker M. (1975), Response calculations for light scattering aerosol particle counters, *Appl. Opt.*, 14, 734 –745.

Czitrovsky A. and Jani P. (1993), New design for a light scattering airborne particle counter and its applications, *Opt. Eng.*, 32, 2557–2582.

Dick W.D., Sachweh B.A., and McMurry P.H. (1996), Distinction of coal dust particles from liquid droplets by variations in azimuthal light scattering, *Appl. Occup. Environ. Hyg.*, 11(7), 637–646.

Dick W.D., McMurry P.H., and Bottiger J.R. (1994), Size and composition dependent response of the DAWN-A multi-angle single particle detector, *Aerosol Sci. Technol.*, 20, 345–355.

Garvey D.M. and Pinnick R.G. (1983), Response characteristics of the particle measuring systems active scattering aerosol spectrometer probe, *Aerosol Sci. Technol.*, 2, 477–488.

Hanusch T. and Jaenicke R. (1993), Simulation of optical particle counter FSSP-100. Consequences for size distribution measurements, *J. Aerosol Sci.*, 23, 112–120.

Hering S. and McMurry P.H. (1991), Response of a PMS LAS-X laser optical particle counter to monodisperse atmospheric aerosols, *Atmos. Environ.*, 25A, 463–461.

Hinds W.C. and Kraske G. (1986), Performance of PMS model LAS-X optical particle counter, *J. Aerosol Sci.*, 17, 67–72.

Hogan A., Ahmed N., Black J., and Barnard S. (1985), Some physical properties of black aerosol, *J. Aerosol Sci.*, 16, 391–396.

Kerker M. (1997), Light scattering instrumentation for aerosol studies and historical overview, *Aerosol Sci. Technol.*, 27, 522–535.

Liu B.Y.H. and Pui D.Y.H. (1975), On the performance of the electrical aerosol analyzer, *J. Aerosol Sci.*, 6, 249–264.

Liu B.Y.H., Szymanski W.W., and Pui D.Y.H. (1983), Response of laser optical particle counter to transparent and light absorbing particles, *ASHRAE Trans.*, 92, 518–527.

Liu Y. and Daum P.H. (2000), The effect of refractive index on size distributions and light scattering coefficients derived from optical particle counters, *J. Aerosol Sci.*, 31, 945–956.

Liu B.Y.H., Szymanski W.W., and Ahn K.H. (1985), Aerosol size distribution measurement by laser and white light optical particle counters, *J. Environ. Sci.*, 28, 19–25.

McMurry P.H., Zhang X. and Lee C.-T. (1996), Issues in aerosol measurement for optics assessment, *J. Geophys. Res.*, 101, 19189–19195.

Nagy A., Szymanski W.W., Golczewski A., Gál P., and Czitrovszky A. (2005), Effects of the evaluation table dimensions on the DWOPS sizing accuracy, *In: Proceedings of the European Aerosol Conference EAC2005* Ghent, Belgium.

Pinnick R.G., Pendleton J.D., and Videen G. (2000), Characteristics of the particle measuring systems active scattering aerosol spectrometer probes, *Aerosol Sci. Technol.*, 33, 334–352.

Renliang Xu. (2000), Particle characterization: Light scattering methods, 13, 432 p, Kluwer, Dordrecht.

Szymanski W.W. and Liu B.Y.H. (1986), On the sizing accuracy of laser optical particle counters, *Part Charact.*, 3, 1–8.

Szymanski W.W., Nagy A., Czitrovszky A., and Jani P. (2002), A new method for the simultaneous measurement of aerosol particle size, complex refractive index and particle density, *Meas. Sci. Tech.*, 13, 303–307.

Szymanski W.W., Ciach T., Podgorski A., and Gradon L. (2000), Optimized response characteristics of an optical particle spectrometer for size measurement of aerosols, *J. Quant. Spectrosc. Radiat. Transf.*, 64, 75–85.

Wyatt P.J. (1998), Submicrometer particle sizing by multi-angle light scattering following fractionation, *J. Colloid Interface Sci.*, 197, 9–20.

Chapter 14
Remote Sensing of Aerosols by Sunphotometer and Lidar Techniques

Anna M. Tafuro, F. De Tomasi, and Maria R. Perrone

Abstract Active and passive remote sensing devices such as lidars and sunphotometers, respectively, are peculiar tools to follow the spatial and temporal evolution of aerosol loads and get complementary data to properly characterize aerosol optical and microphysical properties. A XeF-based Raman lidar is routinely used at the physics department of Lecce's University (40° 20′ N, 18° 6′ E), to monitor aerosol vertical distributions and characterize aerosol optical properties by the vertical profiles of the backscatter and extinction coefficient, lidar ratio, and depolarization ratio. In addition, a sun/sky radiometer operating within AERONET is used to supplement lidar measurements and better infer aerosol types and properties by columnar values of the particle size distribution, the real and imaginary refractive index, the single scattering albedo and the Angstrom exponent (Å). The main objective of this paper is to provide some results on the spatial and temporal evolution of the aerosol properties over south-east Italy, in the central-east Mediterranean basin, by using lidar and sunphotometer measurements. Specifically, results on the characterization of the aerosol load from July 18 to July 21, 2005 are reported and particular attention is devoted to the Sahara dust outbreak that has occurred over south-east Italy on July 18 and 19, 2005.

Keywords: Aerosols, lidar, remote sensing, sunphotometer

14.1 Introduction

Several studies have shown that both natural and anthropogenic aerosols have important effects on the climate of the earth-atmosphere system (Haywood and Shine 1997). Aerosol particles affect the climate directly by scattering and absorbing solar radiation and indirectly by modifying cloud microphysical properties. But, as a consequence of their high spatial and temporal variability, these effects can

---

CNISM, Dipartimento di Fisica, Università di Lecce, via per Arnesano, Lecce, Italy

Y.J. Kim and U. Platt (eds.), *Advanced Environmental Monitoring*, 179–189.

be strongly regional (Nakajima et al. 2003) and the current uncertainty in the quantitative assessment of the Earth's radiative balance, is a direct consequence of the variable nature of aerosols on regional and seasonal scales. A multiple-measurement approach is currently used to assess aerosol impacts on global climate. Specifically, long-term continuous observations from satellites, networks of ground-based instruments and dedicated field experiments in clean and polluted environments are used to feed global aerosol and climate models (Kaufman et al. 2002). The aerosol robotic network (AERONET), which is an international network coordinated by the NASA Goddard Space Flight Center, has been established to assess aerosol optical properties from ground-based sun/sky radiometer measurements (Holben et al. 1998). The need of assessing a vertical resolved aerosol climatology has led to the establishment of the European aerosol research lidar network (EARLINET) that at present consists of a network of 22 lidar stations spread over whole Europe (Bösenberg et al. 2003).

In this paper, an elastic-Raman lidar operating within EARLINET and a sun/sky photometer operating within AERONET are used to characterize the aerosol load from July 18 to July 21, 2005. Particular attention is devoted to the Saharan dust event occurred in the south-east of Italy on July 18–19. A main paper's objective is to show that complementary measurements are required to infer optical and microphysical properties of aerosols of different type. Measurements have been performed at the physics department of Lecce's University that is located on a flat peninsula of south-east Italy, nearly 50 km away from large industrial areas. The site is well suited to contribute to the aerosol characterization of the east Mediterranean basin, which represents a unique area in terms of suspended particulate matter. Aerosols from different sources converge to the Mediterranean basin: urban/industrial aerosols and seasonal biomass burning from Europe (Zerefos et al. 2000; Lelieveld et al. 2002), maritime and long-range transported polluted air masses from the Atlantic Ocean, mineral dust from North Africa (Gobbi et al. 2000; Tafuro et al. 2006), and sea spray from the Mediterranean sea itself (Perrone et al. 2005). A brief description of the XeF lidar and of the sunphotometer is given in Section 14.2. The advection pattern characterization over south-east Italy from July 18 to July 21, 2005 is provided in Section 14.3 by using 7-day analytical backtrajectories and satellite MODIS images. Lidar and sunphotometer measurements are analyzed in Sections 14.4 and 14.5, respectively. Summary and conclusion are reported in Section 14.6.

14.2 Experimental Facilities

An elastic-Raman lidar employing a XeF excimer laser (Lambda Physik LPX 210 I) as radiation source has been used in this study for the aerosol vertical monitoring. The lidar system is located at the physics department of Lecce's University (40° 20′ N, 18° 6′ E), on a flat area of the Salento peninsula in Italy. The XeF laser emits light pulses of 30 ns duration at 351 nm, with a maximum energy and repetition rate of 250 mJ and 100 Hz, respectively. Collection of the backscatter radiation is obtained by a Newtonian telescope, whose primary mirror has 30 cm-diameter and 120 cm-focal length. The lidar system has been designed to measure as a function of the altitude, the aerosol backscatter coefficient, the extinction coefficient, the extinction-to-backscatter ratio, known as lidar ratio, the depolarization ratio, and the water vapor mixing ratio. More details on the experimental apparatus are reported in De Tomasi and Perrone (2003).

A sun/sky radiometer made by CIMEL (France) and operating within AERONET has been used to derive columnar values of the aerosol size distribution, the real and imaginary parts of the complex refractive index, the single scattering albedo, and the aerosol optical thickness. The CIMEL sunphotometer (CE318-1) is an automatic sun/sky radiometer, with a 1.2° field of view and two detectors, to measure direct Sun and sky radiances at eight spectral channels: 340, 380, 440, 500, 670, 870, 940, and 1020 nm. Holben et al. (1998, 2001) have provided a detailed description of the instrument, the data acquisition system and the data inversion procedure. The instrument is located on the roof of the physics department of Lecce's University.

14.3 Characterization of Advection Patterns by Backtrajectories and Satellite Images

The 7-day analytical backtrajectories and MODIS satellite images have been used to infer the source regions of the air masses advected over south-east Italy from July 18 to July 21, 2005. The 7-days backtrajectories, for seven distinct arrival height levels (950, 850, 700, 500, 300, 250, and 200 hPa) and for two arrival times (00:00 and 12:00 UTC), are provided by NASA for each AERONET site (http://croc.gsfc.nasa.gov/aeronet/index.html). Figure 14.1 shows the 7-day analytical backtrajectories of the air masses reaching the monitoring site on (a) July 18 at 12:00 UTC, (c) July 19 at 00:00 UTC, and (e) July 21 at 00:00 UTC, respectively. Figure 14.1a indicates that on July 18, at 12:00 UTC, the 950 and 850 hPa air masses have been advected to Lecce from Europe, while the 700 hPa air mass has been advected from the Atlantic Ocean. Figure 14.1b shows as function of time, the change of height of the 950, 850, and 700 hPa air masses for the 12:00 UTC arrival time.

Figure 14.1c, which displays the July 19 backtrajectories, indicates that north-west Africa was the source region of the 700 hPa air mass advected to Lecce on July 19, while Fig. 14.1d shows as function of time, the change of height of the 950, 850, and 700 hPa air masses, for the 00:00 UTC arrival time, and we observe that the air mass reaching Lecce on July 19 at about 700 hPa, was at ~900 hPa over north-west Africa. Hence, this air mass is expected to be affected by Sahara dust particles. Last conclusion is supported by satellite images.

Figures 14.2a and 14.2b show two MODIS images of the central Mediterranean provided by the Aqua satellite (http://rapidfire.sci.gsfc.nasa.gov/) for July 18 at 12:00 UTC and for July 19 at 12:40 UTC, respectively. It is worth observing from Fig. 14.2a that a dust plume, moving from Sahara, was present over the south-west Mediterranean on July 18, at midday. Then, according to the backtrajectories of

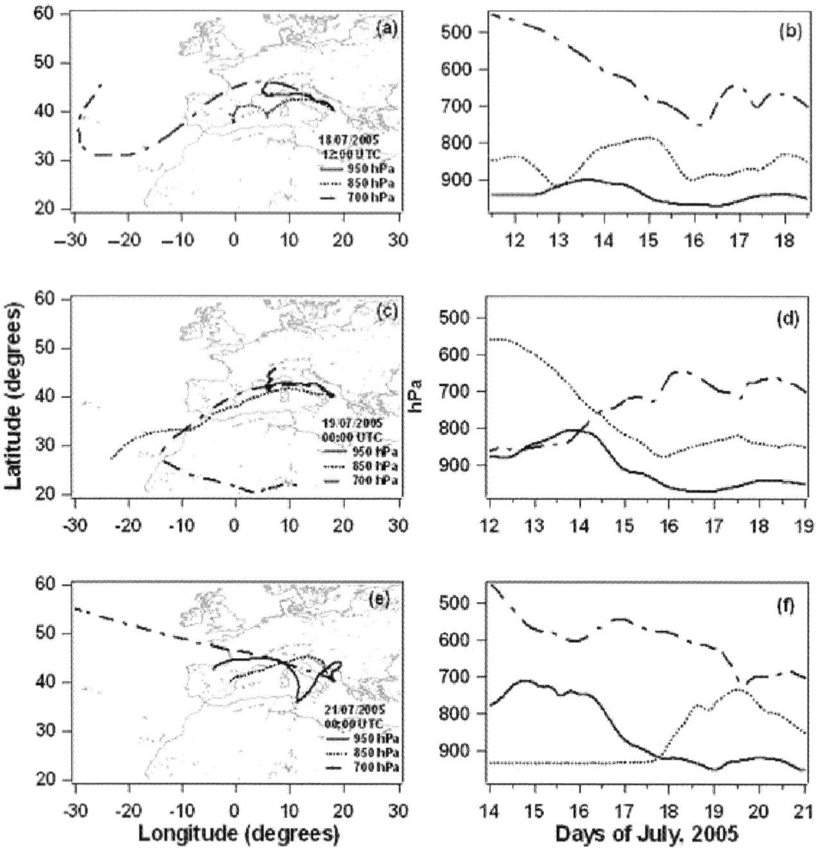

Fig. 14.1 7-days analytical back trajectories for the air masses reaching Lecce on (**a**) July 18 at 12:00 UTC, (**c**) July 19 at 00:00 UTC, and (**e**) July 21 at 00:00 UTC. The change of the height of the 950, 850 and 700 hPa air masses as function of time is also shown for the arrival time of (**b**) 12:00 UTC of July, 18, (**d**) 00:00 UTC of July 19, (f) 00:00 UTC of July 21

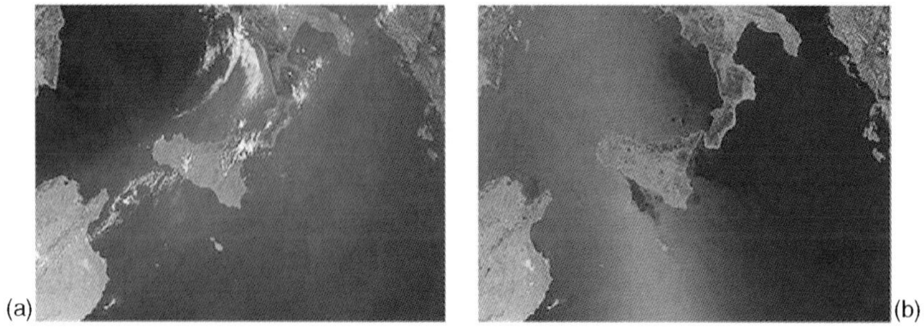

Fig. 14.2 MODIS image of the Mediterranean from Aqua satellite on (**a**) 18 and (**b**) 19 July, 2005 at 12:00 and 12:40 UTC, respectively

Fig. 14.1d, Fig. 14.2b indicates that the dust plume was over south-east Italy on July 19 at 12:40 UTC. Finally, it is also worth noting from Fig. 14.1e that north-west Europe was the source region of the 950 and 850 hPa air masses that have reached Lecce on July 21, while the Atlantic Ocean was the source region of the 700 hPa air mass.

In conclusion, the analysis of backtrajectories and satellite images indicates that aerosols of different type and from different source regions have been advected to Lecce from July 18 to July 21, 2005.

14.4 Aerosol Characterization by Lidar Measurements

Lidar techniques have the unique feature of retrieving vertical resolved aerosol properties. This is particularly useful because passive remote sensing techniques can give only column-averaged aerosol quantities. Aerosol characteristics have been monitored almost continuously on July 18 and July 21. The daily evolution of the backscatter coefficient retrieved from lidar measurements performed from 8:20 to 19:20 UTC of July 18, is shown on Fig. 14.3. The backscatter coefficient is retrieved by the Klett inversion technique (Fernald 1984), that requires a hypothesis on the extinction-to-backscatter coefficient ratio. The lidar ratio depends on size distribution, shape and chemical composition of particles and as a consequence it allows characterizing aerosols of different origin and type (Ackermann 1998; Matthias and Bösenberg 2002). It generally increases with decreasing particle size and increasing contribution to light extinction (Ansmann et al. 2001). Literature values of lidar ratios range from about 10 to 150 sr (Barnaba et al. 2004). According to published works, lidar ratios in the range 50–80 sr can be representative of

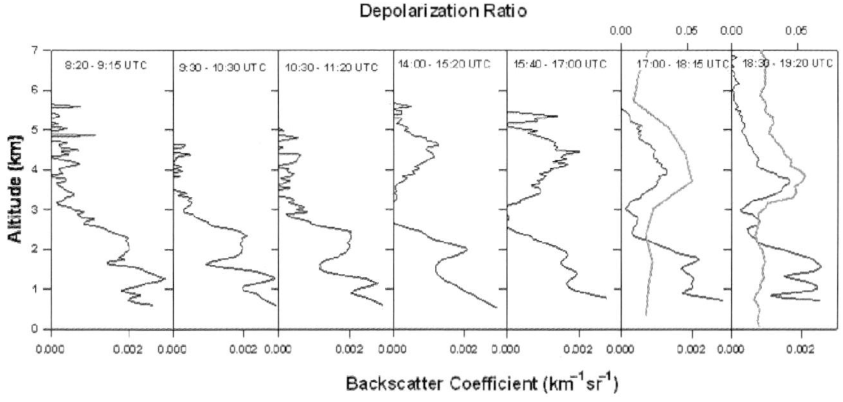

Fig. 14.3 Backscatter coefficient vertical profiles retrieved by lidar measurements on July 18, 2005. Grey lines provide depolarization ratio vertical profiles

non-spherical dust particles and of small absorbing particles (Mishchenko et al. 1997; Mattis et al. 2002; Perrone et al. 2004).

We have set the lidar ratio to 80 sr, to retrieve the backscatter coefficient plotted in Fig. 14.3 This choice will be explained later in. Fig. 14.3 shows that the aerosol vertical distribution varies significantly during July 18. Aerosols are confined at altitudes below 3 km in the morning, while a separate aerosol layer appears from ~14:00 UTC until the evening above 3 km. According to backtrajectories (Fig. 14.1) and satellite images (Fig. 14.2), it is very likely that the higher layer is made of Saharan dust that reaches south-east Italy on the afternoon of July 18. This conclusion is supported by the depolarization ratio measurements (Fig. 14.3, grey lines). The depolarization ratio is the ratio between the total cross-polarized backscatter coefficient and the total polarization preserving backscatter coefficient; it has a value of 0.014 in a pure molecular atmosphere and generally higher values in presence of aerosols. Specifically, spherical particles produce low depolarization ratios, while depolarization ratios larger than 10–15% are associated to non spherical particles such as desert dust particles (Tafuro et al. 2006). We observe from Fig. 14.3 (grey lines) that the aerosol layer above 3 km of altitude is characterized by a depolarization ratio larger than that of the lower altitude aerosol layer. Hence, latter results may indicate either a large presence of non-spherical dust particles on the upper aerosol layer and that the lowermost aerosol layer is mostly made of spherical particles. Our lidar system allows depolarization ratio measurements only on late afternoons, when the solar background is not too high compared to the weak depolarized signal.

Figure 14.4a shows the vertical profiles of the backscatter and extinction coefficient, and of the lidar ratio that have been retrieved by Raman lidar measurements performed on July 18 from 20:50 to 22:15 UTC. The backscatter and extinction coefficient vertical profiles of Fig. 14.4a also reveal the presence of two aerosol layers: one between 2.5 and 5.5 km and the other below 2 km of altitude. However, it is worth noting from Fig. 14.4a that the lidar ratio takes value of ~80 sr from about 1 to 5 km: the upper and lower aerosol layers are characterized by rather close lidar ratio values, even if in accordance to depolarization ratios, the two aerosol layers are expected to be made of particles of different microphysical properties. As we have mentioned, large non-spherical dust particles and small absorbing particles can both be characterized by high lidar ratios.

The nighttime Raman lidar measurements of July 18 have been taken into account to set the lidar ratio to 80 sr and retrieve diurnal backscatter coefficient profiles (Fig. 14.3). In particular, we have assumed that the aerosol layer monitored by the lidar during the day hours is characterized by a lidar ratio value equal to that of the lowermost aerosol layer of Fig. 14.4a. According to depolarization ratio vertical profiles (Fig. 14.3, grey lines), the lowermost layer appears to have not been affected by the dust plume located above ~2 km of altitude.

Lidar measurements have also been performed on July 21, but are not available for July19 and 20. Figure 14.5 shows the backscatter coefficient profiles retrieved on July 21 at different day hours. Grey lines on Fig. 14.5 represent depolarization ratio vertical profiles. Nighttime Raman lidar measurements performed on July, 21

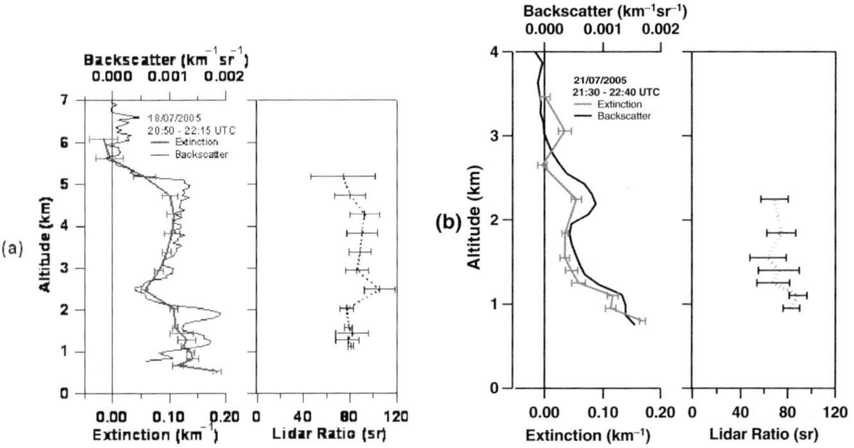

Fig. 14.4 Vertical profiles of the backscatter and extinction coefficient, and of the lidar ratio retrieved by Raman lidar measurements (**a**) on July 18 and (**b**) on July 21 at night

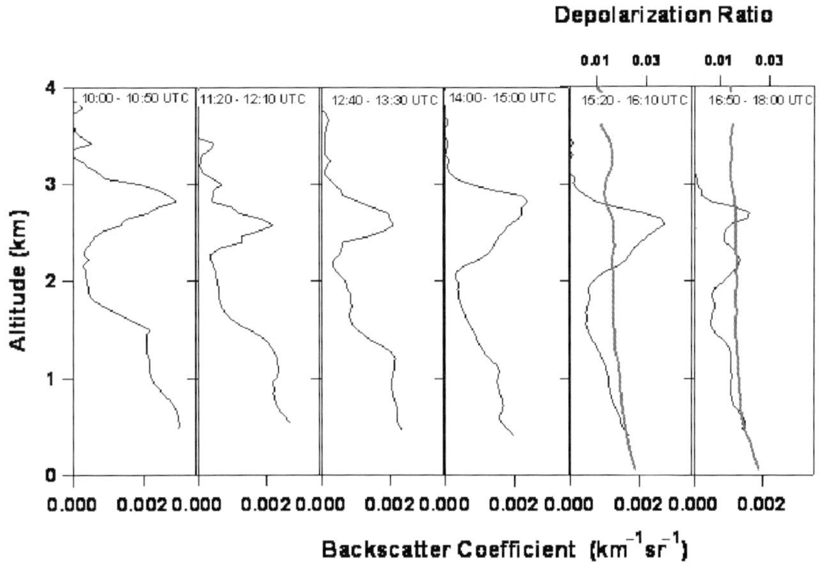

Fig. 14.5 Backscatter coefficient vertical profiles retrieved by lidar measurements on July 21, 2005. Grey lines provide depolarization ratio vertical profiles

from 21:30 to 22:40 UTC are shown on Fig. 14.4b. The lidar profiles of Figs 14.5 and 14.4b reveal the presence at least of two aerosol layers during all measurement hours: one between 2 and 3 km and the other below 2 km of altitude. It is worth noting from Fig. 14.5 that depolarization ratios take, above 2 km of altitude, values that are significantly smaller than those of Fig. 14.3. Latter results may indicate that the layer monitored on July 21 between 2 and 3 km is likely not due to the presence of non-spherical dust particles, in accordance to backtrajectories (Fig. 14.1) even if the lidar ratio values are about 70 sr from ~1 to 2.5 km of altitude (Fig. 14.4b).

14.5 Aerosol Characterization by Sunphotometer Measurements

AERONET sunphotometer retrievals provide column-averaged aerosol quantities that are rather useful to define aerosol microphysical and optical properties. Figure 14.6 shows the daily evolution of the aerosol optical thickness (AOT) at 440 nm (black symbols) and of the Angstrom coefficient (Å), computed from AOT values at 440 and 870 nm (grey symbol) from July 18 to July 21. Both aerosol parameters vary significantly from July 18 to July 21. The AOT increases from ~0.25 to 0.4 on the afternoon of July 18, takes values between 0.4 and 0.5 on July 19, and decreases up to ~0.3 on the following day. In addition, Å takes values within the 1.7–1.8 range up to ~11:00 UTC and values spanning the 0.8–0.9 range from 15:00 to 18:00 UTC (Fig. 14.6, grey symbols). The Angstrom coefficient temporal evolution allows inferring that aerosols of different type have been advected over the monitoring site on the afternoon of July 18. Å is the best marker to infer changes of columnar microphysical aerosol properties (Tafuro et al. 2006): typical values range from 1.5 for particles dominated by accumulation mode aerosols, to nearly zero for large dust particles. Thus, the Å daily evolution further more indicates that large size particles as those coming from the Sahara desert have been advected over the central-east Mediterranean basin in the afternoon of July 18, in accordance to lidar measurements (Fig. 14.3). This interpretation is also supported by the evolution of the AOT_{fine} to AOT_{coarse} ratio plotted on Fig. 14.7, where AOT_{fine} and AOT_{coarse} represent the aerosol optical thickness due to particles with radius

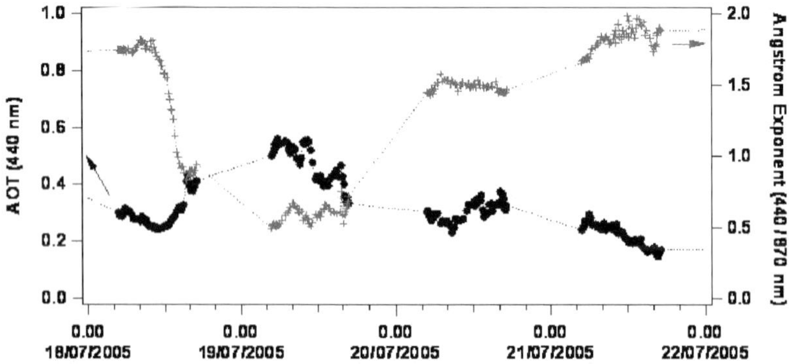

Fig. 14.6 Daily evolution of the optical thickness at 440 nm and of the Angstrom coefficient (440/870 nm).

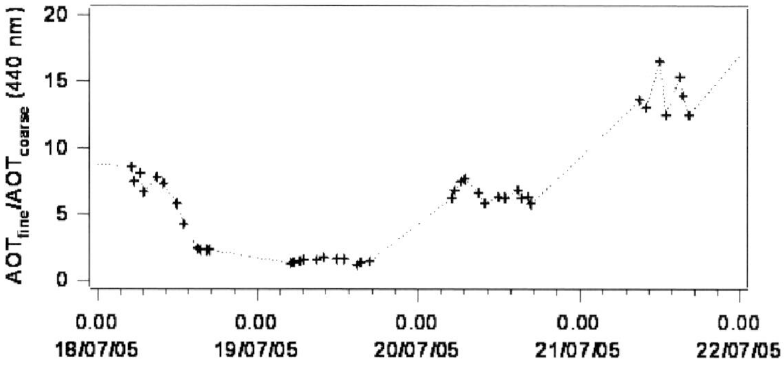

Fig. 14.7 Daily evolution of the AOT_{fine} to AOT_{coarse} ratio

smaller and larger than 0.6 μm, respectively. We observe from Fig. 14.7 that the contribution of fine mode particles that decreases on July 18, increases on July 20 and gets quite large on July 21, as a consequence of the change of the columnar aerosol properties. Hence, sunphotometer retrievals also indicate the advection over south-east Italy of Sahara dust particles from the afternoon of July 18. In addition, they reveal that the Sahara dust outbreak interested Lecce for about 24 h: Å, AOT, and AOT_{fine}/AOT_{coarse} again take values close to those before the dust outbreak on July 20.

A significant increase of the AOT during the dust outbreak is also revealed by Fig. 14.6 (black symbols). It is worth mentioning that the aerosol optical thickness at 351 nm, calculated from the aerosol extinction profile, retrieved by the lidar measurements performed on July 18, from 20:50 to 22:15 UTC, which is equal to 0.58 ± 0.02, is in satisfactory accordance with the AOTs retrieved by the sunphotometer measurements on July 18 at 16:20 UTC, and on July 19 at 05:30 UTC, which are 0.50 ± 0.01 and 0.61 ± 0.01, respectively at 440 nm.

It is now interesting to compare lidar and sunphotometer measurements retrieved after the dust event of July 18–19. Figure 14.6 (grey symbols) reveals that, on July 21, the columnar Angstrom coefficient has recovered to the values 1.5–2 observed before the dust event. According to Fig. 14.7, latter result indicates that the size distribution is again dominated by fine mode particles. However, lidar measurements again show the presence of a separate aerosol layer between 2 and 3 km (Figs 14.5 and 14.4b) whose vertical distribution varies during the day. The depolarization ratio is very small in this layer (Fig. 14.5, grey line) and as we have mentioned, this result suggests that the aerosol layer is likely not due to the presence of non-spherical dust particles. Figure 14.1e indicates that air masses from the Atlantic Ocean arrive to Lecce at about 3 km of altitude (700 hPa) on July 21. However, the particle layer revealed by the lidar at 2–3 km can have been lofted inside Europe, in accordance with Fig. 14.1f. Then, this aerosol layer could be due to the presence of small absorbing particles coming from west European industrial areas or from North

America. Lidar ratio values of Fig. 14.7 support the hypothesis that the aerosol layer is made of small absorbing particles (e.g. De Tomasi and Perrone 2003).

We can see from this example that care must be taken when trying to attribute an origin to aerosol layers in the free troposphere: in this case, if only lidar measurements would have been available, the layer at 2–3 km observed by the lidar on July 21, could easily be confused with a residual layer due to the Saharan outbreak occurred two days before. But, the use of different sources of experimental and modeling information has allowed us to better define the origin and the microphysical proprieties of the investigated aerosol layer.

14.6 Summary and Conclusion

A lidar and a sun/sky radiometer have been used to follow the spatial and temporal evolution of the aerosol load over south-east Italy and get complementary data to properly characterize optical and microphysical properties and the vertical distribution of the aerosol particles. It is rather important to characterize, besides aerosol vertical distributions, microphysical properties of the main aerosol layers, because the aerosol radiative forcing is quite dependent on location and properties of the different aerosol layers. Paper's results have shown that the dust transport from Sahara is, in many cases, characterized by a defined layer above the boundary layer. In addition, it has been shown that particularly during Sahara dust outbreaks, the aerosol vertical distribution can change significantly within few hours and as consequence the aerosol radiative forcing may also vary significantly within few hours.

Specifically, the temporal and spatial evolution of the optical and microphysical aerosol proprieties have been investigated from July 18 to July 21, 2005 and particular attention has been devoted to the Sahara outbreak that has occurred over the central-east Mediterranean from July 18 to July 19. It has been shown that the Angstrom coefficient and the AOT_{fine}/AOT_{coarse} ratio represent the best parameters to trace changes of the aerosol columnar content: their respective values have significantly decreased during the advection of Sahara dust particles. Lidar depolarization ratio measurements have allowed inferring the vertical distribution of non-spherical dust particles that are also responsible of the high lidar ratio values that characterize dust layers (Mishchenko et al. 1997). It has also been shown that fine mode particles can lead to high lidar ratios, but in the latter case depolarization ratios much smaller than those observed during dust outbreaks have been found. The significant support of bactrajectories and satellite images to infer the source region of the monitored aerosols has also been demonstrated.

Acknowledgements This work has been supported by Ministero dell'Istruzione dell'Università e della Ricerca of Italy (Programma di Ricerca 2004. Prot. 20004023854) and by the European Project EARLINET-ASOS (2006–2010, contract n. 025991). Results presented in this paper have been obtained using data from the Aerosol Robotic Network (AERONET). The authors kindly thank the AERONET team.

References

Ackermann J. (1998), The extinction-to-backscatter ratio of tropospheric aerosol: A numerical study, *J. Atmos. Ocean. Technol.*, 15, 1044–1050.

Ansmann A., Wagner F., Althausen D., Muller D., Herber A., and Wandinger U. (2001), European pollution outbreaks during ACE 2, Part I: Alofted aerosol plumes observed with Raman lidar at the Portuguese coast, *J. Geophys. Res.,* 106 D, 20723–20733.

Barnaba F., De Tomasi F., Gobbi G.P., Perrone M.R., and Tafuro A. (2004), Extinction versus backscatter relationships for lidar applications at 351 nm: Maritime and desert aerosol simulations and comparison with observations, *Atmos. Res.*, 70, 229–259.

Bösenberg J., et al. (2003), A European aerosol research lidar network to establish an aerosol climatology, MPI-Report 348, Max-Planck-Institut für Meteorologie, Hamburg, Germany.

De Tomasi F. and Perrone M.R. (2003), Lidar measurements of tropospheric water vapor and aerosol profiles over south-eastern Italy, *J. Geophys. Res.*, 108, 4286–4297.

Fernald F.G. (1984), Analysis of atmospheric lidar observations: Some comments, *Appl. Opt.,* 23, 652–653.

Gobbi G.P., Barnaba F., Giorgi R., and Santacasa A. (2000), Altitude-resolved properties of a Saharan-dust event over the Mediterranean, *Atmos. Environ.*, 34, 5119–5127.

Haywood J.M. and Shine K.P. (1997), Multi-spectral calculations of the direct radiative forcing of the tropospheric sulphate and soot aerosols using a column model, *Q. J. R. Meteorol. Soc.,* 123, 1907–1930.

Holben B.N., et al. (1998), AERONET – A federate instrument network and data archive for aerosol characterization, *Remote Sens. Environ.,* 66, 1–16.

Holben B.N., et al. (2001), An emerging ground-based aerosol climatology: Aerosol optical depth from AERONET, *J. Geophys. Res.,* 106, 12067–12097.

Kaufman Y.J., Tanré D., and Boucher O. (2002), A satellite view of aerosols in the climate system, *Nature.,* 419, 215–223.

Lelieveld J., et al. (2002), Global air pollution crossroads over the Mediterranean, *Science*, 298, 794–799.

Matthias V. and Bösenberg J. (2002), Aerosol climatology for the planetary boundary layer derived from regular lidar measurements, *Atmos. Res.*, 63, 221–245.

Mattis I., Ansmann A., Müller D., Wandinger U., and Althausen D. (2002), Dual-wavelength Raman lidar observations of the extinction-to-backscatter ratio of Saharan dust, *Geophys. Res. Lett.*, 29, No. 9, 20.1–20.4.

Mishchenko M.I., Travis L.D., Kahn R.A., and West R.A. (1997), Modeling phase functions for dust-like tropospheric aerosols using a shape mixture of randomly oriented polydisperse spheroids, *J. Geophys. Res.*, 102, 16, 831–16,847.

Nakajima T., Sekiguchi M., Takemura T., Uno I., Higurashi A., Kim D., Sohn B-.J., Oh S. N., Nakajima T.Y., Ohta S., Okada I., Takamura T., and Kawamoto K. (2003), Significance of direct and indirect radiative forcings of aerosols in the East China Sea region, *J. Geophys. Res.*, 108, 8658, DOI 10.1029/2002JD003261.

Perrone M.R., Barnaba F., De Tomasi F., Gobbi G.P., and Tafuro A.M. (2004), Imaginary refractive-index effects on desert-aerosol extinction versus backscatter relationships at 351 nm: Numerical computations and comparison with Raman lidar measurements, *Appl. Opt.*, 29, 5531–5541.

Perrone M.R., Santese M., Tafuro A.M., Holben B., and Smirnov A. (2005), Aerosol load characterization over South-East Italy by one year of AERONET sun-photometer measurements, *Atmos. Res.,* 75, 111–133.

Tafuro A.M., Barnaba F., De Tomasi F., Perrone M.R., and Gobbi G.P. (2006), Saharan dust particle properties over the central Mediterranean, *Atmos. Res.*, 81, 67–93.

Zerefos C.S., Ganev K., Kourtidis K., Tzortziou M., Vasaras A., and Syrakov E. (2000), On the origin of SO_2 above northern Greece, *Geophys. Res. Lett.*, 27, 365–368.

Chapter 15
Retrieval of Particulate Matter from MERIS Observations

Wolfgang von Hoyningen-Huene, Alexander Kokhanovsky, and John P. Burrows

Abstract Environmental control of pollution uses concentrations of particulate matter (PM) for evaluation of the pollution load. Retrievals of PM from satellite observations are supplementary information to ground-based national observation networks. A method of PM determination using retrievals of spectral aerosol optical thickness is described. The method has been applied to MERIS L1 data over Germany. PM retrievals from satellite observations have been compared with ground-based PM10 measurements of the Federal Environmental Agency, Umweltbundesamt (UBA).

Keywords: MERIS, PM10, BAER, AOT, aerosol

15.1 Introduction

The determination of particulate matter (PM) from space-borne aerosol observations in terms of spectral aerosol optical thickness is required to fill gaps between ground-based stations of the national air quality networks and to get information on PM for regions with no or poor access to ground-based network data. It is relevant information for environmental control.

The importance of control and observations of PM mass concentrations increases with the relevance of national and European regulations on various air pollution species. National ground-based observation networks of air pollutants deliver information at local stations. Since satellites give normally columnar observation of the whole atmosphere, and PM data are valid only for the atmospheric boundary layer, the satellite-derived data for columnar PM must be reprocessed to account for conditions at the ground (e.g., at 2 m height as observed by ground stations).

University of Bremen, Institute of Environmental Physics, Otto-Hahn-Allee 1, 28334 Bremen, Germany

Y.J. Kim and U. Platt (eds.), *Advanced Environmental Monitoring*, 190–202.

Therefore, the retrieval of PM requires the integration of very different information: (a) aerosol optical thickness (AOT), (b) aerosol type and composition, (c) vertical profile and distribution of aerosol and (d) humidity.

The most approaches, exploring this task in the past, use empirical correlations between AOT and PM observations or simple linear relationships between number concentration and AOT. (Fraser 1974; Fraser et al. 1984; Kaufman and Fraser (1990); Gasso and Hegg 1997, 2003). Since this neglects the inherent physical relations between the AOT, the aerosol type (size distribution, main shape of particles and composition), vertical structure and ambient meteorological conditions, correlations of AOT and 'dry' PM data are relatively poor, (Uhlig and von Hoyningen-Huene 1993). The reason is, that while the ground-based aerosol sampling for PM concentrations are performed for 'dry' measurement conditions in the atmospheric surface layer, the AOT is a columnar aerosol parameter of the whole atmosphere, containing all ambient air influences. For this purpose, the boundary layer fraction of the aerosol must be separated and transformed into a 'dry' reference status. First estimations of columnar aerosol concentrations from satellite observations of AOT considering variations in effective radius (r_{eff}) are made by Kokhanovsky et al. (2006).

Al-Saadi et al. (2005) presented an integrated complex approach for the American air quality forecast.

The present contribution integrates spectral AOT retrievals using MERIS L1 data made with the Bremen AErosol Retrieval (BAER) approach (von Hoyningen-Huene et al. 2003) with the estimation of r_{eff}, number concentration and finally mass load within an atmospheric column. Estimates of planetary boundary layer (PBL) height and average relative humidity yield the transfer of these information into concentrations of PM10 within the PBL. The approach is demonstrated using a MERIS L1 scene over Germany.

15.2 Theory

The retrieval of particulate matter from satellite observation uses spectral properties of AOT, as derived by BAER for clear sky conditions. The retrieval of AOT provides the spectral behaviour of AOT for seven shortwave channels of the MERIS instrument, which are used for the determination of the spectral slope in terms of the Angström α-parameter. Angström α is obtained using AOT, retrieved from the seven MERIS channels with wavelength $\leq 0.665\,\mu m$. The BAER approach is described by von Hoyningen-Huene et al. (2003, 2006) and is used in different applications (Kokhanovsky et al. 2004; Lee et al. 2004, 2005). AOT for a certain reference wavelength, here MERIS channel 1 or 2 with 0.412 or $0.443\,\mu m$ is used, and the Angström α-parameter is the basis for the PM retrieval.

The PM retrieval requires an assessment of a size distribution model to convert spectral AOT into columnar aerosol volume, respectively mass. Kokhanovsky et al. (2006) used a mono-modal logarithmic size distribution, characterized by the r_{eff} and

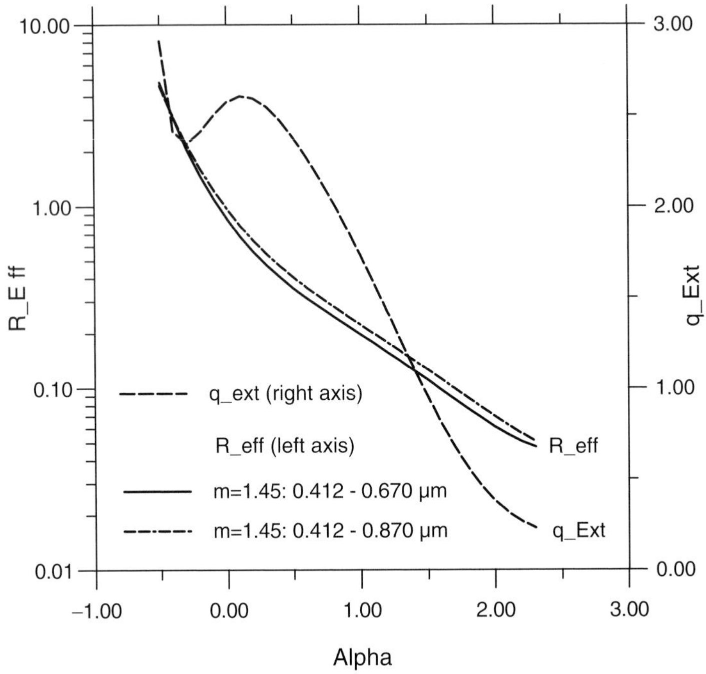

Fig. 15.1 Effective radius (r_{eff}) (black and blue curve) and extinction factor (red curve) as functions of Angström α-parameter

a fixed mode width σ = 0.8326. Mie theory is used to derive parameterisations for r_{eff} and extinction factor as a function of spectral slope of AOT $\delta_{Aer}(\lambda)$, expressed by the Angström α-parameter: $r_{eff}=f_1(\alpha)$, $q_{ext}=f_2(\alpha)$. The relationships for $r_{eff}=f_1(\alpha)$ and $q_{ext}=f_2(\alpha)$ derived are presented in Fig. 15.1.

Both quantities give together with the AOT a columnar number concentration of aerosol:

$$n_{Aer} \approx 8 \cdot \delta_{Aer}(0.412\mu m) \frac{1}{\pi \cdot r_{eff}^2 \cdot q_{ext}} \tag{1}$$

at σ = 0.8326, which is an adequate mode width for PM10. Thus a dynamical link between the spectral AOT and columnar number concentrations is obtained.

The simple monomodal lognormal size distribution can be substituted later by a more complex bimodal model to distinguish between fine and coarse aerosol fraction. The selected monomodal size distribution fits to the size range, relevant for PM10.

An assessment of aerosol density ρ_{Aer} relates the columnar number concentration to an estimate of the columnar aerosol mass:

$$M_{Col} \approx \frac{\pi}{6} \rho_{Aer} \cdot n_{Aer} r_{eff}^3 \cdot \tag{2}$$

For the estimation of PM concentrations the columnar aerosol mass needs to be related to the planetary boundary layer (PBL) conditions. Under clear sky conditions about 90% of aerosol is within the PBL. The observed aerosol exists under ambient humidity conditions with the relative humidity (rh). Therefore, a correction for humidity effects is required, giving $r_{eff}(dry) = r_{eff}(rh) f(rh)$, where $f(rh)$ is given by Hänel (1984). Finally PM concentration can be estimated by

$$PM10 \approx a \frac{M_{Col}(r_{eff}(dry))}{h_{PBL}}. \tag{3}$$

The parameter a gives the fraction of total aerosol, which is within the PBL and h_{PBL} characterizes the thickness of PBL. The approach is described in detail by Kokhanovsky et al. (2006) together with applications to the Sea-viewing Wide Field-of-view Sensor on the SeaStar spacecraft (SeaWiFS).

15.3 PM10 Retrieval from a Satellite: A Case Study

The retrieval of PM10 requires scenes with clear sky conditions only. Cloudy parts of the scenes need to be excluded by a rigorous cloud screening. For the purpose of PM10 retrieval, the MERIS L1 scene with reduced resolution ($1.2 \times 1.2\,km^2$), October 13, 2005, 09:45:11 UTC, over Central Europe is selected. It shows the most parts of Germany as cloud free, thus retrievals of PM10 should not be disturbed by cloud influences. The RGB image of the selected scene is presented in Fig. 15.2.

Over land surface, the spectral AOT has been retrieved by BAER for MERIS channels 1–7 (0.412–0.665 μm). Using these channels Angström α-parameter is determined. The AOT and Angström α-parameter of the scene above are presented in Figs. 15.3 and 15.4, respectively.

The most parts of the scene are suitable for a retrieval of PM10 concentrations. Unless a cloud screening is applied, some cloud disturbances, such as thin *Ci* or contrails, are still visible in the AOT results. Over Germany, AOT at 0.443 μm is ranging between 0.2 and 0.3. Pollution is clearly visible in northern Italy with values of AOT in the range 0.4–0.65. The Angström α-parameter over land surface ranges between –0.2 and 1.1. Clearly a change in aerosol type can be seen from west to east of the scene, with low values in the west and the highest values in the east. Over sea, Angström α-parameter has not been determined. Therefore, PM retrieval has been performed only over land.

The results of AOT retrieval over land have been used for the determination of the columnar size distribution parameters, columnar number concentration and columnar r_{eff}, using equations given above. The size distribution is the basis for the determination of PM10 concentrations.

Figure 15.5 gives the regional pattern of r_{eff}. Since the average relative humidity from ECMWF reanalysis for 1000 hPa is about 50%, a humidity correction

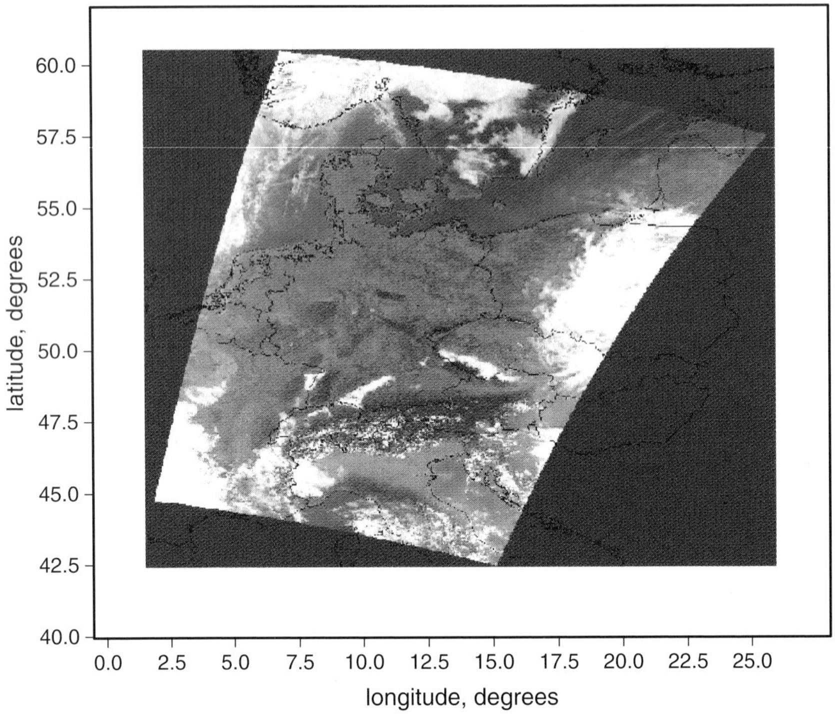

Fig. 15.2 RGB image of the MERIS RR scene of October 13, 2005 09:45:11 UTC over Central Europe

of r_{eff} was not required. Regional differences in humidity are therefore not considered.

One can see an increase of r_{eff} form east to north-west part of Germany, comparable with the change of Angström α-parameter. The region in the north-west, however, has relatively low AOT. The change of Angström α and r_{eff} seems to be connected with a change in aerosol type.

For the calculation of the PM10 concentration, according to Eq. 3, PBL height h_{PBL} and aerosol fraction a in the boundary layer are required.

We assumed a boundary layer height of 1 km. This is based on observations of Meteorological Observatory Lindenberg, giving PBL = 1.2 km at 10:00 UTC, and the PBL height of ECMWF. ECMWF PBL underestimates the values of Lindenberg by about 20%. Although regional differences in PBL height from ECMWF model predictions exist, for a first assessment we used a fixed h_{PBL} for the whole scene.

Aerosol fraction within the PBL is estimated from backscatter LIDAR at Lindenberg. From the vertical profile, one can conclude that 90% of the aerosol expressed by the AOT is within the PBL. This aerosol fraction is used for the whole scene. With these assumptions PM10 concentration has been derived. Figure 15.6 presents the cloud screened PM10 concentration of the scene.

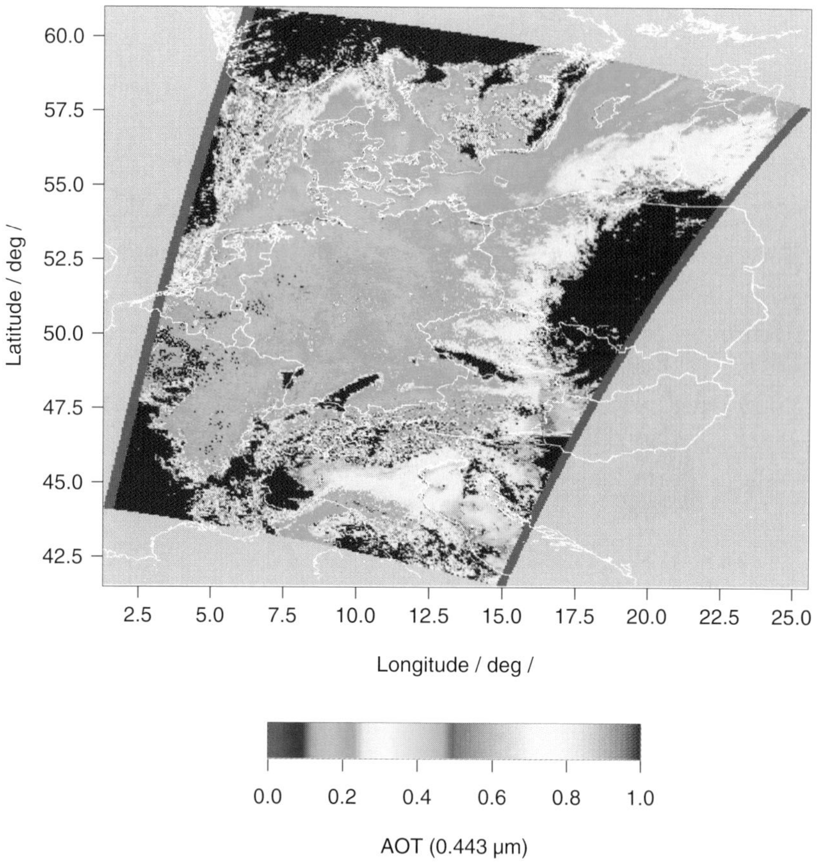

Fig. 15.3 Aerosol optical thickness for MERIS channel 2 (0.443 μm) of the scene of October 13, 2005. Black parts of the scene are excluded because of cloud or snow

15.4 Cloud Screening

The first application of the method showed multiple disturbances by various effects, mainly caused by an insufficient cloud screening. Locally, single very high, partly unrealistic, PM10 concentrations are obtained. For these spots no indication of clouds in the RGB data and in cloud mask products for MERIS could be found. However, the unrealistic high spots occurred in regions close to cloud fields. Thus the cloud screening applied before has not been effective enough for the task of PM retrieval.

The cloud screening applied in BAER determines clouds from radiance boarders for different spectral channels. This removes thick clouds of significant cloud

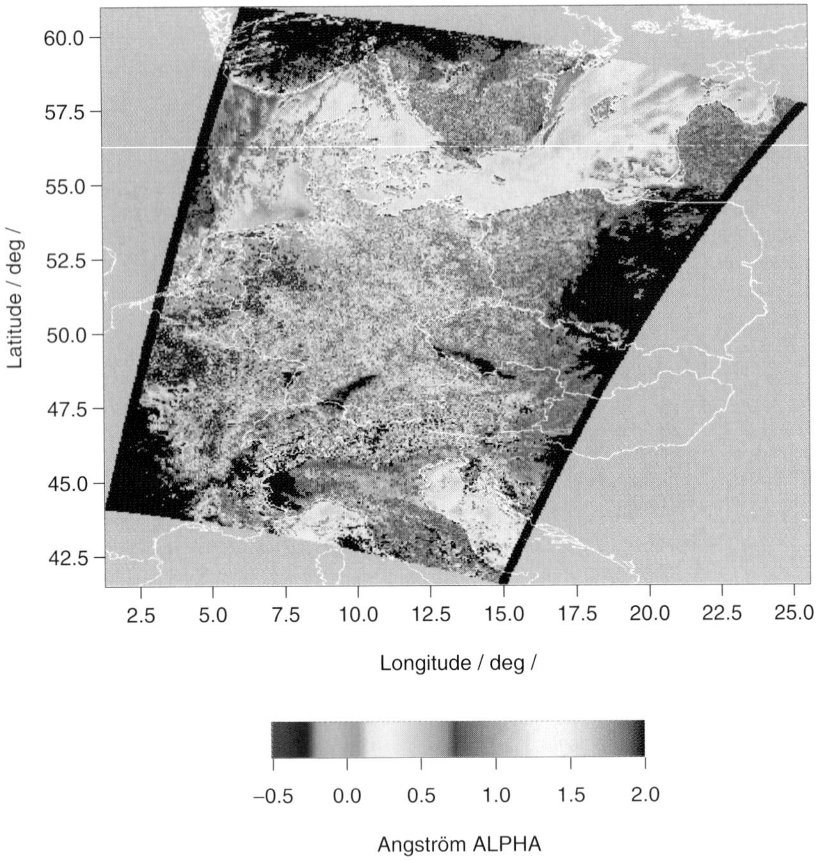

Fig. 15.4 Angström α-parameter

fractions within the scene (>10%). Thin cirrus or sub-pixel clouds with low cloud fractions will not be recognized well by these radiance boarders.

A cloud disturbance results in lowering Angström α significantly and increasing the r_{eff}. This can be caused by both aerosol and cloud disturbances. Assuming that aerosols are more homogeneous distributed than convective clouds, a high standard deviation within a sub-mask of 5 × 5 pixels, indicates for cloud disturbance in this area.

For the purpose of cloud screening we calculated average PM_{Av} and standard deviation σ within a moving 5 × 5 pixel matrix and removed all PM10 results, if the ratio $R = \sigma/PM_{Av} < 0.04$. Thus regions of high PM10 variability will be excluded to avoid cloud disturbance in PM10 retrievals. All unrealistically high PM10 concentrations disappeared.

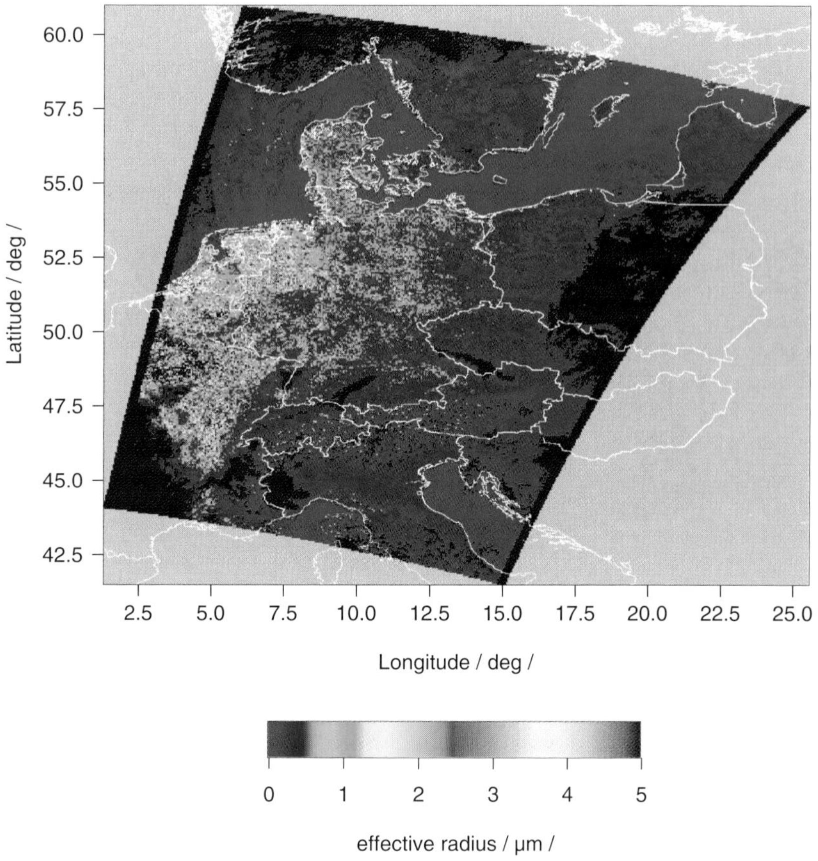

Fig. 15.5 Effective radius

Now three different criteria for the cloud screening are used:

1. The top-of-atmosphere MERIS reflectance ρ_{TOA} for three shortwave channels is spectrally neutral and larger than 0.2.
2. The ratio $\rho_{TOA}(0.412\,\mu m)/\rho_{TOA}(0.443\,\mu m)$ is smaller than 1.07. This indicates reduced Rayleigh- and aerosol scattering and, therefore, a contribution of elevated clouds.
3. The standard deviation of PM within a moving 5×5 pixel mask is lesser than 4%.

All criteria together enable an effective cloud screening for the purpose of PM retrieval. Thus, effects of small sub-pixel clouds with a cloud fraction <0.1, cirrus and contrails could be reduced significantly. A real cloud free scene is a pixel fulfilling all three criteria. All other pixels are rejected from the PM10 retrieval and are masked in the figures with black colour.

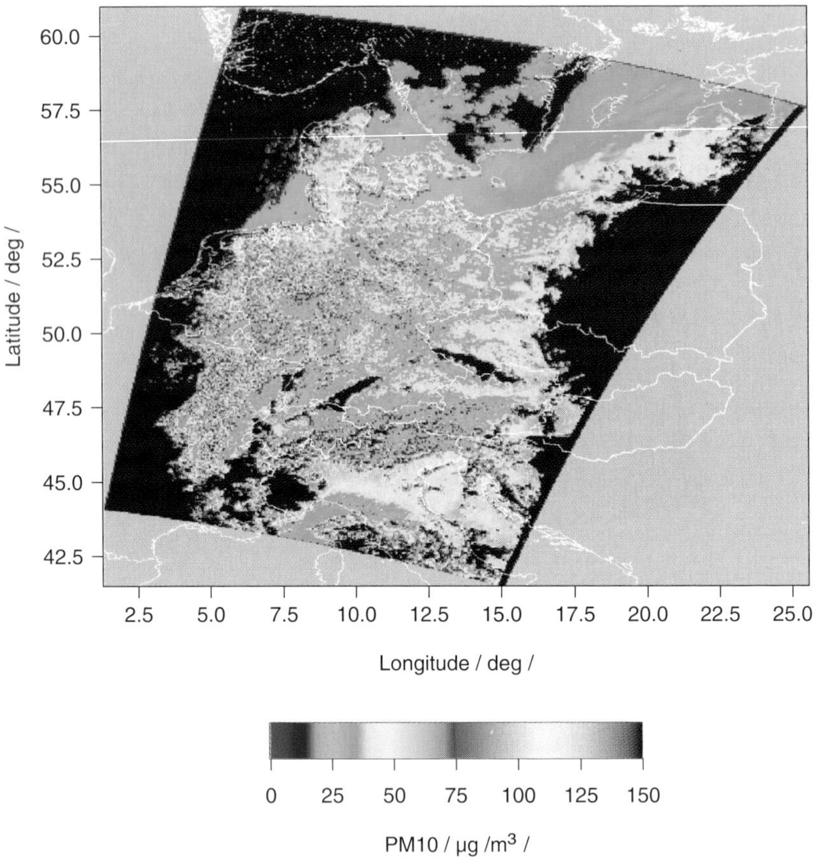

Fig. 15.6 Retrieval of PM10 concentration for the scene of October 13, 2005

15.5 Discussion and Validation of Results

After the additional rigorous cloud screening, the regional pattern of PM10 concentrations over central Europe is obtained for the MERIS RR scene of October 13, 2005. In the vicinity of thick clouds, still effects of thin-high clouds, like cirrus and contrails remain. These clouds do not have a high spatial variability like convection events at the PBL.

Neglecting these effects, the regional pattern of PM10 concentration shows high pollution in northern Italy. Surprisingly, increased pollution can be seen in north-west Germany, where AOT was relatively low. The increased pollution is due to larger effective radii and is confirmed by the ground-measurements of PM10 too. The pattern of PM10 concentration, derived from MERIS observations, follows the

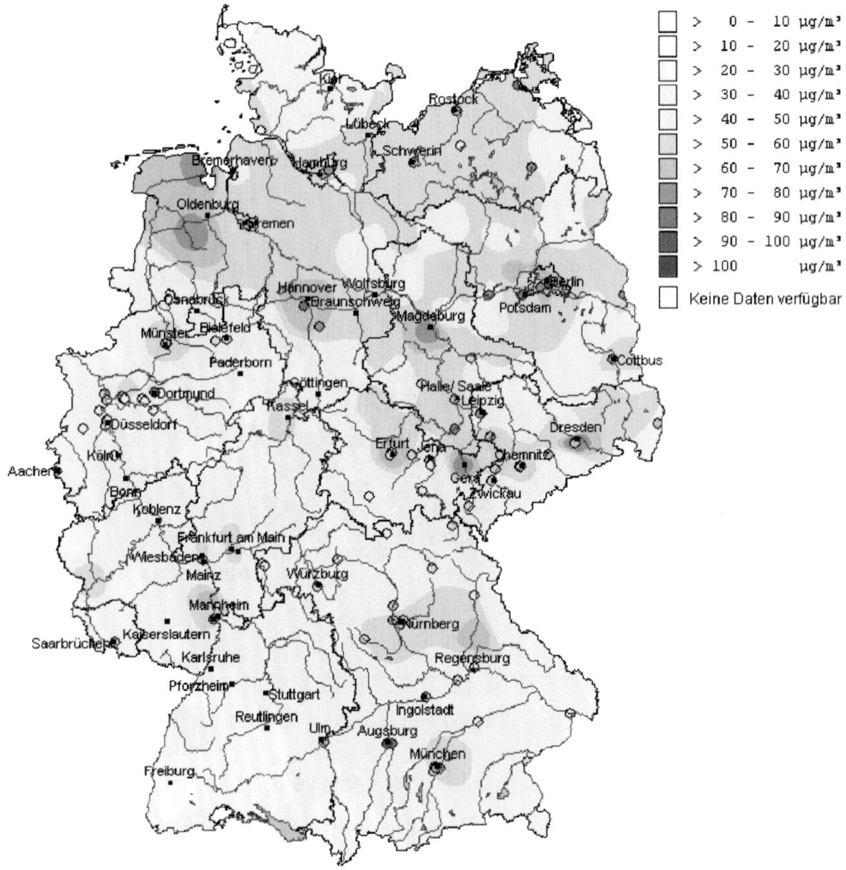

Fig. 15.7 Average daily PM10 concentration over Germany for October 13, 2005, published by Umweltbundesamt, http//www.uba.de

general PM10 distribution obtained from ground based measurements (see Fig. 15.7), with high values in north-west Germany and low values in south-west. AOT and PM10 are nearly uncorrelated over Germany. The correlation coefficient between the AOT (0.443 µm) and retrieved PM10 for the investigated scene is 0.02. An exception is the pollution in northern Italy, where increased AOT is connected with increased PM10 concentrations. Therefore, a monochromatic AOT alone do not display the pollution pattern.

The consideration of the spectral properties of AOT and the retrieval of the r_{eff} gives PM10 values comparable with ground based data as it is presented in Fig. 15.8.

Cloud-screened PM10 retrievals have been used for comparisons with ground-based PM10 measurements of the overflight time of ENVISAT. Ground-based PM10 measurements are obtained within the measurement networks of the 16

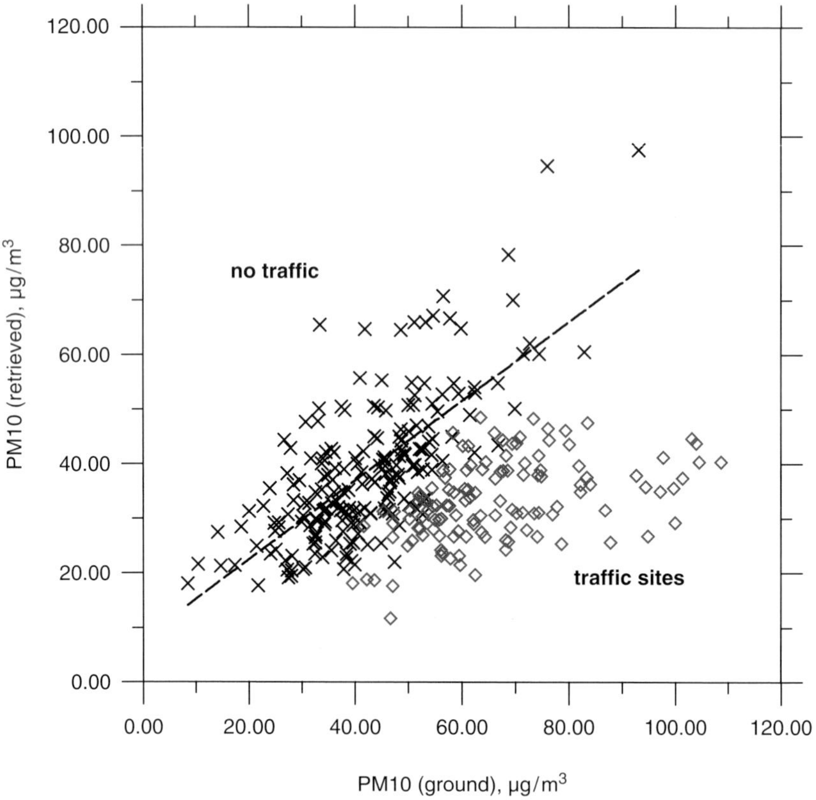

Fig. 15.8 Comparison of cloud-screened retrievals of PM10 from MERIS scene of October 13, 2005 and ground-based measurements of PM10 concentrations for the MERIS overflight time, provided by the German Federal Environmental Agency (Umweltbundesamt). For the traffic sites, no correlation between satellite derived and ground data is found

German federal countries, provided by the Federal Environmental Agency (Umweltbundesamt-UBA) of Germany. The results of the comparison are presented in Fig. 15.8.

Associating all cloud screened data to ground stations of the networks, on a first view it seems, that the retrieval from satellite observations does not reflect really PM10 from ground. (see all points [crosses + open circles] in Fig. 15.8). The reasons need a deeper analysis of the character of the ground stations.

A large number of them, mostly with high PM10 concentrations, are operating in urban areas and measure the pollution of traffic along main road connections. The MERIS RR data, however, give an average estimation for this parameter on a scale of 1.2 × 1.2 km, mostly above the urban canopy layer. This could explain why the satellite observations will underestimate these conditions. Comparisons with stations, affected strongly by urban traffic are indicated in Fig. 15.8 with open circles. This part of station gives no correlation with the PM10 concentrations, derived from satellite observations.

If one removes all stations affected strongly by urban traffic, one obtains a scatter plot, which gives a better correlation with ground data (crosses in Fig. 15.8). We performed a linear fit of PM10 derived using satellite measurements with those on the ground: PM (satellite) = 0.725 PM10 (ground) + 7.98 with a correlation coefficient of 0.71. The average standard deviation is 11 $\mu g/m^3$. Considering the fact, that the regional variability of the meteorological conditions is treated for the whole scene as constant, the scattering of the data is in an acceptable range. Further improvements can be expected, if the real regional meteorological conditions (rh and h_{PBL}) will be taken into account. The data shows, that the retrieved PM10 concentrations give the average pollution by particulate matter on the larger scale of the MERIS satellite pixel and not local peaks.

15.6 Conclusions

For real clear sky scenes, aerosol size distribution parameters, like r_{eff} and number concentration of a simple mono-modal lognormal distribution, is retrieved from spectral AOT measurements. On this basis, PM10 concentrations are obtained.

The cloud-screened PM10 concentration can give the general regional distribution of aerosol pollution.

For the derivation of PM information, the following are important:

1. Spectral AOT should be retrieved with several spectral channels (to obtain the spectral slope of AOT from measurements). For a retrieval over land, instruments are required, like MERIS, which operate with several spectral channels below the red edge wavelength of green vegetation. The retrieval approach needs to consider the whole available spectral information.
2. Improved cloud screening for especially sub-pixel cloud effects is required. Sub-pixel clouds bias the results on PM significantly. Sub-pixel cloud screening for this purpose is not a solved problem so far.
3. The presented results are promising, although regional meteorological influences are not considered in detail. This needs an integration of regional meteorological information, like rh and h_{PBL}.
4. The information seems to be limited by the spatial resolution of the satellite instrument. Thus MERIS RR data do not provide locally high pollution peaks in urban areas. For this purpose, investigations with higher spatial resolutions, like MERIS FR, will be required.

Acknowledgements We like to mention, that ESA was supporting the development of the AOT retrieval over land described in this work for the purpose of atmospheric correction of MERIS land surface data. Further we like to acknowledge the contribution W. Bräuniger, W. Garber and M. Wichmann-Fiebig for providing ground-based PM10 data to us. J. Güldner of Lindenberg Meteorological Observatory of German Weather Service (DWD) contributed PBL information from microwave radiometer. J. Bösenberg of MPI Hamburg provided information on LIDAR profile. P. Glanz contributed to EMCWF relative humidity analysis for the scene studied. The authors are grateful to ESA for providing MERIS data.

References

Al-Saadi J., Szykman J., Pierce R.B., Kitaka C., Neil D., Chu D.A., Remer L., Gumley L., Prins E., Weinstock L., McDonald C., Wayland R., Dimmick F., and Fishman J. (2005), Improving national air quality forecast with satellite aerosol observations, *Bul. Am. Meteorol. Soc.*, 9, 1249–1261.

Fraser R.S. (1976), Satellite measurements of mass of Sahara dust in the atmosphere, *Appl. Optics*, 15, 2471–2479.

Fraser R.S., Kaufman Y.J., and Mahoney R.L. (1984), Satellite measurements of aerosol mass and transport, *Atmos. Environ.*, 18, 2577–2584.

Hänel G. (1984), Parameterization of the influence of relative humidity on optical aerosol properties. (In: *Aerosols Clim. Eff.* A. Deepak Publ. Hampton 117–122.)

Gasso S. and Hegg D.A. (1998), Comparison of columnar aerosol optical properties measured by MODIS airborne simulator with in-situ measurements: A case study, *Remote Sens. Environ.*, 66, 138–152.

Gasso S. and Hegg D.A. (2003), On the retrieval of columnar aerosol mass and CCN concentration by MODIS, *J Geophys. Res.*, D108, D1, 4010, doi: 10.1029/2002JD002382.

Kaufman Y.J., Fraser R.S., and Ferrare, R.A. (1990), Satellite measurements of large scale air pollution. Methods, *J. Geophys. Res.*, D95, 9895–9909.

Kokhanovsky A.A., von Hoyningen-Huene W., and Burrows J.P. (2006), Atmospheric aerosol load from space, *Atmos. Res.*, 81, 176–185.

Uhlig E.-M., and von Hoyningen-Huene W. (1993), Correlation of atmospheric extinction coefficient with the concentration of particulate matter in a polluted urban area, *Atmos. Res.*, 30, 181–195.

von Hoyningen-Huene W., Freitag M., and Burrows J. P. (2003), Retrieval of aerosol optical thickness over land surfaces from top-of-atmosphere radiance, *J. Geophys. Res.*, 108, D9 4260, doi: 10.1029/2001JD002018.

von Hoyningen-Huene W., Kokhanvosky A.A., Burrows J.P., Bruniquel-Pinel V., Regner P., and Baret F. (2006), Simultaneous determination of aerosol- and surface characteristics from top-of-atmosphere reflectance using MERIS on board of ENVISAT, *Adv. Space Res.*, 37, 2172–2177.

Kokhanovsky A.A., von Hoyningen-Huene W., Bovensmann H., and Burrows J.P. (2004), The determination of the atmospheric optical thickness over Western Europe using SeaWiFS imagery, *IEEE T. Geosci. Remote Sensing*, *TGSR* 42(4), 824–832

Lee K.H., Kim Y.J., and von Hoyningen-Huene W. (2004), Estimation of regional aerosol optical thickness from satellite observations during the 2001 ACE-Asia IOP, *J. Geophy. Res.*, 109, D19S16, DOI 10.1029/2003JD004126.

Lee K.H., Kim J.E., Kim Y.J., and Kim J, von Hoyningen-Huene W. (2003), Impact of smoke aerosol from Russian Forest Fires on atmospheric environment over Korea during May, *Atmos. Environ.*, 39, 85–99.

Chapter 16
Bioaerosol Standoff Monitoring Using Intensified Range-Gated Laser-Induced Fluorescence Spectroscopy

Sylvie Buteau[1], Jean-R. Simard[1], Pierre Lahaie[1], Gilles Roy[1], Pierre Mathieu[1], Bernard Déry[1], Jim Ho[2], and John McFee[2]

Abstract The biological aerosol threat has become a major military and civilian security challenge, primarily due to the increased accessibility to biological technologies, and perhaps partially due to technical difficulties in developing effective detection systems. Defence Research and Development Canada (DRDC) has investigated various technologies, including point and standoff systems for environmental aerosol monitoring, to enhance readiness for such threats. Standoff bio-aerosol systems were based on infrared techniques and laser-induced fluorescence (LIF) approaches. These LIDAR systems were designed to monitor the atmosphere from a standoff position, measuring light scatter or fluorescence signals originating from particle-based biological molecules. In the case of LIF, the signal is spectrally resolved by a combination of grating elements and a range-gated intensified charge couple device (ICCD) that records the spectral information within a range-selected atmospheric volume. Multivariate data analysis techniques may be used to achieve real time detection. Advanced data processing techniques combined with the sensitive sensor have demonstrated the potential to detect and discriminate a mixture of several biological species. Instrument detection limits were determined to be within the target range specified for military and civilian scenarios. The potential of this innovative sensor to measure spectral data of various biological agent simulants, interferants and ambient bio-aerosols of natural and anthropogenic origins will be discussed. The detection limits obtained with this new sensor during open air releases for several given materials, cloud depths and ranges were assessed.

Keywords: Agent simulants, biological aerosol, LIF, range-gated CCD, standoff detection

[1]*Defence R & D Canada Valcartier, 2459 Boul. Pie-XI Nord, Québec, QC, Canada, G3J 1X5*

[2]*Defence R & D Canada Suffield, Box 4000, Medicine Hat, AB, Canada, T1A 8K6*

Y.J. Kim and U. Platt (eds.), *Advanced Environmental Monitoring*, 203–216.

16.1 Introduction

Bioaerosol threats have been around for several decades and have been mostly associated with war activities of terrorist organisations or rogue nations. The importance of these bioaerosol threats has considerably increased in the past few years. Consequently, Canada has adopted different measures to increase its readiness facing this potential threat, one of which is the improvement of atmospheric bioaerosol monitoring.

Point and standoff detection are the two main approaches to monitor the bioaerosol contents in the atmosphere. Remote detection is an extension of point detector technologies with the added advantage of measuring material at kilometre distances. Technologies associated with point detection systems have progressed rapidly, mainly because of their lower cost. Proximity to particles confer to point detectors the capability of deducing different types of information such as particle size and the fluorescent spectral signature characteristics (Faris et al. 1997; Hairston et al. 1997; Fell et al. 1998). However, they must be positioned within the bioaerosol cloud in order to be efficient for detection and the information gathered is only relevant for a single location (or multiple discrete locations for a network of point detectors). Alternatively, standoff detection does not have these drawbacks but the information produced by those sensors is limited (detection capability inversely proportional to distance and particle concentration). Nevertheless, similar types of information, such as fluorescent spectra (Grant 1997; Hargis et al. 1998; Simard et al. 2004) and aerosol size statistics (Roy et al. 1999), can also be obtained using standoff sensor approaches.

Light detection and ranging (LIDAR), a technology that transmits laser pulses and detects the returned signal, has been successfully applied in aerosol standoff sensing. Using the strong elastic signal returns, the LIDAR technique can perform demanding tasks such as cloud mapping (Houston and Evans 1987), long-range low-concentration aerosol detection (Condatore et al. 1998), and with a multiple field of view LIDAR, determine the particle size distribution (Hutt et al. 1994) from a standoff position. In these cases, the returned signal is at the same frequency as the laser source, which is defined as the elastic or Rayleigh scattering. However, in order to distinguish clouds with similar distribution in aerosol size but of different types, inelastic LIDAR techniques must be used. These specific techniques are either based on Raman scattering or fluorescence emission, both involving emission of photons having frequencies different than the incident radiation. The main difference between Raman scattering and fluorescence emission is the time scale involved in the process, which is of the order of picosecond versus nanosecond, respectively. The Raman LIDAR return is characterized by spectrally narrow signals at specific shifted wavelengths. Measured in wave numbers (energy), the magnitude of the shift is independent of the irradiation wavelength and specific to the scattering molecule, in addition, the intensity of the Raman band is proportional to the scattering concentration. The fluorescence LIDAR return, which is referred as LIF, can either be spectrally narrow or distributed over a wide spectral interval depending on the complexity of the irradiated molecular structure. In the case of a spectrally distributed return signal, it can either be collected with a wide bandwidth filter (Gelbwachs and Birnbaum 1973) or with a combination of a dispersive element and an efficient photon counting instrument to simultaneously collect the detailed spectrum (Chen et al. 1997). The choice of the irradiating wavelength is critical when using techniques based on resonant effects, such as LIF.

In the mid 1980s, Defence Research and Development Canada (DRDC) initiated LIDAR investigations in biological aerosol detection (Evans et al. 1989). This initiative came about from a serendipitous meeting between curious physicists and microbiologists who wondered if LIDAR systems could detect micron size biological particles. Much later, it was discovered that live biological aerosols exhibit UV-induced fluorescence characteristics and the signal source has metabolic significance if the relevant wavelength is used (Ho et al. 1999). In the late 1990's, DRDC initiated a three-year program called SINBAHD (standoff integrated bioaerosol active hyperspectral detection). SINBAHD assessed the potential of detecting and characterizing bioaerosols from a standoff position by Ultraviolet (UV) LIF and intensified range-gated spectrometric detection techniques. First, a model was developed to predict the performance of such a device as a function of its numerous design parameters (Simard et al. 1999). This model combined the classical LIDAR equation for multi-spectral returns and previous knowledge from laboratory fluorescence measurements using flow cytometry and fluorescence aerodynamic particle sizing (FLAPS) techniques. Following this initial modeling, an exploratory prototype was completed by December 2000. After an extensive optical characterization, SINBAHD was tested at DRDC Suffield (May and September 2001) during open air releases of two different simulants of biological agents: *Baccilus globigii* (BG), a spore-based aerosol, and *Erwinia herbicola* (EH), a vegetative type aerosol, [9]. Other testing conducted since then included two data acquisition campaigns at Dugway Proving Ground (DPG), Utah, USA, on interferant and simulants (July 2002, June 2005) (Buteau et al. 2006), and two others at DRDC Valcartier on natural bioaerosols, essentially pollens (August 2004 and September 2005) (Buteau et al. 2005). The goal pursued was always the same: acquire spectral signatures of various bioaerosol simulants and interferants of natural and anthropogenic origin in order to build a pattern set (signature library). This signature library is the key element of the chosen data processing method, based on multivariate analysis (MA) techniques.

This paper begins by reviewing the performance model and by identifying the associated important technical and phenomenological parameters. In the following section, the technical design of the SINBAHD prototype is described and the general testing procedures is presented. Different results are then shown and briefly analyzed. Finally, a conclusion highlights the important aspects of the presented work. The present work intends to give a global overlook of the entire project and will not go into too specific details especially in the data processing algorithmic which would be a whole topic in itself.

16.2 Theory

A fundamental step to properly evaluate the capabilities of remotely sensed LIF generated by bioaerosols using spectral analysis is to establish a detailed model describing the physical phenomenon. To achieve this task, the general LIDAR equation for the case of a fluorescent target proposed by Measures (Measures 1984) has been reformulated. This new formulation has the advantage of clearly defining the contribution from a single type of bioaerosol as a function of the optics, atmospherics, and the bioaerosol's specific characteristics, and is given as:

$$\frac{dE}{d\lambda}_{ba}(\lambda,\lambda_0,r) = \overbrace{n_{pu}E_0(\lambda_0)}^{\text{Delivered}} \; \overbrace{\xi(r)\frac{A_0}{r^2}}^{\text{Geometry}} \; \overbrace{t_{oe}(\lambda_0)\,t_{oc}(\lambda)}^{\text{Optic}} \; \overbrace{t_{ae}(\lambda_0,r)\,t_{ac}(\lambda,r)}^{\text{Atmospherics}} \; \overbrace{\Delta r N_{ba}(r)}^{\text{Column thickness}} \; \overbrace{\frac{d^2\sigma^F_{ba}}{d\Omega\, d\lambda}(\lambda_0,\lambda)}^{\text{Bioaerosol}} \tag{1}$$

In this equation, $dE/d\lambda_{ba}$ is the radiated energy per unit wavelength interval, collected at wavelength λ, from a single type of bioaerosol located at a range r, when excited by laser pulses at a wavelength λ_0. The range r is fixed by the time delay between firing of the laser and the detection of fluorescence. The depth of this atmospheric cell results from the spatial gate Δr, which is defined as the convolution of the time interval during which the ICCD intensifier is active, the laser pulse duration and the time decay of the fluorescing aerosol. The radiated energy is the result of an integrated number of laser pulses n_{pu}, each of energy E_0, fired at the probed cell. The overlap function ξ multiplied by the aperture of the telescope A_0 divided by the squared range defines the geometric configuration of the collector. The optical parameter is the product of the overall spectral transmission of the emitter t_{oe} and of the collector t_{oc} optics. Similarly, the atmospheric parameter is the product of the spectral transmission of the atmosphere over the emission path t_{ae} and over the collection path t_{ac}. The column thickness is defined as the product of the depth of the atmospheric probed cell Δr multiplied by the averaged density, N_{ba}, of the specific bioaerosol in that cell. Finally, assuming isotropic fluorescence, the differential fluorescent cross-section of bioaerosol, $d^2\sigma^F_{ba}/d\Omega d\lambda$, can be expressed as (Simard et al. 2003)

$$\frac{d^2\sigma^F_{ba}}{d\Omega d\lambda}(\lambda_0,\lambda) = \frac{\psi_{ba}(\lambda_0)\,A_{ba}}{4\pi} P_{ba}(\lambda_0,\lambda), \tag{2}$$

where $\psi_{ba}(\lambda_0)$ and A_{ba} are the effective quantum yield and the averaged projected area of the bioaerosol. P_{ba} is the spectral probability distribution of the inelastically induced photons. This latter parameter provides the classification potential of the bioaerosol from a standoff position and can be obtained once a proper optical calibration of the sensor is performed. Indeed, with the following assumptions: use of a stable white source with a known emission profile (an integrating sphere for example), an atmospheric spectral transmission equal to 1 (short-range measurements), and an ideal spectral resolution for the spectrometer, the spectral probability distribution of a bioaerosol can be expressed as (Simard et al. 2003):

$$P_{ba}(\lambda) = \frac{P_{ba}^{'}(\lambda)}{\int P_{ba}^{'}(\lambda) d\lambda}, \text{ where } P_{ba}^{'}(\lambda) = \frac{ne_{ba}(\lambda) R_{sphere}(\lambda)}{ne_{sphere}(\lambda)} \quad (3)$$

In this equation, ne_{ba} and ne_{sphere} are the detected electronic signals from the bioaerosol and the calibrated integrating sphere, respectively, and R_{sphere} is the integrating sphere calibrated radiance. This spectral probability distribution P_{ba} is intrinsic to the bioaerosol and hence independent of the sensor used to acquire the signals.

Equations 1–3 refer to a single type of bioaerosol; however, the collected spectra during a measurement usually originates from different sources, these being biofluorescence E_{ba} and Raman scatters E_{Raman} of different sources. In order to separate the signal associated with the inelastic scatter E_{is} from the Raman scatters (or any other spectrally distributed signals present in the collected spectra, such as background), a MA technique is used. This latter technique represents the collected inelastic spectra $\vec{E}_{\lambda}$ as a linear combination of normalized spectral signatures $\vec{s}_{is}^{\lambda}$ in the multivariable space from which individual energetic contributions, E_{is}, can be derived:

$$\vec{E}_{\lambda} = E_{ba1}\vec{s}_{ba1}^{\lambda} + E_{ba2}\vec{s}_{ba2}^{\lambda} + \ldots + E_{N_2}\vec{s}_{N_2}^{\lambda} + E_{H_2O}\vec{s}_{H_2O}^{\lambda} + \ldots, \text{ where} \sum_{\lambda} \left(\vec{s}_{is}^{\,i}\right) = 1 \quad (4)$$

A model-based least squares analysis method is used to solve Eq. 4, for the amplitudes associated with the component spectra of a given collected spectrum. The disadvantage of a model-based least squares fitting is that it can be more computationally intensive and is only possible for simple spectra having a few component materials. In the present application, this technique could be used without major drawback since there is a long time interval between acquisitions of the target spectra. In addition, the spectra are simple and they can be accurately described by simple physics-based Raman and fluorescent models. Singular value decomposition (SVD) (Bronson 1989) was chosen for the linear least squares fitting. SVD fitting provides a very good estimate of the coefficients if the problem is over determined, as do other linear least squares methods. However, when the problem is underdetermined (two signatures fit the data equally well), SVD finds a solution that minimizes, in a least squares sense, the coefficients corresponding to those signature functions and will make a least squares choice for setting a balance between them (Press et al. 1992). From Eq. 4, it can be seen that the bioaerosol spectral signatures $\vec{s}_{ba}^{\lambda}$, which are sensor dependent, can be obtained during well-controlled short-range releases of the bioaerosol of interest.

16.3 Sensor Description

SINBAHD, an exploratory prototype, was used to evaluate the accuracy of the model proposed in the previous section. The complete LIDAR system, which includes the transmitter, receiver, electronics and cooling systems, is integrated

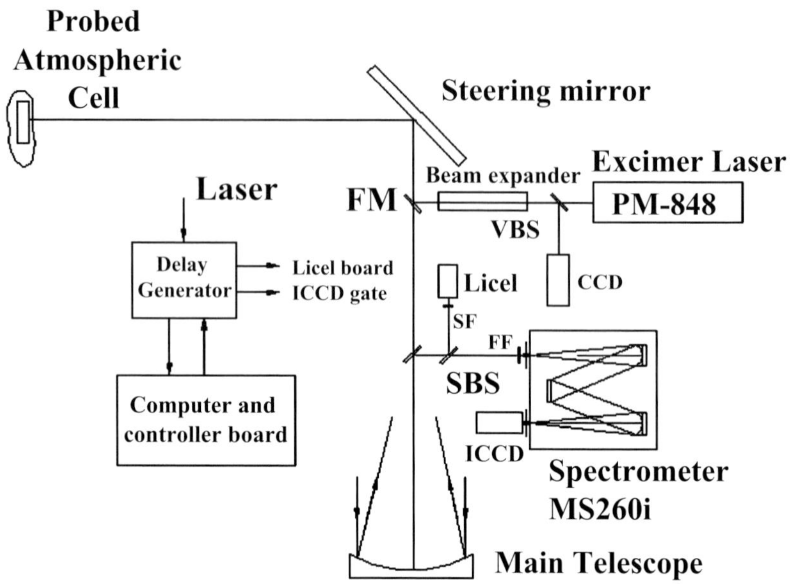

Fig. 16.1 Schematic representation of SINBAHD prototype

within a 12 m-long modified tow-able trailer and, with a diesel-electric generator, constitutes a completely self-sufficient system. The selection of the different components of the prototype was based mainly on maximizing the LIDAR output power and collection sensitivity while staying within the available budget allocation. The prototype schema is presented in Fig. 16.1.

The laser source is an UV Xenon-Fluoride excimer laser (GSI Lumonix, model PM-848) emitting between 150 mJ and 200 mJ per pulse at 351 nm. 15.1-ns long pulses are emitted at a pulse repetition frequency (PRF) of 125 Hz (limited by the acquisition camera), exit the unstable optical cavity of the laser and pass via a 3.65x beam expander providing a final measured beam divergence (FWHM, including pointing stability) of 147 μrad (width) × 308 μrad (height). A visual channel is inserted between the laser and the beam expander with a beam splitter (VBS). With the help of a Charged Couple Device (CCD) camera equipped with a zoom lens, this visual channel allows the laser beam to be directed precisely towards the target of interest. The expanded laser beam is then made co-axial with the collecting optical axis using an adjustable 45° folding square mirror (FM) placed at the center of the telescope-collecting aperture. The combined emitter and collector optical axes are directed at the monitored atmospheric cell using a 50 cm by 33 cm elliptical steering mirror mounted on motorized gimbals providing 10 μrad pointing accuracy (model AOM130M-400-2-41 made by Industrial Automation). A 30-cm diameter, 127-cm focal length, Newtonian telescope (custom made by Space Optics Research Labs) collects the radiation returned by the monitored cell and focuses it at the entrance slit of the imaging spectrometer (Oriel, model MS260i), after passing through two UV high-pass (wavelength) filters (FF) to block the elastic scattered radiation. The 300 line/mm grating of this spectrometer in combination with a 200 μm wide entrance slit confers a spectral resolution of 4.8 nm and a span of 230 nm, optimized between 300 and

600 nm. The entrance slit defines a 157 μrad wide collecting field of view (FOV). The height of the slit, which has no impact on the spectral resolution but fixes the background contribution, is 3.3 mm (2.6 mrad FOV in height).

An intensified CCD camera, ICCD, (Andor model DH501-18F-01, option W) detects the dispersed radiation at the exit window of the spectrometer. The 128 × 1024-pixel CCD array is binned vertically (over the 128-pixel column) to keep only the spectrally dependent signal. From the 1024 pixel array, 675 pixels are in the intensified region and define the 230-nm spectral span of the inelastic scattering collector. The intensifier gate is synchronized with each fired laser pulse with a delay defining the range of the probed atmospheric cell. Furthermore, the natural radiant contribution present in the atmospheric cell, and collected simultaneously with the laser scatter, is sampled between each laser pulse. This requires operating the intensifier gate and the CCD readout at a frequency twice the laser PRF (250 Hz). The combination of the intensifier sensitivity, the 16-bit dynamic range of the camera, and the spectral distribution of the collected signal over the CCD columns give to this detector configuration the capability of counting individual detected photons while keeping a very large dynamic range. This makes the actual sensor very attractive for detecting very low signal levels while retaining the spectral information. In parallel with the fluorescence channel, the elastic scattering is sampled from the collector axis with an anti-reflection coated quartz beam splitter (SBS) and is directed at a photomultiplier (Licel) tube after being filtered by a 10 nm-wide band pass filter centered at 350 nm (SF). This photo-multiplier is connected to a transient recorder and provides elastic scatter returns as a function of range. This information is used to configure the width and position of the intensified range-gate.

16.4 Testing Procedures

All the results presented herein were obtained during the Joint Biological Standoff Detection System increment II field demonstration trial (JBSDS Demo II). This international trial took place at Dugway Proving Ground (DPG), Utah, USA, in June 2005 and regrouped 12 systems, each using their specific technology. This trial was separated into three phases: (1) standoff ambient breeze tunnel (sABT), (2) crosswind testing and, (3) excursion/on the move tests. The latter will not be mentioned further since SINBAHD did not participate in this trial phase.

The sABT phase represented a controlled measurement portion of the trial. The sABT is a tunnel in which bioaerosols are released and the evolution of their concentration with time is measured by six calibrated aerosol particle sizers (APS) placed at even incremental distances in the tunnel. SINBAHD did not participate in this portion of the trial but used the results of DPG west desert LIDAR (WDL) referee system that has obtained calibrated concentrations over 63 releases of simulants or interferants.

The crosswind testing phase consisted of open air releases from a given predefined target location. This type of release represents with reasonable fidelity the operational condition of a real biological attack event except for one detail: the target location is known in advance. An overall total of 48 crosswind trials, some with multiple releases of simulants/interferants, were performed. In most cases, the WDL provided reference

Table 16.1 Simulants/interferants disseminated during JBSDS Demo II trial, Utah, USA, June 2005

Type	Material	Informations
Bacteria spore	Killed-Ba	Dry-killed vaccine strain, Bacillus anthracis Sterne
	BG	Dry, new *Bacillus subtilis* var. *niger* (Denmark) milled at the same size as the old BG
Vegetative cell	Killed-Yp	Wet-killed vaccine strain KIM of *Yersinia pestis* (5×10^9 cfu/ml)
	EH	Wet unwashed *Erwinia herbicola* in the spent media, from ATCC # 33243
Protein	OV	Dry, ovalbumin, the white of an egg (toxin simulant)
Virus simulant	MS2	Wet male-specific bacteriophage type 2 (1×10^{10} pfu/ml)
Interferant	Kaolin	$Al_2Si_2O_5(OH)_4$
	Cabosil	Anti-caking agent
	Diesel	Exhaust from a 100 kW generator and release from HMMWV
	Yellow smoke	M18 colour smoke grenade (for signalling and screening)
	HC smoke	AN-M8 grenade (white smoke for signalling and screening)
	Pollen, top soil, burning brush, burning tires, burning cotton	

data. Three different dissemination mechanisms were used for this type of testing: point, puff or aerial. The positioning of the point and puff mechanisms, which were mounted on a mobile platform, was determined as a function of the wind direction and speed in order to obtain the bio-cloud on the target location. SINBAHD participated in 34 releases during the crosswind testing period. First of all, testing and alignment period was planned at the beginning of each night of testing. For SINBAHD, this period consisted first of the alignment and focussing of the emitting and collecting channel and then, the adjustment of the sensor line of sight with target location. After this testing period, time synchronization would follow and the dissemination initiated depending on wind conditions. Participating detection systems commenced data recording at a given time followed by aerosol release 4 min later. In most cases, the dissemination was ramped down for the last portion of the release. Information on the disseminated materials (sABT and/or crosswind phase) can be found in Table 16.1.

16.5 Experimental Results

The LIF contribution of the background and that emitted by the different studied species must be adequately characterized in order to use these fluorescence signatures subsequently in the MA. Before presenting the processed spectral signatures obtained from the numerous acquisitions performed by SINBAHD, the background

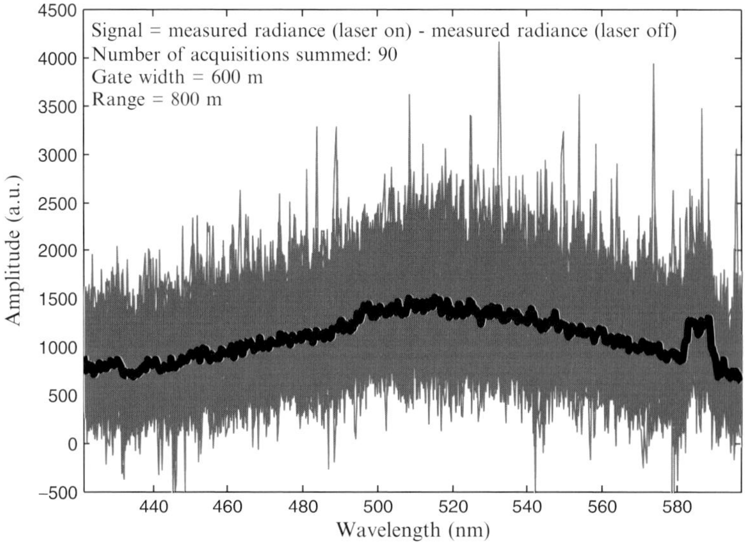

Fig. 16.2 Background laser-induced fluorescence (LIF): 90 single acquisitions (red) and the mean (black)

LIF will be globally characterized. This fluorescent component must also be included into the MA. The background normalized spectral signature was obtained by filtering and normalizing the spectral average of many induced background radiance acquisitions, from which the measured naturally occurring radiance signal was subtracted since it is included in the former. Figure 16.2 presents 90 background fluorescence acquisitions (red) that were averaged (black) and from which the background normalized signature was calculated. The spectral feature around 585 nm is a detector artefact. The background spectral signature measured by SINBAHD was globally quite constant over the attended trial period of 5 days. The background level is globally proportional to the gate width and to the inverse of the squared gate range as predicted by the LIDAR equation (Eq. 1). The spectral shape and level of the LIF originating from the background depends on many factors such as the specific nature of this aerosol background, illumination conditions and gate width.

The representative spectral signatures of the different studied materials were extracted by summing all the spectra still showing some signal once the naturally occurring fluorescence and background LIF components were subtracted. This spectral average is then filtered and normalized. Among the 16 different studied materials, cabosil, kaolin and HC-white smoke did not show any detectable fluorescence. The burning tires had some signal but not enough to extract a signature, the burning cotton and top soil releases had sufficient signals to obtain a signature for the MA utility, but not enough to clearly characterize them. Clear signatures could be obtained for all the others (killed-Ba, killed-Yp, BG, EH, MS2, OV, yellow smoke, fog oil, diesel, burning brushes). Figure 16.3 presents these

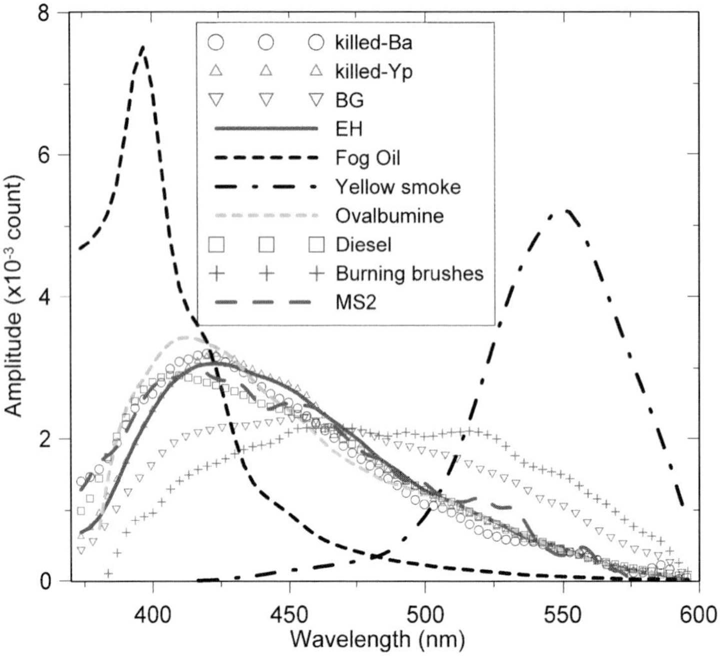

Fig. 16.3 Normalized fluorescence signatures extracted from the numerous acquisition performed by SINBAHD at JBSDS Demo II trial, 2005

sensor-dependent processed signatures, which can be considered as representing, to a different degree depending on the signal/noise (S/N) ratio, the specific LIF for these materials as measured by SINBAHD.

The fog oil and yellow smoke signatures are the most easily discernable among the ten clear signatures. In addition to having clear distinct spectral features, their fluorescence signal levels were much higher than the others (high S/N ratio). This implies that these interferants would be easily categorized by a classifier algorithm but could however, interfere with bioagent detection due to their relatively high signal intensity. Signatures of the vegetative cells EH and irradiated Yp have quite similar spectral features, which shows how difficult the discrimination may be between different materials of the same type. The last six signatures show discrepancies and differences between them to various degrees. The first interesting fact that must be underlined from these results is the obvious spectral differences between killed-Ba and BG, which are two species of *Bacillus* bacteria spores. In contrast to the vegetative cells studied, the two *Bacillus* species can easily be discriminated even though they are both bacteria spores. Note that the killing processes utilised for both the Ba spores and the YP vegetative cell (killed vaccine strain) probably had observable effects on their LIF signatures. Ovalbumin (OV) and MS2, which are respectively toxin and virus simulants, have spectral features much closer to killed-Ba than BG has. Discrimination between killed-Ba, MS2 and OV may be challenging.

16.6 Data Processing

Once the spectral signatures of all different materials have been obtained, the MA can be performed on trial data acquired during the open-air crosswind releases. The MA will evaluate the amplitudes of some pre-selected reference signatures in order to obtain the best possible fit, in a least squares sense, of the acquired spectra. Comparing the amplitudes obtained for different reference spectra permits the classification of the detected signal, up to a certain point. However, this is behind the scope of the present work. Depending on the specificity of the collected spectra and the 'goodness' of the reference spectra, the degree of confidence of this classification process may vary. In bio-medical science, using the term 'classification' as compared to 'identification' convey different meanings. The latter term implies that very specific microbiological and biochemical techniques have been followed to obtain taxonomic genus and species information.

The MA results obtained can then be converted into equivalent concentrations to obtain the corresponding detection limit. Since SINBAHD did not participate in the absolute measurement portion of the trial (sABT phase), its MA results will have to be correlated to the calibrated concentration data measured by the Dugway west desert reference LIDAR (WDL). This latter process will hence facilitate the evaluation of SINBAHD concentration detection limit for the different released materials. This process is limited in efficiency due mainly to the spatial correlation between the WDL and SINBAHD collected data. Indeed, they did not have the exact same line of sight through the target cloud. In spite of that, valid linear correlation between the two systems should be able to be obtained.

Figure 16.4 presents the WDL and SINBAHD MA results for a release of 20 g of killed *Bacillus anthracis* (BA) at about 1.2 km from sensors location on June 13, 2005. The WDL concentration length (solid line) was obtained by integrating the concentration over the region where the cloud appeared and multiplying by the length of a bin. SINBAHD MA results presented in Fig. 16.4 include the background (triangles) and killed-Ba (dashed line) signature amplitude. The concentration detection threshold, namely 4sigma, corresponds to four times the standard deviation of the studied material signature amplitude during the non-presence of this material. This 4sigma directly provides the SINBAHD detection limit for that particular material at a given range. Once the correlation between WDL and SINBAHD MA results is performed and the off-signal standard deviation is calculated, the detection limit can then be determined. In this particular case, for a cloud of 10 m along the LIDAR line of sight at 1.2 km, the detection limit in cloud concentration would be of 17 kppl (kppl = 1×10^3 particles per litre of air). One interesting point in this result is the ability to detect the LIF signature (dashed line) at lower signal levels than the background (full triangles). This result demonstrates the efficiency of the MA to exploit the acquired data even at detected signal levels lower than the background levels.

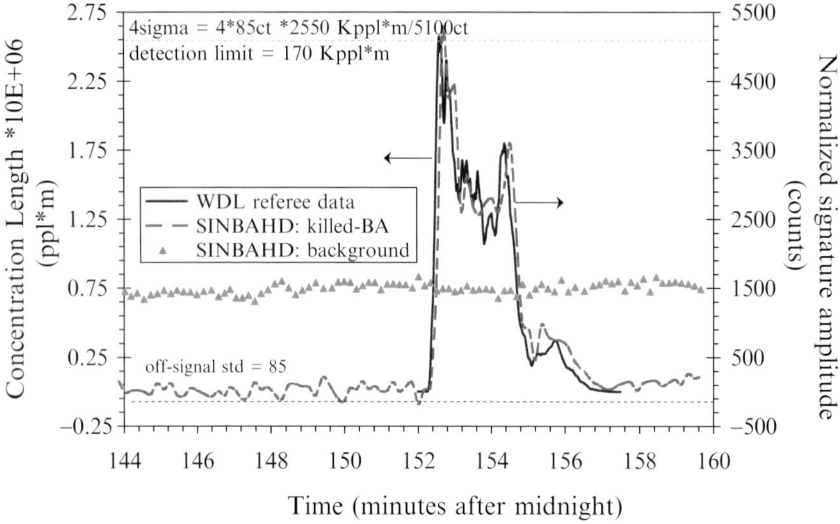

Fig. 16.4 WDL calibrated concentration (solid line) and SINBAHD (600 m gate width, around 10 m cloud depth) MA results for a release of killed-Ba at about 1.2 km (TDA015)

16.7 Conclusion

SINBAHD participation within the JBSDS Demo II field trial has provided further opportunities to explore its detection potential and also to obtain new spectral signatures in the pattern set (library). SINBAHD participated in 34 test releases for crosswind open air trials. From these acquisitions, 19 signatures could be extracted, nine did not have any fluorescence signal and one was lost due to a computer failure. The five other acquisitions, from which no fluorescence signal could be visually observed in real time, were processed with SINBAHD multivariate analysis. In these five cases, a detection signal could be extracted but with a fairly low S/N ratio. Signature differentiation was fairly obvious in most cases but not between the two vegetative cells, irradiated-Yp and EH. Indeed, the killed vaccine strain of *Yersinia pestis* (Yp) had about the same signature as *Erwinia herbicola* (EH), even though the first had been irradiated but not the latter. Spectral differences between irradiated-Ba and BG were noted (the two species of *Bacillus* bacteria spores). The signature of OV and MS2 showed spectral features much closer to killed-Ba. The background fluorescence was also properly characterized in order to optimise the MA results. The background fluorescence signature has a fairly stable level over an acquisition trial, and a quite stable spectral shape for different acquisition gate widths. All the SINBAHD results, once processed by the MA, were correlated with the data of DPG west desert LIDAR (WDL) referee system. From these correlations, the 4sigma sensitivity could be evaluated for a given cloud depth and range.

Acknowledgements We are grateful to the Biological Standoff Detection Field Demo trial team and to DPG personnel for their support throughout the trial. Special thanks to John Strawbridge, JPM Bio Defence, to have given us the opportunity to attend this trial.

References

Bronson R. (1989), *Matrix Operations*, Schaum's Outline Series, McGraw-Hill, New York.

Buteau S., Simard J.-R., and Roy G. (2005, October), *Standoff detection of natural bioaerosol by range-gated laser-induced fluorescence spectroscopy.* (Paper presented at Optics East SPIE conference—Chemical and Biological Standoff Detection III, #5995, Boston, USA).

Buteau S., Simard J.R., McFee J., Ho J., Lahaie P., Roy G., and Mathieu P. (2006), *Joint Biological Standoff Detection System Increment II:Field Demonstration—SINBAHD performances.* DRDC Valcartier TM2006-140, 83, UNCLASSIFED.

Chen C.L., Heglund D.L., Ray M.D., Harder D., Dobert R., Leung K.P., Wu M.,and Sedlacek III A.J. (1997), *Application of Resonance Raman Lidar for Chemical Species Identification,* SPIE Vol. 3065, pp. 279–285.

Condatore L.A.Jr., Guthrie R.B., Bradshaw B.J., Logan K.E., Lingvay L.S., Smith T.H., Kaffenberger T.S., Jezek B.W., Cannaliato V.J., Ginley W.J., and Hungate W.S. (1998, October), *U.S. Army Chemical and Biological Defense Command Counter-Proliferatioin Long-Range Biological Standoff Detection System*, (paper presented at the 4th Joint Workshop on Standoff Detection for Chemical and Biological Defense).

Evans B.T.N., Roy G., and Ho J. (1989), *The Detection and Mapping of Biological Simulants, Part 2: Preliminary Lidar Results*, DREV R-4480/89, UNCLASSIFIED.

Faris G.W., Copeland R.A., Mortelmans K., and Bronk B.V. (1997), Spectrally resolved absolute fluorescence cross sections for bacillus spores, *Appl. Opt.*, 36(4), 958–967.

Fell N.F.Jr., Pinnick R.G., Hill S.C., Videen G., Niles S., Chang R.K., Holler S., Pan Y., Bottiger J.R. and Bronk B.V. (1998, November), *Concentration, Size, and Excitation Power Effects on Fluorescence from Microdroplets and Microparticles Containing Tryptophan and Bacteria*, (paper presented at the Conference on Air Monitoring and Detection of Chemical and Biological Agents, Boston). SPIE Vol. 3533, pp. 52–63.

Gelbwachs J. and Birnbaum M. (1973), Fluorescence of atmospheric aerosols and lidar implications, *Appl. Opt.*, 12(10), 2442–2447.

Grant W. B. (1977), *Optical Bases for Remote Biological Aerosol Detection*, Edgewood Arsenal Contractor Report No. ED-CR-77025, Standford Research Institute.

Hairston P.P., Ho J.and Quant F.R. (1997), Design of an instrument for real-time detection of bioaerosols using simultaneous measurement of particle aerodynamic size and intrinsic fluorescence, *J. Aerosol Sci.*, 28(3), 471–482.

Hargis P.J. Jr., Lang A.R., Schmitt R.L., Henson T.D., Daniels J.W., Jordan J.D., Shcroder K.L. and Shokair I.R. (1998, October), *Sandia Multispectral Airborne Lidar for UAV Deployment*, (paper presented at the 4th Joint Workshop on Standoff Detection for Chemical and Biological Defense).

Ho J., Spence M., and Hairston P. (1999), Measurement of biological aerosol with a fluorescent aerodynamic particle sizer (FLAPS): Correlation of optical data with biological data, *Aerobiologia*, 15, 281–291.

Houston J.D. and Evans B.T.N. (1987), Monitoring Aerosols Optically, *Photo. Spec., 93.*

Hutt D.L., Bissonnette L.R., and Durand L. (1994), Multiple field of view lidar returns from atmospheric aerosols, *Appl. Opt.*, 33, 2338–2348.

Measures R.M. (1984), *Laser Remote Sensing: Fundamentals and Applications*, Wiley Interscience, New York.

Press W.H., TeukolskyS.A., Vetterling W.T. and Flannery B.P. (1992), *Numerical Recipes in C: The Art of Scientific Computing, Second Edition*, Cambridge University Press, Cambridge.

Roy G., Bissonnette L., Bastille C., and Vallée G. (1999), Retrieval of droplet-size density distribution from multiple-field-of-view cross-polarized lidar signals: Theory and experimental validation, *Appl. Opt.*, 38(24), 5202–5211.

Simard J.-R., Mathieu P., Theriault J.-M., Larochelle V., Roy G., McFee J.E., and Ho J. (1999). *Active Range-Gated Spectrometric Standoff Detection and Characterisation of Bioaerosols: Preliminary Analysis*, DREV TM1999–056, 34 pp, UNCLASSIFED.

Simard J.-R., Roy G., Mathieu P., Larochelle V., McFee J.E. and Ho J. (2003). *Standoff Integrated Bioaerosol Active Hyperspectral Detection (SINBAHD): Final report*, DREV TR 2002–125, 118 pp, UNCLASSIFED.

Simard J.R., Roy G., Mathieu P., Larochelle V., McFee J.E. and Ho J. (2004), Standoff sensing of bioaerosols using intensified range-gated spectral analysis of laser-induced fluorescence. *IEEE Trans. Geosci. Remote Sens.*, 42(4), 865–874.

Chapter 17
MODIS 500 × 500-m^2 Resolution Aerosol Optical Thickness Retrieval and Its Application for Air Quality Monitoring

Kwon H. Lee[1], Dong H. Lee[1], Young J. Kim[1*], and Jhoon Kim[2]

Abstract Atmospheric aerosol optical thickness (AOT) was retrieved over urban areas using moderate resolution imaging spectroradiometer (MODIS) 500-m resolution calibrated radiance data. A modified Bremen Aerosol Retrieval (BAER) algorithm was used to retrieve AOT over Seoul, Korea. Since the surface reflectance of an urban area is typically brighter than that of vegetation or soil areas in the visible wavelength region, the error associated with aerosol retrieval is higher. Surface reflectance determination using the linear mixing model (LMM) produced values that were smaller (~0.02) than those obtained using the minimum reflectance technique (MRT). This difference would lead to an overestimated AOT when using the LMM approach. Retrieved AOT data using MRT and standard MODIS 10-km aerosol products (MOD04) were compared to Rotating Shadow-band Radiometer (RSR) observations at Yonsei University, Seoul, Korea. Regression analysis showed that the root mean square error (RMSE) of MODIS AOT and MOD04 AOT was 0.05 in both cases. In addition, MODIS AOT data were compared with ground-based particulate matter data (PM10) from air quality monitoring networks of the National Institute of Environmental Research (NIER). MODIS AOT data showed a relatively low correlation ($r = 0.41$) with surface PM10 mass concentration data due to differences in ground-based and column-averaged data, variability of terrain, and MODIS cloud mask. This result is similar to that of other investigations. The application of fine-resolution satellite data supports the feasibility of local and urban-scale air quality monitoring.

Keywords: AOT, LMM, MODIS, MRT, PM10

[1] *Advanced Environmental Monitoring Research Center (ADEMRC), Gwangju Institute of Science & Technology (GIST), 1 Oryong-dong, Buk-gu, Gwangju 500-712, Republic of Korea*

[2] *Department of Atmospheric Sciences, Yonsei University, Shinchon-dong 134, Seodaemun-gu, Seoul 120-749, Republic of Korea*

**Corresponding author: Tel: + 82-62-970-3401, Fax: + 82-62-970-3404*

Y.J. Kim and U. Platt (eds.), *Advanced Environmental Monitoring*, 217–230.

17.1 Introduction

Satellite remote sensing has provided quantitative information on aerosols with an accuracy comparable to that of surface measurements. The large-scale distribution of aerosol concentration and characteristics, aerosol radiative forcing, and property changes of clouds interacting with aerosols are among the important observations that can be provided by satellite remote sensing. Due to the growing recognition of the importance of aerosol properties for studies of climate and global change, it is indeed fortunate that a number of very significant and greatly enhanced satellite systems are being developed for launch in the next few years. These systems will also provide information on the global distribution, including seasonal and interannual variation, of aerosol sources (forest fires, desert dust and aerosol from the oxidation of SO_2 emissions from industrial regions), aerosol loading and optical properties, as well as direct and indirect radiative forcing.

Satellite measurements can also be inverted to yield information on aerosol optical thickness (AOT), angular scattering properties, and size distribution. Satellite measurements clearly have the advantage of being the only set of measurements that provide a wide coverage. The disadvantage of the presently used passive sensors is that they can only be applied under clear-sky conditions, and their application over land is difficult. Furthermore, a number of additional assumptions have to be made regarding atmospheric conditions. In spite of these disadvantages, many researchers have studied global and local-scale aerosol remote sensing using satellites. Techniques associated with satellite observation in the field of atmospheric science are advancing rapidly (Kaufman et al. 1997a; Gordon et al. 1997; King et al. 1999). Total ozone mapping spectrometer (TOMS) measurements are designed to monitor ozone and provide valuable information on aerosol geographical distributions (Herman et al. 1997). Such measurements were used to detect an increase in biomass burning smoke in African savanna regions during the 1990s (Hsu et al. 1999). The Sea-viewing Wide Field-of-view Sensor (SeaWiFS), developed to provide ocean color data products for the study of marine biogeochemical processes, produces an aerosol data product from its atmospheric correction algorithm (Gordon and Wang 1994) through a separate algorithm (von Hoyningen-Huene et al. 2003), and has been used to investigate transport of Asian dust plumes (Lee et al. 2004). The launch of polarization and directionality of the earth's reflectance (POLDER) on ADEOS II adds greater capabilities (Leroy et al. 1997). Extracting the full potential from the satellite data requires correlative suborbital data to not only validate the satellite measurements (Clark et al. 1997; Fraser et al. 1997; Vermote et al. 1997), but to supply information not obtainable from space. The EOS/Terra and EOS/AQUA satellites and their new instruments, such as the moderate resolution imaging spectroradiometer (MODIS) and multi-angle imaging spectroradiometer (MISR), provide improved measurement capabilities and unprecedented volumes of data regarding the atmosphere, land, ocean, and radiative processes (Tanré et al. 1997; Wanner et al. 1997).

AOT is a key property used in the characterization of atmospheric aerosols from satellite data. The radiant energy obtained from an earth-observing satellite sensor is converted into reflectance or apparent reflectance using information about the atmospheric condition along the path between the ground and the satellite, as well as the viewing geometry and solar geometry. Aerosols also block the view for systems observing changes in land use. Basic aerosol retrieval techniques are therefore employed to remove the aerosol signal, and can also be used to highlight or enhance the aerosol-only signal. King et al. (1999) summarized the major retrieval methods well.

Aerosol retrieval from remote sensing data has many important applications, including a role in climate studies and the atmospheric correction of satellite imagery. Remote sensing of atmospheric aerosol using MODIS satellite data has been shown to be very useful in global/regional-scale aerosol monitoring. Recently, satellite-retrieved AOT has also been used to estimate local/regional air quality (Engel-Cox et al. 2004a,b; Hutchison 2003; Hutchison et al. 2004). Due to the large spatial resolution of $10 \times 10\text{-km}^2$ for the MODIS standard aerosol product (MOD04), AOT data have limitations for local/urban-scale air quality monitoring applications. Its large spatial resolution is incapable of reflecting various types of urban-scale (Seoul ~ 600 km^2) pollution with a complicated topography. Therefore, a modified Bremen Aerosol Retrieval (BAER) algorithm (von Hoyningen-Huene et al. 2003) as detailed by Lee et al. (2005) was used in this study to retrieve AOT for fine resolutions of $500 \times 500\,\text{m}^2$ over Seoul, Korea. Figure 17.1 shows the study area.

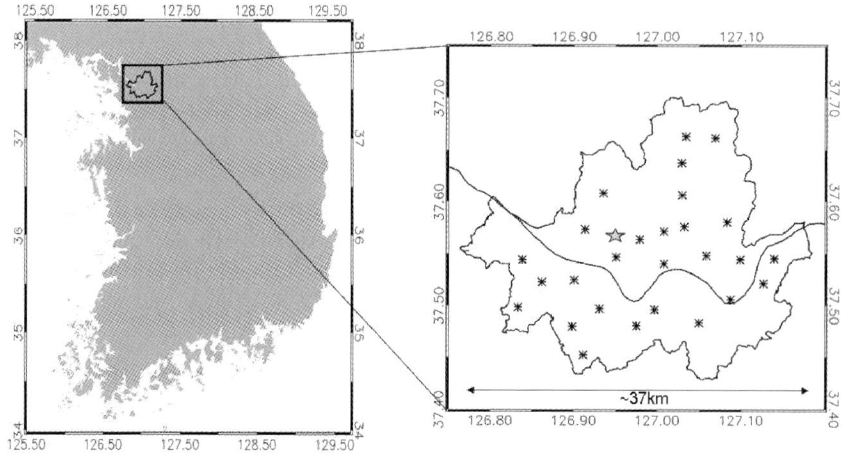

Fig. 17.1 Map of the Korean peninsular and the Seoul metropolitan area. Symbols on the right represent the 27 air quality monitoring stations (asterisks) and the radiometer measurement site (large star)

17.2 Methodology

MODIS is a multispectral (36 bands), multi-resolution (1 km, 500 m, 250 m) sensor dedicated to the observation of the Earth. It is a key instrument aboard the Terra (EOS AM) and Aqua (EOS PM) satellites. Terra's orbit around the Earth is timed so that it passes from north to south across the equator in the morning, while Aqua passes south to north over the equator in the afternoon. Two bands are imaged at a nominal resolution of 250 m at nadir, with five bands at 500 m and the remaining 29 bands at 1 km. A ±55-degree scanning pattern at the EOS orbit of 705 km achieves a 2,330-km swath and provides global coverage every 1–2 days. Global/regional-scale aerosol distribution information has been provided by MODIS measurements since the launch of the satellites. However, the operational MODIS AOT from the NASA algorithm is not appropriate for local/urban-scale aerosol monitoring as it has a spatial resolution of 10 × 10-km^2. Therefore, we used TERRA/MODIS 500-m resolution data (0.47, 0.55, 0.66 μm) for aerosol retrieval over an urban area for 2004 in this study. Figure 17.2 briefly describes the aerosol retrieval processes using MODIS 500-m resolution data.

The aerosol retrieval algorithm for MODIS 500-m resolution data is a modified version of BAER, which retrieves aerosol properties in greater detail and can be applied to other sensors with relative ease. In order to retrieve AOT, a radiative transfer equation was introduced to determine aerosol reflectance by decomposing the top-of-atmosphere (TOA) reflectance into surface reflectance and Rayleigh path radiance (Kaufman et al. 1997a). The TOA reflectance ρ_{TOA} (θ_0, θ_S, ϕ) is expressed as

$$\rho_{TOA}(\theta_0,\theta_S,\phi) = \rho_{ATM}(\theta_0,\theta_S,\phi,\tau_{Aer},\tau_{Ray},p(\theta),\varpi_0) + \frac{T_{Tot}(\theta_0)\cdot T_{Tot}(\theta_S)\cdot \rho_{Surf}(\theta_0,\theta_S)}{1-\rho_{Surf}(\theta_0,\theta_S)\cdot r_{Hem}(\tau_{Tot},g)} \quad (1)$$

where θ_0 is the solar zenith angle, θ_S the satellite zenith angle, and ϕ the azimuth angle. τ_{Aer}, τ_{Ray} and τ_{Tot} represent aerosol, Rayleigh, and total optical thickness, respectively. $p(\theta)$ is the phase function, ϖ_0 a single-scattering albedo, g the asymmetry parameter, ρ_{ATM} the atmospheric path reflectance, $T_{Tot}(m_0)$ the total transmittance, $\rho_{Surf}(m_0, m_S)$ the surface reflectance, and r_{Hem} (τ_{Tot}, g) the hemispheric reflectance. The atmospheric contribution consists of the Rayleigh and aerosol fractions. The Rayleigh path reflectance can be determined through the use of a relevant wavelength method (Bucholtz 1995). The Rayleigh optical thickness ($\tau_{Ray}(\lambda)$) can then be determined from the surface pressure $p(z)$ in each pixel through the following equation:

$$\tau_{Ray}(\lambda) = A\cdot\lambda^{-\left(B+C\cdot\lambda+\frac{D}{\lambda}\right)}\cdot\frac{p(z)}{p_0} \quad (2)$$

where A, B, C and D are the constants for the total Rayleigh scattering cross-section and total Rayleigh volume scattering coefficient for a standard atmosphere. p_0 is the mean sea level surface pressure.

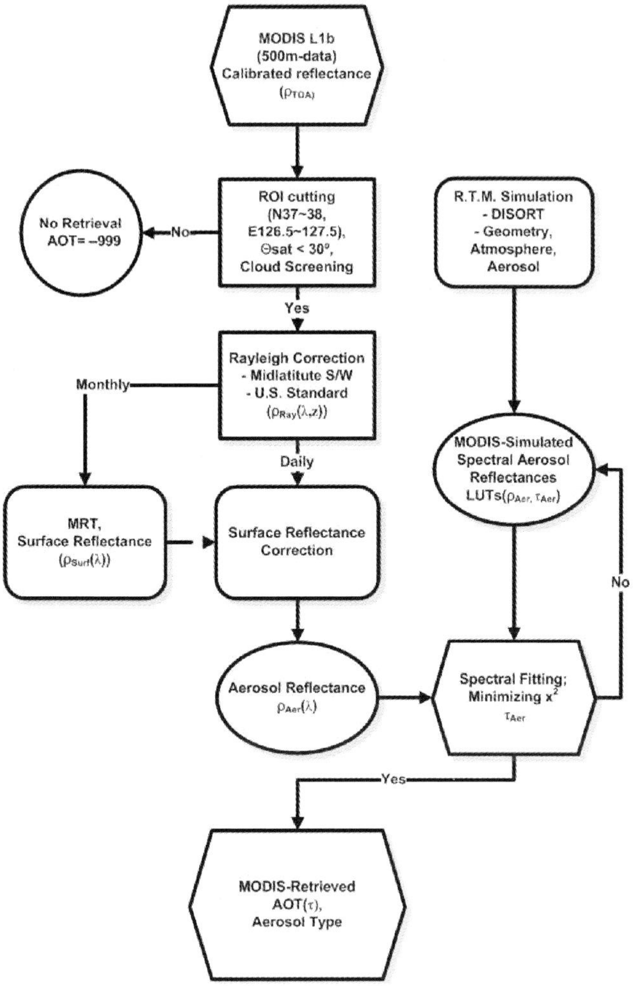

Fig. 17.2 Schematic diagram for aerosol retrieval in this study

There exists an inherent problem in determining surface reflectance for aerosol retrieval, since photons reflected by the surface reach the satellite after they transmit through the atmosphere. However, aerosol retrieval over a bright surface such as an urban area is problematic because of the large uncertainties concerning the contribution of surface reflectance to satellite data. Although Kaufman et al. (1997c) used the 2.1-um channel reflectance for estimation of surface reflectance at visible wavelengths, it is mainly suitable over vegetation or dark soils. The original BAER algorithm used a linear mixing model (LMM) between vegetation and soil spectra for surface reflectance determination in a pixel. Surface reflectance (ρ_{Surf}) can be calculated by portion of vegetation and bare soil spectra in the following manner:

$$\rho_{Surf}(\lambda) = C_{Veg} \cdot \rho_{Veg}(\lambda) + (1 - C_{Veg}) \cdot \rho_{Soil}(\lambda) \tag{3}$$

where λ is wavelength, and $\rho_{Veg}(\lambda)$ and $\rho_{Soil}(\lambda)$ the spectral reflectances of 'green vegetation' and 'bare soil', respectively. C_{Veg} is the vegetation index representative of the vegetation fraction in each pixel. LMM also possesses a problem since it uses only two surface composition parameters. Recently, new techniques like 'Deep Blue' as introduced in Hsu et al. (2004) can retrieve AOT over bright surfaces using the minimum reflectivity technique (MRT). In this study, we used MRT to determine the surface reflectance by finding the clearest scene during each month over an urban area. The MOD35 cloud screening method (Ackerman et al. 1998) was used to determine cloud contamination. Near nadir data (viewing angle < 35°) were only used to minimize angular effects due to bidirectional reflectivity by inhomogeneous surfaces. Since cloud shadows can also lead to lower reflectance in a pixel, the second minimum reflectance was taken.

Physical and optical characteristics of aerosol particles in the atmosphere also greatly affected satellite aerosol retrieval. Since the original BAER method uses one Look-up Table (LUT) for aerosol retrieval to one whole satellite scene, this method should be improved. In this study, a few aerosol models for LUT construction were used to determine the best aerosol model selection method for a pixel. LUTs were constructed from the discrete ordinate radiative transfer (DISORT) code (Stamnes et al. 1988). MODIS-received aerosol reflectances were simulated with the four aerosol models (1, 2, 3 and 4) presented in Table 1 from Lee et al. 2007, comprising four standard aerosol models (rural [RR], urban [UB]) from Shettle and Fenn (1979) and two AERONET-derived aerosol models (Asian dust [AD] and biomass burning [BB] aerosol models). The non-sphericity of dust particles should be considered in scattering calculations, which could cause changes in the phase function signature. Therefore, non-spherical dust phase function data were used to minimize the effects of non-sphericity of aerosol in our retrieval method. For LUT construction, TOA reflectances were computed for various input conditions that included solar and sensor geometries, Rayleigh scattering, the six aerosol models, a dark surface, and an AOT ranging from 0 to 3. An optimum spectral shape-fitting procedure was used for best aerosol model selection by comparing the MODIS-measured spectral aerosol reflectances with corresponding calculated values (Kaufman and Tanré 1998; Costa et al. 1999; Torricella et al. 1999). Minimizing the error term (x^2) that describes the agreement between measurement and simulation involved the use of the following equation:

$$x^2 = \frac{1}{n} \sum_{i=1}^{n} \left(\frac{\rho^m_{Aer}(\lambda_i) - \rho^c_{Aer}(\lambda_i)}{\rho^m_{Aer}(\lambda_i)} \right)^2 \tag{4}$$

where n is the number of selected wavelengths (λ_i), and $\rho^m{}_{Aer}$ and $\rho^c{}_{Aer}$ are the measured and calculated aerosol reflectances, respectively. Fitting was done at the selected wavelengths of 0.47, 0.55, 0.66 μm.

Retrieval errors under various aerosol loadings were estimated by comparing MODIS AOT data with broadband rotating shadowband radiometer (RSR) observations at Yonsei University (N 37.565°, E 126.935 °), Seoul, Korea. Furthermore, the application of MODIS AOT data for an assessment of urban air quality is also discussed through comparisons with PM10 mass concentrations recorded at a ground air quality monitoring network in Korea.

Daily aerosol product (Tanré et al. 1997; Kaufman et al. 1997a,b) of MODIS level 2 aerosol datasets (MOD04 L2; MODIS aerosol product, Version 4.1.3) were also collected from the National Aeronautics and Space Administration (NASA) Distributed Active Archive Center (DAAC) for comparisons with MODIS 500-m resolution AOT data obtained in this study. The MOD04 data include various aerosol physical and optical parameters with a $10 \times 10\text{-km}^2$ spatial resolution. MODIS AOT has been validated with ground-based sunphotometer AOT using a spatiotemporal approach (Ichoku et al. 2002). MODIS aerosol retrievals over land surfaces, except in coastal zones, possess retrieval errors of $\Delta\tau_a = \pm 0.05 \pm 0.2\ \tau_a$ (Chu et al. 2002).

17.3 Result

17.3.1 Surface Reflectance Assessment

In order to retrieve AOT using MODIS 500-m resolution data, surface reflectance should first be determined and then the temporal and local variability of surface properties within the study area should be estimated. Our surface reflectance determination by MRT was compared with that of LMM in Figure 17.3. The MODIS RGB (red = 0.66 μm, green = 0.55 μm, and blue = 0.47 μm) color composite image shows green vegetation areas and vegetation index values that are relatively higher (>0.6) than the urban area (<0.4), which mainly consists of artificial construction materials. In LMM analysis, a low vegetation index leads to a high soil reflectance portion for mixed reflectance in a pixel. However, this value is less than the real reflectance of an urban area. The difference in surface reflectance between LMM and MRT was about 0.02. This underestimated surface reflectance estimation could lead to a larger AOT (~0.2).

17.3.2 Validation

MODIS-retrieved AOT results were compared with ground-based measured AOT in Figure 17.4. The RSR measurement site is located within Yonsei University (see Figure 17.1). Since MODIS scans the study area at around 02~03 UTC, RSR data closest in time (±30 min) and MODIS data from the nearest pixel to the measurement site were used for the comparison. Although the comparison was done for a limited number of cases, there was good agreement between RSR-measured

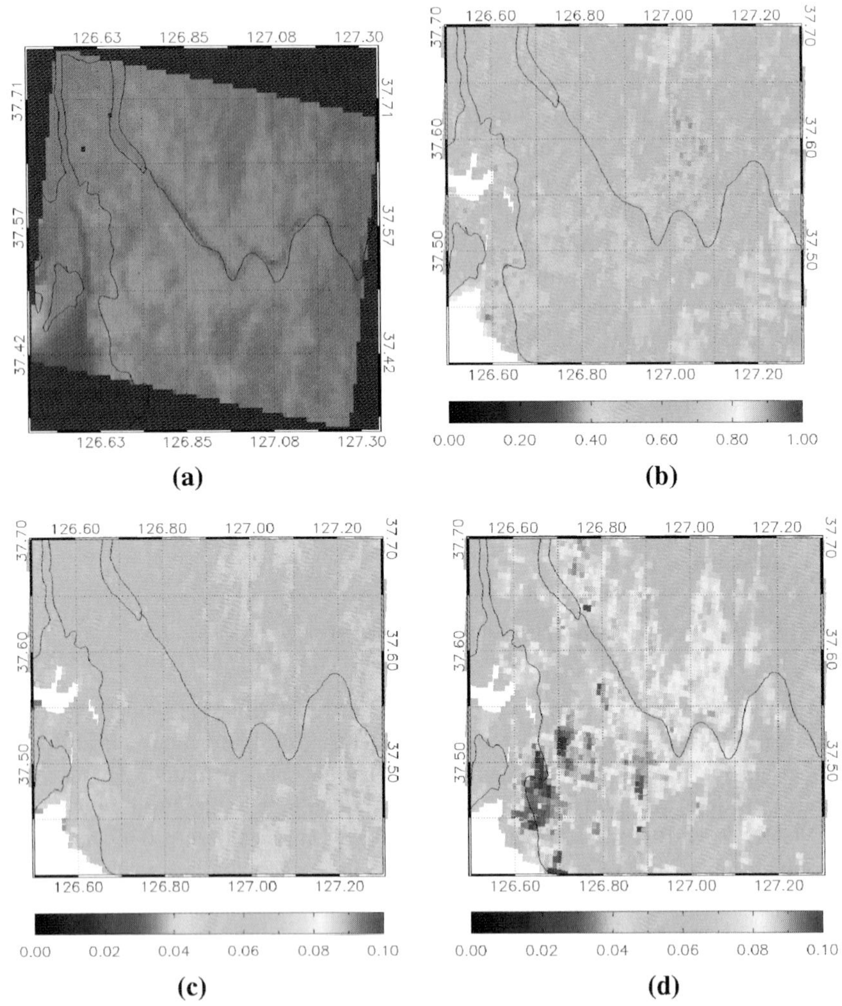

Fig. 17.3 MODIS RGB composite image (**a**) aerosol-free vegetation index (**b**) (AFRI, Karnieli et al. 2001), and surface reflectance ($\lambda = 0.46$) determined by (**c**) LMM and (**d**) MRT

and MODIS-retrieved AOT, with a linear-fitting correlation coefficient (r) of 0.80 and 0.74, respectively. Moreover, when compared with MOD04 AOT data, MODIS 500-m resolution AOT data showed a much better correlation with RSR AOT data over Seoul, suggesting a lower level of accuracy for 10-km resolution data because of its coarse spatial resolution.

Figure 17.5 shows the AOT map for different aerosol loading cases on a hazy and clear day, with MODIS RGB images and MODIS AOT showing the different aerosol loading states. On April 24, 2004, the sky was relatively clear and there was

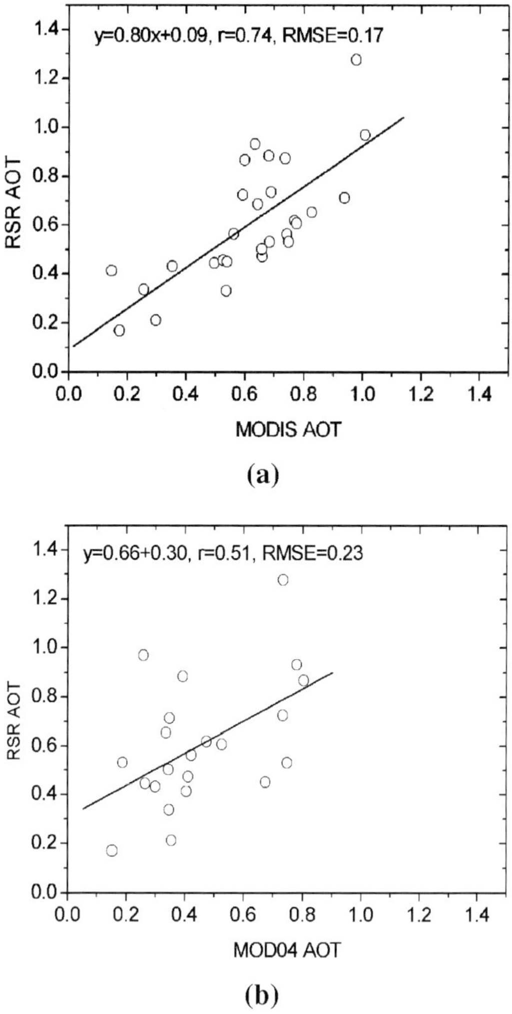

Fig. 17.4 Comparison between (**a**) MODIS AOT and (**b**) RSR AOT

a northerly wind blowing from North Korea towards the Seoul metropolitan area. AOT was very low (~0.3) in Seoul. On December 8, 2004, a Chinese haze plume was transported to the Korean peninsular by a westerly wind and an increased AOT value (~2.0) was the result of aerosol mass increases in the local atmosphere. Note that the white areas indicate no-retrieval pixel due to clouds and a high satellite viewing angle.

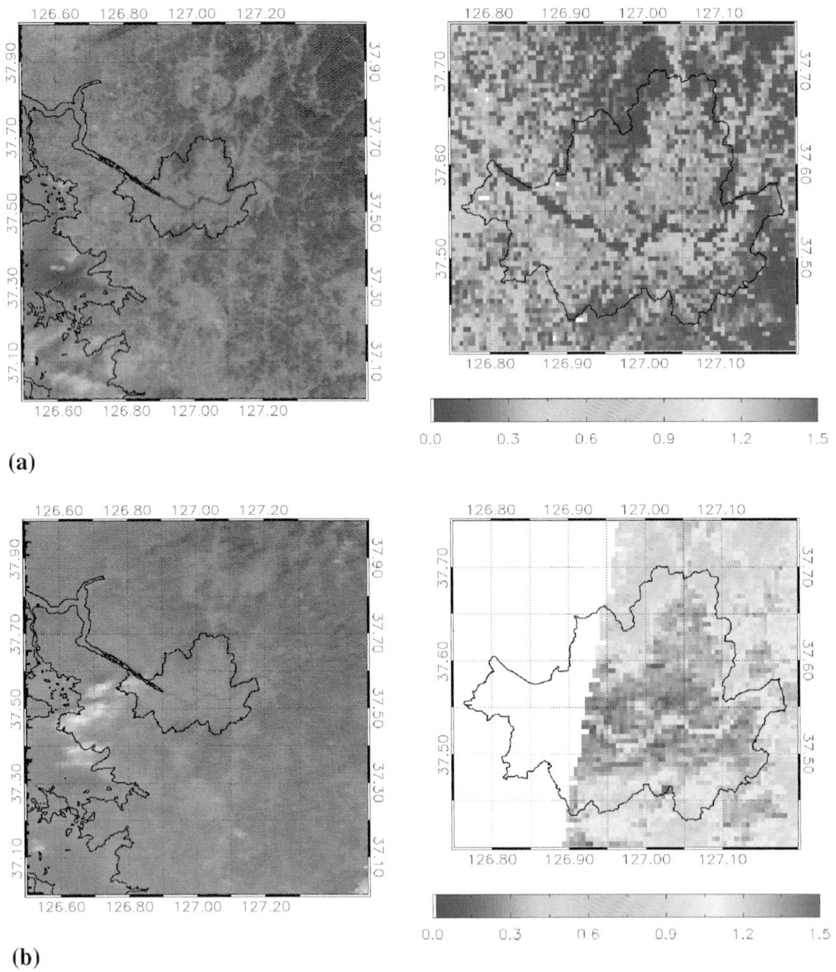

Fig. 17.5 MODIS RGB images and AOT over Seoul, Korea on (**a**) April 24, 2004 (clear day), and (**b**) December 8, 2004 (hazy day)

17.3.3 Application to Air Quality

In this section, MODIS-retrieved AOT data were compared to data pertaining to PM monitoring in Seoul, Korea. Figure 17.6 shows the comparison between PM10 and MODIS AOT for the 27 locations in Seoul, Korea. The linear correlation coefficient (r) is 0.41, suggesting that the PM10 mass concentration is indicative of near surface values that are not as well reflected by the MODIS column AOT data. The correlation between satellite AOT and PM increases with the use of PM2.5 mass concentration data. There was a better correlation between MODIS AOT and surface PM10 concentration when compared to other published values

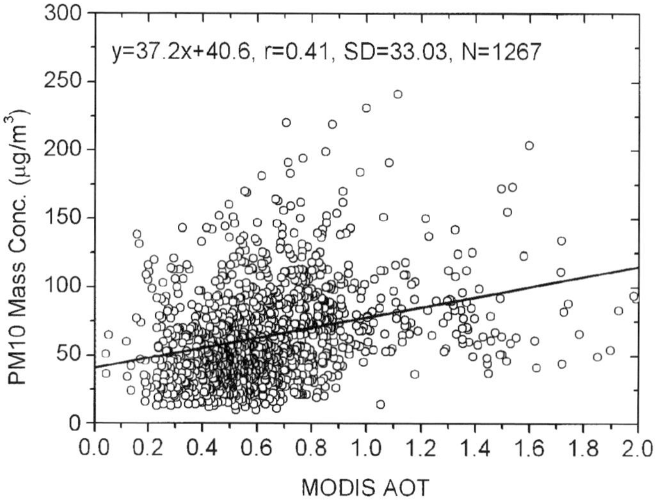

Fig. 17.6 Comparison between MODIS AOT and PM10 mass concentrations measured from 27 air quality monitoring stations in Seoul, Korea

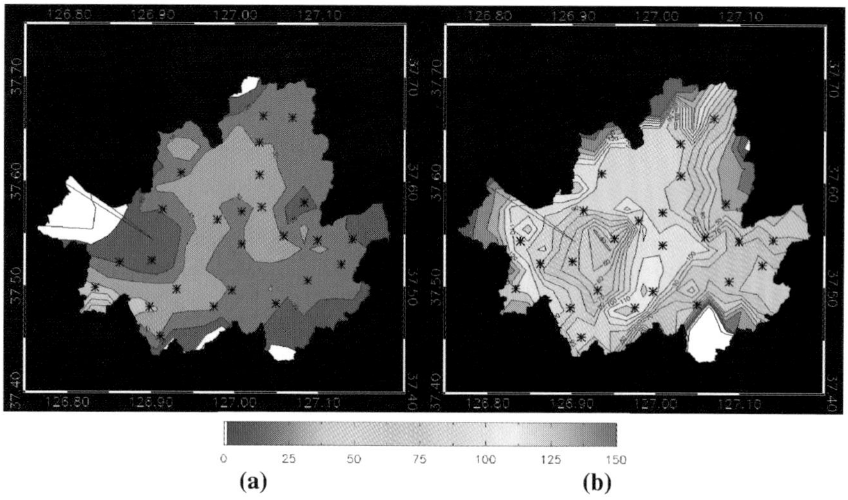

Fig. 17.7 Contoured PM10 mass concentration map on (**a**) April 24, 2004 and (**b**) December 8, 2004 in Seoul, Korea

(Engel-Cox et al. 2004b), which supports the feasibility of using high-resolution MODIS AOT for local and urban-scale air quality monitoring.

For an easier comparison between satellite-retrieved AOT and ground-based PM10 mass concentrations, the contour map of MODIS AOT and PM10 in Seoul for the days detailed in Figure 17.5 are shown in Figure 17.7(a) and 17.7(b). These PM10 distributions clearly depict a high PM10 loading case with large AOT values on December 8, 2004 and a low PM10 loading case with low AOT on other day.

17.4 Conclusion

This study has demonstrated the retrieval of MODIS 500-m resolution AOT products and their potential application to urban air quality monitoring. MRT was used successfully for fine-resolution aerosol retrieval over bright land surfaces. Furthermore, 500-m resolution AOT was estimated with a better accuracy of MODIS standard aerosol products when compared with ground-based radiometer observation data. Fine-resolution aerosol retrieval data can be used to study spatial aerosol loading distribution and the impact of transient pollution on local/urban air quality. The initial application of MODIS AOT for a study of air quality with ground-measured PM10 mass concentration data is encouraging and indicates that fine spatial resolution satellite scenes can provide air quality information. Despite the moderate correlation between MODIS AOT and PM10 mass concentrations, this case study demonstrates the usefulness of satellite-retrieved AOT data for air quality monitoring.

Acknowledgement This work was supported in part by the Korea Science and Engineering Foundation (KOSEF) through the Advanced Environmental Monitoring Research Center (ADEMRC) at the Gwangju Institute of Science and Technology (GIST) and a research project from the Korea Aerospace Research Institute (KARI). PM data used in this study were acquired from the Korea National Institute of Environmental Research (NIER). MODIS data used in this study were acquired from the NASA Information Services Center (DISC) Distributed Active Archive Center (DAAC).

References

Ackerman S.A., Strabala K., Menzel P., Frey R., Moeller C., Gumley L., Baum B., Seeman S.W., and Zhang H. (1998), Discriminating clear-sky from cloud with MODIS: Algorithm theoretical basis document (MOD35), *J. Geophys. Res.*, 103, 96, 2857–2864.

Bucholtz A. (1995). Rayleigh-scattering calculations for the terrestrial atmosphere. *Appl. Opt.*, 34, 2765–2773.

Chu D.A., Kaufman Y.J., Ichoku C., Remer L.A., Tanré D., and Holben B.N. (2002), Validation of MODIS aerosol optical depth retrieval overland. *Geophys. Res. Lett.*, 29, DOI 10.1029/2001GL013205.

Clark D.K., Gordon H.R., Voss K.J., Ge Y., Broenkow W., and Trees C. (1997), Validation of atmospheric correction over the oceans *J. Geophys. Res.-Atmos.*, 102(D14), 17209–17217.

Costa M.J., Cervino M., Cattani E., Torricella F., Levizzani V., and Silva A.M. (1999), Aerosol optical thickness and classification: Use of METEOSAT, GOME and modeled data. EOS-SPIE International Symposium on Remote Sensing, Proc. SPIE Vol. 3867, Satellite Remote Sensing of Clouds and the Atmosphere IV, J. E. Russell (Ed.), 268–279.

Engel-Cox J.A., Hoff R.M., and Haymet A. (2004a), Recommendations on the use of satellite remote-sensing data for urban air quality, *J. Air Waste Manag.*, 54, 1360–1371.

Engel-Cox J.A., Holloman C.H., Coutant B.W., and Hoff R.M. (2004b), Qualitative and quantitative evaluation of MODIS satellite sensor data for regional and urban scale air quality, *Atmos. Environ.*, 38(16), 2495–2509.

Fraser R.S., Mattoo S., Yeh E.N., and McClain C.R. (1997) Algorithm for atmospheric and glint corrections of satellite measurements of ocean pigment, *J. Geophys. Res.*, 102, (D14), 17107–17118.

Gordon H.R. and Wang M. (1994), Retrieval of water-leaving radiance and aerosol optical thickness over the oceans with SeaWiFS: A preliminary algorithm, *Appl. Opt.*, 33, 443–452.

Gordon H.R., Du T., and Zhang T. (1997), Atmospheric correction of ocean color sensors: Analysis of the effects of residual instrument polarization sensitivity, *Appl. Opt.*, 36, 6938–6948.

Herman J.R., Bhartia P.K., Torres O., Hsu C., Seftor C., and Celarier E. (1997), Global distribution of UV-absorbing aerosols from Nimbus 7/TOMS data, *J. Geophys. Res.*, 102, 16911–16922.

Hsu N.C., Herman J.R., Torres O., Holben B.N., Tanré D., Eck T.F., Smirnov A., Chatenet B., and Lavenu F. (1999), Comparison of the TOMS aerosol index with Sun-photometer aerosol optical thickness: Results and application, *J. Geophy. Res.*, 23, 745–748.

Hsu N.C., Tsay S.C., King M.D., and Herman J.R. (2004), Aerosol retrievals over bright- reflecting source regions, *IEEE Trans. Geosci. Remote Sens.*, 42, 557–569.

Hutchison K.D. (2003), Applications of MODIS satellite data and products for monitoring air quality in the state of Texas, *Atmos. Environ.*, 37(17), 2403–2412.

Hutchison K.D., Smith S., and Faruqui S. (2004) The use of MODIS data and aerosol products for air quality prediction, *Atmos. Environ.*, 38(30), 5057–5070.

Ichoku C., Chu D.A., Mattoo S., Kaufman Y.J., Remer L.A., Tanré D., Slutsker I., and Holben B. (2002), A spatio-temporal approach for global validation and analysis of MODIS aerosol products, *Geophys. Res. Lett.*, 29, DOI 10.1029/2001GL013206.

Stamnes K., Tsay S., Wiscombe W., and Jayaweera K. (1988), Numerically stable algorithm for discrete-ordinate-method radiative transfer in multiple scattering and emitting layered media, *Appl. Opt.*, 27, 2502–2509.

Kaufman Y.J., Tanré D., Gordon H.R., Nakajima T., Lenoble J., Frouin R., Grassl H., Herman B.M., King M.D., and Teillet P.M. (1997a), Passive remote sensing of tropospheric aerosol and atmospheric correction for the aerosol effect, *J. Geophys. Res.*, 102(D14), 16815–16830.

Kaufman Y.J., Tanré D., Remer L., Vermote E.F., Chu A., and Holben B.N. (1997b). Operational remote sensing of tropospheric aerosol over the land from EOS-MODIS, *J. Geophys. Res.*, 102(14), 17051–17068.

Kaufman Y.J., Wald A.E., Remer L.A., Gao B.C., Li R.R., and Flynn L. (1997c), The MODIS 2.1-μm channel correlation with visible reflectance for use in remote sensing of aerosol, *IEEE Trans. Geosci. Remote Sens.*, 35, 1286–1298.

Kaufman Y.J. and Tanré D. (1998), Algorithm for Remote Sensing of Tropospheric Aerosol from MODIS, Algorithm theoretical basis document, ATBD-MOD-02, NASA Goddard Space Flight Center.

King M.D., Kaufman Y.J., Tanré D., and Nakajima T. (1999), Remote sensing of tropospheric aerosols from space: Past, present and future, *Bull. Am. Meteorol. Soc.*, 80(11), 2229–2259.

Lee K.H., Kim Y.J., and Hoyningen-Huene W. (2004), Estimation of aerosol optical thickness over Northeast Asia from SeaWiFS data during the 2001 ACE-Asia IOP, *J. Geophys. Res.*, 109, D19S16, DOI 10.1029/2003JD004126.

Lee K. H., Kim J. E., Kim Y. J., Kim J., and Hoyningen-Huene W. (2005), Impact of the smoke aerosol from Russian forest fires on the atmospheric environment over Korea during May 2003, *Atmos. Environ.*, 39(1), 85–99.

Lee K.H., Kim Y.J., von Hoyningen-Huene W., and Burrow J.P. (2007), Spatio-Temporal variability of atmospheric aerosol from MODIS data over Northeast Asia in 2004, *Atmos. Environ.*, 41(19), 3959–3973.

Leroy M., Deuzé J.L., Bréon F.M., Hautecoeur O., Herman M., Buriez J.C., Tanré D., Bouffies S., Chazette P., and Roujean J.L. (1997), Retrieval of atmospheric properties and surface bidirectional reflectances over land from POLDER/ADEOS, *J. Geophys. Res.*, 102, 17023–17038.

Shettle E.P. and Fenn R.W. (1979) Models for the aerosols of the lower atmosphere and the effects of humidity variations on their optical properties. AFGL-TR-79-0214, U.S. Air Force Geophysics Laboratory, Hanscom Air Force Base, MA.

Tanré D., Kaufman Y.J., Herman M., and Mattoo S. (1997), Remote sensing of aerosol properties over oceans using the MODIS/EOS spectral radiances, *J. Geophys. Res.*, 102, 16971–16988.

Torricella F., Cattani E., Cervino M., Guzzi R., and Levoni C. (1999), Retrieval of aerosol properties over the ocean using global ozone monitoring experiment measurements: Method and applications to test cases, *J. Geophys. Res.*, 104 (D10), 12085–12098.

Vermote E.F., El Saleous N., Justice C.O., Kaufman Y.J., Privette J.L., Remer L., Roger J. C., and Tanré D. (1997), Atmospheric correction of visible to middle infrared EOS-MODIS data over land surface, background, operational algorithm and validation, *J. Geophys. Res.*, 102(14), 17131–17141.

von Hoyningen-Huene W., Freitag M., and Burrows J.B. (2003), Retrieval of aerosol optical thickness over land surfaces from top-of-atmosphere radiance, *J. Geophys. Res.*, 108(D9), 4260, doi:10.1029/2001JD002018.

Wanner W., Strahler A.H., Hu B., Lewis P., Muller J.P., Li X., Schaaf C.L. Barker, and Barnsley M.J. (1997), Global retrieval of bidirectional reflectance and albedo over land from EOS MODIS and MISR data: Theory and algorithm, *J. Geophys. Res.*, 102, 17143–17162.

Section 3
Contaminant-Control Process Monitoring

Chapter 18
Aquatic Colloids: Provenance, Characterization and Significance to Environmental Monitoring

Jae-Il Kim

Abstract Aquatic colloids are ubiquitous in all kinds of natural water and in general found to be small in size (<100 nm) and low in number density (<10^{14} particles per liter). Colloids of such properties may play a significant role for the aquifer migration of environmentally hazardous contaminants: radioactive elements as well as other trace chemical composites. Insightful knowledge on aquatic colloids is therefore perceived as indispensable for monitoring the environmental behavior of hazardous trace constituents.

This chapter describes the chemical process of generating aquatic colloids, e.g. their kernels like hydroxy aluminosilicate (HAS) colloids, as well as the incorporation of radionuclides into such colloid formation. Likely processes are characterized in particular by a combination of different nanoscopic approaches. The colloid formation is monitored radiochemically in conjunction with the highly sensitive spectroscopic speciation, e.g. time-resolved laser fluorescence spectroscopy (TRLFS), which facilitates the chemical characterization of trace actinides in particular. Colloids thus generated are quantified for their average size and number density by laser-induced breakdown detection (LIBD) upon optical plasma monitoring. Exemplary illustrations are summarized for the formation of colloid-borne trivalent actinides (Am, Cm), which become incorporated into HAS-colloids. Discussion is extended to the migration behavior of radionuclides as colloid-borne species in natural aquifer systems, for which a field experiment is chosen as a case in point.

Keywords: Aquatic colloids, speciation, actinides, migration, environmental monitoring

---

Institut für Nukleare Entsorgung (INE), Forschungszentrum Karlsruhe (FZK)
76021 Karlsruhe, Germany

Y.J. Kim and U. Platt (eds.), *Advanced Environmental Monitoring*, 233–247.

18.1 Introduction

Aquatic colloids are ubiquitous in all kinds of natural water (Yariv 1979; Bolt et al. 1991; Kim 1986, 1991, 1994). Their concentrations and size range vary widely depending on the geochemical surrounding of each given aquifer. Normally the particle size ranges from 1 nm up to 400 nm in average diameter but the particles of predominant number density are found as less than 100 nm (Kim et al. 2002). Number density (particles per liter water) varies from 10^{11} upwards over 10^{14} (Kim et al. 2002). Chemical composition of aquatic colloids is wide-ranging as maintained by their provenance (Yarif 1979; Bolt et al. 1991; Kim 1993). In general, they can be categorized into two different composites: inorganic colloids composed by heterogeneous polynucleation via oxo-bridging of different metal ions and organo–inorganic colloids produced by aggregation of inorganic colloids via complexation with organic molecules, e.g. humic acid (Bolt et al. 1991; Kim 1991; Malcolm and Bryan 1998). By nature they are hydrophilic, as exposed with the negatively charged surface, and thus play a carrier role for contaminant trace metal ions, e.g. radioactive elements like actinides, in aquifer systems (Artinger et al. 2002a,b; Hauser et al. 2002; Geckeis et al. 2004). Organic contaminants of polarized nature can also be carried on migration by inorganic colloids (Bolt et al. 1991).

Colloid-facilitated migration, especially, of trace actinide ions is of cardinal importance for the environmental monitoring in particular with respect to the radioecological safety aspect (Kim and Grambow 1999). It is to note that the radiochemical toxicity is attributed to radioelement concentrations that are much lower than concentration limitations of given elements for the chemical toxicity. For this reason, the long-term safety assessment of nuclear waste disposal entails the well-founded knowledge on aquatic colloids, on their interaction with trace radionuclides, above all long-lived actinides, and eventually on their migration behavior (Kim and Grambow 1999; Kim 2000).

This chapter is a brief summary review of recent investigations dealing with the provenance of aquatic colloids, the characterization as regards their interaction with trace actinide ions and the migration behavior of colloid-borne actinides. For this purpose, notable examples are selected to demonstrate how to characterize aquatic colloids with modern instrumentation, how to speciate the actinide interaction with colloids and how to appraise the colloid-facilitated migration of actinides in a given aquifer system. The subject matter under discussion may certainly be applicable to the environmental monitoring of other contaminants as well.

18.2 Aquatic Colloids in Nature

As natural water is always in contact with various mineral surfaces in a given geological formation, weathering products of surfaces are dispersed, dissolved and coagulated into new chemical composites, either hydrophobic states to precipitate

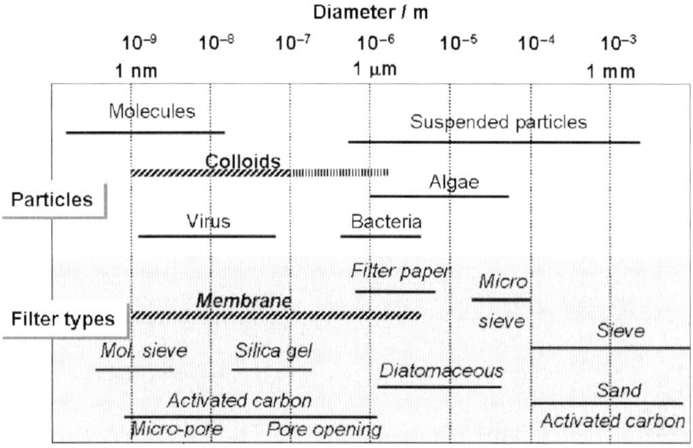

Fig. 18.1 Size ranges of various aquatic particles and pore openings

or hydrophilic states as colloids (Yarif 1979; Bolt et al. 1991). A colloidal fraction generated in such a process is often mixed with suspended particles of relatively large size (up to a few μm), which are prone to sedimentation with time; as a result, a stable fraction of colloids appears to be small in number density (10^{11}–10^{14} particles per liter, or over) as well as in average size (<100 nm) (Kim 1993; Kim et al. 2002). The particle size range of aquatic colloids can be compared with other aquatic particulates as shown in Fig. 18.1 (Stumm and Morgan 1981; Kim 1993). The size range of aquatic colloids is comparable with that of virus but evidently smaller than that of bacteria and microalgae. Sampling of subsurface water or groundwater induces often oxidation (e.g. Fe^{2+}) and decomposition of carbonate, thus leading to hydroxide of metal ions, as a result, produces suspended particles of larger size (Kim 1991). This kind of artifacts confuses in fact the particle size range of actual aquatic colloids. Therefore, the colloid size range is marked in Fig. 18.1 in two regions: actual range of <100 nm and uncertain range up to a few μm. Pore openings of various filter types (Stumm and Morgan 1981) are also given in this figure for the purpose of comparison. Modern membrane filters of different pore openings facilitate the size characterization of aquatic colloids, once careful precaution is attended to experimental handling.

An example of characterizing aquatic colloids in deep groundwater interwoven in various aquifer areas in Gorleben, northern Germany, is shown in Fig. 18.2 (Kim 1993). Colloids in this groundwater are composed of organo–inorganic composites, which contain a vast array of trace metal ions. The average size of predominant number density (10^{11}–10^{14} particles per liter) ranges <100 nm with a minor fraction up to 450 nm (Kim 1993). Colloid-borne concentrations of selected trivalent and tetravalent elements (homologues of actinides) are found to be a proportional function of the DOC (dissolved organic carbon) concentration, made mostly of fulvic and humic acids. All these elements are incorporated quantitatively into colloids. Age determination by ^{14}C-dating of humic components indicates these aquatic colloids

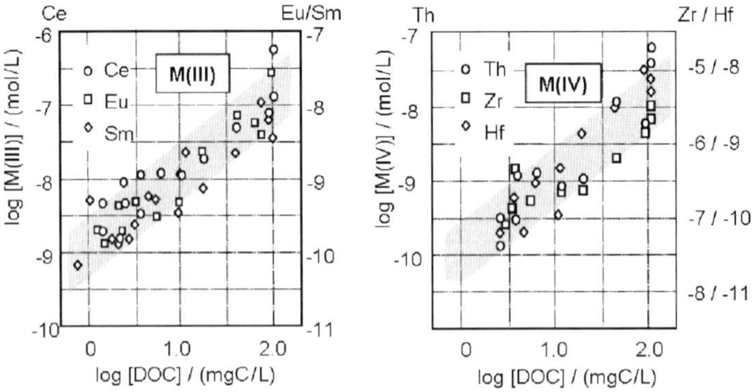

Fig. 18.2 Concentrations of water-borne trivalent and tetravalent trace elements associated with aquatic colloids as a function of the DOC (dissolved organic carbon, mostly humic and fulvic acids) in deep groundwater at Gorleben, Germany

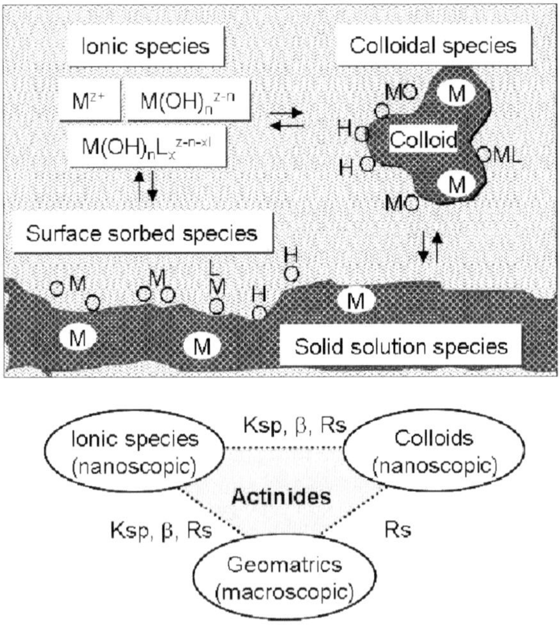

Fig. 18.3 A three phase system: ionic, colloid and solid, overruling the distribution of trace metal ions in aquifer systems

older than 20,000 years (Buckau et al. 2000). The fact suggests that aquatic colloids can remain stable for a long period.

Knowing that mineral-water interface systems add always in aquatic colloids (Bolt et al. 1991; Kim 1991), the environmental monitoring entails the appraisal of contaminant distributions into three different phases: ionic, colloid and solid (minerals) phases. Illustrations given in Fig. 18.3 demonstrate typical three phase interactions of given metal ions that can be anticipated in aquifer systems (Kim and Grambow 1999). Specific reactions can be quantified separately between the two phases; ionic–colloid, ionic–solid and colloid–solid by assessing solubility products (Ksp), complexation constants (β) and partition ratios (Rs) for individual elements concerned. Such quantifications are, however, difficult, if not unfeasible, since necessary parameters to be taken into account are not always accessible straightforwardly (Kim 2000). Main difficulty arises in the characterization and quantification of aquatic colloids involved in a given system, namely particle size, number density and chemical composition. As aquatic colloids are relatively small in size and low in number density, conventional methods, like ultrafiltration, ultracentrifugation, light-scattering measurement etc., often fail in their proper quantification. As a response to such difficulty, a noble method has been introduced recently (Scherbaum et al. 1996; Bundschuh et al. 2001b), which is capable of quantifying aquatic colloids for their average size and number density in parallel. Discussion on this approach is briefly made here without commenting on other conventional methods, since they are handled multiply in the open literature (Ross and Morrison 1988).

18.3 Quantification of Aquatic Colloids by Laser-Induced Breakdown Detection

Principle of laser-induced breakdown detection (LIBD) can be appreciated by illustrations depicted in Fig. 18.4 (Kim and Walther 2006). Modulated laser light of high-energy intensity (irradiance) sorbed into a colloid particle incites ionization of

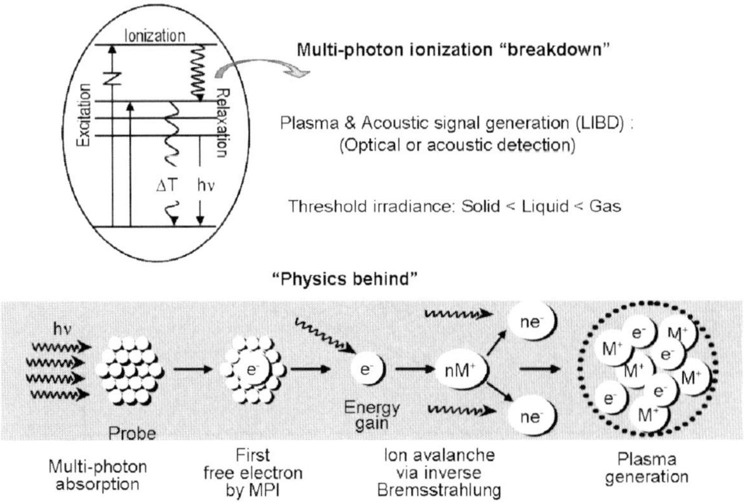

Fig. 18.4 An illustration of the laser-induced breakdown process for the determination of aquatic colloids

elements embraced therein. Energetic free electrons thus generated relax via giving off Bremsstrahlung that in turn creates further ionization and hence ion avalanche via so-called inverse Bremsstrahlung (Scherbaum et al. 1996; Walther et al. 2002; Kim and Walther 2006). Ignition of plasma takes place above the threshold irradiance for the physical state of a given prove, which follows: gas > liquid > sold. This principle makes selective detection of colloids in water possible. Consequently, the colloid particle undergoes "breakdown", leading to a nanoscopic plasma bundle. Whereas acoustic waves generated at the event of breakdown can be monitored by acoustic detection for determining number density (Scherbaum et al. 1996), the indiscrete relaxation of plasma can be monitored optically for a two-dimensional localization of plasma within the laser focus volume to ascertain an average size of colloids (Bundschuh et al. 2001b). Calibration of the latter process as a function of the laser beam irradiance enables monitoring both the average particle size and number density.

The particle detection sensitivity of LIBD, as calibrated by well-defined polystyrene reference particles, is compared with that of light scattering methods in Fig. 18.5: static and dynamic (PCS: photon correlation spectrometry) modes (Bundschuh et al. 2001b). As is apparent from this figure, the light scattering method is basically not sensitive enough for detecting colloids of small in size and low in number density, which is generally the case for aquatic colloids. On the other hand, LIBD is capable of detecting small colloids of very dilute concentrations, for which it is superior to light scattering by a factor of 10^4–10^7 in the predominant size range of actual aquatic colloids (indicated by gray shade).

Potential of LIBD is visualized, as shown in Fig. 18.6 (Bundschuh et al. 2001a), by monitoring colloids present in a variety of potable waters together with those in laboratory water from a Milli-Q apparatus. Non-processed tap water of the author's

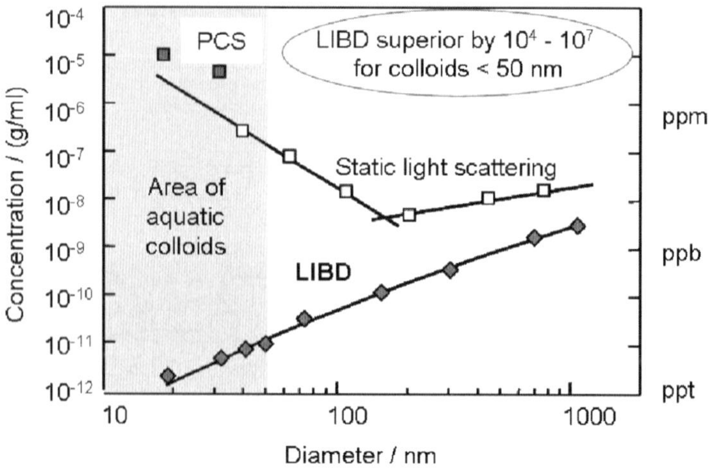

Fig. 18.5 Detection sensitivities of laser-induced breakdown detection (LIBD) in comparison with those of conventional light scattering methods

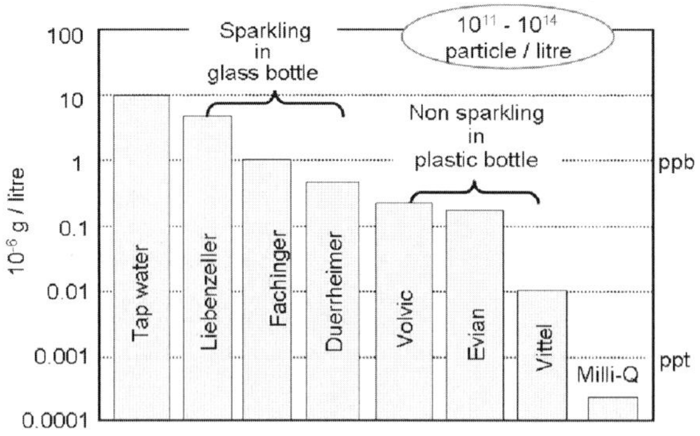

Fig. 18.6 Aquatic colloids present in various potable water

laboratory shows the highest colloid concentration compared to other bottled waters, reaching 10 µg/L (10 ppb). With a dominant particle size of <50 nm, the number density approaches to 10^{14} particle per liter. Non sparkling waters in plastic bottles contain less colloids than sparkling waters in glass bottles. Surface of glass bottles may be leached out with time, as a result, nanoscopic particles dispersed as colloids. The number density of colloids in potable water ranges in general from 10^{11} to 10^{14} particle per liter (Kim and Walther 2006). Fig. 18.6 corroborates simply the omnipresence of aquatic colloids. Provenance of such aquatic colloids is of keen interest to comprehend, above all, for appraising how the environmental monitoring can be advanced for trace actinides, of which the migration is eventually facilitated by aquatic colloids.

18.4 Provenance of Aquatic Colloids

Sound perception of how aquatic colloids are generated entails fundamental knowledge on prime kernels of their composition (Iler 1997; Kim et al. 2002). Following this trait, the major components of abundant aquifer minerals are sought out for the preliminary investigation. Aluminum and silicon oxides are dominant components in lithosphere (Pettijohn 1948; Mason 1952) and therefore, aluminosilicate composites become dominating aquifer minerals, much of which is clayey composition (Stumm and Morgan 1981; Bolt et al. 1991). Besides amorphous silicon oxide, most aluminosilicate minerals are sparingly soluble in the neutral pH range of water; the low solubility shores up on the other hand-dissolved species becoming colloidal. Some aluminosilicate minerals, together with amorphous silicon and aluminum oxides, are selected for illustrating their solubilities (Lindsay 1979; Stumm and Morgan 1981) as a function of pH in Fig. 18.7 (Kim et al. 2002). Relatively good soluble sillimanite and poorly soluble kaolinite and low albite are chosen for illustration. The near neutral pH range (indicated by gray shade), where solubilities

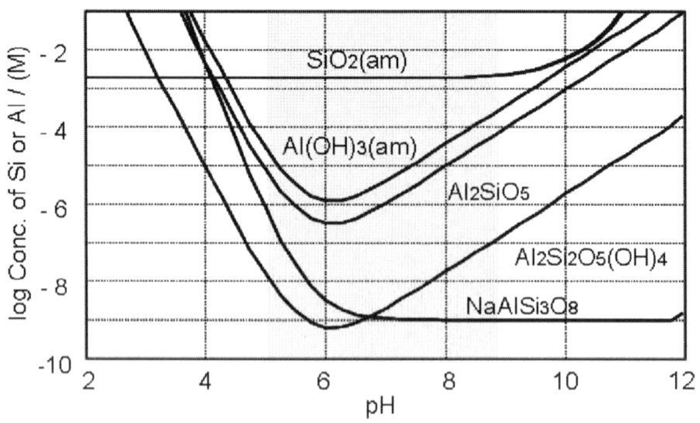

Fig. 18.7 Solubilities of amorphous silica and aluminum hydroxide, together with aluminosilicate composites: sillimanite (Al_2SiO_5), kaolinite ($Al_2Si_2O_5(OH)_4$) and low albite ($NaAlSi_3O_8$)

are low, appears to be the favorable condition for generating colloids in a composition of hydroxy aluminosilicates (HAS). They are kernels of aquatic colloids primarily. How such colloids incorporate trace actinides is an essential insight into the provenance of colloid-borne actinides in a given nuclear waste repository. Knowing that natural aquatic conditions are complex, owing to multiple interactions of a wide range of waterborne trace components: metal ions, either inorganic or organic anions, hydrophilic molecules etc. (Lindsay 1979; Stumm and Morgan 1981), this chapter concentrates only on a key feature for the formation of aquatic HAS-colloids as well as of colloid-borne actinides.

18.5 Generation of Aquatic Colloid-Borne Actinides

Hydroxy aluminosilicate (HAS) colloids are generated by mixing of acidic Al with alkaline Si to predisposed neutral pH in the presence of trivalent actinides, Am or Cm, in a trace concentration of 5×10^{-8} M (Kim et al. 2002). Am and Cm (chemical homologues) are alternately used for the reason that the former is on hand for a large-scale experiment, while the latter available only in a trace concentration is favored for the spectroscopic speciation (Panak et al. 2003). The Al concentration is varied from 10^{-3} M to 10^{-7} M (from over to under saturation, cf. Fig. 18.7), while keeping the Si concentration constant at 10^{-2} M (over saturation) and at 10^{-3} M (under saturation) for delivering polysilicic acid in the former and monosilicic acid in the latter (Kim 2005). Radiochemical measurements, after phase separation by a sequential ultrafiltration at two different pore openings: at 450 nm followed at 1 nm, make the evaluation of colloid fraction available, according to an empirical convention: *ionic species < 1 nm < colloids < 450 nm < precipitate* (Kim et al. 2002). In parallel, colloids are monitored directly by LIBD and visually by AFM (atomic force microscopy) (Kim et al. 2002). Both approaches give a comparable result:

an average size range of 10–50 nm for predominant particles. The number density determined by LIBD ranges 10^{11}–10^{14} particle per liter depending on the experimental condition applied, pH and initial concentrations of Al and Si.

Speciation of colloid-borne trivalent actinides is performed by time-resolved laser fluorescence spectroscopy (TRLFS) (Klenze et al. 1991), which provides the possibility of appraising excitation and relaxation spectroscopy as well as measuring the fluorescence lifetime of a probe elements concerned. Application of the three optical characteristics in parallel leads to chemical speciation with high sensitivity. To attain the high spectroscopic sensitivity, Cm(III) is chosen for the purpose of demonstrating the trivalent actinide behavior (Panak et al. 2003). Formation of colloid-borne Cm is surveyed as a function of pH along with the generation of HAS-colloids in a mixed solution containing 10^{-2} M Si (polysilicic acid prevails), 10^{-4} M Al and 5×10^{-8} M Cm. Three colloid-borne Cm species are identified (Kim et al. 2005), as illustrated in Fig. 18.8, which are named: Cm-HAS(I), Cm-HAS(II) and Cm-HAS(III). Increasing pH converts Cm-HAS(I) to CmHAS(II) and further to Cm-HAS(III) progressively. At pH 6 and beyond, Cm-HAS(III) becomes gradually prevailing, which remains stable in a period of over 60 days of observation. The similar experiment with 10^{-3} M Si (monosilicic acid prevails) results in the formation of Cm-HAS(I) and Cm-HAS(II) species only under the same experimental conditions (Panak et al. 2003).

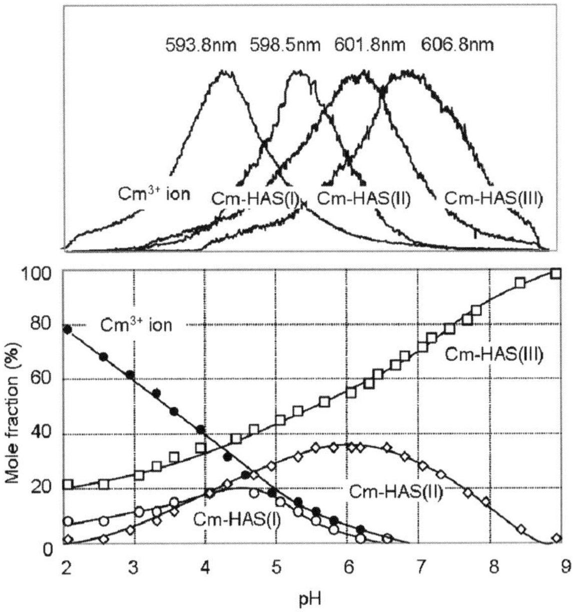

Fig. 18.8 Spectroscopic speciation of different HAS-colloid-borne Cm(III) species as a function of pH by time-resolved laser fluorescence spectroscopy (TRLFS). HAS denotes hydroxy aluminosilicate

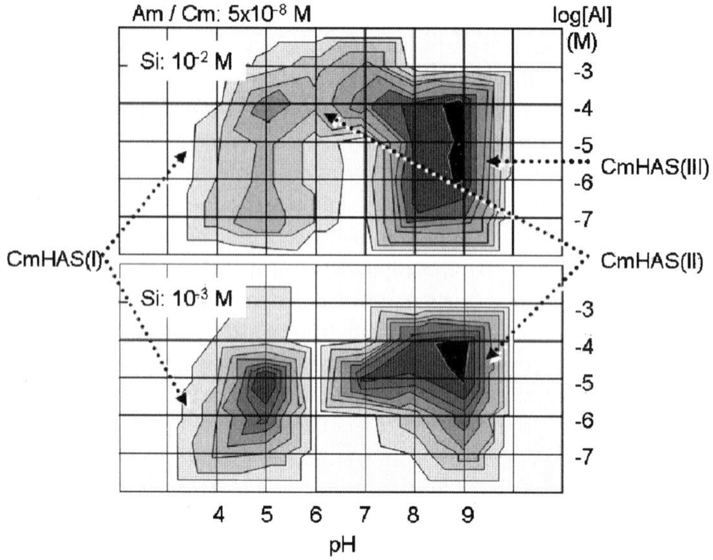

Fig. 18.9 Formation of HAS-colloid-borne Cm (or Am) at 10^{-2} M Si (polysilicic acid) and at 10^{-3} MSi (monosilicic acid) as a function of the Al concentration and pH. Contour areas are divided at a 10% interval of the colloid-borne Cm (or Am) fraction in solution (increasing gradually with darker shade)

A broad screening experiment is carried out radiochemically by adding 5×10^{-8} M Am in order to ascertain favorable conditions for the formation of colloid-borne actinides(III). The results are summarized in Fig. 18.9 as contour diagrams for the fractions of colloid-borne Am (Kim et al. 2005). Since Am and Cm are chemically homologues in solution, Figs. 18.8 and 18.9 can be correlated with each other. Correlation of the two figures, together with the speciation results of Cm-HAS formation at 10^{-3} M Si (Kim et al. 2002: not shown here), makes distinction of favorable experimental conditions for generating each of HAS-colloid-borne Am (or Cm). Colloidal species of Cm-HAS(I) is formed only at low pH, Cm-HAS(II) in 10^{-3} M Si at high pH or in 10^{-2} M Si around pH 6 and Cm-HAS(III) only in 10^{-2} M Si at pH >7.

The fluorescence relaxation time of each colloid-borne Cm species is found to be different from one another, following the order of Cm-HAS(I) < Cm-HAS(II) < Cm-HAS(III) (Kim et al. 2005). Resolving the relaxation time, the hydration number of water molecules bound to each Cm in HAS-colloids is found: 7 H_2O to Cm-HAS(I), 6 H_2O to Cm-HAS(II) and 0/1 H_2O to Cm-HAS(III) in relation to the reference value of 8/9 H_2O coordinated to the aqueous Cm^{3+} ion (Kim et al. 2005). According to these results, the chemical structure of each HAS-colloid-borne Cm species can be postulated as shown in Fig. 18.10. Based on the results discussed hitherto, it is possible to draw a conclusion that, in the formation process of HAS-colloids, trace trivalent actinides are incorporated by surface sorption

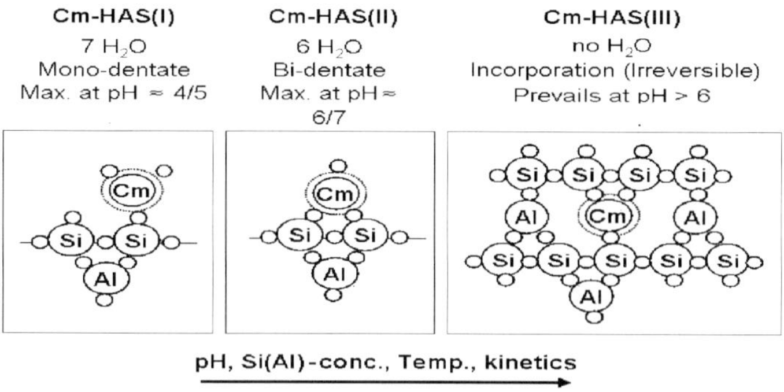

Fig. 18.10 Chemical states postulated based on the fluorescence lifetime of each HAS-colloid-borne Cm species: a conversion of the species takes place with increasing pH, the concentration of Si (Al) and temperature from left to right

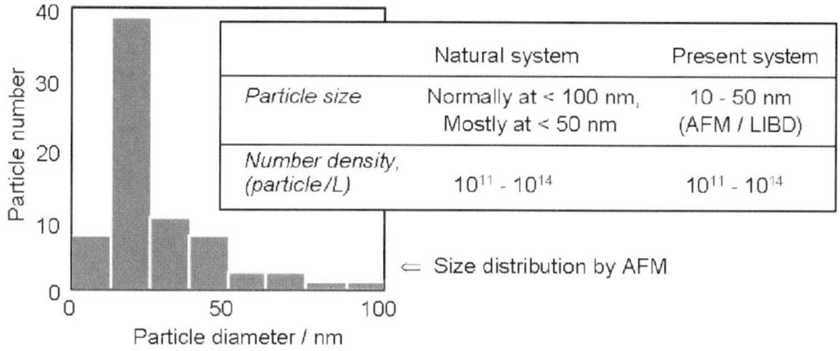

Fig. 18.11 A particle size distribution of HAS-colloids determined by AFM; average size and number density ranges ascertained by LIBD for natural aquatic colloids and HAS-colloids

as well as by coalescing into the colloidal structure, substituting Cm (or Am) at the Al site. The latter process generates a stable HAS-colloid-borne Cm (or Am), i.e. Cm-HAS(III) colloids. Characterization of such stable colloids made by LIBD and AFM leads to the results given in Fig. 18.11 (Kim et al. 2002). The average size and number density of HAS-colloids vary obviously with experimental conditions, e.g. pH, the concentration of each component involved.

Summarizing the so far discussed results and according the atomic ratio found as approximately 1 for Al/Si (Kim et al. 2002) in the HAS composition, the formation of HAS-colloids can be described as follows:

$$nAl(OH)_{3-y}^{\ y+} + Si_{n-1}(OH)_{2n-2} \Leftrightarrow Al_nSi_nO_{4n-2-x}(OH)_x^{\ x-n-4} + (7+ny-x)H^+ + (2n-ny-1)H_2O$$

In this reaction an isomorphic substitution of trivalent trace actinides to Al may take place as depicted in Fig. 18.10 (right side). The reaction is pH reversible and hence a conversion of HAS-colloid-borne Cm species takes place as shown in Fig. 18.8.

18.6 Migration Behavior of Colloid-Borne Actinides

Aquatic colloids generate evidently colloid-borne actinides as exemplified by the aforementioned example, which are called pseudocolloids of actinides (Kim 1991). Further, the migration behavior of colloid-borne actinides is of cardinal importance for the environmental monitoring, in other words, for the safety assessment of nuclear (or other) waste disposal (Kim and Grambow 1999; Kim 2000). In natural aquifer systems, pore-openings are in general large enough for unconstrained waterway of aquatic colloids as can be appreciated from Fig. 18.12. The migration behavior of actinides (as well as other contaminants) in a given aquifer system can be assessed by monitoring the retardation coefficient (R_f) of given ions, molecules or colloids. R_f represents a ratio of the migration velocity of a waterborne component vs. the water flow velocity (Kim and Grambow 1999). The migration process is depicted in Fig. 18.12 for waterborne components (e.g. ionic radionuclides [RN])

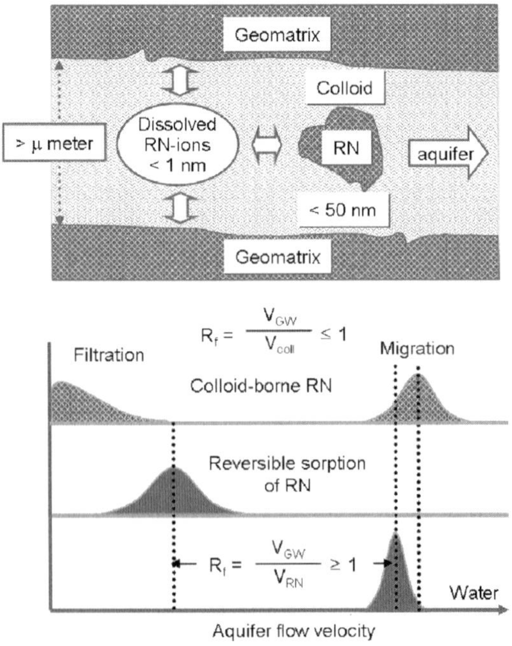

Fig. 18.12 An illustration of the migration behavior of radionuclides in ionic species interacting with mineral surfaces and in colloidal form devoid of interaction

undergoing surface interactions as well as for colloid-borne RN, which appears to be least interacting. Surface interactions result in the R_f value larger than one always, whereas the migration of colloids, devoid of interaction, is enhanced by advection to a certain extent and hence R_f becomes slightly less than one (Artinger et al. 2002a,b). Consequently, aquatic colloids facilitate the migration of actinides only. On the other hand, non-colloidal actinide ions are prone to sorption onto aquifer mineral surfaces, thus immobilized easily, as noticed by significantly large R_f values (Mckinley and Scholtis 1993; Vandergraaf et al. 1993).

The colloid-facilitated migration of actinides is observed in a field experiment. The field-laboratory in Grimsel, Switzerland, which has been built up for the investigation of the nuclear waste disposal safety, is chosen for the migration experiment of colloid-borne actinides (Hauser 2002, Geckeis 2004). Bentonite colloids (average diameter <200 nm; smectite type) dispersed in in situ granite water are spiked by trace concentrations of trivalent and tetravalent actinide homologue ions: Th(IV), Hf(IV) and Tb(III) (Hauser 2002). To a water passage of 5 m granite fracture distance, aquatic colloids are injected by air pressure at one end and extracted at another end. The experiment is visualized in Fig. 18.13. The colloid migration is monitored in-situ by LIBD for their concentration and average size in input and output waters. Colloid-borne elements are analyzed afterward by ICP-MS in laboratory. Migration profiles summarized in Fig. 18.13 (inset) indicate that the

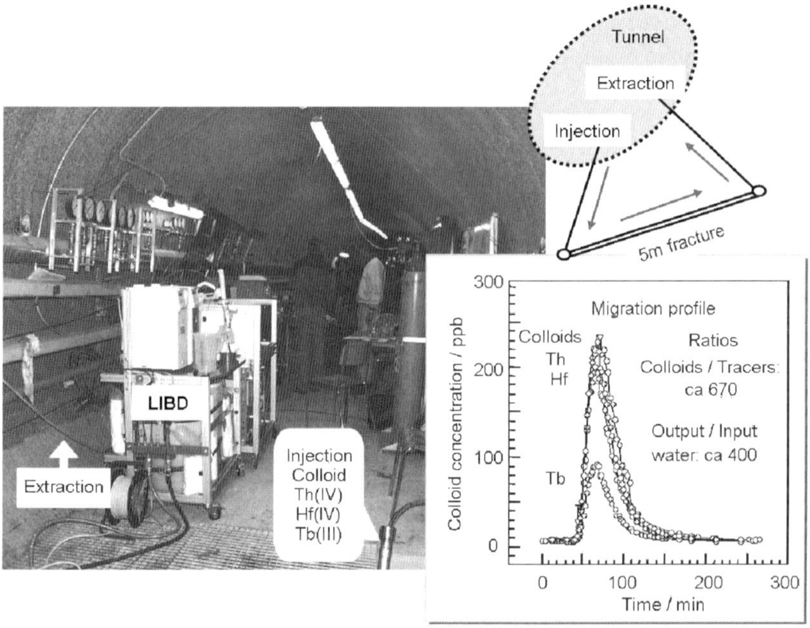

Fig. 18.13 A field experiment of the colloid-borne actinide (homologues) migration in the Grimsel, Switzerland underground laboratory (tunnel) for a 5 m distance of the granite fracture. Trace metal ions are accompanied by bentonite colloids of <200 nm

migration of trivalent and tetravalent elements is facilitated by aquatic colloids. Recovery of tetravalent Th and Hf amounts to about 75% and of trivalent Tb to about 35% (Hauser et al. 2002). Improvement of spiking metal ions onto colloids in the subsequent experiment results in a near quantitative recovery of all trace elements introduced (Geckeis et al. 2004). Both laboratory and field experiments (Artinger et al. 2002a,b; Hauser et al. 2002; Geckeis et al. 2004) substantiate that the actinide migration is facilitated by aquatic colloids, which thus play as a carrier role.

18.7 Final Remarks

The colloid-facilitated migration of actinides has been investigated by laboratory experiments as well as by field experiments. All these endeavors have corroborated that actinides become migrational only in accompany with aquatic colloids. Therefore, fundamental knowledge on aquatic colloids and their provenance is an essential prerequisite for the radioecological monitoring and also for the environmental monitoring of other contaminants. This is particularly true for the long-term safety assessment of nuclear waste disposal. However, not much is known for the provenance of colloid-borne actinides, i.e. geochemical processes of such colloid formation, because of experimental difficulties involved in the appraisal of nanoscopic reactions. Limitation of space allocated for this chapter renders summarizing only some notable examples from the recent investigations in this subject field.

References

Artinger R., Schuessler W., Scherbaum F., Schild D., and Kim J.I. (2002a), Am-241 migration in a sandy aquifer studied by long-term column experiments. *Environ. Sci. Technol.*, 36, 4818–4823.

Artinger Rabung T., Kim J.I., Sachs S., Schmeide K., Heise K.H., Bernhard G., and Nitsche H. (2002b), Humic colloid-borne migration of uranium in sand columns, *J. Contam. Hydrol.*, 58, 1–12.

Bolt G.H., De Boodt M.F., Hayes M.H.B., and McBride M.B. (1991), *Interact. Soil Colloid-Soil Solut.*, Kluwer Academic Publishers, Dordrecht, The Netherlands.

Buckau G., Artinger R., Geyer S., Wolf M., Fritz P., and Kim, J.I. (2000), C-14 dating of Gorleben groundwater. *Appl. Geochem.*, 15, 583–597.

Bundschuh T., Knopp R., and Kim, J.I. (2001a), Laser-induced breakdown detection of aquatic colloids with different laser systems. *Colloids and Surfaces A: Physicochem. Eng. Asp.*, 177, 47–55.

Bundschuh T., Hauser W., Kim J.I., Knopp R., and Scherbaum F.J. (2001b), Determination of colloid size by 2-D optical detection of laser-induced plasma. *Colloids and Surfaces A: Physicochem. Eng. Asp.*, 180, 285–293.

Geckeis H., Schäfer T., Hauser W., Rabung T., Missana T., Degueldre C., Mori A., Eikenberg J., and Fierz T. (2004), Results of the colloid and radionuclide retention experiment (CRR) at the

Grimsel test site, Switzerland: Impact of reaction kinetics and speciation on radionuclide migration. *Radiochim. Acta*, 92, 765–774.
Hauser W., Geckeis H., Kim J.I., and Fierz T. (2002), A mobile laser-induced breakdown detection system and its application for the in situ monitoring of colloid migration. *Colloids and Surfaces A: Physicochem. Eng. Asp.*, 203, 37–45.
Iler R.K. (1997), *The Chemistry of Silica*. Wiley-Interscience, New York.
Kim J.I. (1986), Chemical behavior of transuranic elements in natural aquifer systems. In *Handbook on the Physics and Chemistry of the Actinides*, A. J. Freeman and C. Keller, (eds.), Elsevier Science Publishers, B. V., Amsterdam, New York, chap. 8, 413–455.
Kim J.I. (1991), Actinide colloid generation in groundwater. *Radiochim. Acta*, 52–53, 71–81.
Kim J.I. (1993), The chemical behavior of transuranium elements and barrier functions in natural aquifer systems. *Mat. Res. Symp. Proc.*, 294, 3–21.
Kim J.I. (1994), Actinide colloids in natural aquifer systems. *MRS Bull.*, 19(12), 47–53.
Kim J.I. and Grambow B. (1999), Geochemical assessment of actinide isolation in a German salt repository environment. *Eng. Geol.*, 52, 221–230.
Kim J.I. (2000), Is the thermodynamic approach appropriate to describe natural dynamic systems (status and limitations). *Nucl. Eng. Des.*, 202, 143–155.
Kim M.A., Panak P.J., Yun J.I., Kim J.I., Klenze R., and Köhler K. (2002), Interaction of actinides with aluminosilicate colloids in statu nascendi, Part 1: generation and characterization of actinide(III) pseudocolloids. *Colloids and Surfaces A: Physicochem. Eng. Asp.*, 216, 97–108.
Kim M.A., Panak P.J., Yun J.I., Priemyshev A., and Kim J.I. (2005), Interaction of actinides(III) with aluminosilicate colloids, in statu nascendi, Part III: Colloid formation from monosilanol and polysilanol. *Colloids and Surfaces A: Physicochem. Eng. Asp.*, 254, 137–145.
Kim J.I. and Walther C. (2007), Laser-induced breakdown detection. In *Environmental Colloids and Particles: Behaviour, Separation and Characterisation.* Wilkinson K.J. and Lead J.R (Eds) chap. 12, Wiley & Sons, London, 555–612.
Klenze R., Kim J.I., and Wimmer H. (1991), Speciation of aquatic actinide ions by pulsed laser spectroscopy. *Radiochim. Acta*, 52–53, 97–103.
Lindsay W.L. (1979), *Chemical Equilibria in Soils*, Wiley, New York.
Malcolm N. and Bryan N.D. (1998), Colloidal properties of humic substances. *Adv. Colloid Interfac.*, 78, 1–48.
Mason B. (1952), *Principles of Geochemistry*, Wiley, New York.
McKinley I.G. and Scholtis A. (1993), A comparison of radionuclide sorption databases used in recent performance assessments. *J. Contam. Hydrol.*, 13, 347–363.
Panak P.J., Kim M.A., Yun J.I., and Kim J.I. (2003), Interaction of actinides with aluminosilicate colloids in statu nascendi, Part 2: spectroscopic speciation of colloid-borne actinides(III). *Colloids Surf. A: Physicochem. Eng. Asp.*, 227, 93–103.
Pettijohn F.J. (1948), *Sedimentary Rocks*, Harper and Brothers, New York.
Ross S. and Morrison I.D. (1988), *Colloidal Systems and Interfaces*, Wiley, New York.
Scherbaum F.J., Knopp R., and Kim J.I. (1996), Counting of particles in aqueous solutions by laser-induced photoacoustic breakdown detection. *Appl. Phys. B: Lasers Opt.*, 63, 299–306.
Stumm W. and Morgan J.J. (1981), *Aquatic Chemistry*, Wiley, New York.
Yariv S and Cross H. (1979), *Geochemistry of Colloid System*, Springer-Verlag, Berlin.
Vandergraaf T.T., Ticknor K.V., and Melnyk T.W. (1993), The selection of a sorption database for the geosphere model in the Canadian nuclear fuel waste management program. *J. Contam. Hydrol.*, 13, 327–345.
Walther C., Bitea C., Hauser W., Kim J.I., and Scherbaum F.J. (2002), Laser induced breakdown detection for the assessment of colloid mediated radionuclide migration. *Nucl. Instr. Meth. Phys. Res. Section B*, 195, 374–388.

Chapter 19
Progress in Earthworm Ecotoxicology

Byung-Tae Lee, Kyung-Hee Shin, Ju-Yong Kim, and Kyoung-Woong Kim

Abstract Earthworms are regarded as one of the most suitable animals for testing the toxicity of chemicals in soils and have been adopted as standard organisms for ecotoxicological testing. In several guidelines concerning earthworm toxicity tests, *Eisenia fetida/andrei* (*E. fetida/andrei*) was chosen because it can be easily cultured in the laboratory and an extensive database on the effects of all classes of chemicals exists for this species. Acute and chronic toxicity tests have been used traditionally to assess the toxicity of contaminants, with mortality and changes in biomass, reproduction rates and behavioral responses representing endpoints. Moreover, the avoidance behavior test (AVT) using earthworm is under development and standardization, which records the ability of earthworms to choose or avoid a certain soil. Recent studies have shown that neutral red retention time (NRRT) has the potential for a rapid assessment of the toxic effects for earthworms of soils contaminated with heavy metals and metalloids. Toxicity is the apparent expression of the metal accumulation in earthworm body. The uptake, accumulation and elimination properties of metals by earthworm are the major part of toxicology, which is called toxicokinetics. Geochemical factors may have significant effects on metal transport or bioavailability. Prediction models for metal accumulation and toxicity in soils are being developed based on metal bioavailability. In this study, methodologies and research trends in earthworm toxicity are reviewed for understanding metal bioavailability. Toxicity prediction models are introduced for terrestrial environment and several studies are referred to understand the role of geochemistry in toxicology.

Keywords: Bioavailability, earthworm toxicity, metal, prediction model, toxicokinetics

Department of Environmental Science and Engineering, Gwangju Institute of Science and Technology (GIST), Gwangju 500-712, Republic of Korea; Tel: + 82-62-970-3391, Fax: + 82-62-970-3314

Y.J. Kim and U. Platt (eds.), *Advanced Environmental Monitoring*, 248–258.

19.1 Earthworm Toxicity Tests

19.1.1 Standard Laboratory Test

Earthworms are one of the most suitable animals for use as key bioindicator organisms for testing the toxicity of chemicals in soils because of certain characteristics, such as large size, specific behavior and high biomass. (Callahan 1988; Goats and Edwards 1988; BoucheÂ 1992). As a consequence, they have been adopted as standard organisms for ecotoxicological testing by the European Union (EEC 1984) and the OECD (1984). In addition, the International Standards Organization (ISO 1993, 1998) and the U.S. Environmental Protection Agency (US EPA 1988) have guidelines regarding the earthworm toxicity test. In these studies, the species *Eisenia fetida/andrei* (*E. fetida/andrei*) was chosen because it can be easily cultured in the laboratory (Tomlin and Miller 1989) and an extensive database on the effects of all classes of chemicals exists for this species (Edwards and Bohlen 1996).

Acute and chronic toxicity tests have been used traditionally to assess the toxicity of contaminants; test methods often operate at different ranges of sensitivity (lethal, sublethal) using end points such as mortality, changes in biomass, reproduction rates and behavioral responses. The mortality endpoint is relatively sensitive at high pollutant concentrations and indicates the maximum damage likely to an organism. However, sublethal concentrations of a pollutant can also have dramatic ecological effects on populations, such as a decrease in reproduction rate or emigration.

The key characteristics of several standard methods are summarized in Table 19.1. One main difference between the acute and reproduction test is the method of application of the test substance. In the acute test, all substances are mixed thoroughly with the soil substrate before introducing it into a test container. In the reproduction test, the test substances such as a pesticide may be applied to the soil surface by spraying in preparation for the test. This method of application makes the conditions of the laboratory test closer to those associated with agricultural practices. In the acute test, test animals are not fed and body weight loss is recorded in control boxes. However, there are no validity criteria regarding acceptable body weight in the acute test, even though body weight loss of control organisms is sometimes high. This problem typically does not occur in the reproduction test because the worms are fed during the test.

19.1.2 Modified Methods for Field Test

The earthworm toxicity test can be used to detect and quantify the toxicity of chemical contaminants in soil. For example, soil contamination by heavy metals near mine tailings can be determined using an earthworm toxicity test. The bioassays are more useful tools in assessing the bioavailability of metals, and earthworms appear to be more sensitive to heavy metals than other soil invertebrates

Table 19.1 Summary of test conditions of laboratory tests on earthworms

	Acute test (OECD 207)	Reproduction test (OECD 222)	Reproduction test (ISO 11268–2)
Species	*Eisenia fetida* *Eisenia andrei*	*Eisenia fetida* *Eisenia andrei*	*Eisenia fetida* *Eisenia andrei*
Test duration	2 weeks	8 weeks	8 weeks
Test substrate	Artificial soil	Artificial soil	Artificial soil
Conditions	20 ± 2°C In the dark	20 ± 2°C Light–dark cycles (16–8 h)	20 ± 2°C Light–dark cycles (12~16 h–12~8 h)
Food	–	Oatmeal, cow or horse manure	Cow manure
Validity criteria	Mortality ≤10%	Mortality (initial 4 weeks) ≤10%	Mortality ≤ 10%
	–	Number of juveniles ≥30	Number of juveniles ≥30
	–	Coefficient of variation <30	Coefficient of variation <30
Endpoints	Mortality body weight of adults	Mortality body weight of adults	Mortality body weight of adults
	–	Number of juveniles	Number of juveniles,
	Others (e.g. behavior)	Others (e.g. behavior)	Others (e.g. behavior)
Introduction of test substances	Mixed into the soil	Mixed into the soil Application to the soil surface	Mixed into the soil for pesticides, as close to field conditions as possible (Annex D)

(Ma et al. 2002; Boularbah et al. 2006). To prepare test soils, field soil samples could be mixed thoroughly with air-dried artificial soil at several dilution ratios. In general, artificial soil consisted of 10% coarse ground sphagnum peat, 20% kaolinite clay and 70% fine sand, depending on OECD or ISO guidelines. If a field soil is suspected or known to be highly toxic, a lower range of mixing ratios should be used and additional dilutions could be added after observing the level of mortality during the first 1–2 h of the test. In this modified method for field testing, the soil type (size distribution) may differ depending on the dilution ratio, and it is important to consider whether this difference can affect earthworm survival and growth irrespective of the toxicants in the soil.

19.1.3 Toxicity Endpoints

Mortality: Mortality is the main end point in the acute toxicity test and is expressed by means of LC50 after 7 and 14 days of exposure. The mortality is assessed by emptying the test medium onto a glass tray or plate, sorting worms from the medium and testing their reaction to a mechanical stimulus at the front end

(OECD 1984). This end point is valid only when the mortality in control treatments is lower than 10% for both the acute and reproduction tests.

Although the relevance of mortality as an endpoint for the field is high, the calculated toxicity is not identical due to a number of factors. For example, the pH and organic content of the soil could affect the toxic concentration of metals in soils and these could be altered in the test (Davies et al. 2003).

Reproduction: Reproduction is the main end point in the sublethal test and the number of juveniles is chosen as the parameter measuring reproduction. The results are considered valid only if the rate of production of juveniles is at least 30 per control container and the coefficient of variance of reproduction in the control treatment does not exceed 30% (ISO, 1998). The reference compounds for the reproduction test, carbendazim or benomyl, should be tested and significant effects should be observed between 1 and 5 mg of active ingredient/kg dry mass or 250–500 g/ha (ISO, 1998; OECD, 2004). Other parameters such as the number of cocoons or number of cocoons hatching could be introduced. These parameters have disadvantages /advantages and appropriate parameters should be selected on the basis of the purpose of the study. For example, cocoon production destroys the substrate during the washing procedure even though it is a well-defined and sensitive parameter. Moreover, determining the weight of juveniles could provide valuable information concerning reproduction despite the problem of handling small juveniles.

Growth: Weight change is a useful end point in both the acute and reproduction tests, although there are no validity criteria in OECD Guideline 207 or the ISO reproduction test guideline. However, the ISO acute test guideline (ISO, 1993) suggests that results are valid if the average loss of biomass of worms in control treatments does not exceed 20%.

A problem of potential bias arises when calculating the average change in body weight per adult when a relatively high mortality occurs in a test. Therefore, changes in body weight should be evaluated only when less than 15% mortality occurs in a test. In addition, body weight could be expressed as overall biomass instead of mean body weight, thus including the mortality endpoint (Kula, 1997).

Avoidance behavior test: The acute and chronic toxicity tests for earthworms have several disadvantages: acute tests do not consider the effect on population dynamics (OECD, 1984; ISO, 1993), while chronic tests are labor intensive and need quite a long duration ranging from 4 to 7 weeks, all of which delay the making of decisions concerning risk management and remediation strategy for a contaminated area (ISO, 1998; OECD, 2004). To overcome these weaknesses, a rapid toxicity assessment in the form of an Avoidance behavior test (AVT) using earthworms is under development and standardization that assesses the ability of earthworms to choose or avoid a certain soil (ISO/CD, 2003; Natal da Luz et al. 2004; Stephenson et al. 1997). The test has the potential to provide quick answers concerning the chronic effects for earthworms of contaminated soils.

Several studies have suggested that the results of AVT can increase the sensitivity of an evaluation of contaminated soil, quickly assessing an ecological endpoint which is not measured by any other test using the soil matrix (Yeardley et al.

1996; Stephenson et al. 1997). An abandoned mine area was evaluated ecologically by AVT as a real situation, and the results showed that AVT can be regarded as a valuable tool in the screening level evaluation of soil contamination (Loureiro et al. 2005).

Neutral red retention time: At the sub-cellular level, the lysosomal system has been identified as a particular target for the toxic effects of contaminants (Moore 1990). Lysosomal membrane stability is affected by chemical and nonchemical factors, and may be useful as an integrative biomarker of multiple stressors (Weeks and Svendsen, 1996). Lowe et al. (1994) developed an assay employing neutral red for the study of lysosomal injury of marine mussel digestive cells caught in clean and contaminated sites. Weeks and Svendsen (1996) employed the response of lysosomal membrane to earthworm toxicity assessment. In earthworm NRRT, coelomocytes extracted from the coelomic cavity of earthworms into an isotonic earthworm ringer solution were allowed to adhere to a microscope slide for 30 s before the application of a neutral red dye. The red dye rapidly accumulated within the lysosomes. Observation of the loss of this dye from these lysosomes into the surrounding cytosol has enabled the quantification of the degree of lysosomal damage caused to earthworms from exposure to soil contaminants.

Many studies have shown that NRRT has the potential for a rapid assessment of the toxic effects for earthworms of soils contaminated with heavy metals and metalloids. Hankard et al. (2004) showed that NRRT detected significant biological effects, while no effect on earthworm survival was found when earthworms were exposed to contaminated soils. Even though soil contaminant concentrations did not show any chemical exposure or toxic effects for earthworms living in the soils, NRRT successfully identified significant exposure and biological effects caused by the contaminants. Moreover, NRRT indicated toxic stress due to Cu exposure at an early stage of the earthworm life cycle, and was closely linked to the effects on certain life cycle traits (Reinecke et al. 2002). For acetochlor-polluted soils, NRRT showed a sensitive response even when earthworms were exposed to a concentration as low as 10 mg/kg (Xiao et al.). Several studies have shown that the lysosome is the sub-cellular target for the action of toxic metals and other contaminants in soils, and that the use of NRRT provides the potential for recognition of the damaging effects of soilborne heavy metals and metalloids. Though NRRT has the disadvantages of being time consuming and possessing an element of subjectivity regarding the initial quantification of staining, computer imaging and microscopic instruments can be applied to the technique to obtain a high degree of accuracy and automation (Weeks and Svendsen 1996).

19.2 Toxicokinetic Study

In traditional toxicology, the toxic effects on earthworms are usually related to total contaminant concentrations in soils. However, the results showed a limited relationship between the positive or negative effects on earthworms and the total

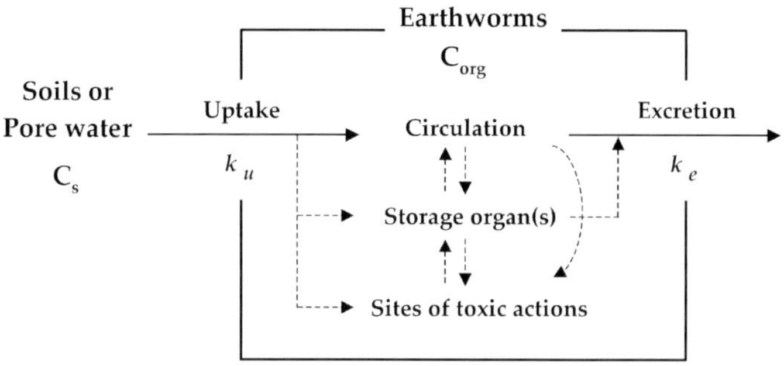

Fig. 19.1 Contaminant transport process from the terrestrial environment into earthworms. Contaminants can be circulated through the earthworm body, stored at certain organs, or cause adverse effects at the site of toxic actions

concentration of contaminants in soils. Several abiotic factors play a key role in the transport of contaminants from soils to earthworms. Environmental conditions (e.g. pH, temperature, organic carbon, competitive ions and ligands) and the chemical properties of contaminants in soils or pore water have a significant effect. Additionally, biotic factors (e.g. earthworm metabolism, metallothionein and lifestyle) have an effect on the toxicity.

Body residues are better estimates of the amount of a contaminant at the sites of toxic actions in earthworms than soil concentrations to which earthworms are exposed (Peijnenburg 1997). Toxicokinetic studies address the rate and extent to which contaminants are transported between soils and earthworms. Toxicokinetics depend not only on the biological features of the earthworm, but also on the chemical properties of the soil to which earthworms are exposed. The main objective of current study is to predict the physiological fate of contaminants in earthworms. Contaminants which are taken up may be partitioned into either biologically available or biologically unavailable fractions (Fig. 19.1).

Biologically available fractions are metabolized to less- or nontoxic forms, or may cause toxic effects at the site of toxic actions. Biologically unavailable fractions circulate through the earthworm body and are finally excreted without causing any toxicity.

19.2.1 One-Compartment Model

Many studies have applied kinetic models to describe the uptake, accumulation and elimination of toxicants by earthworms. The one-compartment model is generally used to characterize the contaminant process between the environment and organisms. The governing equation of the model is the contaminant mass balance; it assumes that the contaminant transport between soils and earthworms is governed by passive diffusion, and is therefore expressed as Fick's first law. A least square fitting is applied for the experimental data concerning contaminant concentration with exposure time to model uptake or eliminate changes in earthworm burden. The kinetic parameters (uptake rate and elimination rate) can be calculated through the statistical analysis. The parameters depend not only on the organisms, but also on the contaminants. Kinetic analysis shows the properties of the biological metabolism for the contaminants. Contaminant accumulation is given by

$$Q_t = C_o + \frac{k_u C_s}{k_e}(1 - e^{-k_e t}) \tag{1}$$

where

Q_t = total amount of metal in the earthworm
C_o = amount of contaminant in the earthworm at $t = 0$
C_s = total concentration in soils (or porewater), mg kg^{-1} (or mg l^{-1})
k_u = uptake rate constant, day^{-1}
k_e = elimination rate constant, day^{-1}
t = time, day

The steady- and non-steady-state kinetics of metals in earthworms suggested that bioconcentration depends on the metal concentrations in the soils, with bioconcentration being greater at lower soil concentration (Neuhauser et al. 1995). Toxicokinetic studies have shown that uptakes in earthworms were contaminant dependent (Spurgeon and Hopkin 1999). The uptake kinetics of metals or metalloids in earthworms has been studied in field soils (Peijnenburg et al. 1999a,b). Internal metal concentrations varied less than corresponding external levels. Zn and especially Cu provided the most extreme examples of this general behavior, which suggests regulation by the organism. Several studies suggest that earthworms regulate metal concentrations in their tissues. Spurgeon and Hopkin (1996) showed that Zn is regulated in earthworms and that tissue Zn concentrations did not show a good correlation with total soil Zn concentrations, thereby supporting the notion of biological regulation. Neuhauser et al. (1995) suggested that Cd, Cu, Pb and Zn were regulated, while no such regulation existed for Ni uptake.

Janssen et al. (1997a) tested the field-based partition coefficients for metals or metalloids by calculating the ratio of the amount of metal extracted by concentrated HNO_3 to the metal concentration in pore water. Regression equations based on soil characteristics were obtained and predicted partition coefficient values were in good agreement with measured values. The study suggests that the most influential factor determining the distribution of Cd, Cr and Pb between the soil solid phase and pore water is pH. The most important component determining As and Cu is the Fe content, whereas for Ni it is DOC. An approach using partition coefficients was applied to predict metal accumulation in earthworms (Janssen et al. 1997b). The study showed that the bioconcentration of metals in earthworms was governed by the same soil characteristics that determine equilibrium partition coefficients between the soil solid phase and pore water.

Many studies focused on the development of models to predict toxicity of soils for earthworms in field soils. Soil properties and contaminants were considered key parameters for contaminant uptake, and consequently, the control of toxicity.

Scientists are becoming interested in the prediction of toxicity and metal accumulation in earthworms. Janssen et al. (1997b) showed that the most important soil property affecting metal accumulation in earthworms was pH, which agreed with the results of partition coefficient factors. Bradham et al. (2006) supported the role of pH in metal uptake of earthworms in soils. They showed that pH was the most important soil property affecting earthworm mortality and internal Pb. Soil pH was inversely correlated to mortality and internal Pb, soil solution Pb and Pb bioavailability. The results confirmed that soil properties are important factors modifying metal bioavailability and toxicity, and should be considered during an ecological risk assessment of metals in contaminated soils.

19.2.2 Biotic Ligand Model

The general assumption that the major uptake route in earthworms is dermal contact has been tested and verified (Vijver et al. 2003, 2005). Analysis of the dermal route of metal uptake in earthworms revealed that pore water is the main metal pool for earthworms in soils. This finding was used to develop prediction models of toxicity for earthworms in a terrestrial environment. Steenbergen et al. (2005) applied a biotic ligand model (BLM) and regression model to predict acute copper toxicity in earthworms. The study used the De Schamphelaere and Janssen method (De Schamphelaere and Janssen 2002) (Eq. 2). The final BLM (Eq. 3) was obtained and the model showed that pH and Na play a significant role in the toxicity of copper for earthworms.

$$LC50_{Cu^{2+}} = \frac{f_{CuBL}^{50\%L}}{(1-f_{CuBL}^{50\%L})\cdot K_{CuBL}} \cdot \{1+K_{NaBL}\cdot(Na^{+})+K_{HBL}\cdot(H^{+})\} \qquad (2)$$

where

$LC50_{Cu^{2+}}$ = the free copper activity, mol L^{-1}

$f_{CuBL}^{50\%L}$ = the fraction of binding sites occupied by copper at 50% lethality

K_{CuBL} = the stability constants for binding to the biotic ligand of Cu^{2+}, L mol^{-1}

K_{NaBL} = the stability constants for binding to the biotic ligand of Na^{+}, L mol^{-1}

K_{HBL} = the stability constants for binding to the biotic ligand of H^{+}, L mol^{-1}

$$LC50_{Cu^{2+}} = 3.02\cdot10^{-7} + 1.24\cdot10^{-2}(H^{+}) + 2.79\cdot10^{-4}(Na^{+}) \qquad (3)$$

The regression models of the study showed that DOC plays an independent protective role, in addition to decreasing copper toxicity by complexation of the metal and thus reducing the free copper ion activity at the same total copper concentration. The study concludes by suggesting that terrestrial BLMs provide a new tool for environmental risk analysis. To develop more predictive tBLMs, it should be noted that the endpoint of toxicity bioassays is representative and generalized.

A toxicity test should minimize the uncertainty resulting from differences between laboratory and field conditions. However, few studies concentrate on this area of research because of limited data and the difficulty of conducting necessary experiments. The development of powerful prediction models is the main objective, which will provide useful tools to conduct ecological risk assessment and establish reasonable criteria for the management of soil environments. The models will consider and explain the toxicokinetic properties of contaminants based on metal- and species-dependent aspects. Furthermore, factors that play significant roles in metal transport or accumulation (including metal biotransformation and toxicity) will receive an in-depth analysis as a result of model development.

References

BoucheÂ A. (1992), Earthworm species and ecotoxicological studies. In: Greig-Smith, P.W., Becker, H., Edwards, P.J., Heimbach, F. (Eds.), *Ecotoxicology of Earthworms.* Andover, Hants, U.K., pp. 20–35.

Boularbah A., Schwartz C., Bitton G., and Morel J.L., (2006), Heavy metal contamination from mining sites in South Morocco: 1. Use of a biotest to assess metal toxicity of tailings and soils. *Chemosphere,* 63, 802–810.

Bradham K.D., Dayton E.A., Basta N.T., Schroder J., Payton M., and Lanno R.P. (2006), Effect of soil properties on lead bioavailability and toxicity to earthworms, *Environ. Toxicol. Chem.*, 25, 769–775.

Callahan C.A. (1988), Earthworms as ecotoxicological assessment tools. In: Edwards, C.A., Neuhauser, E.F. (Eds.), *Earthworms in Waste and Environmental Assessment.* SPB Academic Publishing, Hague, Netherlands, pp. 295–301.

Davies N.A., Hodson M.E., and Black S. (2003), Is the OECD acute worm toxicity test environmentally relevant? The effect of mineral form on calculated lead toxicity. *Environ. Pollut.*, 121, 49–54.

De Schamphelaere K.A.C. and Janssen C.R. (2002), A biotic ligand model predicting acute copper toxicity for *Daphnia magna.* The effects of calcium, magnesium, sodium, potassium and pH. *Environ. Sci. Technol.,* 36 (1), 48–54.

Edwards C.A., and Bohlen P.J. (1996). *Biology and Ecology of Earthworms*, 3rd ed. Chapman and Hall, London, U.K.

European Economic Community (1984). EEC Directive 79/831/ EEC, Annex V, part C. *Method for the determination of ecotoxicity.* Level 1. Earthworms: artificial soil test. Commission of the European Communities, DGXI/128/82 Rev. 5, Brussels.

Goats G.C. and Edwards C.A. (1988), The prediction of field toxicity of chemicals to earthworms by laboratory methods. In: Edwards, C.A., Neuhauser, E.F. (Eds.), *Earthworms in Waste and Environmental Assessment.* SPB Academic Publishing, Hague, Netherlands, pp. 283–294.

Hankard P.K., Svendsen C., Wright J., Wienberg C., Fishwick S.K., Spurgeon D.J., and Weeks J.M. (2004), Biological assessment of contaminated land using earthworm biomarkers in support of chemical analysis. *Sci. Tot. Environ.*, 330, 9–20.

ISO 11268–1 (1993), Soil quality—effects of pollutants on earthworms (*Eisenia fetida*). Part 1: Determination of acute toxicity using artificial soil substrate. Geneva, Switzerland.

ISO 11268–2 (1998), Soil quality—effects of pollutants on earthworms (*Eisenia fetida*). Part 2: Determination of on reproduction. Geneva, Switzerland.

ISO/CD 17512 (2003), Soil quality—avoidance test for testing the quality of soils and the toxicity of chemicals—test with earthworms (*Eisenia fetida*). Geneva, Switzerland.

Janssen R.P.T., Peijnenburg W.J.G.M., Posthuma L., and Hoop M.A.G.T. (1997a), Equilibrium partitioning of heavy metals in Dutch field soils. I. Relationship between metal partition coefficients and soil characteristics. *Environ. Toxicol. Chem.*, 16, 2470–2478.

Janssen R.P.T., Posthuma L., Baerselman R., Hollander H.A., Van Veen R.P.M., and Peijnenburg W.J.G.M. (1997b), Equilibrium partitioning of heavy metals in Dutch field soils. II. Prediction of metal accumulation in earthworms. *Environ. Toxicol. Chem.*, 16, 2479–2488.

Kula C. (1998), Endpoints in laboratory testing with earthworms: Experience with regard to regulatory decisions for plant protection products. In: Sheppard S., Bembridge J., Holmstrup M., Phostuma L. (Eds.), *Advances in Earthworm Ecotoxicology*. SETAC, Amsterdam, Netherlands, pp. 3–24.

Loureiro S., Soares A., and Nogueira A. (2005), Terrestrial avoidance behavior tests as screening tool to assess soil contamination. *Environ. Pollut.*, 138, 121–131.

Lowe D.M., and Pipe R.K. (1994), Contaminant induced lysosomal membrane damage in marine mussel digestive cells: An in vitro study. *Aquat. Toxicol.*, 30, 357–365

Ma Y, Dickinson N M, and Wong M H. (2002), Toxicity of Pb/Zn mine tailings to the earthworm *pheretima* and the effects of burrowing on metal availability. *Biol. Fertil. Soils*, 36, 79–86.

Moore M.N. (1990), Lysosomal cytochemistry in marine environmental monitoring. *Hystochemistry*, 22, 187–191.

Natal da Luz T., Ribeiro R., and Sousa J.P. (2004), Avoidance tests with collembola and earthworms as early screening tools for site specific assessment of polluted soils. *Environ. Toxicol. Chem.*, 24, 2188–2193.

Neuhauser E.F., Cukic Z.V., Malechi M.R., Loehr R.C., and Durkin P.R. (1995), Bioconcentration and biokinetics of heavy metals in the earthworm. *Environ. Pollut.*, 89, 293–301.

OECD, (1984), OECD guidelines for testing of chemicals: Earthworm acute toxicity test. OECD Guideline No. 207. Paris, France.

OECD, 2004. OECD guidelines for testing of chemicals: Earthworm reproduction test. OECD Guideline No. 222. Paris, France.

Peijnenburg W.J.G.M. (1997), Bioavailability of metals to soil inbertebrates. In: Allen, H.E. (Eds.), *Bioavailability of Metal in Terrestrial Ecosystems: Importance of Partitioning for Bioavailability to Invertebrates, Microbes, andPlants*. SETAC, Florida, USA, pp. 89–112.

Peijnenburg W.J.G.M., Posthuma L., Zweers P.G.P.C., Zbaerselman R., de Groot A.C., Van Veen R.P.M., and Jager T. (1999a), Prediction of metal bioavailability in Dutch field soils for the oligochaete *Enchytraeus crypticus. Ecotox. Environ. Safety*, 43, 170–186.

Peijnenburg W.J.G.M., Baerselman R., de Groot A.C., Jager T., Posthuma L., and Can Veen R.P.M. (1999b), Relating environmental availability to bioavailability: Soil-type dependent metal accumulation in the oligochaete *Eisenia Andrei. Ecotox. Environ. Safety*, 43, 294–310.

Reinecke S.A., Helling B., and Reinecke A.J. (2002), Lysosomal response of earthworm (*Eisenia fetida*) coelomocytes to the fungicide copper oxychloride and relation to life-cycle parameters. *Environ. Toxicol. Chem.*, 21, 1026–1031.

Spurgeon D.J., and Hopkin S.P. (1996), Effects of variations of the organic-matter content and pH of soils on the availability and toxicity of Zn to the earthworm *Eisenia fetida. Pedobiologia*, 40, 80–96.

Spurgeon D.J., and Hopkin S.P. (1999), Comparisons of metal accumulation and excretion kinetics in earthworms (*Eisenia fetida*) exposed to contaminated field and laboratory soils. *Appl. Soil Ecol.*, 11, 227–243.

Steenbergen N.T.T.M., Iaccino F., Winkel M., Reijnders L., and Peijnenburg W.J.G.M. (2005), Development of a biotic ligand model and a regression model predicting acute copper toxicity to the earthworm *Aporrectodea caliginosa. Environ. Sci. Technol.*, 39, 5694–5702.

Stephenson G.L., Kaushik A., Kaushik N.K., Solomon K.R., Sttele T., and Scroggins R.P. (1997), Use of an avoidance–response test to assess the toxicity of contaminated soils to earthworms. In: Sheppard S., Bembridge J., Holmstrup M., and Phostuma L. (Eds.), *Advances in Earthworm Ecotoxicology*. SETAC, Amsterdam, Netherlands, pp. 67–81.

Tomlin A.D. and Miller J.J. (1989), Development and fecundity of the manure worm, *Eisenia fetida* (Annelida: Lumbricidae) under laboratory conditions. In: Dindal D.L. (Ed.), *Soil*

Biology as Related to Land Use Practices. Proceedings of Seventh International Soil Zoology Colloquium, Syracuse, New York, USA, pp. 673–678.

U.S. Environmental Protection Agency. 1988. Protocols for short- term toxicity screening of hazardous waste sites, EPA/600/3–88/029.

Vijver M.G., Vink J.P.M., Miermans C.J.H., and van Gestel C.A.M. (2003), Oral sealing using glue: A new method to distinguish between intestinal and dermal uptake of metals in earthworms. *Soil Biol. Biochem.*, 35, 125–132.

Vijver M.G., Wolterbeek H.Th., Vink J.P.M., and van Gestel C.A.M. (2005), Surface adsorption of metals onto the earthworm *Lumbricus rubellus* and the isopod *Porcellio scaber* is negligible compared to absorption in the body. *Sci. Tot. Environ.*, 340, 271–280.

Weeks J.M. and Svendsen C. (1996), Neutral red retention by lysosomes from earthworm (*Lumbricus rubellus*) coelomocytes: A simple biomarker of exposure to soil copper. *Environ. Toxicol. Chem.*, 15, 1901–1805.

Yeardley Jr. R.B., Lazorchak J.M., and Gast L.C. (1996), The potential of an earthworm avoidance test for evaluation of hazardous waste sites. *Environ. Toxicol. Chem.*, 15, 1532–1537.

Xiao N.W., Song Y., Ge F., Liu X.H., and Ou-Yang Z.Y. (2006), Biomarkers responses of the earthworm *Eisenia fetida* to acetochlor exposure in OECD soil. *Chemosphere,* 65, 907–912.

Chapter 20
Differentiating Effluent Organic Matter (EfOM) from Natural Organic Matter (NOM): Impact of EfOM on Drinking Water Sources

Seong-Nam Nam[1], **Stuart W. Krasner**[2], **and Gary L. Amy**[3]

Abstract In this study, the characteristics of effluent organic matter (EfOM) were investigated and differentiated from natural organic matter (NOM) using several analytical methods: XAD resin fractionation, size-exclusion chromatography in combination with a dissolved organic carbon detector (SEC–DOC), fluorescence spectroscopy (excitation-emission matrix [EEM] and fluorescence index [FI]), biodegradable dissolved organic carbon (BDOC) testing, ultraviolet absorbance at 254 nm (UVA_{254}), and DOC measurements. Also, the impact of EfOM on drinking water sources was evaluated in terms of treatability of the EfOM by a drinking water treatment process. Treatability experiments were carried out with various mixtures of NOM and EfOM by a coagulation process. Characterization results indicated that the distinct properties of EfOM were lower specific UVA (SUVA), more hydrophilic organic matter, increased FI values, higher polysaccharides peak in SEC, and clear protein-like peak in EEM, as compared to NOM. Removal of DOC in EfOM-dominated waters by coagulation was not as high (~38%) as those of NOM samples (~57%), which was attributed to the higher hydrophilicity of EfOM. BDOC of EfOM occurred primarily via degradation of microbially derived organic constituents, such as proteins and polysaccharides.

Keywords: Natural organic matter, effluent organic matter, drinking water treatment

20.1 Introduction

Wastewater reuse has been increasing as an alternative source for regions with limited water supplies. Indirect potable reuse of wastewater occurs when the treated wastewater is discharged into rivers or lakes that are used downstream as

[1] *Civil and Environmental Engineering, University of Colorado, Boulder, Colorado USA*

[2] *Metropolitan Water District of Southern California, La Verne, California USA*

[3] *UNESCO-IHE Institute for Water Education, Delft, The Netherlands*

Y.J. Kim and U. Platt (eds.), *Advanced Environmental Monitoring*, 259–270.

drinking water sources. However, it is not well understood what impact EfOM present in wastewater will have on drinking water treatment plants. EfOM may impact the removal efficiency of NOM, increase disinfectant or coagulant dose, and increase the formation of disinfection by-products (DBPs) of health and regulatory concern.

EfOM present in biologically treated wastewater consists of refractory NOM derived from drinking water sources, soluble microbial products from bacteria in the activated sludge, and trace levels of synthetic organic compounds or DBPs from domestic and/or industrial use (Crook et al. 1999; Drewes et al. 2003, Namkung et al. 1986). However, despite increased studies on EfOM, it is not fully understood compared to the understanding of NOM, due to its complex and heterogeneous composition and varied sources, such as domestic, industrial, or agricultural origins.

Therefore, a better understanding of EfOM will enable the drinking water treatment plant receiving EfOM-impacted or -dominated waters to improve and/or select other treatment approaches to efficiently remove EfOM. The aims of this research are to investigate the characteristics of EfOM, differentiated from NOM, and to evaluate the impacts of EfOM-receiving waters on the drinking water treatment process.

20.2 Methods

20.2.1 Water Sources

This study was part of a research project funded by the American Water Works Association Research Foundation and the U.S. Environmental Protection Agency. Although the original project collected samples from ~20 wastewater treatment plants (WWTPs) and drinking water treatment plants, including river water upstream and downstream of WWTPs, representative results from an EfOM-impacted river in a Northeast USA watershed will be discussed. In addition, some results from an EfOM-dominated stream in a Southwest USA watershed will be shown.

Wastewater effluent was collected from a wastewater treatment plant (WWTP) which has a 2-stage aeration activated sludge process for secondary treatment and filtration for tertiary treatment. Treated wastewater (after chlorination/plant outfall) was discharged into a river. A sample was taken upstream (UPST) of the WWTP outfall site on the river. Note, there were other WWTPs that discharged into this watershed upstream of the subject WWTP. The river converged with another river and downstream of the confluence was used as an influent (INFL) to a drinking water treatment plant. Finally, samples were collected in an effluent-dominated stream from downstream of a WWTP in a Southwest USA watershed.

20.2.2 Analytical Parameters

NOM and EfOM were characterized by the following measurements: UVA at 254 nm, DOC, and SUVA (= UVA_{254} × 100/DOC). XAD-8 and XAD-4 resins were used for fractionating hydrophobic (HPO), transphilic (TPI) and hydrophilic (HPI) organic matter. High-pressure SEC (LC600 Shimadzu Corp., Japan) with a TSK HW-50S column (26 mm × 300 mm), in combination with an online DOC detector (modified Sievers 800 Turbo, Ionics Instruments, USA) or a UV detector, was used to separate organic matter by molecular size distribution (polysaccharides, humic substances, and low molecular weight organic acid peaks) (Her et al. 2002).

A fluorescence EEM was developed by scanning over an excitation range of 240–450 nm by 10-nm increments and an emission range of 290–530 nm by 2-nm increments using a FluoroMax-3 spectrofluorometer (HORIBA Jobin Yvon, Inc., USA). The fluorescence samples were adjusted to 1 mg/L of DOC and pH 2.8 with an acidified 0.01-M KCl solution. After subtracting the blank, EEMs were corrected with excitation and emission correction files generated by the FluoroMax manufacturer. Excitation and emission bandwidths were each 5 nm, respectively. Intensities were normalized to the blank Raman peak area measured the same day as the sample EEMs.

A fluorescence index (FI) was calculated by the ratio of fluorescence intensity at emission 450 and 500 nm at excitation 370 nm (McKnight et al. 2001). Total dissolved polysaccharides were measured using the phenol–sulfuric acid colorimetric method (Dubois et al. 1956). The biodegradability of EfOM was measured by a 5-day aerobic BDOC test (Allgeier et al. 1996).

20.2.3 Coagulation of NOM/EfOM Mixtures

In order to investigate the impact of EfOM on drinking water treatment plants, treatability experiments using jar tests were performed with coagulation–flocculation, which is a representative drinking water treatment process for the removal of NOM. Mixtures of water upstream of the WWTP (i.e., NOM) and the WWTP wastewater (WW) effluent (i.e., EfOM) were prepared with 100/0, 75/25, 50/50, 25/75, and 0/100 (NOM %/EfOM %) ratios (for samples collected in August 2005). Predetermined doses of ferric chloride ($FeCl_3$) were added in the range of 2–40 mg/L. Jar testing was carried out for 1 min at 100 rpm for rapid mixing, followed by 15 min at 20 rpm for flocculation, and 20 min for (quiescent) settling.

Removal efficiency was evaluated in terms of DOC. Table 20.1 shows the characteristics of various NOM/EfOM waters.

Table 20.1 Characteristics of NOM/EfOM mixture samples (Nam, 2007)

Mixture samples (NOM/EfOM)	pH	UVA_{254}, cm^{-1}	DOC mg/L	SUVA L/mg-m	Conductivity, µS/cm	TDS, mg/L	Alkalinity, mg/L as $CaCO_3$
100/0	8.26	0.077	3.01	2.56	451	216	80
75/25	7.94	0.088	3.80	2.31	511	245	60
50/50	7.50	0.099	4.59	2.15	552	264	40
25/75	7.53	0.109	5.38	2.03	678	326	28
0/100	7.43	0.12	6.17	1.94	822	397	27.5

20.3 Results and Discussion

20.3.1 Characteristics of NOM and EfOM

Table 20.2 summarizes trends showing the characteristics of NOM, EfOM and EfOM-*impacted* waters in the Northeast USA watershed. Compared to UPST (upstream NOM sample), WWTP (EfOM sample) exhibited relatively low SUVA values (1.63–2.13 L/mg-m, average: 1.90 L/mg-m), increased fraction of hydrophilic organic matter (32.5–38.6%, average: 34.8%), and higher FIs (1.39–1.43, average: 1.41). The low SUVA of wastewater implies that the DOC of EfOM is comprised a more nonaromatic (or less aromatic) organic carbon than NOM.

Organic matter derived from allochthonous (terrestrial plant origin) sources have lower FI values and organic matter derived from autochthonous (microbially derived, such as algae and bacteria) sources show higher FI values. As a point of

Table 20.2 Characterization results of upstream water, wastewater and downstream water (Nam, 2007)

Sample ID	Sample type	Sampled date	UVA_{254} cm^{-1}	DOC mg/L	SUVA L/mg-m	HPO (mg/L as DOC)	TPI (mg/L as DOC)	HPI (mg/L as DOC)	FI
UPST	upstream of WWTP	February 2004	0.074	2.58	2.87	1.61	0.58	0.38	1.207
		August 2004	0.149	5.22	2.85	2.34	1.46	1.42	1.160
		August 2005	0.077	3.01	2.56	1.47	0.53	1.01	1.319
		Average	**0.100**	**3.60**	**2.76**	**1.81**	**0.86**	**0.94**	**1.229**
WWTP	WW effluent	February 2004	0.124	7.61	1.63	3.03	1.64	2.93	1.422
		August 2004	0.115	5.40	2.13	2.36	1.29	1.75	1.397
		August 2005	0.120	6.17	1.94	2.50	1.62	2.05	1.433
		Average	**0.120**	**6.39**	**1.90**	**2.63**	**1.52**	**2.25**	**1.418**
INFL	downstream of WWTP	February 2004	0.088	3.10	2.84	2.02	0.62	0.45	1.173
		August 2004	0.158	5.12	3.09	2.66	1.14	1.32	1.150
		August 2005	0.119	4.72	2.52	2.43	0.98	1.31	1.459
		Average	**0.122**	**4.31**	**2.82**	**2.37**	**0.91**	**1.03**	**1.261**

Results from a Northeast USA watershed

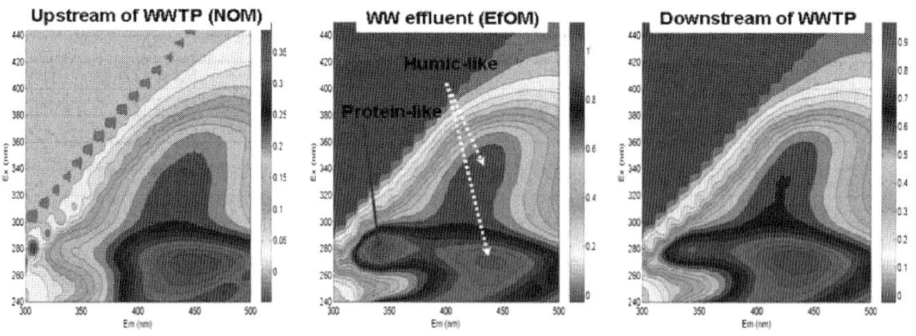

Fig. 20.1 EEM contour maps of samples (February 2004) collected upstream of WWTP (left), wastewater effluent (middle), and downstream of WWTP (right) from a Northeast USA watershed. (*x*-axis: emission [Em] wavelengths of 290–500 nm, *y*-axis: excitation [Ex] wavelengths of 240–450 nm) (Nam, 2007)

reference, FIs of the Suwannee River humic and fulvic acids, and pure EfOM (which was generated from a laboratory bench-scale bioreactor and isolated by an electrodialysis/reverse osmosis desalting process) were determined to be 0.864, 0.969, and 1.459, respectively. The FIs of EfOM were higher than those of NOM, which means that the properties of EfOM that distinguish it from NOM are mainly microbial—in origin. The same trends of FIs were also observed in other wastewaters in this study (results not shown).

Fluorescence EEM measurements for upstream water, wastewater, and downstream of WWTP (influent of drinking water treatment plant) clearly showed different peaks between NOM and EfOM (Fig. 20.1). Previous researchers working with fluorescence spectroscopy identified tryptophan-like, protein-like peaks at Ex/Em = 275/340 nm and humic-like peaks at 260/380–460 nm and 350/420–480 nm (Coble 1996; Coble et al. 1998; Stedmon et al. 2003). EEMs of wastewater and wastewater-impacted water exhibit the presence of protein-like substances at the range of excitation wavelength 260–290 nm and emission wavelength 320–370 nm, which were at a similar location observed in other studies, and this protein-like peak may have originated from soluble microbial products (SMPs) present in biologically treated wastewater.

In general, the SEC–DOC of surface waters and wastewaters is separated into three main peaks according to their molecular weight distributions, as shown in Fig. 20.2: the zone 1 peak represents large molecules, such as polysaccharides, proteins, and organic colloids, the zone 2 peak is attributed to humic substances, and the zone 3 peak corresponds to low molecular weight organic acids. Corresponding UVA_{254} chromatograms showed low absorbance in the zone 1 peak, and greatly increased absorbancy in zone 2, with intermediate absorbance in the zone 3 peak, which indicates that macromolecular constituents in the zone 1 peak do not consist of aromatic rings or conjugated bonds, but saturated structures or linear ring systems that may be more labile and hydrophilic, although they are macromolecules. On the other hand, components (humic substances) in the zone 2 peak are likely to be

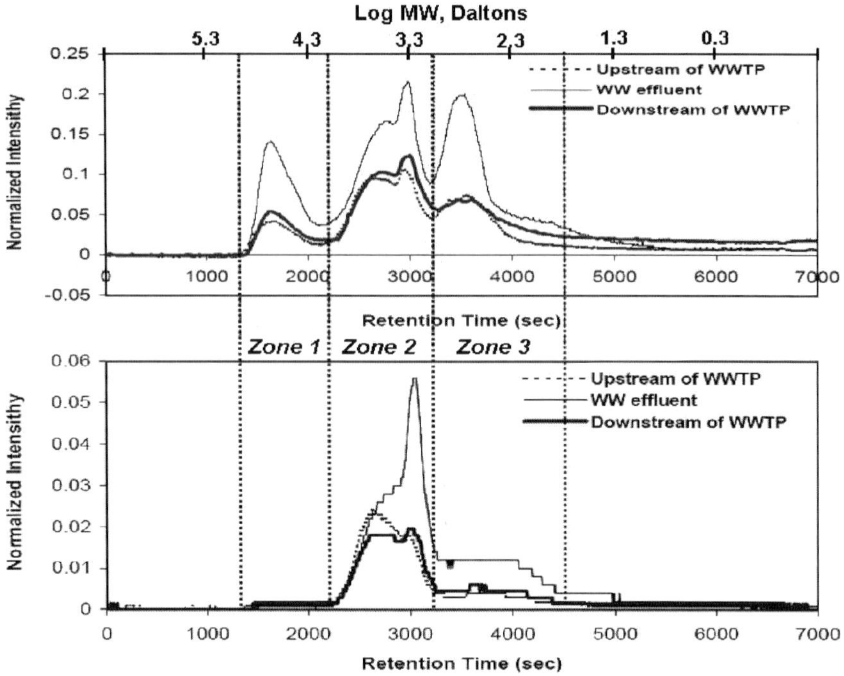

Fig. 20.2 SEC of DOC fractions (upper) and UVA_{254} fractions (lower) for upstream water, wastewater, and downstream water samples (February 2004) from a Northeast USA watershed, and their molecular weight (MW) distributions (WWTP sample diluted approximately threefold relative to upstream and downstream samples) (Nam, 2007)

hydrophobic and microbially resistant in spite of somewhat smaller molecules (relative to zone 1).

The SEC–DOC fractions for each peak from the SEC are summarized in Table 20.3. The percentage for each peak was determined by the ratio of integrated peak area to total peak area, and each percentage was then converted to DOC (relative to the bulk water DOC). As can be seen in Table 20.3, EfOM had a higher DOC content of polysaccharides, indicating a higher contribution to the DOC of the EfOM by microbial sources. In addition, the EfOM also showed a higher DOC content of both humic substances and low molecular weight organic acids relative to the NOM; given that humic substances are predominately derived from drinking water NOM, this higher EfOM humic substance content may reflect a diverse drinking water source encompassing more than the upstream sample.

Figure 20.3 shows the results of total dissolved carbohydrates measurements using the phenol–sulfuric acid method. Higher levels of polysaccharides were present in EfOM, which was consistent with SEC–DOC data. Polysaccharides are one of the components of bacterial cell walls and are released during the

Table 20.3 Characterization results of upstream water, wastewater, and downstream water from a Northeast USA watershed (SEC–DOC data from samples taken in February 2004) (Nam, 2007)

		DOC (mg/L)		
Samples	DOC (mg/L)	PS	HS	LMA
UPST (Upstream water)	2.58	0.27	1.19	1.12
WWTP (WW effluent)	7.61	0.58	4.39	2.63
INFL (Downstream water)	3.10	0.22	1.21	1.67

PS, HS, and LMA denote polysaccharides, humic substances, low molecular weight organic acids

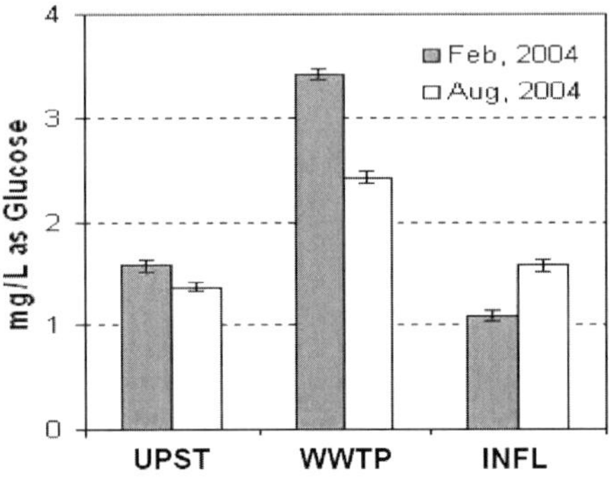

Fig. 20.3 Total dissolved carbohydrates (polysaccharides) for upstream water (UPST), wastewater (WWTP), and downstream water (INFL) (n = 3) (Nam, 2007)

endogenous cell decay. Therefore, the primary characteristics of EfOM mostly reflect microbial sources.

20.3.2 Treatability of EfOM-Impacted or -Dominated Water by Coagulation

Figure 20.4 shows the DOC removal efficiencies of NOM/EfOM mixtures by coagulation. Here, EfOM-dominated waters are defined as waters mixed with 50/50 or 25/75 or 0/100 (NOM %/EfOM %) ratios, and EfOM-impacted water will be the 75/25 mixture. As shown, the DOC removals from adding 40 mg/L of coagulant ferric chloride were 54 and 57% in the upstream sample and the EfOM-impacted water, respectively, whereas the DOC in the EfOM-dominated waters was removed by 38–46%.

Based on previous researchers' studies related to coagulation of NOM, it is known that the coagulation process preferentially removes hydrophobic organic matter over hydrophilic organic matter. The fluorescence EEM spectra of samples through the coagulation process support this trend (Fig. 20.5). As the NOM water was increasingly mixed with more of the EfOM, the protein-like peak became

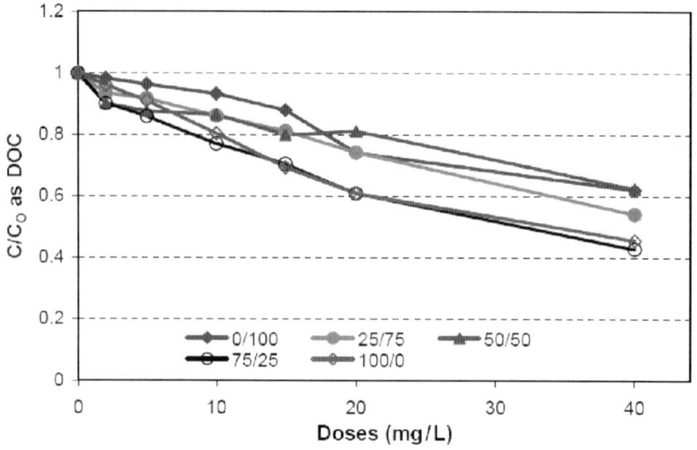

Fig. 20.4 Removal of DOC in NOM/EfOM mixtures by coagulation (Nam, 2007)

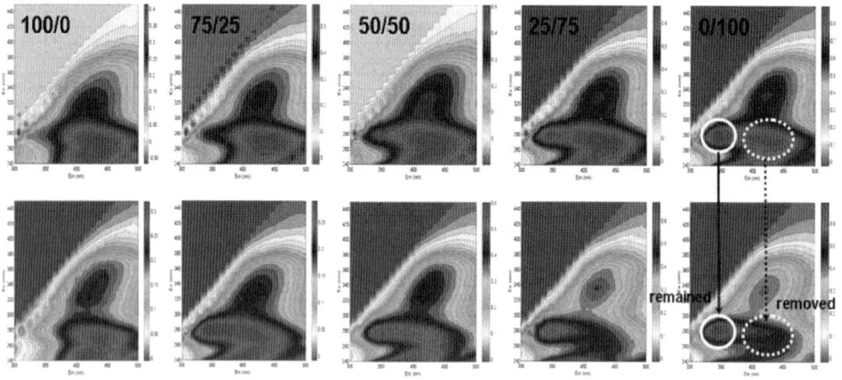

Fig. 20.5 EEM contour maps before coagulation (upper) and after coagulation with 40 mg/L dose (lower) (*x*-axis: emission wavelength of 290–500 nm, *y*-axis: excitation wavelength of 240–450 nm) (Nam, 2007)

clearer and its fluorescence intensity increased as well (upper Figures in Fig. 20.5). From the coagulation process, humic-like materials were significantly removed, but the protein-like materials showed lower removals.

Considering that EfOM is more hydrophilic than NOM according to previously presented characteristics of EfOM, these results are reasonable. In addition, as the upstream water was mixed with more of the EfOM, the bulk concentration of DOC increased (Table 20.1). So when 40 mg/L of coagulant was added, the coagulant to DOC ratio was 13 mg/mg for the upstream (100/0) sample, whereas the ratio was only 6.5 mg/mg for the EfOM (0/100) sample. So the presence of EfOM can also increase the coagulant demand. On a constant coagulant/DOC ratio (e.g., ~6.5 mg/mg),

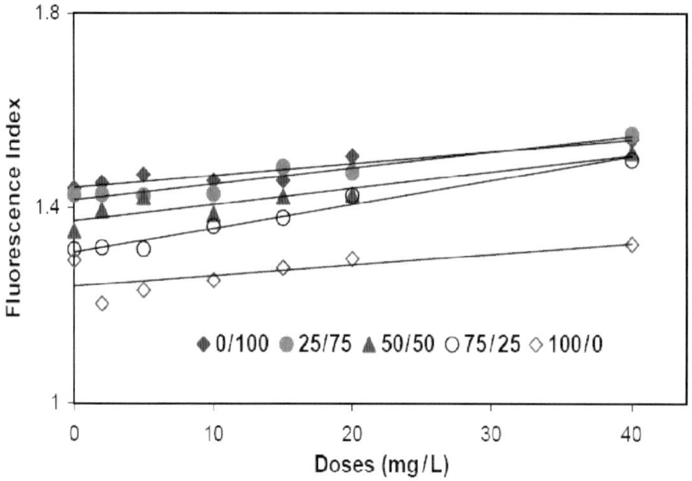

Fig. 20.6 Changes in FIs of the NOM/EfOM mixtures by the coagulation process (Nam, 2007)

coagulant doses of 20 and 40 mg/L would be needed for the upstream and EfOM samples, respectively. The use of 20 mg/L of coagulant on the upstream sample or 40 mg/L of coagulant on the EfOM sample both resulted in similar DOC removal efficiencies (~40%, Fig. 20.4).

These results mean that the presence of EfOM in a drinking water source may have an impact on the efficiency of the drinking water treatment process due to the different properties of EfOM with respect to NOM, especially in terms of DOC (DBP precursor) removal and/or coagulant demand.

Coagulation resulted in increased FIs of the NOM/EfOM mixtures. Allochthonous materials, which are sources of humic-like materials, were preferentially removed over autochthonous organic matter, which caused an increase in the relative portion of organic matter from autochthonous sources (Fig. 20.6).

20.3.3 Biodegradation of EfOM

Figures 20.7 and 20.8 shows the EEM contour plots and SEC–DOC of EfOM through aerobic BDOC tests. Fig. 20.7 portrays a WWTP sample from the Northeast USA watershed while Fig. 20.8 corresponds to a sample taken from a wastewater-dominated stream in a Southwest USA watershed. In contrast to the coagulation process, biodegradation preferentially removed protein-like substances. In Fig. 20.8, degradation of organic carbon in the polysaccharide peak was more complete and faster, whereas humic substances were slowly and incompletely biodegraded.

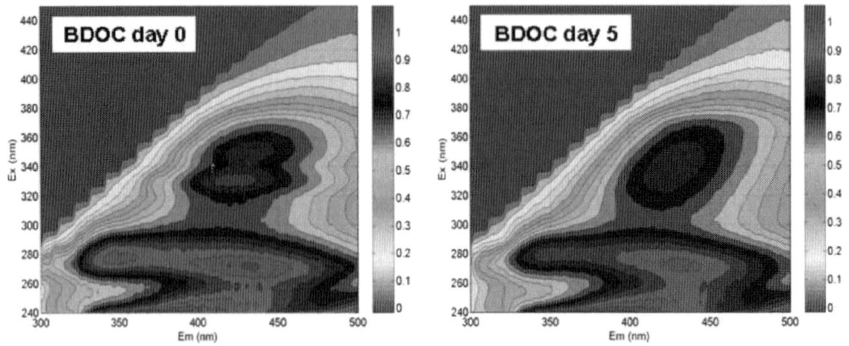

Fig. 20.7 Changes in EEM contour maps by aerobic $BDOC_5$ test of EfOM (data from wastewater sample taken from a Northeast USA WWTP in August 2005) (*x*-axis: emission wavelength of 290–500 nm, *y*-axis: excitation wavelength of 240–450 nm) (Nam, 2007)

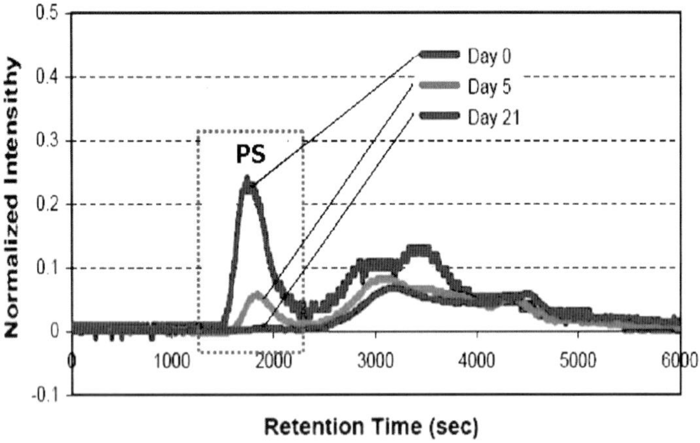

Fig. 20.8 SEC-DOC of aerobic BDOC test of EfOM for 5 and 21 days (data used here were for a sample collected from a Southwest USA watershed in February 2005, which was EfOM-dominated) (Nam, 2007)

After a 21-day BDOC test, the polysaccharide peak almost completely disappeared. Because large-sized components of the polysaccharide peak are saturated or linear structures, with fewer double bonds, bacteria were more easily able to assimilate these structures compared to humic substances with unsaturated, branched or ring structures.

Through this research project, aerobic $BDOC_5$ showed a linear relation with DOC, the hydrophilic organic matter fraction, the amount of polysaccharides of raw waters, as well as the concentration of dissolved organic nitrogen (DON). However, it was inversely related to the SUVA values of samples (results not shown).

20.4 Conclusions

NOM and EfOM collected from a WWTP and river sites upstream and downstream of the WWTP were characterized with analytical techniques. Also, treatability of various NOM/EfOM mixtures was evaluated with the coagulation process as a representative drinking water treatment process. The experimental results from this work are as follows:

- EfOM exhibited lower SUVA, higher fractions of hydrophilic organic matter, and higher FIs compared to NOM, indicating that the EfOM source is from autochthonous and microbially derived sources.
- A distinct difference in EEMs between NOM and EfOM was the presence of a clear protein-like peak, which was in EfOM, but not in NOM.
- In SEC–DOC results, EfOM contained a higher DOC content polysaccharide (or protein) groups, and those constituents exhibited nonaromatic and hydrophilic characters.
- Total dissolved polysaccharide measurements were also higher in EfOM.
- DOC removals of EfOM with the same coagulant dose were less efficient than with NOM, and these results were attributed to higher hydrophilic character of EfOM (where the coagulation process removed hydrophobic organic matter more favorably) and due to an increase in coagulant demand.
- In contrast with coagulation, biodegradation of EfOM preferentially (and faster) removed hydrophilic substances (protein-like and polysaccharide groups).

Through these results, it is elucidated that EfOM has distinguishable properties that are different from NOM and which may have a negative impact on drinking water treatment plants using conventional (coagulation) processes. Therefore, when EfOM-impacted waters are used as a potable source, other additional treatment processes or environmental buffers (e.g., biological filtration, river bank filtration, soil aquifer treatment, etc.) may need to be considered in order to remove hydrophilic components (proteins and polysaccharides).

Acknowledgements This study has been performed through funding from the American Water Works Association (Awwa) Research Foundation (project # 2948). The authors gratefully acknowledge that the Awwa Research Foundation is the joint owner of the technical information upon which this publication is based. The authors thank the Foundation and the U.S. government, through the Environmental Protection Agency (USEPA) for its financial, technical, and administrative assistance in funding and managing the project through which this information was discovered. The comments and views detailed herein may not necessarily reflect the views of the Awwa Research Foundation, its officers, directors, affiliates or agents, or the views of the U.S. Federal Government. The project manager is Alice Fulmer.

The authors also acknowledge our coinvestigators at Arizona State University (Paul Westerhoff and Baiyang Chen) who provided the DON data for this study. Finally, the participating utilities and agencies and their staff are acknowledged for their invaluable assistance and financial support.

References

Allgeier S.C., Summers R.S., Jacangelo J.G., Hatcher V.A., Moll D.M., Hooper S.M., Swertfeger J.W., and Green R.B. (1996), *A Simplified and Rapid Method for Biodegradable Dissolved Organic Carbon Measurement*. Proceedings: AWWA Water Quality Technology Conference. Boston, MA.

Coble P.G. (1996), Characterization of marine and terrestrial DOM in seawater using excitation-emission matrix spectroscopy, *Mar. Chem.*, 51(4), 325–346.

Coble P.G., Castillo C.E.D., and Avril B. (1998), Distribution and optical properties of CDOM in the Arabian Sea during the 1995 Southwest Monsoon, *Deep-Sea Res. Part II*, 45, 2195–2223.

Crook J., MacDonald J.A., and Trussell R.R. (1999), Potable use of reclaimed water, *J. Am. Water Works Assoc.*, 91(8), 40–49.

Drewes J.E., Reinhard M., and Fox P. (2003), Comparing microfiltration-reverse osmosis and soil-aquifer treatment for indirect potable reuse of water, *Water Res.*, 37(15), 3612–3621.

Dubois M., Gilles K.A., Hamilton J.K., Rebers P.A., and Smith F. (1956), Colorimetric method for determination of sugars and related substances, *Anal. Chem.*, 28(3), 350–356.

Her N., Amy G., Foss D., Cho J., Yoon Y., and Kosenka P. (2002), Optimization of method for detecting and characterizing NOM by HPLC-size exclusion chromatography with UV and on-line DOC detection, *Environ, Sci. Technol.*, 36(5), 1069–1076.

McKnight D.M., Boyer E.W., Westerhoff P.K., Doran P.T., Kulbe T., and Andersen D.T. (2001), Spectrofluorometric characterization of dissolved organic matter for indication of precursor organic material and aromaticity, *Limnol. Oceanogr.*, 46(1), 38–48.

Nam S.-N. (2007), Characterization and Differentiation of Wastewater Effluent Organic Matter (EfOM) versus Drinking Water Natural Organic Matter (NOM): Implications for Indirect Potable Reuse, PhD Dissertation, University of Colorado, Boulder, Colorado USA.

Namkung E. and Rittmann B.E. (1986), Soluble microbial products (SMP) formation kinetics by biofilm, *Water Res.*, 20, 795–806.

Stedmon C.A., Markager S., and Bro R. (2003), Tracing dissolved organic matter in aquatic environments using a new approach to fluorescence spectroscopy, *Mar. Chem.*, 82, 239–254.

Chapter 21
An Advanced Monitoring and Control System for Optimization of the Ozone-AOP (Advanced Oxidation Process) for the Treatment of Drinking Water

Joon-Wun Kang[1], Byung Soo Oh[1], Sang Yeon Park[1], Tae-Mun Hwang[2], Hyun Je Oh[2], and Youn Kyoo Choung[3]

Abstract This study was conducted to illustrate an innovative method for the optimization and control of the ozone-AOP (advanced oxidation process) to achieve the target oxidation objective for the removal of hazardous micropollutants, while minimizing the formation of harmful by-products, bromate (BrO_3). The control method was based on a specifically conceived analytical setup with FIA (flow injection analysis), which can accurately isolate and measure an individual oxidant in the presence of other oxidants (ozone, hydrogen peroxide, free chlorine, etc). Three auto-analyzing units (Ozone Kinetic Analyzing Unit, Hydrogen Peroxide Analyzing Unit and Bromate Analyzer) were specially devised for this study, which were successfully tested on both in lab- and demo-scales for in-plant operation.

Keywords: Bromate, ozone, ozone/hydrogen peroxide, hydroxyl radical (OH•), optimization

21.1 Introduction

Ozone has been continuously used in the treatment of drinking water for close to 100 years. Recently, the number of ozone installations has rapidly grown world wide for the effective removal of toxic chemicals and inactivation of pathogenic microorganisms. On ozonation, pollutants can be oxidized by molecular ozone or

[1] *Department of Environmental Engineering, YIEST, Yonsei University at Wonju, 234, Maeji, Wonju, KOREA (220–710)*

[2] *Korea Institute of Construction Technology, 2311 Daehwa-Dong, Ilsan-gu, Kyonggi-do, KOREA (411–712)*

[3] *School of Civil & Environmental Engineering, Yonsei University, Seoul, KOREA*

Y.J. Kim and U. Platt (eds.), *Advanced Environmental Monitoring*, 271–281.

the hydroxyl radical (OH•), from the decomposition of molecular ozone in water (Staehelin and Hoigné 1982). For efficient disinfection, the minimization of ozone decomposition is desirable, but for the oxidation of micropollutants, a process modification to increase the proportion of OH• is much favored since OH• is more reactive than ozone molecule. Ozonation alone does not appear to generate a sufficiently high concentration of OH•. Therefore, many attempts have been made to explore AOPs (advanced oxidation processes) to enhance the production of OH•. Of the ozone-based AOPs, the ozone/hydrogen peroxide (peroxone) process is one of the most effective processes for the production OH• (Duguet et al. 1985).

However, in ozone-related processes, i.e. ozone alone or the peroxone process, the formation and control of bromate has been regarded as the most important issue since it was first implicated as a potential carcinogen (Buffel et al. 2004). The maximum permissible level of bromate in drinking water has been set at 10 µg/L in both the U.S. and the European Union (Fed. Resist. 1994; Amtsblatt der Euopäischen Gemeinschaften L 330 1998). During oxidation, it has been found that two mechanisms are responsible for bromate formation: molecular ozone and OH•. The contribution to bromate formation via these two oxidants is affected by water quality variables and operational conditions, such as pH, temperature, bromide level, ozone dosage and contact time, as well as other matrix conditions. In the peroxone process, the ratio of hydrogen peroxide to ozone could also affect bromate formation, as the molecular ozone concentration, residual hydrogen peroxide and OH• are sensitive to fluctuations in the water quality. Therefore, the aim of this study was to propose an automated real-time control system for optimization of the ozone-AOP process in relation to bromate formation.

21.2 Methods

21.2.1 Ozone Analysis System

Figure 21.1 is the schematic diagram of an ozone analysis system (O_3-k_c device) designed for the continuous monitoring of residual ozone, as well as the decomposition rate, using a highly sensitive detective analytical method employing an FIA technique. The design of this measuring system, incorporating several peristaltic pumps, solenoid valves and a UV/Vis detector, was based on the indigo method (Hoigné and Bader 1981). The device was designed to be controlled by an ICU equipped with a data acquisition system (data acquisition interval was less than 0.05 s) (Park et al. 2000). Figure 21.1(b) shows a typical decomposition pattern of ozone in raw water. Two apparent ozone decay phases, which consists with the instantaneous ozone demanded phase (IOD) followed by a rather slower decay rate with the first order (k_c) in $[O_3]$, were successfully measured with this online monitoring setup.

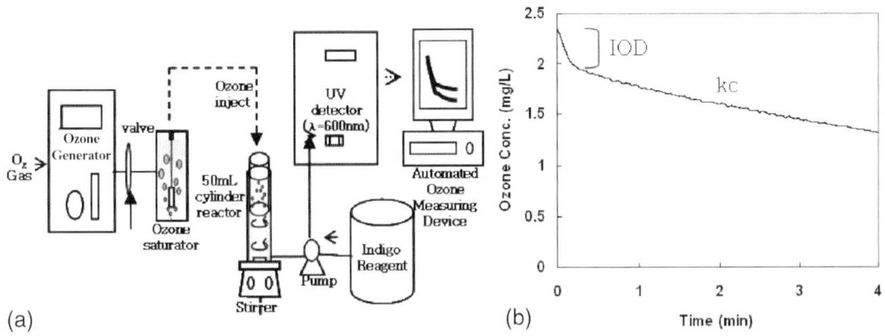

Fig. 21.1 **(a)** Schematic diagram of the ozone analysis system. **(b)** Typical ozone decomposition pattern of ozone in raw water obtained from this setup.

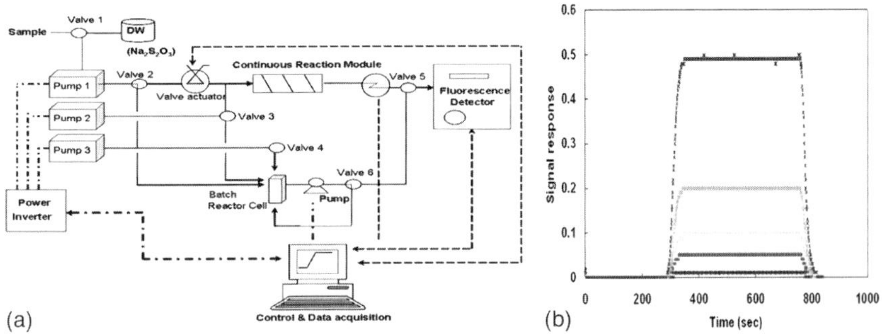

Fig. 21.2 **(a)** Schematic diagram of the hydrogen peroxide analysis system. **(b)** Signal response.

21.2.2 Hydrogen Peroxide Analysis System

The hydrogen peroxide analyzing unit was based on a fluorometric detection method using the reaction of *p*-hydroxyphenyl acetic acid and residual hydrogen peroxide in the presence of a peroxidase enzyme (Lazrus et al. 1985). Figure 21.2(a) shows a scheme for the on-line residual peroxide detection system, which was carefully devised to exclude interference from ozone. Three DC drive pumps (Cole-Parmer Peristaltic pump, USA) continuously injected the following reagents; fluorometric reagent (0.35 M potassium hydrogen phthalate (KHP), 8.0×10^{-3} M *p*-hydroxyphenyl acetic acid (POPHA), 2 purpurogallin units of peroxidase/ml reagent), buffered solution (0.5 M NaOH) and the sample, into the hydrogen peroxide analyzer, with tubing connected to both the inlet and outlet of the instrument. The POPHA dimer formed in the reaction module was stabilized raising pH >10, and analyzed using fluorescence detection (SOMA, Japan), with excitation and emission at 320 and 400 nm, respectively. Data signals were converted into actual hydrogen peroxide concentration units (µg/l), and recorded by a computer system via an interface card (Labview, USA). Figure 21.2(b) shows the stability and sensitivity of data signal obtained with this method, which enabled detection to the ~ppt level (MDL < 1 µg/l) acquiring two signals per second.

21.2.3 Bromate Analysis System

The bromate analysis was based on the EPA 326.0 method, using a suppressor acidified post-column reagent (Salhi and Gunten 1999). As shown in Fig. 21.3, the bromate analyzing system was composed of two main parts: ion chromatograph

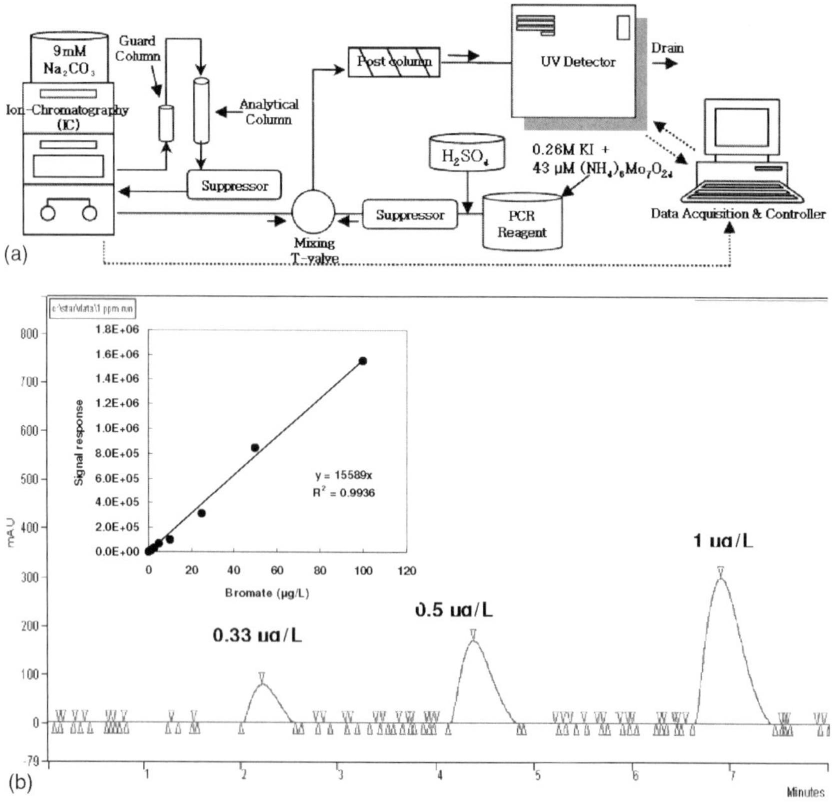

Fig. 21.3 **(a)** Schematic diagram of the bromate analysis system. **(b)** Shapes of the bromate peaks (MDL <0.3 µg/l)

and post-column reaction (PCR). The ion chromatograph system (DX-500, Dionex) was comprised of a guard column (AS9-HC, Dionex), an analytical column (AG9-HC, Dionex), a suppressor devices (ASRS-I) and a conductivity detector. The apparatus for the PCR were composed of a post-column reagent delivery system (PC-10, Dionex), a heated post-column reaction coil (Dionex) and an ultraviolet/visible (UV/Vis) detector (Prostar325, Varian). The effluent of bromate ions separated by the ion chromatograph mixed with an acidic solution of potassium iodide containing an ammonium molybdate and heated at 80 °C to form tri-iodide ions, which were measured by UV/Vis detection at 352 nm. With this setup, bromate could be precisely detected at levels as low as 0.3 μg/l.

21.3 Results and Discussion

21.3.1 Peroxone Process for the Removal of Pharmaceuticals

The peroxone process is one of the most effective ways of producing OH• for the oxidation of micropollutants. The stoichiometric optimum for this reaction is a hydrogen peroxide:ozone ratio of 0.5 (mole: mole) or 0.35 (weight: weight), but since other ozone-consuming species are usually present in water, the required dose ratio is usually less than this value. In fact, the actual ratio is to be directly determined with experimental studies. Figure 21.4 shows the effect of hydrogen peroxide on the percentage of the substrate removed; the optimum hydrogen peroxide:ozone ratio was 0.3 (w/w).

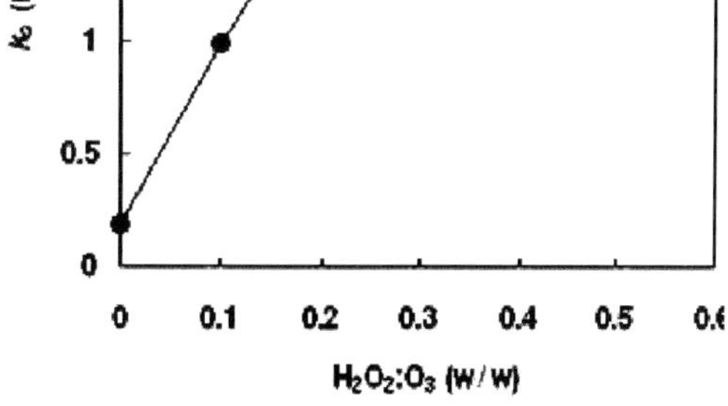

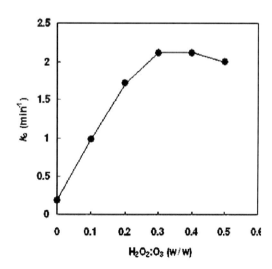

Fig. 21.4 Effect of hydrogen peroxide on Ibuprofen removal efficiency. Initial Ibuprofen: 2 μM. Ozone dose: 2 mg/l, hydrogen peroxide:ozone ratio = 0 ~ 0.8 (w/w)

21.3.2 Bromate Formation Characteristics in Ozone Alone

For the future application of ozone, the characteristics of bromate formation during ozonation were studied by testing water samples collected from six different water treatment plants located in Korea. With ozone doses of 1 ~ 2 mg/l, the formation of bromate ranged between 2 and 12 μg/l, which in some cases exceeded the MCL (= 10 μg/l). The level of bromide in the water samples tested were relatively low (11 ~ 42 μg/l), but the conversion rate of bromide to bromate appears to be high under certain conditions. The key factors that could increase the production of bromate are the ozone doses (Fig. 21.5), bromide level (Fig. 21.6), and pH. Increasing the ozone dose from 1 to 3 mg/l and the contact time from 2 to 10 min, the level of

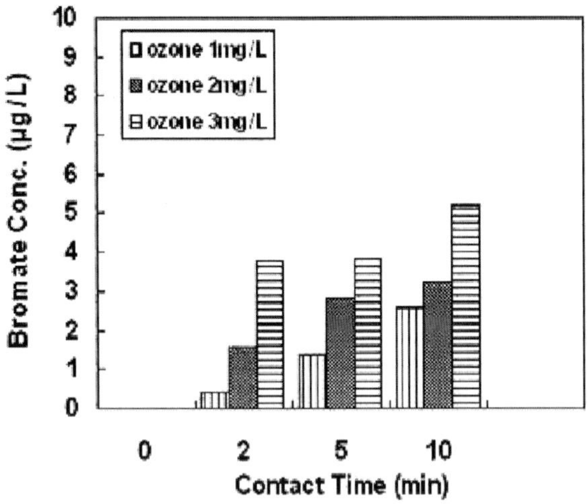

Fig. 21.5 Effect of ozone dose and contact time on the formation of bromate. pH: 7.2. Br^-: 40 μg/l

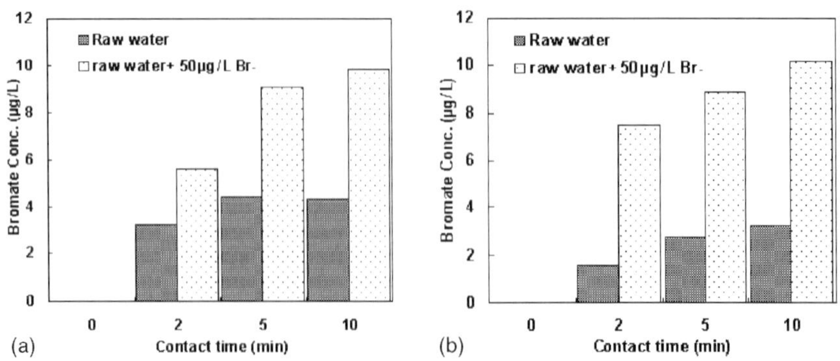

Fig. 21.6 Effect of contact time and bromide level on the formation of bromate. **(a)** Ozone dose: 2 mg/l. pH: 7.0. Initial Br^-: 40 μg/l. **(b)** Ozone dose: 2 mg/l. pH: 7.6. initial Br^-: 14 μg/l

bromate formation was found to increase from 0.4 to 5.2 μg/l (Fig. 21.5). Figure 21.6 compares the effect of the bromide level on the formation of bromate in the two different raw water sources with that of raw waters collected from S and M water treatment plants. In each raw water sample, 50 μg/l of bromide was intentionally added and ozonated. In results, about twofold increase of bromate formation was observed in the bromide spiked water as compared to the case in raw water. The effect of pH was found to be the most significant factor affecting the formation of bromate. Increasing the pH from 7 to 8.5, the conversion yield (%) of bromate was increased from 6.8 to 58%, far exceeding the MCL guideline (data not shown).

21.3.3 Characteristics of Bromate Formation in the Peroxone Process

The peroxone process has been proved to enhance the degradation of various micropollutants by converting aqueous ozone into the more reactive OH•. However, the formation of bromate cannot be avoided as OH• is also strongly involved in the peroxone process, bromate could also be formed, as the amount of molecular ozone would be small, but would still be involved in the bromate formation pathways. In this process, less ozone would be utilized in the formation of bromate due to its competitive reaction with the conjugate base of H_2O_2 ($O_3 + HO_2^- \rightarrow O_3^{-\bullet} + HO_2^{\bullet}$); therefore, less bromate production would expected in the peroxone process. However, the opposite was observed in our experiments. With regard to the effect of hydrogen peroxide on the formation of bromate in the peroxone process, the results reported in a reference survey were inconsistent (Gunten et al. 1988). Furthermore, it is hard to assess the effect of hydrogen peroxide on the formation of bromate without performing experimentation. Figure 21.7 shows the bromate formation profiles vs. contact time for a raw water initially spiked with bromide (= 100 μg/l) by varying the hydrogen peroxide dose between 0 and 0.6 (w/w), with

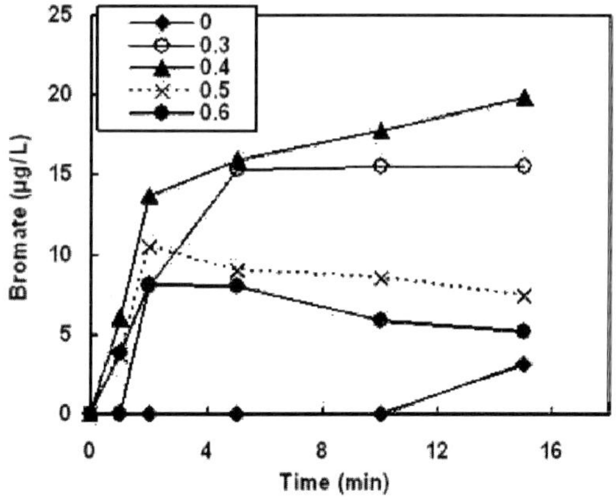

Fig. 21.7 Effect of hydrogen peroxide on the rate of bromate formation. Ozone: 2 mg/l. pH: 7.0. Br^-: 100 μg/l

the ozone dose fixed at 2 mg/l. On the addition of hydrogen peroxide, the observed level of bromate production compared to ozone alone was much higher. The highest level of bromate production was observed with a hydrogen peroxide:ozone ratio (w/w) of 0.4, but the rate of bromate production decreased with a ratio of 0.5, which is the desired peroxide dose for the control of bromate.

21.3.4 Scheme for Oxidant Dosage Control in the Ozone and Peroxone Processes for the Control of Bromate Formation

A real-time method for the effective control of ozone and hydrogen peroxide dosages has been developed, based on a specifically devised analytical setup, as previously shown (Fig. 21.1 and 2), and an optimum ozone dosage control method proposed (Oh et al. 2003; Oh et al. 2005). This control method achieved both the required disinfection and oxidation objectives. For the purpose of disinfection, ozone was dosed to the corresponding *CT* (*C*: residual disinfectant concentration (mg/l); *T*: contact time [min]) criteria required to meet the target inactivation goal. With respect to the oxidation objectives, which have recently increased in emphasis due to the contamination of water by various toxic chemicals, ozone was dosed to take into account the yield of the more reactive species, OH•, by measuring the ozone decay rate constant, k_c, and R_{ct} value (ratio of the OH• to residual ozone). For this purpose, the *kCT*-control method, which optimizes the ozone dosage on a real-time basis (Oh et al. 2003), is to be further developed and utilized to achieve the target water treatment objectives, the oxidation and/or disinfection goal. The *kCT* is defined as

$$kCT = \int [O]_3 dt$$

where, *kCT* is the area under the curve for the time varied residual ozone concentration (Fig. 21.1(b) to be precisely measured by the setup, as shown in Fig. 21.1 (a). The k_c value varies according to variations in the water quality, but could be continuously monitored by the automated real time monitoring/control system, and the amount of ozone adjusted to the set *kCT* value. The recommended target value, which was tested in a prior study on H river water, was 2 ~ 3 mg/l-min (Oh et al. 2003).

Since the oxidation of toxic micropollutants, such as pesticides and pharmaceuticals, and taste and order causing compounds, has now become a more centered issue in water treatment, several municipal drinking water treatment plants that used ozone have now been converted to the peroxone process by the addition of a hydrogen peroxide feeding system. In this process, the control of the optimum hydrogen peroxide dose is regarded as an important factor in the optimization of the process. The stoichiometric optimum for the reaction between ozone and hydrogen peroxide is a hydrogen peroxide:ozone ratio of 0.35 (w/w), but due to other demands on ozone or reacted compounds usually present in water, the required dose ratio would usually be less, but this can be determined experimentally. The *kCT* analytical devise enables

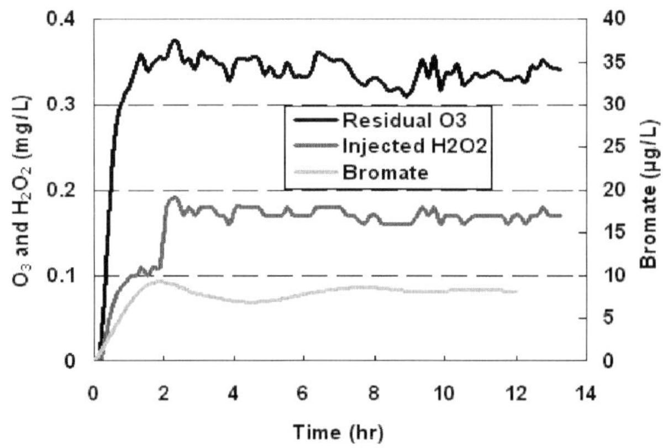

Fig. 21.8 Optimum oxidant dose control for the peroxone process using the *kCT* control method

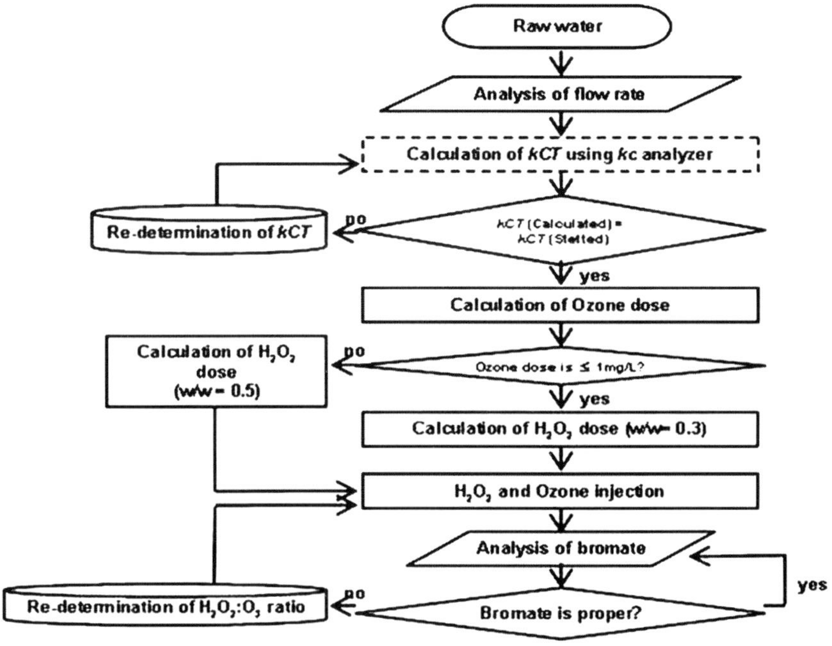

Fig. 21.9 Optimum process control logic to meet bromate MCL in ozone and peroxone AOPs

calculation of the optimum hydrogen peroxide dosage based on the measured residual ozone, which is controlled by the set *kCT* value. Figure 21.8 shows the residual ozone output obtained with the input *kCT* set at 2 mg/l-min vs. the calculated hydrogen peroxide dosage. The residual ozone output response to the set *kCT* was ~0.35. In a previous study, the recommended hydrogen peroxide dose was a hydrogen peroxide:ozone (w/w) ratio of 0.5, which corresponds to a hydrogen peroxide dose of

0.16 ~ 0.18 mg/l, as shown in Fig. 21.7. With this *kCT* based peroxide dose control, the bromate production could be successfully suppressed below the MCL guide line (10 μg/l). Figure 21.9 shows the flow chart of the *kCT* based control logic for the determination of the dosages of ozone and hydrogen peroxide to meet the bromate MCL guideline. The overall control system incorporated with the specifically-conceived analytical devices and a control unit was able to attain the best water treatment objectives for minimizing the formation of bromate.

21.4 Conclusion

The following conclusions were drawn from this study:

1) Online monitoring devices for the analyses of residual ozone and hydrogen peroxide were devised, not only for real time monitoring, but also for optimum dosage control.
2) The online monitoring device for the key ozone related by-product, bromate, was also devised, and was able to detect bromate at a level as low as 0.3 μg/l.
3) For the future application of ozone and ozone related AOPs, the characteristics of bromate formation have been studied. Even at low bromide levels, the conversion rate of bromide to bromate could be high, but this will depend on fluctuation in the water quality.
4) In the peroxone process, the formation of bromate was sensitive to the change in the hydrogen peroxide:ozone ratio. At a ratio of 0.5 (w/w), the minimum bromate formation was attained.
5) The overall control system, incorporated with the specifically conceived analytical devices and control logic, could attain the best water treatment objectives, with the suppression of the formation of the harmful by-product, bromate.

Acknowdegements This subject was supported by Ministry of Environment as "The Eco-technopia 21 project," and also by "The Brain Korea 21 project (BK21)" of Ministry of Education & Human Resource Development.

References

Amtsblatt der Euopäischen Gemeinschaften L 330: Richtlinie 98/83/EG, 1998.
Buffle M-O., Galli S., and von Gunten U. (2004), Enhanced bromate control during ozonation: The chlorine-ammonia process, *Environ. Sci. Technol.*, 38, 5187–5195.
Duguet J.P., Brodard E., Dussert B., and Mallevialle J. (1985), Improvement in the effectiveness of ozonation of drinking water through the use of hydrogen peroxide, *Ozone Sci. Eng.*, 7, 241–258.
Hoigné J. and Bader H. (1981), Determination of ozone in water by the indigo method, *Water Res.*, 15, 449–456.

Lazrus A.L., Kok G.L., Gitlin S.N., Lind J.A., and Mclaren S.E. (1985), Automated fluorometric method for hydrogen peroxide in atmospheric precipitation, *Anal. Chem.*, 57, 917–922.

Oh B.S., Kim K.S., Kang M.G., Oh H.J., and Kang J.W. (2005), Kinetic study and optimum control of the ozone/UV process measuring hydrogen peroxide formed in-situ, *Ozone Sci. Eng.*, 27, 421–430.

Oh H.J., Kim W.J., Choi J.S. Gee C.S. Hwang T.M., Kang J.G. and Kang J.W. (2003), Optimization and control of ozonation plant using raw water characterization method, *Ozone Sci. Eng.*, 25, 383–392.

Park H.S. Hwang T.M., Kang J.W., Choi H.C., and Oh H.J. (2000), Characterization of raw water for the ozone application measuring ozone consumption rate, *Water Res.*, 35, 2607–2614.

Staehelin J. and Hoigné J. (1982), Decomposition of ozone in water: Rate of initiation by hydroxide ions and hydrogen peroxide, *Environ. Sci. Technol.*, 16, 676–681.

Salhi E. and von Gunten U. (1999), Simultaneous determination of bromide, bromate and nitrate in low $\mu g l^{-1}$ levels by ion chromatography without sample pretreatment, *Water Res.*, 33, 3239–3244.

USEPA, Fed. Resist. (1994) National primary drinking water regulations; disinfectants and disinfection byproducts; proposed rule59:145:38668.

von Gunten U., Bruchet A., and Costentin E. (1988), Bromate formation in conventional and advanced oxidation processes (O_3/H_2O_2): Theoretical and empirical evaluation of full-scale experiments, *J. Am. Water Works Assoc.*, 88, 53–65.

Chapter 22
Monitoring of Dissolved Organic Carbon (DOC) in a Water Treatment Process by UV-Laser Induced Fluorescence

Uwe Wachsmuth[1], Matthias Niederkrüger[2], Gerd Marowsky[2], Norbert Konradt[3], and Hans-Peter Rohns[3]

Abstract Results of online investigations of water quality during a water treatment process by ultraviolet-laser-induced fluorescence (UV-LIF) are presented. In the first part the integrated fluorescence intensity is correlated to the classically determined concentration of dissolved organic carbon (DOC). Decision, detection and determination limits are evaluated for this procedure and online DOC measurements conducted with the presented LIF system are compared to reference analysis. In the second part the ozone demand in the water treatment process is derived from the LIF-signal directly. The calibration was done by correlating LIF-signals with the ozone doses for different conditions and a definite residual ozone concentration in the processed water.

Keywords: Dissolved organic carbon (DOC), drinking water, laser induced fluorescence, online process control, real-time measurements

22.1 Introduction

Large cities situated at big streams often fulfill their demand of drinking water from these streams. Often the water is not obtained from the river directly, but after natural filtering from the river banks. This raw water is still enriched with aromatic organic substances which have to be removed to avoid the growth of bacteria in the water supply pipe systems and for health reasons. In water treatment processes, this is often realized by oxidation of the aromatic components and by additional filtering of the hydrophilic reaction derivates on active carbon. The amount of the oxidant added to the raw water is controlled by the residual concentration of this substance after the process.

[1] *Laser-Laboratorium Göttingen GmbH, Hans-Adolf-Krebs-Weg 1, 37077 Göttingen, Germany*

[2] *Laser-Laboratorium Göttingen e.V., Hans-Adolf-Krebs-Weg 1, 37077 Göttingen, Germany*

[3] *Stadtwerke Düsseldorf AG, Qualitätsüberwachung Wasser (OE 423), Postfach 101136, 40002 Düsseldorf, Germany*

Y.J. Kim and U. Platt (eds.), *Advanced Environmental Monitoring*, 282–294.

The concentration of dissolved organic carbon (DOC) in the raw water seems to be the wanted control parameter. The problem is that the general DOC-analysis setups are not able to monitor the process in a short-time interval and that they require intensive service due to obligatory analysis gases and chemicals.

The UV–laser induced fluorescence method seems to be a suitable tool to overcome these problems. The Laser-Laboratorium Göttingen (LLG) and others (Lewitzka et al. 2004 and literature cited therein) have shown the potential of this technique for the detection of organic substances like polycyclic aromatic hydrocarbons (PAH), mineral oil contamination in water and in soil.

In a joint project of the Stadtwerke Düsseldorf AG and the LLG, the potential of this technique is tested as an online method for the real-time regulation for the oxidation step in the water purification process (Niederkrüger et al. 2004).

22.2 The LLG-Laserfluorimeter

Over the past few years several different, compact optical analyzing systems have been developed at the LLG using laser-induced fluorescence (LIF). The laserfluorimeters are able to acquire nanosecond time-resolved fluorescence spectra of the organic aromatic substances directly in water after excitation with a UV laser pulse. This information is used for qualitative and — after suitable calibration — for quantitative determination of the present organic substances without any purification or preparation step in-between.

The laserfluorimeter used in this project and its schematic is shown in Fig. 22.1a and b respectively. The mobile system ($40 \times 40 \times 55\,cm^3$, weight 30 kg, electric power supply 12 V/3.5 A) consists of a miniaturized frequency quadrupled Nd:YAG-laser emitting UV laser pulses at 266 nm (pulse energy <50 µJ, pulse width 8 ns, repetition rate 50 Hz). The laser pulses are guided via a quartz fiber (diameter 600 µm) to the sensor head at the measuring position. Here the UV-pulses excite the aromatic substances and their fluorescence is collected by four quartz fibers (diameter 400 µm each) that guide the light to the detection unit. The geometry of the optical fibers inside the sensor head is optimized for maximum fluorescence light collection and minimal stray light impact (Bünting et al. 1999).

The detection unit is basically an optical multichannel analyzer consisting of a polychromator, an intensified CCD camera and fast gating electronics (the details are given in Marowsky et al. 2000 and literature cited therein). The spectral bandwidth of the system ranges from 260 nm to 540 nm with a spectral resolution of 7 nm. Single spectra are obtained from a sequence of single-laser pulses and the software constructs a time-resolved 2D-LIF-spectrum by shifting the detection gate with respect to the emission time of the laser pulse. Such a 2D-fluorescence spectrum taken from the raw water at the water plant in Düsseldorf-Flehe is shown in Fig. 22.2.

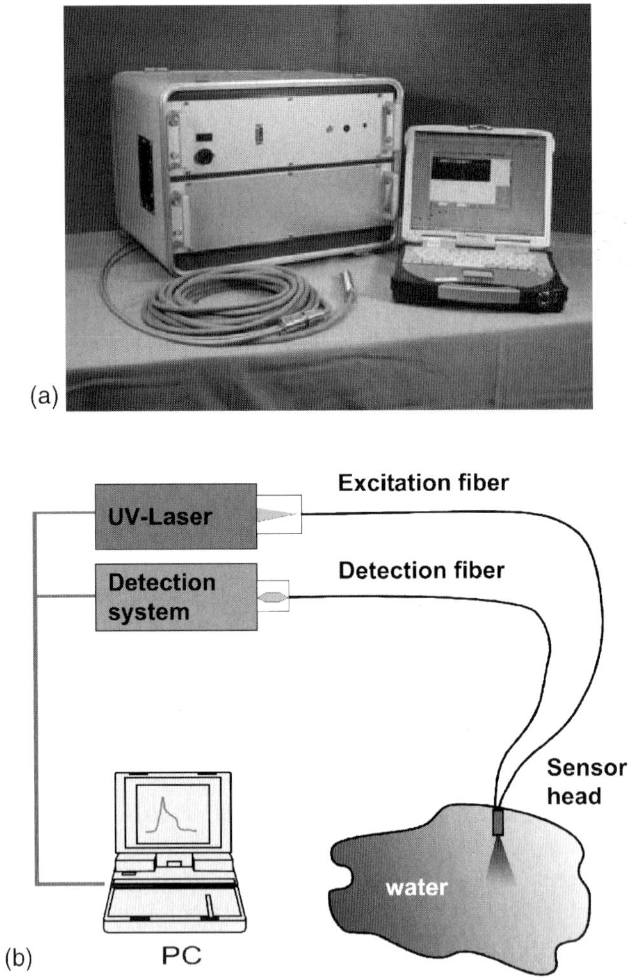

Fig. 22.1 (**a**) LLG laserfluorimeter, (**b**) schematic diagram of LLG laserfluorimeter

The device is controlled by proprietary spectrometer software, which additionally allows the validation of the spectra by calibration routines, which can be deposited in the data base. The result of the analysis can therefore be displayed directly after the measurement.

22.3 DOC Evaluation by LIF in Bank Filtrates

The spectrum shown in Fig. 22.3 is typical for natural water. There are two signals, which are well distinguishable: The Raman stray light signal due to the OH-symmetric stretch vibration of water, which shows up at 293 nm as a sharp peak and a relatively

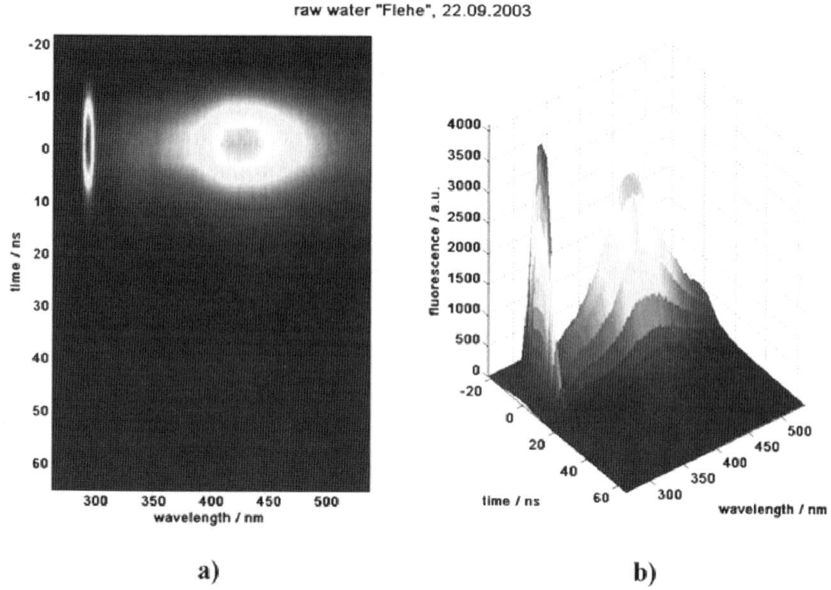

Fig. 22.2 (**a**) Contour plot and (**b**) 3D graph of time-resolved fluorescence spectra from a water specimen from the water plant in Düsseldorf-Flehe

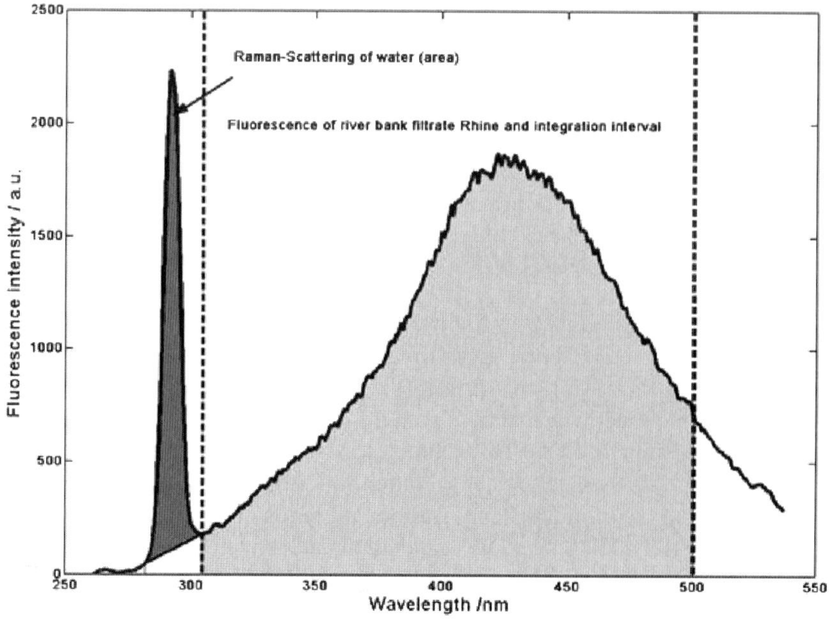

Fig. 22.3 UV-LIF spectra of bank filtrate of river Rhine

broad unstructured fluorescence signal in the range between 300 and 530 nm. The second signal can be attributed to a mixture of homo- and heterocyclic aromatic substances, which are formed by the post mortal degradation of organic materials, generally called humic substances. Humic substances are a mixture of different fluorophores linked through carbon chains. The best description is probably a not well-defined natural oligomer in which different fluorophores are incorporated. Nevertheless, it is well established (e.g. Esteves and Duarte 2001; Baker 2002; Peuravouri et al. 2002) that humic substances from different geological areas are distinguishable by their fluorescence spectra.

Generally, not every organic molecule can be detected by fluorescence spectroscopy, but as long as the origin of the natural organic matter is identical, it is justifiable to assume that the humic substances are in equilibrium with their non-fluorescing accompanying materials and a correlation between the fluorescence and the organic matter content becomes possible.

A calibration on the local conditions should therefore be necessary for generating good DOC results via fluorescence spectra. In the following the acquired fluorescence spectra are generally called DOC-LIF-spectra. During the investigations it was realized that all DOC-LIF spectra exhibit a short time profile, so that the time resolution does not contain further valuable information. Neglecting the time resolution has another advantage: The acquisition for a mean spectrum (out of 40 single spectra) and data evaluation takes less than 30 s. Figure 22.3 shows the spectral regions applied for the data evaluation procedure of a typical DOC-LIF spectrum of raw water. The integral fluorescence intensity is determined in the spectral region between 300 and 500 nm and subsequently normalized by the area of the underground corrected Raman-signal. It is obtained by integrating the signal in the spectral range between 280 to 300 nm and subsequently subtracting the area of the underlying fluorescence signal, which is estimated by a straight line between the intensities at 280 and 300 nm. With this underground correction it was easier to apply the Raman-area for normalization, but there is no difference when applying the underground corrected peak intensity of the Raman signal for normalization. The normalized fluorescence intensity is than plotted versus the classically determined DOC concentration to result in a calibration function for evaluation of the fluorescence signals in the process (Fig. 22.4), which can be deposited in the spectrometer software.

The calibration procedure is as follows: Assuming that the natural organic matter (NOM) in the Rhine is a good approximation for the NOM in the bank filtrate, Rhine water was taken and was filtrated over 0.45 μm mesh. The DOC-value for the filtrated Rhine water was evaluated with the classical method (catalytic oxidation and infrared detection of CO_2). This solution served as stock solution. It was diluted with pure water (after oxidation and filtration [0.45 μm mesh]) for the preparation of ten equidistant concentration steps for the calibration procedure according to the German norm DIN 32 645. The DIN standard dictates an unweighted procedure, so that the error bars in Fig. 22.4 have only informative character. These error bars have been calculated out of the standard deviation of ten single measurements for

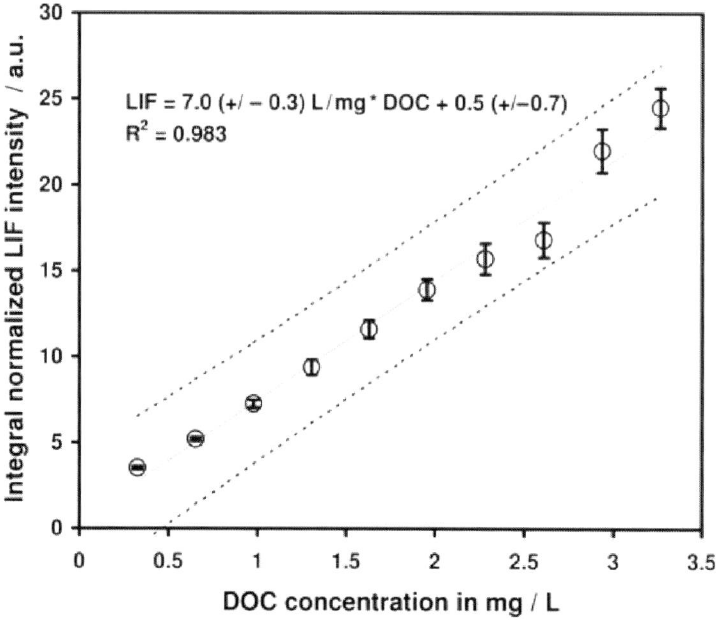

Fig. 22.4 DOC-LIF-calibration for raw water treatment

Table 22.1 Limits according to DIN 32 645, linear calibration method

Decision limit	0.49 mg/l
Detection limit	0.98 mg/l
Determination limit	1.48 mg/l

a 95% confidence interval assuming a normal distribution. The error of the used classical DOC determination of the stock solution was estimated to be about 15%. For clarity this error is not indicated in Fig. 22.4. The lower and upper boundaries specify the 95% confidence interval evaluated by the calibration procedure according to DIN 38 402.

As demonstrated in Fig. 22.4, the normalized LIF intensity shows good linearity with the DOC-values. The slope has a value of seven, which indicates that this new method is by a factor of seven more sensitive than the classically used one. The intercept is slightly positive but zero is within the calculated error. Since the correlation over the investigated region is linear, fluorescence quenching is not observed. The evaluation of the calibration data according to DIN 32 645 is summarized in Table 22.1. The decision, detection and determination limits are in the same magnitude as for the classical methods.

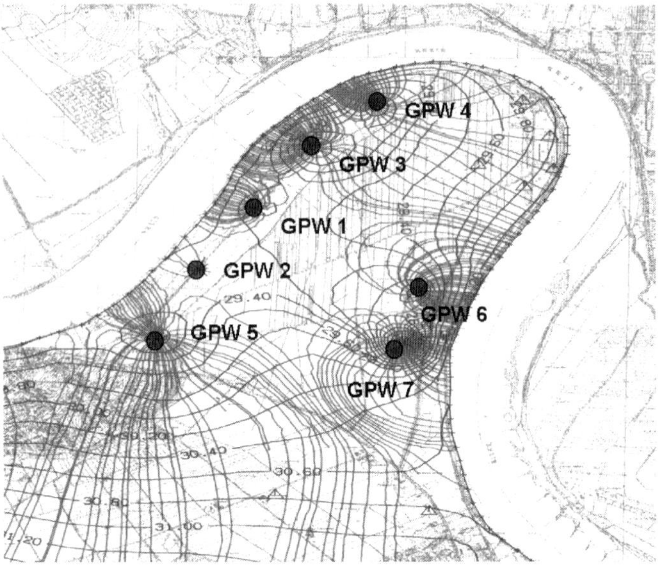

Fig. 22.5 Pumping station on the Grind GPW1 to GPW 7

22.4 Investigation of Bank Filtrates

The laserfluorimeter was used for two different types of investigations: (a) measurements at seven pumping stations at the area of the sinuosity of the Rhine called 'Grind' (Fig. 22.5, position of pumping stations and lines for identical groundwater levels) were performed directly at the stations; (b) measurements in front of the water tank, where the oxidation process (here oxidation by adding ozone) takes place, were performed directly at the joint natural water inlet from several pumping stations. Figure 22.6 shows a schematic drawing of the cleaning process used by the Stadtwerke Düsseldorf AG.

Case (a) The LLG-laserfluorimeter was used as an online system at the pumping stations as shown in Fig. 22.5 during the start and working phase of water withdrawal after a stagnancy period for several days. This phase seems to be interesting, because it is assumed that the quality of the pumped water varies until a steady state in the ground water of the pumping region is reached. Figures 22.7 and 22.8 show typical DOC-profiles for the pumping stations at the 'Grind' (GPW): The dashed line is connecting the discrete analyzing results for conventional DOC measurements (grey diamonds) and the small black diamonds, which represent single values of continuous online measurements, are the results obtained via LIF spectroscopy directly. The whole range of measured DOC concentration is quite narrow, but the change in water quality in three hours for the water of one pumping station is small. This is the experience of the waterworks at Stadtwerke Düsseldorf AG. In all examples, the LIF results have been higher in DOC value than the classical results. However, the trend of the DOC measurements is well reflected in all examples and the changes in water quality can be followed in real time.

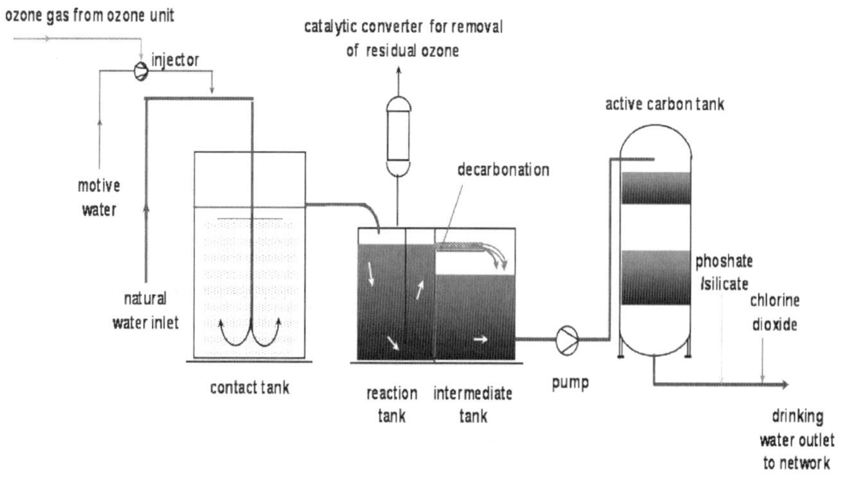

Fig. 22.6 Water cleaning process of the Stadtwerke Düsseldorf AG

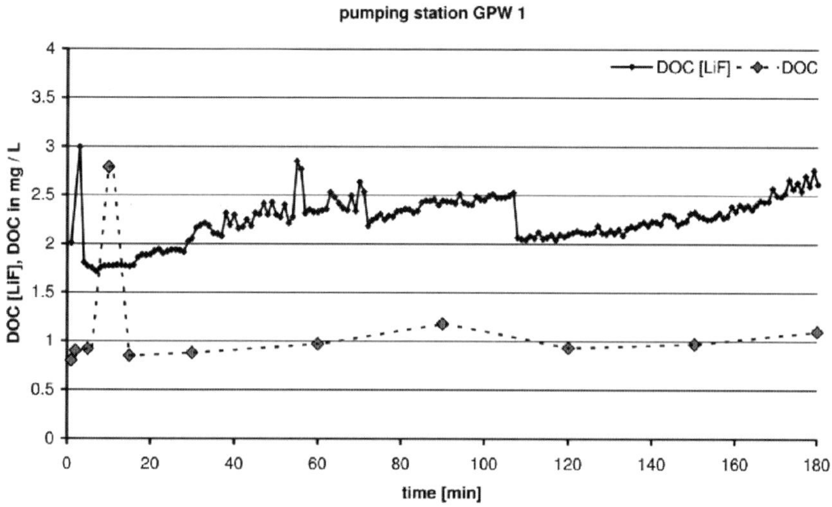

Fig. 22.7 Comparison of conventional DOC (solid line) with DOC values determined via LIF-spectroscopy (dashed line) for pumping station 1 at the Grind

The discrepancy between the two methods can be traced back to two possible explanations: (a) The DOC determined from the LIF measurements rely on a calibration of Rhine water humic acids, whereas in this case the humic acids in the bank filtrate are detected. (b) The raw water was filtered with a mesh of 0.45 μm pore size for the determination of the classical DOC. These removed small particles may still contribute to the fluorescence as well, so that the error is systematic over estimation of the determined DOC. This overestimation is true for all performed measurements at the Grind. Both (a) and (b) are plausible, but the variation in the

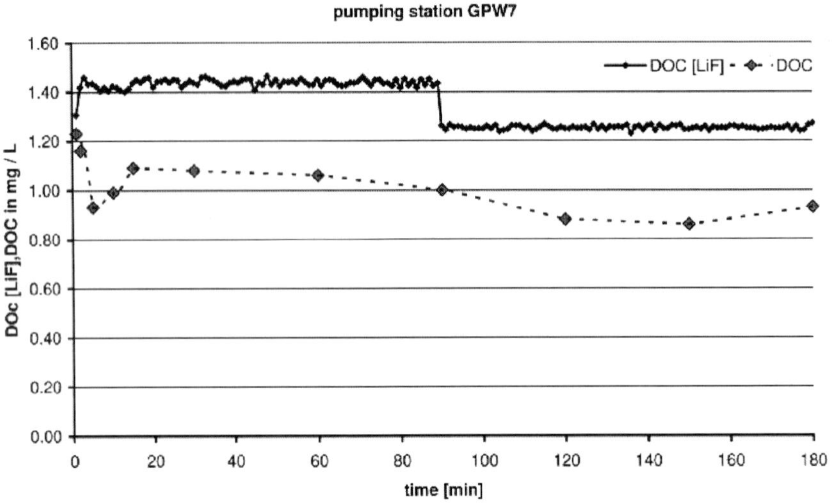

Fig. 22.8 Comparison of conventional DOC (dashed line) with DOC values determined via LIF-spectroscopy (solid line) for pumping station 7 at the Grind

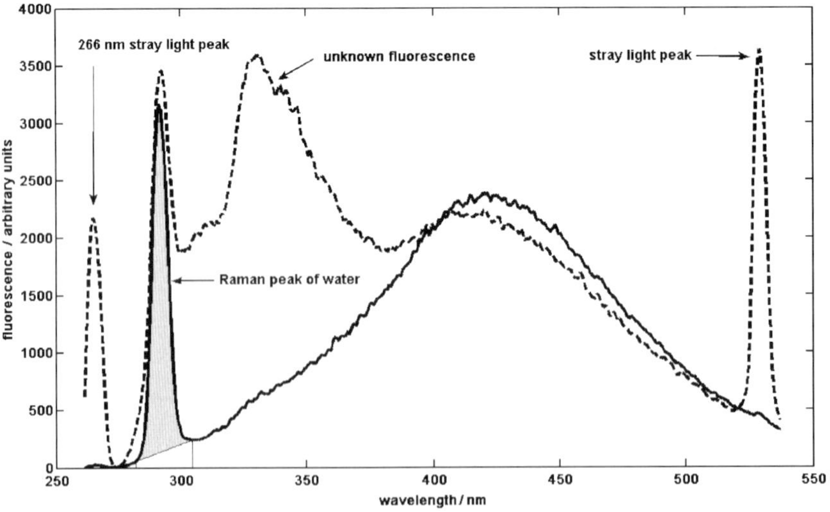

Fig. 22.9 Perturbed fluorescence spectra (dashed), not qualified for DOC-data evaluation, because of stray light interference and additional fluorescence peak in comparison with DOC-fluorescence spectrum (solid black line)

observed spectra between the Rhine and raw waters are small, so this answer (b) seems to be the more dominant effect.

During the measurements perturbations of the humic substance, spectra have been noticed. These perturbations manifest themselves in a growing stray light signal at 266 nm, which allows one to reason the appearance of particles or gas bubbles in front of the sensor. Figure 22.9 shows such perturbed spectra. On the basis of the spectral information, these measurements can be identified as false and can be ignored for the determination of the actual DOC-value. The spectral information therefore is a valuable tool for the quality of the recorded spectra and for the decision whether a true or a false spectrum is present.

Case (b) In this case the LLG-laserfluorimeter was placed in the water plant in Düsseldorf-Flehe directly at the natural water inlet in front of the contact tank. At this measurement position no perturbations as described before have been observed. In the contact tank the oxidation process by ozone takes place (Fig. 22.6). This water cleaning step is necessary to avoid growth of bacteria in the water supply pipe system and for health reasons. Unfortunately hitherto there is no parameter to control the oxidation process in advance. In the actual practiced method, one measures the residual ozone concentration after the oxidation has taken place. This value is then used for re-adjustment of the ozone dosage.

The interest of the Stadtwerke Düsseldorf AG is therefore to use the LIF-signal for the prediction of the ozone dosage. Figure 22.10 (a) shows the recorded LIF-signal over the measurement time. At 7.50 a.m., the raw water from the pumping station named 'Brückerbach' (BB) is replaced by water of pumping station '5' by starting a second pump (PW5 2P). Water with a higher ozone demand is replaced by water with a lower ozone consumption, which is clearly seen in the LIF-trace in Fig. 22.10 (a). However, the ozone dosage is reduced by the technician in the control room for about 20 min (vertical dashed line) after the time indicated by the LIF-signal since the delay between ozone addition and measurement of the residual ozone concentration is about 20–30 min due to the mean residence time of the water in the contact tank. This time delay in adjustment induces a maximum in the residual ozone concentration

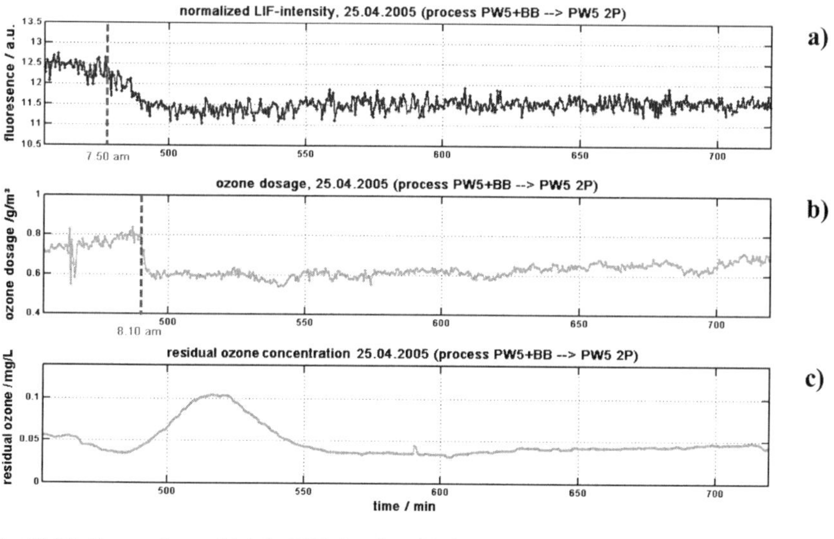

Fig. 22.10 Comparison of (**a**) the LIF-signal to (**b**) the ozone dosage and (**c**) the measurement of residual ozone concentration in the water (25th April 2005)

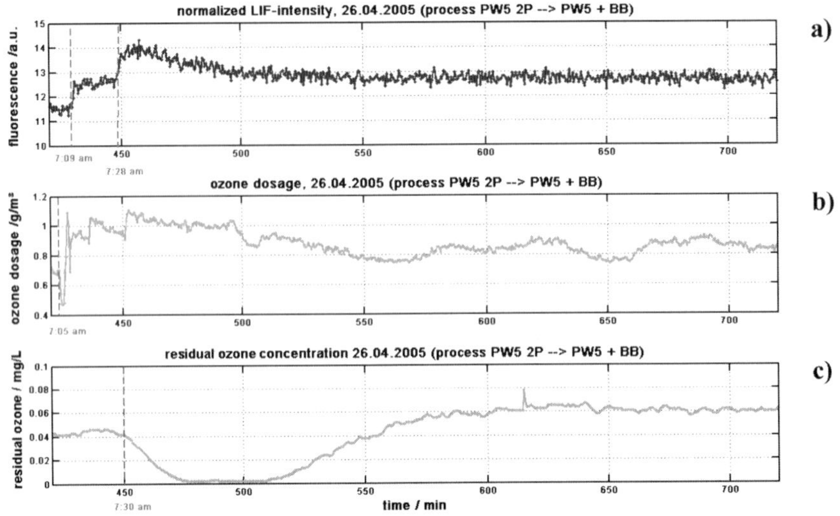

Fig. 22.11 Comparison of (**a**) the LIF-signal to (**b**) the ozone dosage and (**c**) the measurement of residual ozone concentration in the water (26th April 2005)

as seen in Fig. 22.10 (c). High residual ozone concentrations are undesirable because they enhance the bromate concentration which is legally limited.

In Fig. 22.11, the process of switching the pumping lines back to their former state is documented. Before 7.09 a.m., the water was delivered from pumping station PW5 with two pumps. At 7.09 a.m. one pumping line was switched to the pumping station 'Brückerbach' (BB) again. The LIF-signal shows a jump to a higher value, which is constant for 19 min. The volume pumped during this time can be contributed to stagnated water in the pipeline between the well and the cleaning facility. Then at 7.28 a.m., freshwater from the well is arriving and the LIF-trace shows a maximum, which reduces slowly to the constant value registered before again.

In Fig. 22.11 (b) the ozone dosage is depicted. At 7.05 a.m., the ozone generator is changed (two ozone generators are used for ozone production in an alternating way). Since the technician in the control room knows by his experience that the water from pumping station 'Brückerbach' has generally higher ozone consumption, he sets the ozone dosage to a higher value, but in fact, this value was not high enough and the residual ozone concentration (Fig. 22.11 c) drops to zero. Since the technician has-up to now-no actual online information about the incoming water quality, he is not able to recognize the additional maximum peak as it is registered by the LIF-Signal. He can only adjust the ozone dosage with a time delay of 20–30 min according to the residual ozone concentration at the exit of the water processing when the oxidation already has taken place.

In the next step, we have correlated the LIF-signal directly with the ozone dosage by a very simple two-point calibration. In Fig. 22.12, the results for another run at the water plant in Düsseldorf-Flehe are depicted: Fig. 22.12 (a) shows again the

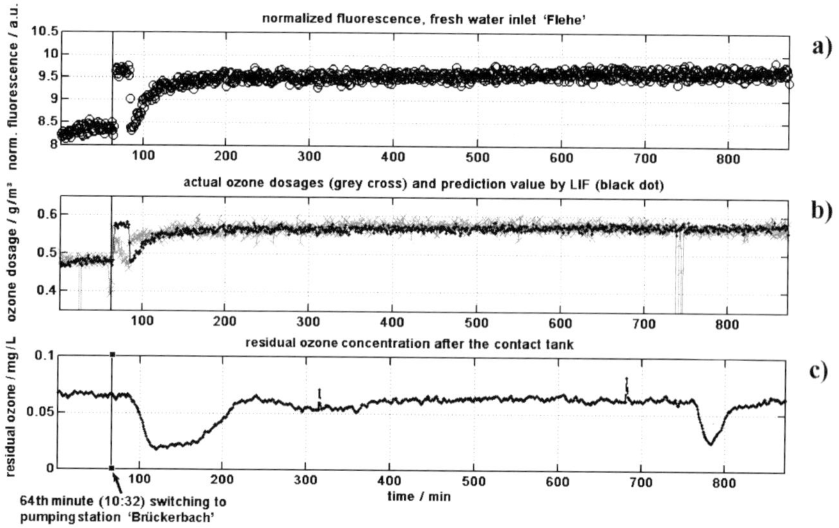

Fig. 22.12 (**a**) Normalized fluorescence, (**b**) actual ozone dosage (grey cross) and prediction value (black dot) determined by LIF and (**c**) residual ozone concentration

LIF-signal recorded at the freshwater inlet at the cleaning facility; the grey crosses in Fig. 22.12 (b) represent the actually applied ozone dosage and the black dots show the recommended ozone dosage determined by LIF-spectroscopy. After switching one pumping line to the pumping station 'Brückerbach,' there is an enhanced need for ozone in the first place. The ozone dosage is adjusted, but the demand is too high resulting in a very low residual ozone concentration. Low residual ozone concentrations are also unfavourable because the disinfection capacity is too low.

At minute 720, there seems to be a short failure at the ozone gas generator, resulting in a difference between the recommended and the actual ozone dosage again. In both cases where the actual ozone dosage differs from the recommended value deduced by LIF spectroscopy, 20 min later a very low residual ozone concentration in the water is recorded. The time delay of 20 min corresponds well to the retention time for the water in the contact tank (see Fig. 22.6).

These examples document well that the LIF-signal can be used for predicting the ozone consumption during the oxidation process using different raw waters from the incoming pumping stations.

22.5 Summary

The investigations have shown that LIF is suitable for the determination of DOC-values of raw waters. Prerequisite is a calibration to the local conditions of the raw water. The demonstrated LIF set-up was successfully used for on-line measurements of DOC at different pumping stations.

Furthermore LIF is a sensitive probe for revealing differences in water quality. It was possible to distinguish between stagnating water and freshwater in the case of waterworks in Düsseldorf-Flehe. The investigations at the 'Grind' show changes of LIF from starting water withdrawal to equilibrium.

The investigations demonstrate also a direct correlation between LIF-signal and ozone consumption. The normalized LIF-signal can be used to determine a prediction value of the ozone demand for oxidation processes. These measurements have been performed directly at the raw water inlet of the cleaning facility at the water works 'Flehe' of the Stadtwerke Düsseldorf AG. Following this prediction value the ozone dose can be adjusted in real time.

In the future, LIF could be applied to control the input and output of drinking water plants to get further information about the water cleaning process which is an important goal in the water safety plans of the suppliers.

References

Baker A. (2002), Spectrophotometric discrimination of river dissolved organic matter. *Hydrol. Process.*, 3203–3213.

Bünting U., Lewitzka F., and Karlitschek P. (1999), Mathematical model of laser-induced fluorescence fiber-optic-sensor head for trace detection of pollutants in soil. *Appl. Spectrosc.*, 53, 49–56.

Esteves V.I., and Duarte A.C. (2001), Differences between Humic substances from riverine, estuarine, and marine environments observed by fluorescence spectroscopy. *Acta Hydrochem. Et Hydrobiol.*, 28, 359–363.

Lewitzka F., Niederkrüger M., and Marowsky G. (2004), Application of two-dimensional LIF for the analysis of aromatic molecules in water. In P. Hering, J.P. Lay and S. Stry (Eds), *Laser in Environmental and Life Sciences, Springer*-Verlag: Berlin, pp.141–161.

Marowsky G., Lewitzka F., Bünting U., and Niederkrüger M. (2000), Quantitative analysis of aromatic compounds by laser induced fluorescence spectroscopy. *Proc. SPIE*, 42, 218–223.

Niederkrüger M., Wachsmuth U., Konradt N., Rohns H.-P., Nelles T., and Irmscher R. (2004), Mobiles Laserfluoreszenzspektrometer zum Monitoring auf gelöste organische Verbindungen bei der Wassergewinnung aus Uferfiltrat. *VDI-Berichte*, 1863, 17–24.

Peuravuori J., Koivikko R., and Pihlaja K. (2002), Characterization, differentiation and classification of aquatic humic matter separated with different sorbents: synchronous scanning fluorescence spectroscopy. *Water Res.*, 36, 4552–4562.

Section 4
Biosensors, Bioanalytical and Biomonitoring Systems

Chapter 23
Biosensors for Environmental and Human Health

Peter-D. Hansen

Abstract Sensors and biosensors as well as biochemical responses (biomarkers) in ecosystems owing to environmental stress provide us with signals (environmental signalling) of a potential damage in the environment. These responses are perceived in this early stage, but in ecosystems, the eventual damage can be prevented. Once ecosystem damage has occurred, the remedial action processes for recovery could be expensive and pose certain logistical problems. Prevention of ecosystem deterioration is always better than curing. Ideally, "early warning signals" in ecosystems using sensing systems and biochemical responses (biomarkers) would not only tell us the initial levels of damage, but these signals will provide us as well with answers to develop control strategies and precautionary measures with respect to the water framework directive. To understand the complexity of the structure of populations and processes behind the health of populations, communities and ecosystems, we have to direct our efforts to promote rapid and cost-effective new emerging parameters of ecological health. New emerging parameters are biochemical effect (biomarker) related parameters in the field of immunotoxicity and endocrine disruption. Environmental effects such as genotoxicity and clastogenicity were detected in organisms from various "hot spots". Vital fluorescence tests are one means that allow us to unmask adverse events (i.e., genetic alterations in field-collected animals or in situ-exposed organisms) by a caging technique. New emerging ecosystem health parameters are closely linked to biomarkers of organisms measured in monitored areas. One problem is always to find the relevant interpretation and risk assessment tools for the environment.

Keywords: Biosensor, effect assessment, ecotoxicological classification in sediment, endocrine effects, assessment of "good ecological status", drug exposure.

Technische Universität Berlin, Faculty VI, Department of Ecotoxicology, Franklin Strasse 29 (OE4), D-10587 Berlin, Germany
(Tel: +493031421463, Fax: +49308318113)

Y.J. Kim and U. Platt (eds.), *Advanced Environmental Monitoring*, 297–311.

23.1 Introduction

For risk assessment and risk assessment tools new recommendations are described in the Technical Guidance Document of EU – Edition 2, in the new EU Chemicals Legislation REACH and in the status report for toxicological methods of the European Centre for the Validation of Alternative Methods. In the EU water framework directive (2000) a general requirement for ecological protection, and a general minimum chemical standard was introduced to cover all surface waters. For the description of "good ecological status" and "good chemical status", effects monitoring tools are needed for the description of the ecological status of river basin systems. In some cases, biomarkers were very helpful to promote an environmentally sensitive and sustainable use in studies with coastal zone samples (Baumard et al. 1999, Hansen et al. 1985, 1990, 1995, 2006). A very promising tool is the scale classification based on toxicity studies and environmental monitoring to classify sediments: i.e., "Ecotoxicological Classification of Sediments by Bioassays" (Krebs 2005a, 2005b). The currently available biomarkers or biochemical responses are used as environmental monitoring tools to assess information on early responses of living organisms to environmental stressors, and to deliver signals of ecosystem damage and pathology owing to both man-made and natural pollutions. Biomarkers are already being applied to the water matrix and to benthic organisms that include new emerging biomarkers such as endocrine effects and immunotoxicity. Usefulness for new emerging effect parameters such as immunotoxicity (phagocytosis) in the context of biotoxins is demonstrated by mussels exposed to sediments in coastal areas and exposure of mussels under controlled conditions to the water matrix. Here, results of immunotoxic effects in mussels and the endocrine load of the sediments are demonstrated. Immunotoxic response of the blue mussel was quantified by phagocytosis activity (Phagocytosis Index) of its hemocytes. It appears that both immunosuppressive and immunostimulative effects are likely to occur at specific sites and that responses will be influenced by the type and intensity of contaminants present. Immune system function in bivalves can be adversely affected by long-term exposure to environmental contaminants. Investigating alterations in immunity can therefore yield relevant information about the relationship between exposure to environmental contaminants and susceptibility to infectious diseases.

23.2 Biochemical Responses and Biosensors for Effects Monitoring in the Environment

Use of biomarkers (biochemical responses) in multi-arrays for environmental monitoring is complementary to chemical analysis since they can alert for presence of toxic compounds that require further instrumental analysis or "bioresponse-linked instrumental analysis". Biosensors are by definition analytical devices incorporating a biological component like micro-organisms, organelles, cell receptors, enzymes, antibodies, nucleic acids and a physicochemical transducer system: optical, electrochemical, thermometric, piezoelectric or magnetic. An enzyme linked recombinant receptor assay like the ELRA is a biosensor according to this definition: the biological component is the enzyme and the transducer is the optical component. Biosensor and biochemical responses for the assessment of environmental health are listed in Table 23.1. It is rather difficult to transfer the monitored biochemical responses or the sensor responses into an operational effect related standard (EQN = environmental quality norm) for environmental monitoring. Sensing systems based on the induction and inhibition of a functional system relating to function, interference and effect related endpoints are listed in Table 23.1 and depicted in Fig. 23.1.

Table 23.1 Environmental monitoring of effect related biochemical responses (biomarkers) and potential sensing systems with their endpoints ("effects at the level of" neurotoxicity etc.) (Modified after Bilitewski et al. 2000.)

Targets	Example	Interference by compounds	Endpoints
Proteins Enzymes	Acetylcholine esterase	Organophosphorus and carbamic compounds	Neurotoxicity
	Protein phosphatase 1 and 2A	Microcystins	Hepatotoxicity
Ion channels	Na+ channel (voltage-gated channel)	Saxitoxin, tetrodotoxin, procaine	Neurotoxicity
Transport protein	SHBG, CBG, TBG	Endocrine disruptors	Growth, reproduction
Receptors	Estrogen receptor	Endocrine disruptors (e.g. o,p-DDT, nonylphenol)	Growth, reproduction
	Nicotinic acetylcholine (ACh) receptor (ligand-gated channel)	Anatoxins	neurotoxicity
Electron carriers	QB protein	Photosynthesis II herbicides (e.g. s-triazines, phenylureas), phytotoxins	Photosynthesis
Nucleic acids DNA	DNA double strands	PAHs, Pesticides, PCBs, EDCs, intercalating polycyclic aromates (ethidium, acridine, caffeine); DNA adducts (metabolites of chloracetamide herbicides)	Genotoxicity
Cytoskeleton	Tubulin	Colchicin, taxol; anti-tubulin herbicides (e.g. trifluralin, oryzalin)	Cytotoxicity
Ribosomes	rRNA (ricin)	Ribotoxins (ricin, abrin, Shga toxin)	cytotoxicity

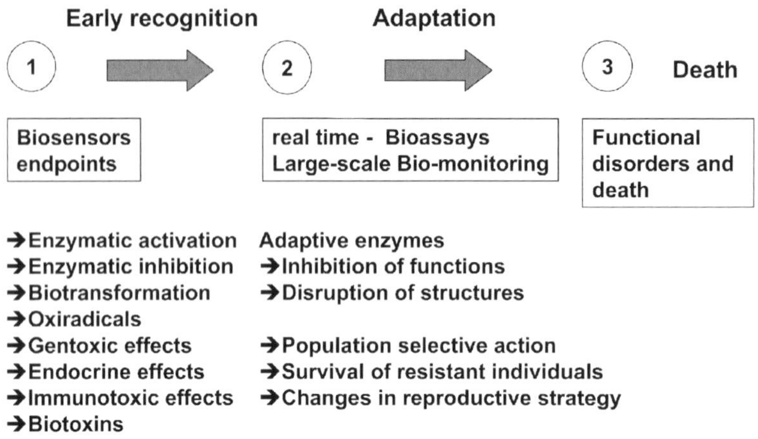

Fig. 23.1 Environmental signalling and diagnosis by biosensors and biochemical responses (biosensors endpoints) In figure: Genotoxic effects; Oxyradicals

Biosensors together with effect related parameters or biochemical responses for environmental monitoring are very complex but they will give a clear picture of the health status of the investigated system. In Fig. 23.1 the biochemical responses are demonstrated by three phases of reaction by the organisms. The enzymes and oxyradicals are helpful tools for a final biosensor. Enzyme biosensors are sensors based on enzymatic bio-recognition elements. Therefore the definition of "biosensors" can be widened: "Biosensors are analytical devices incorporating a biologically derived and/or bio-mimicking material (i.e., cell receptors, enzymes etc.), associated with or integrated within a physicochemical transducer or transducing microsystem, which may be optical, electrochemical, thermometric, piezoelectric or magnetic. Biosensors are distinct from bioassays in that the transducer is not an integral part of the analytical system. Figure 23.1 shows an example of biochemical responses for diagnosis of the health status of environmental systems (Hansen 2003).

The effect related parameters or biochemical responses are complex but they will give a clear picture of the health status of the investigated system. "Ecosystem Health" is synonymous with "environmental integrity", from which it follows that the scope of ecosystem health (EH) research encompasses all the tools and approaches which are efficacious in increasing the cognitive, curative and preventive knowledge as the goal to the preservation of environmental integrity. Ecosystem health research thus directs its attention to the prediction of reversible and irreversible insults which human or other activities could potentially inflict on the environment. For the assessment of ecosystem health, very promising biomarker approaches are centred on quantifying biochemical effects in organisms and populations. One important tool for the acceptance of biomarkers in science, technology and governmental legislature are the so-called inter-laboratory comparison studies of measurements of biomarkers. Laboratory studies have established a strong causal link between exposure of fish to PAHs and co-planar PCBs and the expression of cytochrome P_{450}1A1 and its associated 7-ethoxyresorufin-O-deethylase (EROD) activity. The induction of EROD activity in fish liver has particularly been used as a biomarker for the effects of these organic contaminants in several inter-calibration exercises (see Stagg and Addison 1995). EROD induction is a classical biomarker and well established for MFOs (mixed function oxigenases) and biotransformation in ecotoxicology. So far, however, only phase I of the MFOs is investigated and not phase II with conjugation and the real detoxification process. Additionally, another often used biomarker is cholinesterase inhibition. The basic concept is that organophosphorus pesticides and carbamates inhibit cholinesterase at different levels (Sturm and Hansen 1999; Sturm et al. 1999). For the quantification of neurotoxicity there are two well known cholinesterases (acetyl cholinesterase and butryl cholinesterase) and the methodology in principle is standardised after DIN (German Institute for Norming: DIN 38415-T1 1995). There is an application of validated biomarkers available with endpoints including new emerging biomarkers for endocrine effects and immunotoxicity in addition to genotoxicity. There is a high potential of biochemical responses and the development of fast and reliable biochemical tools (biosensors) for on-site screening in environmental and human health analysis. The principles associated with the different scales of biochemical processes relating to ecosystem and human health are shown in Table 23.2.

23.3 The River System of Berlin and the "Good Ecological Status" Classified by a Biosensor System

In considering the impact of either natural stress or man-made stress we always encounter detoxification, disease defence, regulation and adaptation processes. This situation makes the assessment approach by biomarkers rather complicated. However, in symptoms analysis including functional (behaviour, activity and metabolism) and structural changes in organisms (cells, tissues and organs), biomarkers do have a significant ecological assessment potential. For landscape planning and environmental management it is necessary to get significant data from biochemical responses for relevant and sustainable actions. As an example of transfer of new substances from sediments to groundwater – the River Havel case study in Germany has demonstrated close aspects between environmental monitoring and ecosystem health aspects.

23.3.1 Description of the River System of Berlin and the Biosensor (Receptor Assay–Sensor) Application

The River Havel is an intensively monitored waterway because of its multiple use for beach filtration and finally as a source to produce drinking water. The River

Table 23.2 Selected biochemical responses for assessment of environmental health and the biomarker methods used for their examination (Modified after Bresler et al. 1999)

Method	Characteristic of health
Measurement of blue and green fluorescence of NADH and FAD in living tissues	Metabolic state of mitochondria, cells or tissues respiration and glycolysis
Quantitative fluorescent cytochemistry	DNA, RNA, proteins and lipids content
Using permeable fluorogenic substrates of enzymes, specific inhibitors, and kinetic analysis	Enzyme activity in living cells in situ: a. Non-specific esterases b. Detoxifying enzymes c. Marker enzymes
Using special fluorescent anionic markers	Alterations of permeability of plasma membranes, epithelial layers and histo-hematic barriers
Using specific fluorescent transport substrates, inhibitors and kinetic analysis	State of carrier-mediated transport system for xenobiotics elimination
Using fluorescent xenobiotics or fluorescent analogue of xenobiotics	Xenobiotics distribution, extra- and intracellular accumulation and storage
Using special fluorescent xenobiotics or fluorescent analogues of xenobiotics	State and function of xenobiotic-binding proteins
Vital tests with Acridine Orange or Neutral Red	State of lysosomes and cell viability
Metachromatic fluorescence of intercalated or bound Acridine Orange, 590/530 nm Microfluorometry	Functional rate of nuclear chromatin, DNA denaturation
Complete cyto- and histopathological examination	Early pathological alterations and signs of environmental pathology
Electron microscopy	Cell structures and organoids
Cytogenetic examinations	Detection of environmental genotoxicity and clastogenicity
Mass Spectrometry (MALDI/TOF/MS; ESI-TOF-MS/MS)	Identification and detection of membrane proteins, epitope-binding areas of proteins

Havel is a water stretch of 30 km in length with a surface of more than 20 km^2. It is a slow flowing river with several conflicting uses such as (1) receiving waterway for secondary effluent and rain water, (2) active beach filtration and drinking water production, (3) sport- and professional fisheries, (4) leisure and recreation concerning water sports and the EU water directive for swimming areas with beaches along the river and (5) waterway for leisure boats and commercial cargo vessels. With an extended surface, a low water depth (mean water depth 7 m) and a steady input of nutrients, the river Havel has a potential for algal blooms. Blooms begin in February (Fig. 23.2) and continue until May. During May and June the algae growing are reduced and a second bloom of blue greens with microcystins start at the end of June. There is a combination of increasing temperature (up to 30°C) and decreasing flow of water (approx. 30 m^3/s), which trigger increases in biomass and in the amount of suspended solids. Finally, the latter are eventually deposited as sediments.

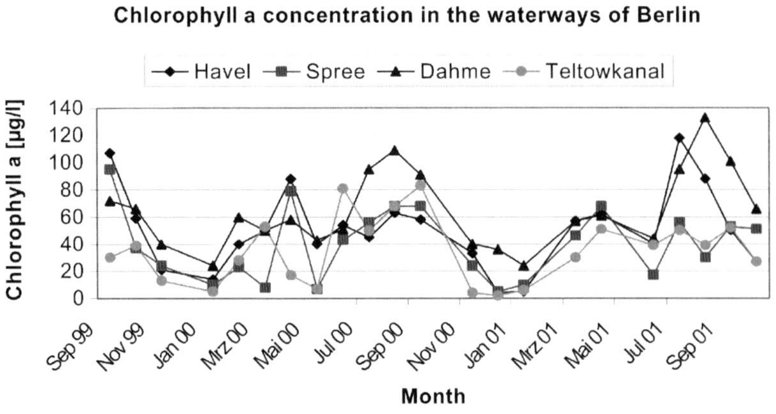

Fig. 23.2 Algal blooms in the Berlin waterways River Havel, Spree, Dahme and the Teltowkanal

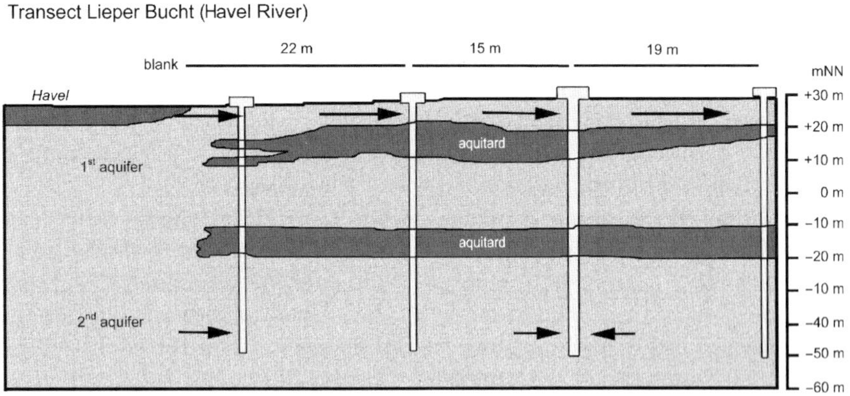

Fig. 23.3 Schematic hydro geologic section of the transect at the River Havel (Lieper Bucht), location of monitoring wells (shallow and deep) and of the raw-water supply well for drinking water (Modified after Heberer and Mechlindki 2003)

Besides algal blooms, there is a second source for suspended solids in the river because of effluent loading by sewage treatment plant emissions. The river consists of 10%, 20%, 30% and 40% treated effluents (tertiary treatment and sometimes membrane filtration). In winter (high water flow = approx. 100 m^3/s) loading corresponds to 10% effluent to 20%–30% in spring and autumn (medium water flow = approx. 30 m^3/s) and to 40% of treated effluent in summer (low water flow = approx. 5 m^3/s).

The second source of suspended solids in the receiving river Havel on top of extreme eutrophication by algae (see Fig. 23.2) is the outlet of the Berlin-Ruhleben sewage treatment plant. The effluent contains 10 mg/l suspended solids in treated sewage sludge. The sewage plant close to the river Havel is one of the main sewage plants in Berlin with an output of 240,000 m^3 of treated effluent per day. Together with the treated effluent, there is an emission of approximately 2.4 t of biological treated sewage sludge to the Berlin River Havel System. A volume of approximately 880 t biologically treated sludge is emitted per year and deposited in the river Havel. The suspended solids of the sludge are carriers of endocrine disrupting compounds (EDCs) and other organic pollutants.

23.3.2 Xenoestrogenic Effects in the River System of Berlin

The concentrations of EDCs like nonylphenol, different phthalates and lower PCBs are rather low. For Nonylphenol the safety factor is approximately100 between the concentration in the effluents and the endocrine effect measured by the enzyme linked recombinant receptor assay (ELRA). The safety factor for Bisphenol A is approximately 3,000. There is no safety factor (Hansen 1998) concerning the lowest effects concentrations (LOEC) of the hormones and the measured concentrations of the hormones in the effluents. Endocrine effects were measured for 17β-Estradiol in the range of 0.1–1 ng/l, for Ethinylestradiol [0.2–3.0 ng/l] and for Estrone [0.5–1.0 ng/l]. These concentrations are already present in the surface water of the River Havel and the river sediments. Our results regarding endocrine disrupters (EDCs) are comparable to investigations conducted under the LOES programme of the Netherlands and cited studies world wide (Vethaak et al. 2002).

Endocrine effects in the sediments of the River Havel, Spree, Dahme and Teltowkanal are calculated and expressed as 17β-Estradiol equivalents [μg/kg] in sediment dry weight and for pore water and elutriates of sediments in [μg/l], see Figs. 23.4 and 23.5. The lower River Havel shows a median concentration of 8.5 μg/kg—17β-Estradiol equivalents, in the upper River Havel 6.7 μg/kg—17β-Estradiol equivalents, 34.2 μg/kg—17β-Estradiol equivalents in the River Spree, 45.7 μg/kg—17β-Estradiol equivalents in the River Dahme and in the Teltowkanal 14.3 μg/kg—17β-Estradiol equivalents. The results in Fig. 23.4 clearly demonstrate that concentration levels up to 175 μg/l 17β-Estradiol equivalents in elutriates of sediments are coming from sites with an extreme suspension rate. In Figs. 23.4 and 23.5 the concentration levels of 17β-Estradiol equivalents were ranged in seven classes. Endocrine effects measurements were done by the ELRA-test (ELRA = enzyme linked recombinant assay, Seifert 1999).

Analysis performed by ELRA (Seifert 1999) and the hER and hAR assay by McDonalds will be soon part of an intercalibration study for endocrine substances after the German Institute for Norming (DIN). Sediment loading by xenoestrogens changes over the year and also because of extreme algal growth (see Fig. 23.1), but they also contain a substantial amount of phytohormones. Apart from estrogenic responses there is only a poor evidence of androgens in River Havel sediments. In Figs. 23.4 and 23.5 the patchiness of the sediments of the River Havel with endocrine responses in the river bed is shown.

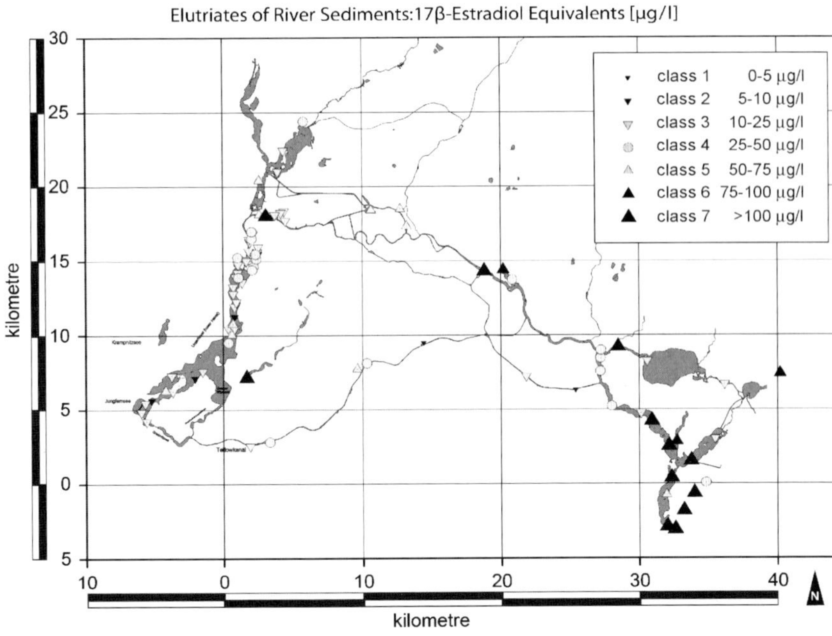

Fig. 23.4 17β-Estradiol equivalents [µg/L] for elutriates of sediments in the Berliner River system using ELRA in spring 2001. Sediments were sampled at the top 0–5 mm layer of the rivers

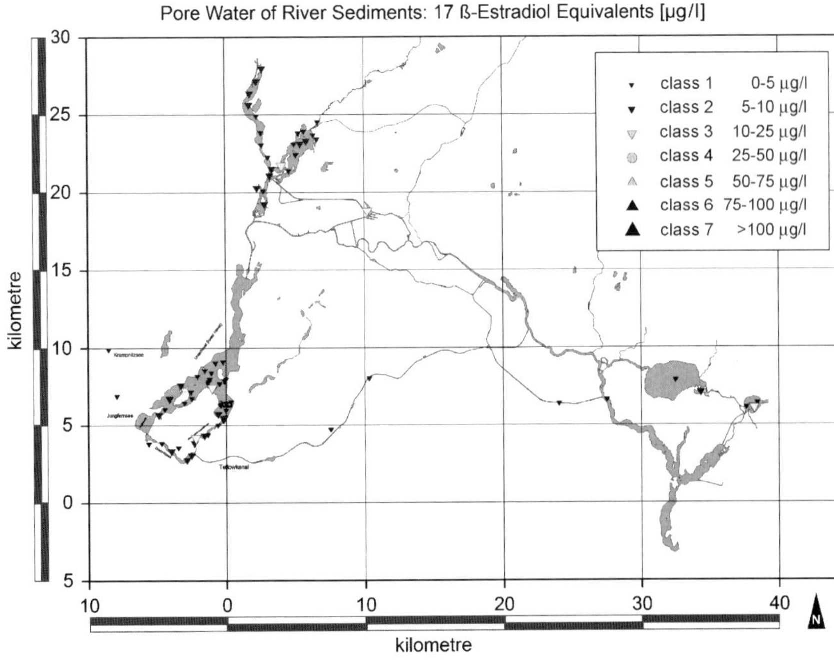

Fig. 23.5 17β-Estradiol equivalents [µg/L] for pore water of sediments in the Berliner River system using ELRA for the year 2001. Sediments were sampled at the top 0–5 mm layer of the rivers

23.3.3 Ecotoxicological Assessment in Relation to the WFD

For an assessment of the endocrine effects reported in Figs. 23.4 and 23.5, a scientific assessment system is necessary. It was thus decided to use the ecototoxicological classification system of sediments (Krebs 2005a, 2005b) developed at the German Federal Institute of Hydrology (BfG). Sediment classification is based on the scientific ecotoxicological approach using so-called pT-values. The method is used to assess the quality of sediments, dredged materials and toxic effects of waste water with standardized bioassays. Different bioassays and different test phases (pore water and elutriate) are considered equal in ranking in this system. The pT-value depends on the dilution factor and the lowest observed effect concentration, (see Krebs 2005). The highest dilution level without effects corresponds to the pT-value, i.e., the dilution step 1:2 represents a pT-value of 1. Toxicity classes are assigned Roman numerals: it is a seven-level-system based on convention. We combined the seven-level toxicity classes system using pT-value of the xenoestrogenic effects measured by the ELRA-test adapted to the five-level ecological classification system of the EU Water Framework Directive (WFD). Figures 23.6 and 23.7 demonstrate the risk assessment and ranking using the pT-value as an ecotoxicological classification concept in correspondence to the WFD.

23.4 Discussion

A relevant question meriting discussion is the following: can biosensors give additional information? For on-line measurements, there are some doubts that "state of the art" biosensors will give additional information and will replace chemical analytes or even bioassays. The case study emphasized the need to develop biosensors which do not only measure "conventional" contaminants but also new emerging parameters like endocrine effects which are needed for compliance with the drinking water directive as well as the water frame work directive (WFD) and which are currently used in monitoring programs. The classification of the contamination of river sediments by xenoestrogens after the pT-values and expressed as 17β-estradiolequivalents contributes to the evaluation by receptor assay—sensors for the ecological risk assessment (ERA), risk communication and risk management. To exploit the principal advantage of biosensors, standardisation and harmonisation is needed for governmental decision making in collaboration with industry and governmental authorities. A promising sign is that ISO (ISO=International Standardization Organisation) already has a working group for Biosensors (ISO TC 147/SC2) and a basic working document by ISO on biosensors is already available. Very promising also is the additional information available especially in the field of bio-effect related sensors like whole cell sensors and receptor sensors. Progress is being made in the development of receptor assays-sensors (Hansen 2003; Hansen et al. 2006). The format of the receptor assays has changed meanwhile to nanotechnology levels, already well established in water and food analysis. For

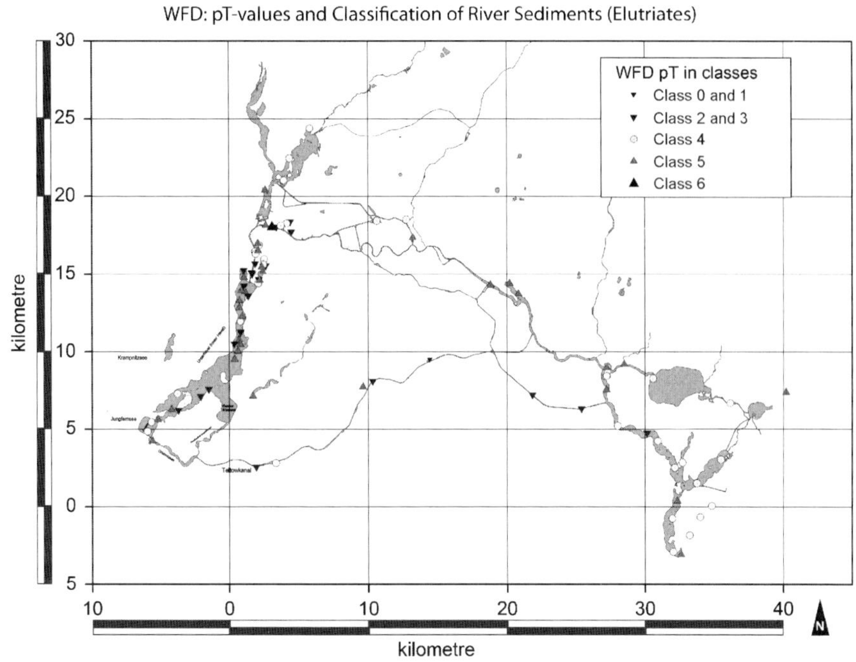

Fig. 23.6 pT-values and classification of River Sediments (Elutriates) in the Berliner River System for the year 2001 as per the WFD classification system (Krebs, 2005a, 2005b). For xenoestrogenic effects, see Fig. 23.4 (Huschek and Hansen 2006)

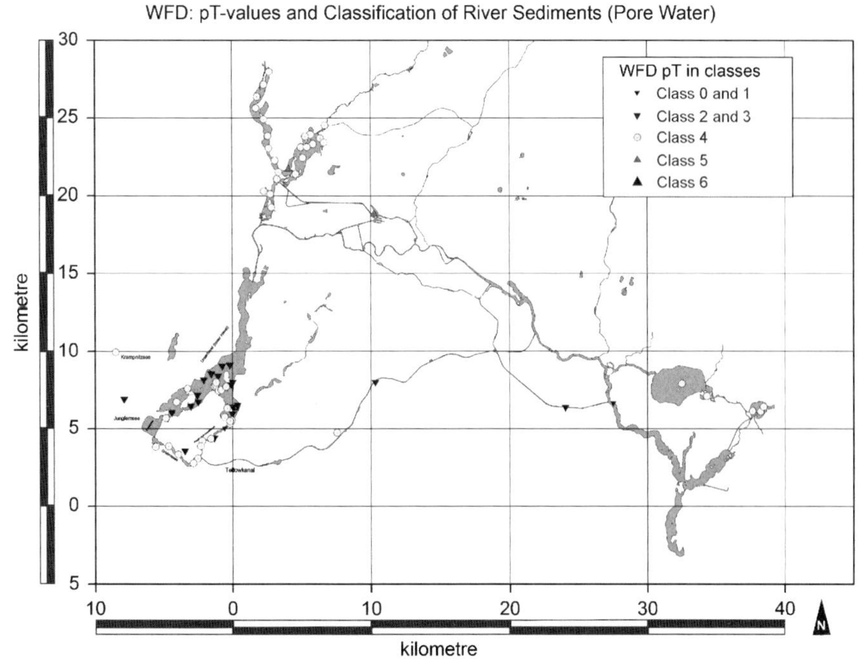

Fig. 23.7 pT-values and classification of pore water of sediments in the Berliner River system for the year 2001 as per the WFD classification system (Krebs, 2005a, 2005b). For xenoestrogenic effects, see Fig. 23.4 (Huscheck and Hansen 2006)

assessment of surface water, the WFD has requested the establishment of "good ecological status" and "good chemical status". The challenge in this study was to develop a strategy to include effect related data of river sediments into WFD concepts. Finally, we proposed a convention to combine the seven level toxicity classes system using pT-value of the xenoestrogenic effects measured by the ELRA-test and adapted these to the five-level ecological classification system of the EU WFD. Bringing in to the WFD the effect related approaches concerning bioassays and biomarkers is only relevant in the context of the "quality norms (QN)" of environmental relevant substances and the "good chemical status". The QN developed after the WFD in correspondence to "good chemical status" has to be enlarged for endocrine disrupting compounds, as it was shown in this study for estrogens (i.e., 17β-Estradiol) and Xenoestrogens (i.e., Nonylphenol and Bisphenol A). Other classes of concern and their potentially hazardous environmental effects should also be investigated.

Besides loading aquatic systems with endocrine disrupting compounds, there are several medicinal products used in large quantities for humans and livestock in industrialised countries (see Table 23.3) that have been frequently found in sewage treatment plants (STPs), surface waters (Huschek et al. 2004), sediments, pore water and groundwater. Because of the low current and loading by treated secondary effluents, there are several hot spots. Most of the beaches of the River Havel are protected areas linked to the drinking water directive. Because of the very active beach filtration (see Fig. 23.3) along the River Havel for the production of drinking water, there is intensive monitoring of sediments and groundwater mainly by instrumental analysis (LC-ESI-MS/MS). The results clearly show that several polar drugs are detected in groundwater and beach filtration. Several compounds are effectively removed by beach filtration but others are still present at higher concentrations (see Table 23.3). In total there are approx. 50–70 substances present.

Apart from drug residues (Huschek et al. 2004) several polyaromatic substances are present in River Havel sediments. In the River Havel sediments, Tributyl-Sn (TBT) concentrations were detected in the range of 1.600 ng/g dry weight. Exposure experiments with mussels showed feminisation and an increase of vitellogenin like proteins (Blaise et al. 2003) and exposed fish displayed an increase of vitellogenin as well as intersex determined by microscopic examination of the gonads (Arab et al. 2006). The results were a clear indication of ecosystem health status (see Table 23.2).

A comprehensive strategy regarding the demands of WFD is necessary and this study is a contribution to the chemistry quality component, but morphology and biology are also of importance. The uniqueness of these three quality components of the WFD has to be discussed in further studies and in relation to ecosystem health. The environmental aspects have to be strongly assessed in the context of public health. Effects in ecological systems are closely linked to health therefore strongly suggesting that biosensors, biomarkers and bioassays should be included in ecosystem health assessment in-line with the WFD. The most successful strategy would be to include bioassays and biosensors standardised by ISO and CEN into the approaches to define a "good chemical status" in the river basin systems. For

Table 23.3 Concentration ranges of active drug substances in µg/L in effluents of two STP's in Berlin (Huschek G and Hansen P.-D. 2006)

Active drug substances	Sales (kg) of active substance in 2001	Effluents of two different in Berlin STP [µg/L] during 2001–2002
Ibuprofen	344884.6	0.054–0.35
Carbamazepine	87604.9	0.9–1.2
Diclofenac	85800.7	1.47–2.1
Metoprolol	68364.4	0.99–1.18
Bezafibrate	33475,6	0.10–1.75
Propyphenazone	28140,2	0.11
Phenazone	24843,2	0.09
Sotalol	26649,2	0.57–0.68
Erythromycin	19199,0	0.15–0.43
Atenolol	13594,4	0.19–0.27
Trimethoprim	11426,6	0.16–0.42
Clarithromycin	7159,1	0.071
Propranolol	6519,0	0.056–0.09
Gemfibrozil	5243,7	0.28
Roxithromycin	9554,5	0.11–0.52
Naproxen	5060,1	0.1
Indomethacine	3720,6	0.14
Bisoprolol	2956,8	0.13
Ethinylestradiol	47,5	n.d. - 0.0003

n.d. = not detectable

the understanding of the status of the environment, a more realistic effect related real time sensor system is required.

The case study on a receptor assay–sensor showed clearly that biosensors are complementary to chemical methods since they can signal the presence of toxic (i.e., Xenoestrogens, Biotoxins) compounds that require further effect related chemical analytical studies. In addition this will allow the correlation of toxicity biosensor data with individual chemical compounds. Because of this cross activity, biosensors can detect chemical compounds which were not target analytes but they are detectable by the cross active biosensors or by the bioeffect related sensor but not by the chemical instrumental system (Alcock et al. 2003).

In summary, biosensors are early recognition systems that indicate the presence of unknown compounds which are responsible for the signals they detect. It is well known and accepted that in addition to chemical analysis effect related parameters are necessary. Sometimes chemical toxicity and related information can not be obtained by chemical instrumental measurements.

For diagnosis concerning human health, new technologies like proteomics are presently under developmental stages. These tools will eventually help to validate effects in humans and in the environment and this could be one relevant direction for the future. Many techniques have been developed to meet the demands of the lucrative biomedical markets and await adaptation for environmental applications, while the demands of continuous on-line monitoring are still hampered with problems that will require unique solutions.

Acknowledgements The author thanks for funding by Senate Department of Urban Development of the City of Berlin—Fischereiamt (Berlin Fishery Board, BFB), the European Commission (SANDRINE: ENV4-CT98–0801 and CITY FISH: ENK1-CT 1999–00009 projects) and the BMBF (MARS 1, BMBF- Az 03F0200A).

References

Alcock S., Barcelo D., and Hansen P.-D. (2003), Monitoring freshwater sediments, *Biosens. and Bioelectron.*, 18, 1077–1083.

Arab N S., Lemaire-Gony E., Unruh P.D., Hansen B.K., Larsen O.K., Andersen N.J., and Narbonne. (2006), Preliminary study of responses in mussels (*Mytilus edulis*) exposed to bisphenol A, diallyl phthalate and tertebromodiphenyl ether. *Aquat. Toxicol.*, 78, Suppl 1, 86–92.

Baumard P., Budzinski H, Garrigues P., Dizer H., and Hansen P.-D. (1999), Polycyclic aromatic hydrocarbons in recent sediments and mussels (*Mytilus edulis*) from the western Baltic Sea: Occurrence, bioavailability and seasonal variation. *Mar. Environ. Res.,* 47, 17–47.

Bilitewski U., Brenner-Weiss G, Hansen P.-D., Hock B., Meulenberg E., Müller G., Obst U., Sauerwein H., Scheller F.W., Schmid R., Schnabl G., and Spener F. (2000), Bioresponse-linked instrumental analysis. *TrAC Trends in Analytical Chemistry*, 19, 7, 428–433.

Blaise C., Gagne F., Salazar M., Salazar S., Trottier S., and Hansen P.-D. (2003), Experimentally-induced feminisation of freshwater mussels after long-term exposure to a municipal effluent. *Fresen. Environ. Bull.,* 12, 8, 865–870.

Bresler V., Bissinger V., Abelson A., Dizer H., Sturm A., Kraetke R., Fishelson L., and Hansen P.-D. (1999), Marine molluscs and fish as biomarkers of pollution stress in littoral regions of the Red Sea, Mediterranean Sea and North Sea. *Helgol. Mar. Res.*, 53:219–243.

DIN 38415–1 (1995), Deutsche Einheitsverfahren zur Wasser-, Abwasser- und Schlammuntersuchung - Suborganismische Testverfahren (Gruppe T) - Teil 1: Bestimmung von Cholinesterase-hemmenden Organophosphat- und Carbamat-Pestiziden (Cholinesterase-Hemmtest).

Hansen P.-D., V. Westernhagen H., and Rosenthal H. (1985), Chlorinated Hydrocarbons and hatching success in Baltic herring spring spawners. *Mar. Environ. Res.*, 15, 59–76.

Hansen P.-D., and Addison R. F. (1990), The use of mixed function oxidases (MFO) to support biological effect monitoring in the sea. *ICES. C.M. 1990/E:33.*

Hansen P.-D. (1995), Assessment of ecosystem health: Development of tools and approaches. In: *Evaluating and Monitoring the Health of Large-Scale Ecosystems*. Papport D., Gaudet C., and Calow P., Series I (Eds.), *Global Environmental Change* (Vol. 28, pp. 195–217). Berlin Heidelberg New York: Springer.

Hansen P.-D., Dizer H., Hock B., Marx A., Sherry J., McMaster M. and Blaise Ch. (1998), Vitellogenin—A biomarker for endocrine disruptors. *Trend. Anal. Chem. (TRAC)*, 17, 7, 448–451.

Hansen P.-D. (2003), Biomarkers. In: Markert B.A., Breure A.M., and Zechmeister H.G. (Eds.), Bioindicators & Biomonitors, *Principles, Concepts and Applications* (pp. 203–220). Amsterdam, Boston, London, Oxford, Paris, San Diego, San Francisco, Singapore, Sydney, Tokyo: Elsevier.

Hansen P.-D., Blasco J., De Valls A., Poulsen V., and van den Heuvel-Greve M. (2006), Biological analysis (Bioassays, Biomarkers, Biosensors) In: Sustainable management of sediment resources,

Volume 2, Sediment quality and impact assessment of pollutants. Eds. Damia Barceló and Mira Petrovic, *Elsevier Publishers Amsterdam, London, New York,* 311 pp.

Heberer T., and Mechlindki A. (2003), Fate and transport of pharmaceutical residues during bank filtration. *Hydrosciences,* 12, 57–60.

Huschek G., Hansen P.-D., Maurer H. H., Krengel D., and Kayser A. (2004), Environmental risk assessment of medicinal products for human use according to European Commission recommendations. *Environ. Toxicol.*, 19, 3, 226–240.

Huschek, G., Hansen, P.-D., (2006), Ecotoxicological classification of the Berlin river system using bioassays in respect to European Water Framework Directive, *Environmental Monitoring and Assessment*, 212, 1–3.

Krebs F. (2005a), The pT-method as a hazard assessment scheme for sediments and dredged material. In Blaise C., and Férard J.-F. (Eds.), *Small-scale Freshwater Toxicity Investigations* (Vol. 2, pp. 281–304). Dordrechtt: Springer.

Krebs F. (2005b), The pT-method as a hazard assessment scheme for wastewaters. In Blaise C., and Férard J.-F. (Eds.), *Small-scale Freshwater Toxicity Investigations* (Vol. 2, pp. 115–137). Dordrechtt: Springer.

Seifert M. (1999), Bestimmung von Östrogenen und Xenoöstrogenen mit einem Rezeptorassay. Dissertation, Fakultät für Landwirtschaft und Gartenbau, Technische Universität München.

Stagg R.M., and Addison R.F. (1995), An inter-laboratory comparison of measurements of ethoxyresorufin O-deethylase activity in dab (Limanda limanda) liver. *Mar. Environ. Res.*, 40, 93–108.

Sturm A., and Hansen P.-D. (1999), Altered cholinesterases and monooxygenase levels in *Daphnia magna* and *Chironimus riparius* exposed to environmental pollutants. *Ecotox. Environ. Safe.*, 42, 9–1.

Sturm A., da Sila de Assis H. C., and Hansen P.-D. (1999), Cholinesterases of marine teleost fish: enzymological characterisation and potential use in the monitoring of neurotoxic contamination. *Mar. Environ. Res.*, 47, 389–398.

Vethaak A.D., Rijs G.B.J., Schrap S.M., Ruiter H., Gerritsen A., and Lahr J. (2002), Estrogens and xeno-estrogens in the aquatic environment of the Netherlands - occurrence, potency and biological effects. *RIZA/RIKZ-report no. 2002.001.*

Chapter 24
Biological Toxicity Testing of Heavy Metals and Environmental Samples Using Fluorescence-Based Oxygen Sensing and Respirometry

Alice Zitova[1], Fiach C. O'Mahony[1], Maud Cross[2], John Davenport[2], and Dmitri B. Papkovsky[1,3]

Abstract A new methodology for simple, rapid, high throughput biological testing of potentially hazardous chemical and environmental samples has been developed, which is based on measurement of oxygen consumption of aquatic test organisms using phosphorescent oxygen-sensitive probes and detection on a fluorescent plate reader. Test organisms are exposed to potential toxicants and then allowed to respire in a sealed measurement compartment in the presence of soluble oxygen probe added to the sample. The resultant depletion of the dissolved oxygen causes an increase in sample fluorescence over time, thus reflecting the organism respiration rate and its alterations. Dedicated low-volume sealable 96-well plates provide improved sensitivity, convenience and miniaturization in such respirometric assays. These assays are carried out using standard laboratory tools and fluorescence plate readers, with multiple samples processed in parallel in 96-well microtitter plates. Oxygen consumption rate is a universal biomarker of general viability and metabolic responses of aerobic organisms, hence, this methodology is applicable to various organisms, including those currently used in toxicity testing. In this study, the new respirometric platform has been demonstrated with different organisms including prokaryotes (*E.coli*), eukaryotes (Jurkat T-cells), invertebrates (*Artemia salina*), and validated with toxicity testing of environmentally relevant chemicals. A panel of water samples discharged from wastewater treatment plants was also analyzed with this panel of test organisms. Organisms were exposed to the samples for a period of time (0–24 h depending on the organism chosen), and then assessed for their respiration rates (1 h assay). Toxicity was recorded as EC_{50}, i.e., the concentration of the toxicant which caused a 50% decrease in the respiration compared to untreated organisms. Responses of various organisms to certain chemicals can

[1] *Biochemistry Department & ABCRF, University College Cork, Cavanagh Pharmacy Building, Cork, Ireland*

[2] *Zoology Ecology and Plants Science Department, University College Cork, Distillery Fields, North Mall, Cork, Ireland*

[3] *Luxcel Biosciences Ltd., Suite 332, BioTransfer Unit, BioInnovation Centre, UCC, Cork, Ireland*

Y.J. Kim and U. Platt (eds.), *Advanced Environmental Monitoring*, 312–324.

be cross-compared and correlated to established toxicity tests (based on LD_{50}). In terms of sample throughput, sensitivity, speed, flexibility and convenience, the new screening platform is seen as being superior to the existing toxicity tests currently used. It provides adequate assessment of biological hazards of complex chemical and environmental samples, allows for the monitoring of sub-lethal effects and provides information-rich data reflecting the mode of toxicity. It is therefore highly suitable for environmental monitoring and screening of potentially hazardous samples, including large scale programs such as EU Water Framework Directive and REACH.

Keywords: Toxicity testing, animal-based testing, oxygen consumption assay, fluorescence-based oxygen sensing, optical oxygen respirometry

Glossary of acronyms: EC_{50} (median effective concentration), LD_{50} (median lethal concentration), PAH (polyaromatic hydrocarbons), PBDEs (polbrominated diphenylethers), RST (respirometric screening technology), Swift (Screening methods for Water data InFormaTion), WWT(wastewater treatment), WFD (water framework directive).

24.1 Introduction

In recent years the European Commission has published two new EU initiatives of particular concern to those involved in environmental monitoring. The first was a regulatory system for chemicals called REACH (EU, 2001b) which stands for Registration, Evaluation and Authorization of Chemicals. Under REACH, each producer and importer of chemicals in volume of 1 tonne or more per year and per producer/importer—around 30,000 substances—will have to register them with a new EU Chemicals Agency, submitting information on properties, uses and safe ways of handling them. Through evaluation each member state or agency is required to look in more detail at registration dossiers and at substances of concern, limiting animal testing to the absolute minimum in addition to prescribing the use of alternative methods wherever possible. Chemicals of particular concern include those which are known to cause cancer, mutations, toxicity to reproduction, including those which are persistent, bio-accumulative and toxic, or very persistent and very bio-accumulative. Based on such studies the Commission can decide to ban certain uses of chemicals or ban the chemicals in question altogether. The second initiative called the Water Framework Directive (EU, 2000) and the associated Swift-WFD (EU, 2003) is concerned with the expansion of the scope of water protection to include all waters, to set clear objectives in order that a "good status" be achieved by 2015 and that water use be sustainable throughout Europe. The directive initially identified 33 substances (EU, 2001a) which had been shown to be of major concern for European Waters. These include existing chemicals, plant protection products, biocides, metals and other groups like polyaromatic hydrocarbons (PAHs) and polbrominated diphenylethers (PBDEs) that are used as flame retardants. Of the 33 substances specified 11 have been identified as being priority hazardous substances which are of particular concern, and are destined to be phased out over a twenty year period. On the 17th of July 2006 the European Commission adopted a proposal to expand this list of 33 to include an additional 8 compounds (EU, 2006).

As a consequence of these new initiatives novel screening techniques capable of sensitive and high throughput analysis while also limiting or replacing the use of animal testing are required. One such technique developed by our group is respirometric screening technology (RST) (Papkovsky 2005). RST-based assays operate by means of fluorescence-based oxygen-sensitive probes (Papkovsky 2004) which are simply added to test samples where the fluorescence of the probe is quenched by sample dissolved oxygen in a nonchemical, reversible manner (collisional quenching of fluorescence (Papkovsky 2004)). The predictable manner in which the probe fluorescence changes in response to the changing concentration of oxygen, allows oxygen consumption rates to be monitored. As molecular oxygen is a key metabolite of all aerobic cells and higher organisms, alterations in cellular oxygen uptake serve as useful markers of their metabolic status, viability, and responses to various stimuli. The simple addition of the water-soluble probe to the sample being tested allows for great flexibility of the assay setup, as the sample can subsequently be pipetted into 96- or 384-well microtitre plates. The compatibility of the oxygen probe with conventional fluorescent plate readers also allows for the rapid and convenient parallel screening of multiple samples using a very simple kinetic assay format.

Monitoring changes in respiration also fulfils the second criteria of these assays in that it allows testing of sublethal toxic effects. One of the tensions in designing toxicity testing strategies is between reducing animal use and suffering and regulatory needs for more information on a wider array of chemicals or more detailed information on a smaller group of chemicals. Traditionally toxicity analysis is measured based on the concentration of a samples that kills 50% of the animals being tested called the LC_{50}. However with certain organisms death can be difficult to determine unequivocally. As a result other effects, such as pharmacological, biochemical or physiological responses, which closely correlate with death such as immobility, are measured instead. As with death for a measurement endpoint, results can be analyzed by comparing percent effect for organisms exposed to the toxicant and those not exposed to the toxicant allowing parameters such as the EC_{50}, the median effective concentration, to be determined. In addition to the aforementioned difficulty of ascertaining mortality in some cases when determining lethal concentration analysis, effective concentration analysis also have the advantage of being superior in terms of sensitivity as unsurprisingly a sublethal effective will occur at a lower toxicant concentration then death (EPA, 1994).

To date RST has proven suitable with a wide range of model systems and *in vitro* assays including the prokaryotes *E. coli* (O'Mahony and Papkovsky 2006) and *S. Pombe* (O'Riordan et al. 2000), the eukaryotic cell lines (Alderman et al. 2004; Hynes et al. 2003; Hynes et al. 2005), and the marine organism *Artemia salina* (O'Mahony et al. 2005) allowing these models to be used in the place of the conventional animal models of toxicity. The current study uses these three different groups of organisms to study their sensitivity and applicability for toxicity monitoring. In particular the current study focuses on heavy metal toxicity as one of the chemicals listed in the REACH program as well as a panel of wastewater samples discharged from treatment plants as would be tested under the Water Framework Directive. These samples are used to evaluate the sensitivity of different test organisms and the effects of exposure time to toxicant on the assay sensitivity.

24.2 Experimental Section

24.2.1 Materials

Phosphorescent oxygen probe (type A65N-1) and low-volume sealable 96-well microplates (MPU96-U1) were from Luxcel Biosciences, Ireland. Standard polystyrene 96-well plates, 20 ml cell culture flasks and 50 ml reagent tube were obtained from Sarstedt (Ireland). Heavy mineral oil, sea salt, heavy metals salts $ZnSO_4$ and $CuSO_4$, dimethyl sulfoxide (DMSO), trypan blue, LB broth and RPMI-1640 growth media were from Sigma-Aldrich (Ireland). All the other chemicals and solvents were of analytical grade. Solutions of heavy metals and dilution of environmental samples were prepared using Millipore grade water. Industrial wastewater samples (discharged from WWT plants after treatment) were obtained from EPA laboratory in Inniscarra, Co. Cork, Ireland.

24.2.2 Organism Culturing

Colonies of *E.coli* (DH5α MCR) were removed from solid agar and suspended in 100 ml LB broth in a 500 ml flask. Bacteria were grown at 37°C on a shaker, until an OD600 of 0.5 was reached. Cells were enumerated using a Neubauer hemocytometer (Assistant) and light microscope Alphaphot-2 YS2 (Nikon).

Jurkat cells (a human leukaemia T-cell line) were a gift from Professor Tom Cotter (University College Cork, Ireland). The cells were cultured in RPMI 1640 medium supplemented with 10% (v/v) fetal bovine serum, 2 mM L-glutamine, 100 U/ml penicillin, and 100 μg/ml streptomycin. Typically, cells were cultured in 75 cm^2 suspension cell flasks (Sarstedt) at 37°C in humidified atmosphere of 5% CO_2 in air to a concentration greater than 1×10^6 cells/ml (stained using trypan blue and counted under a microscope using a hemocytometer) prior to testing.

Artemia salina were hatched in artificial seawater (35 g/l) prepared from sea salts (Sigma-Aldrich). After 48 h at 25°C under continuous illumination by a bulb (no feeding required), *Artemia* molted to the instar II stage, a growth stage when most toxicity testing is conventionally carried out on *Artemia* (Vanhaecke and Persoone 1984).

24.2.3 Respirometric Measurement Setup

Two different platforms were utilised for respirometric measurements, both providing adequate sensitivity and performance with respect to the measurement of respiration of test organisms and relative simplicity of assay setup and execution. The first format employs standard 96-well plates (Sarstedt) which were used for the cell-based assays. The second more sensitive format employs Luxcel sealable microplates (Alderman et al. 2004) used for the *Artemia*-based assays. The MPU96-U1 plates are made of clear polystyrene and contain wide and shallow wells, 6 mm in diameter, 0.5 mm deep, actual volume ~6 μl, which are surrounded by rims and frames that help in sealing the samples, removing air bubbles and excluding ambient oxygen. A small excess of sample (usually 10 μl) containing test organisms, toxicant and oxygen probe is added to each well, which is retained by capillary forces, and the oxygen uptake assay is initiated by applying the adhesive sealing tape to the plate.

A65N-1 phosphorescent oxygen probe (Luxcel) was reconstituted in 1 ml of water to give a 1 μM stock, which was stored in the dark at 4°C and used at the working concentrations described below. *Artemia salina* assays conducted on Luxcel plates required probe working concentration of 100 nM. Fluorescence intensity measurements were carried on a prompt fluorescence reader, SpectraMax Gemini (Molecular Devices), with excitation set at 380 nm and emission at 650 nm. *E.coli* and Jurkat-cell-based assays conducted on 96-well plates required probe concentrations of 3 nM. Time resolved fluorescence (TR-F) measurements were carried out on a Genios Pro fluorescent plate reader (Tecan, Mänerdorf, Switzerland) using standard set of filters of 380 and 650 nm, respectively, with delay and gate times of 60 μs and 100 μs. In order to assess the effect of incubation time on the sensitivity of the organisms to the test samples incubation times of 0 and 24 h were used. Samples were added to test organisms in medium to give a resulting concentrations of 6.4e + 4, 6.4e + 3, 6.4e + 2 μg/l of heavy metals, and wastewaters samples were 10, 100 and 1000 times diluted. Sets of control samples, particularly those containing test organisms, oxygen probe and no toxicant (100% of respiration) and test organisms without the probe and toxicant (blank) were included in each experiment. All concentration points and controls were run in quadruplicates. Protocols for the individual model organisms are outlined below.

24.2.3.1 Artemia Salina

For 0 h and 24 h incubation-based toxicity assays *Artemia salina* were transferred, using Pasteur pipettes (Volac, U.K.), into 1.5 ml Eppendorf tubes (Starstedt) containing the toxicant diluted in seawater in addition to the appropriate concentration of oxygen sensitive probe. Post incubation (0 and 24 h) the animals were pipetted into the wells of the low volume Luxcel plates at a concentration of 5 animals per well (volume ~10 μl). Assays were initiated by sealing the microplate with Mylar sealing film and measuring the plate at constant temperature of 30°C every 5 min over a period of 1–2 h.

24.2.3.2 Jurkat Cells and E. coli

For 0 h incubation, 135 µl sample of Jurkat cells in RPMI-1640 growth medium (10^6 cells/ml) containing the oxygen probe were pipetted into the wells of a standard 96-well plate. 15 µl of toxicant stock(s) were added to the samples to produce the required concentrations of toxicant (may vary for different wells). The samples were then sealed by pipetting 100 µl of heavy mineral oil on top of each and monitored on a plate reader at 37°C as described above. For 24 h incubation cells were split in RPMI-1640 growth medium and dispensed in 4.5 ml aliquots into the 20 ml culture flasks (Sarstedt). 0.5 ml of various concentrations of test sample were added to make up the appropriate concentrations of toxicants. Cells were subsequently cultured for 24 h under standard conditions of 37°C, humidified atmosphere of 5% CO_2 in air. Following the incubation period 150 µl of cell solutions were added to the wells of standard 96-well plate together with appropriate concentration of probe, sealed with 100 µl heavy mineral oil and monitored. Assay with *E. coli* cells in LB broth resembled that conducted with Jurkat cells for the 0 h incubation step. For the 24 h incubation, 10 ml of *E.coli* were added to 100 ml of LB Broth, and subsequently aliquoted in 9 ml volumes into 50 ml reagent tubes (Sarstedt), each containing 1 ml of test compound at required concentration. These samples were cultured at 37°C with shaking and then analyzed in 96-well plates as above.

24.2.4 *Data Analysis*

Measured time profiles of fluorescence were analysed to determine the initial slopes of each sample. The slopes, which reflect oxygen rates by test organisms, were calculated using the following formula:

$$slope = \frac{I_2 - I_1}{(I_1 - I_{blank})(t_2 - t_1)}$$

where I_1 and I_2 = fluorescence signal at time points t_1 and t_2 (usually 10 and 20 min after start of monitoring, selected by operator) and I_{blank} = signal without the probe (when comparable with the signal). To assess respiration rates of test organisms and alterations produced by toxicants, calculated slopes were compared to those of untreated organisms (positive controls, 100% respiration) and to those without organisms (negative controls). EC_{50} values (i.e., the concentration of the toxicant which caused a 50% decrease in the respiration, compared to untreated organisms) were calculated using sigmoidal fits in OriginPro 7.5 G software. Statistical analysis was conducted using a T-test with 95% confidence limits.

24.3 Results

24.3.1 Heavy Metal Toxicity

Typical respiration profiles of *Artemia salina* treated with heavy metal ion (Cu^{2+} in this case) are shown in Fig 24.1a. Oxygen consumption for the treated animal samples is significantly reduced, and samples without animals (negative controls) produce negligible slope. Such slopes can be used to assess the rate of respiration of treated *Artemia* compared to untreated animals. Figure 24.1b shows dose and time dependence of Cu^{2+}

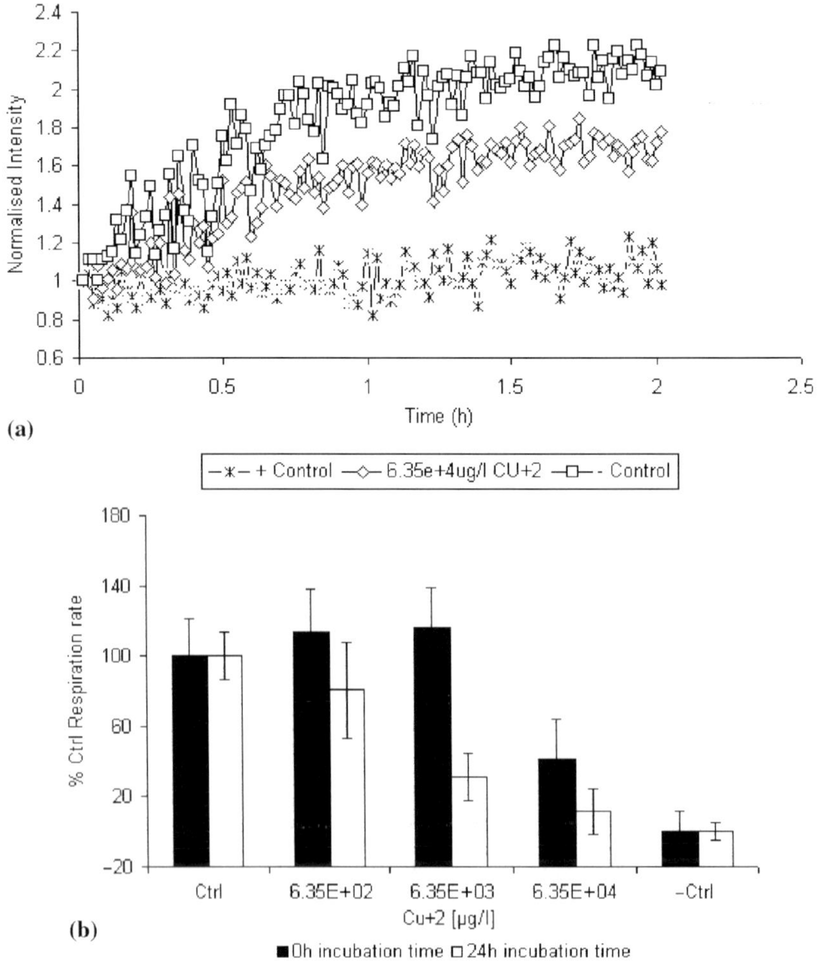

Fig. 24.1 (**a**) Respiration profiles of *Artemia salina* in 6.35e+4 µg/l Cu^{+2} (□), negative (x) and positive controls (◊), at 0 h incubation time, (top): (**b**) Effects of Cu^{2+} on *Artemia* respiration at 0 h and 24 h exposure times, measured with 5 animals per well over 2 h, 30°C (bottom)

on respiration. When no pre-incubation is allowed, a significant effect on respiration is only apparent above 6.4e+3 μg/l with a T value of 5.3292. Whereas with 24 h animal exposure to the toxicant the respirometric assay detects a significance difference at concentrations below 6.4e+3 μg/l of Cu^{2+} with a T value of 10.2365.

Increased sensitivity to heavy metals is also observed with *E.coli* and Jurkat cells, Fig. 24.2. With a 0 h incubation period, *E.coli* was identified as the most sensitive to Cu^{2+} followed by *Artemia* and Jurkat cell, giving EC_{50}−0h values of

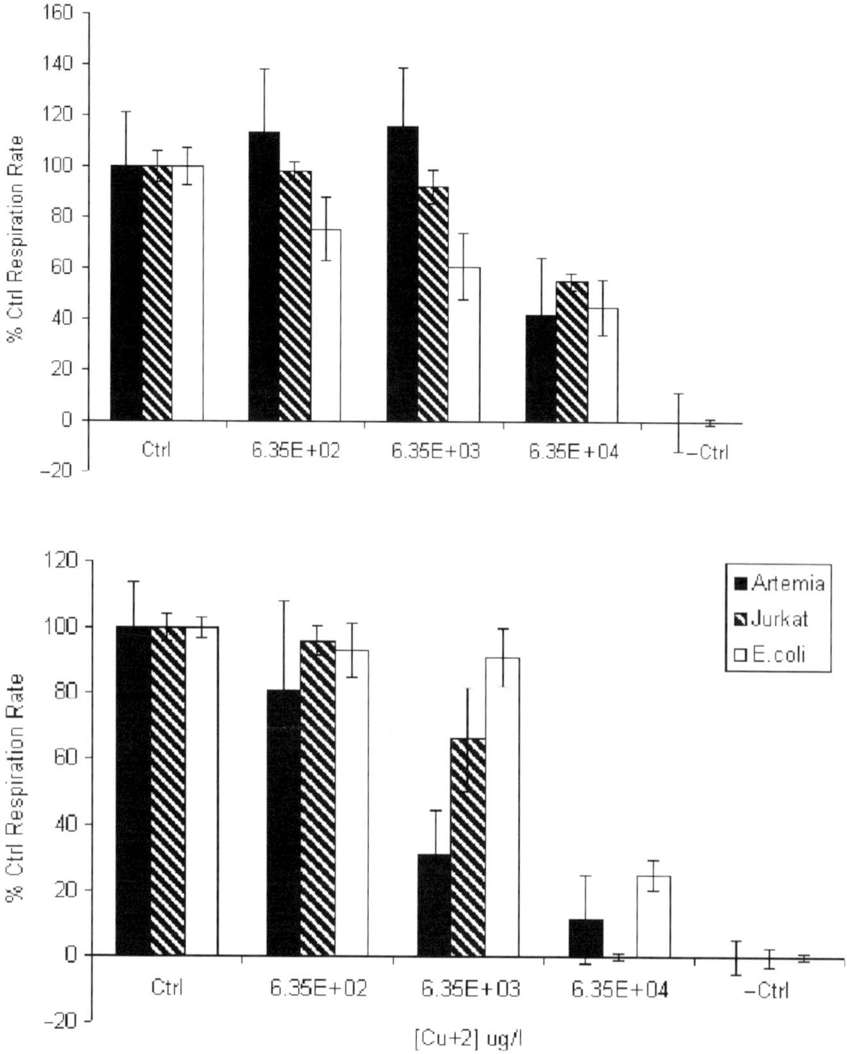

Fig. 24.2 Effects of Cu^{+2} on different animals: *Artemia salina (ϒ), Jurkat* cells (⊄) and *E.coli* (Δ), at 0 h (top), and 24 h (bottom) pre-incubation time

Table 24.1 Heavy metals toxicity on various model organisms measured by oxygen respirometry

		Heavy metal	
		Zn^{+2} [μg/L]	Cu^{+2} [μg/L]
EC_{50} – 0 h	*Artemia*	6.5e+5	6.2e+4
	Jurkat	2.1e+4	9.1e+3
	E.coli	1.0e+4	3.8e+4
EC_{50} – 24 h	*Artemia*	4.9e+2	3.6e+3
	Jurkat	7.7e+3	8.6e+3
	E.coli	9.5e+3	3.0e+4

3.8e+4 μg/L, 6.2e+4 μg/L, 9.1e+3 μgL, respectively (Table 24.1). Increasing the incubation time to 24 h resulted in *Artemia* being the most sensitive followed by Jurkat cells and *E.coli* with EC_{50}–24 h values of 3.6e+3 μg/L , 8.6e+3 μg/L and 3.0e+4 μg/L respectively. Table 24.1 shows that a similar pattern is observed for Zn^{+2}, with *Artemia* being the least sensitive at 0 h incubation and the most sensitive at 24 h incubation periods.

24.3.2 Analysis of Industrial Wastewater Samples

The model organisms were also used to analyse a set of industrial wastewater samples for their residual biological toxicity. All these samples underwent treatment at WWTP and were discharged into the environment. An initial screening of the samples (diluted 1:10 in assay medium) using Jurkat cells and 0 h pre-incubation time revealed that only sample 4 caused a marked reduction in respiration, while the others had practically no effect. Sample 4, which produced a respiration profile similar to that of the negative control (i.e., almost complete inhibition, see Fig. 24.3a). These samples underwent subsequent investigation to determine their dose and time dependant effects on the respiration of Jurkat cells, *E.coli* and *Artemia*.

Sample 4 displayed the greatest level of toxicity with T values of 78.0615 for (Jurkats), 12.4033 for (*E.coli*) and 9.6363 for (*Artemia*) at 0 h incubation. A dilution factor of 79 was required for sample 4 to achieve 50% inhibition of respiration (Table 24.2). Of the three model organisms tested, Jurkats were shown to be the most sensitive when an incubation period of 0 h was used followed by *E.coli* and *Artemia* with 50% of control respiration achieved from dilution factors of 79, 67.1 and 12 respectively. Using the 24 h incubation period, Jurkat cells were still the most sensitive however *Artemia* were shown have developed a greater sensitivity then *E.coli* with dilution factors of 27.5, 13.3 and 3.8 respectively. Samples 1, 2 and 3 showed no measurable effect on the respiration, whilst samples 5 and 6 showed some effects, but no clear pattern of toxicity at both 0 and 24 h incubation times.

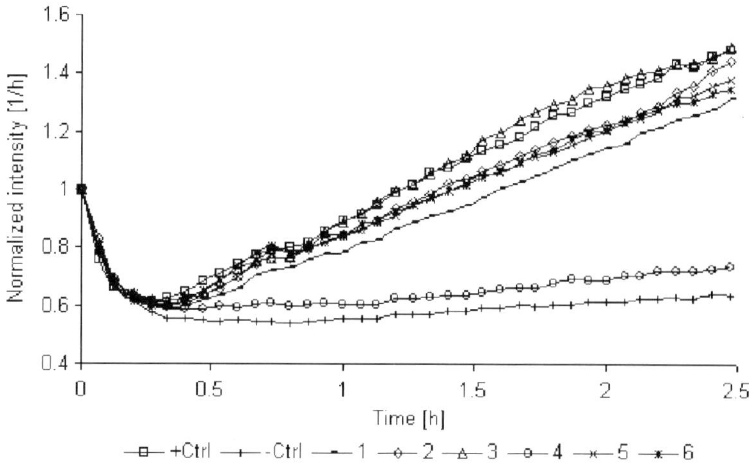

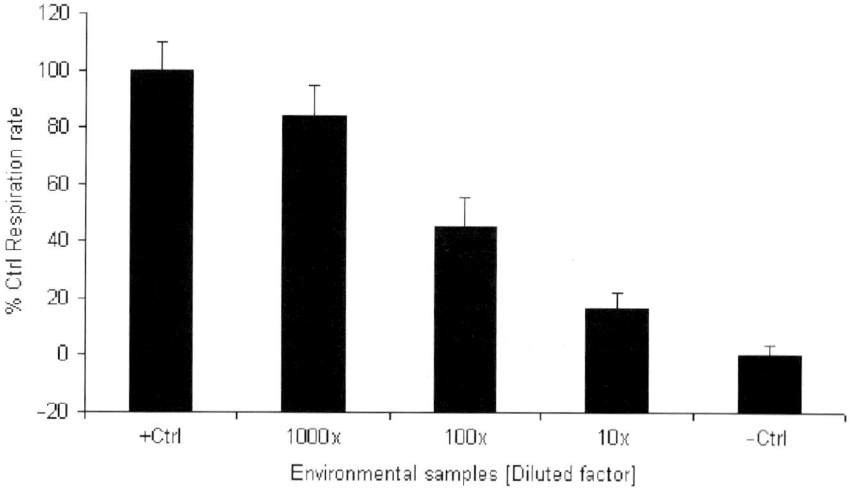

Fig. 24.3 Effects on Jurkat cells respiration of six wastewater samples (top). Effect of increasing concentrations of sample 4 (bottom). Incubation time 0 h, Jurkat cells $5 \cdot 10^6$ cells/ml

Table 24.2 Comparison of wastewaters samples toxicity effect (EC_{50}) on various model organisms

		Environmental sample [dilution factor causing the effect]					
		n.1	n.2	n.3	n.4	n.5	n.6
$EC_{50}-0\,h$	*Artemia*	< 1	< 1	< 1	12.0	< 1	< 1
	Jurkat	< 1	< 1	< 1	79.0	< 1	< 1
	E.coli	< 1	< 1	< 1	67.1	5.2	< 1
$EC_{50}-24\,h$	*Artemia*	< 1	< 1	< 1	13.3	1.5	4.1
	Jurkat	< 1	< 1	< 1	27.5	< 1	< 1
	E.coli	< 1	< 1	< 1	3.8	< 1	3.9

24.4 Discussion

The above results show the suitability and effectiveness of oxygen respirometry and RST for the analysis of chemicals and environmental samples for their biological hazard and toxicity. The ability to monitor the respiration of a panel of model organisms and study changes in their metabolism induced by external toxicants facilitated the assessment and ranking of heavy metals ions for their toxicity. Zn^{2+} was identified by each of the three model systems as being of greater toxicity then Cu^{2+}, at both short (0h) and long (24h) exposure times a result which agrees with similar assays conducted on Microtox system (Sillanpaa and Oikari 1996). In comparison to the same Microtox study the sensitivities of the various organism respirometric-based assays described in this study were less sensitive at both incubation times, however as different organisms were used for RST and Microtox a direct comparison is not possible. For Zn^{2+} and Cu^{2+} respectively the Microtox gave EC_{50} values of 2.6e+2µg/L and 1.02e+3µg/L respectively whereas the RST assays at 0h incubation gave 1.0e+4µg/L (*E.coli*), 2.1e+4µg/L (Jurkats) and 6.5e+5µg/L (*Artemia*) for Zinc and 3.8e+4µg/L (*E.coli*), 9.1e+3µg/L (Jurkats), 6.2e+4µg/L (*Artemia*) for Copper. Increased incubation time was seen to increase the toxicological impact, especially for the multi-cellular organism *Artemia*, where the EC_{50} values dropped from 6.5e+5µg/L to 4.9e+2µg/L for Zn^{2+} and 6.2e+4µg/L to 3.5e+3µg/L for Cu^{2+}, respectively. The lack of a similar large increase in the cell-based assays is assumed to be due to the more simple organization and physiology of the cultured cells compared to multi-cellular organisms. The ability of longer exposure time to increase sensitivity of *Artemia* but decrease it for the cell-based assays was also observed when the environmental samples were tested.

The screening of environmental samples highlighted the power of the RST approach for its ability to assay high numbers of samples simultaneously. The microtitter plate format of the assay, is only limited by the number of wells in the plate and the degree of automation. Even for manual liquid handling used in our case, several hundred samples or assay points can be run each day. Qualitative and objective (instrumental) toxicological assessment allowed samples to be rapidly (1–2h) screened, reliably identifying toxic samples, ranking them and determining EC_{50} values. The ability of the platform to quantify the toxicity of real environmental samples such as wastewater has been demonstrated. The tracking of the toxicity of a sample over time by comparing its effect on the respiration of the chosen test organism and exposure time was shown. The versatility of the RST platform allows the user to choose test organisms and customise the assay based on such factors as the availability of culturing facilities, measurement instrumentation and trained personnel. With this in mind the brine shrimp *Artemia salina* is an attractive model for primary screening of samples for potential biological hazard and toxicity by RST. Its ease of culture, high fecundity, low cost, facilitate the growth of sufficient numbers of test organisms in laboratories with little knowledge of animal handling, without the legal or ethical issues that surround the use of high animal models such as mice. From the cell-based RST assays, test organisms such as *E. coli* look attractive, as they are also robust, easy to culture and use.

24.5 Conclusions

EU initiatives such as REACH and the Water Framework Directive have highlighted the need for high throughput screening for the analysis of toxic chemicals and environmental samples. RST and simple respirometric *in vitro* assays with a panel of model organisms have been shown to provide relevant and information-rich toxicological data and to meet the requirements of such initiatives whilst avoiding the legal and ethical issues of using conventional animal-based testing methods. The flexibility and cost-effectiveness of the assays allows them to be adapted for various test organisms such as bacteria, mammalian cell culture and small marine animals. The microtitter plate platform (96- and 384-well) on which the assay are conducted means that the number of samples amenable for testing is greatly increased and only limited by the level assay of automation. These factors combined with the rapid and sensitive nature of the assay clearly highlight the potential for RST use in these increasingly important areas of environmental research, biochemical and molecular toxicology.

Acknowledgements Financial support of this work by the Marine Institute and Marine RTDI Measure, Productive Sector Operational Programme, National Development Plan 2000–2006 (Grant-aid agreement No AT-04-01-01), is gratefully acknowledged.

References

Alderman J., Hynes J., Floyd S.M., Kruger J., O'Connor R. and Papkovsky D.B. (2004), A low-volume platform for cell-respirometric screening based on quenched-luminescence oxygen sensing, *Biosens. Bioelectron.*, 19, 1529–1535.

EPA, U. (1994). Using toxicity tests in ecological risk assessment (pp. 2–3). Office of emergency remedial response, hazardous site evaluation division, US EPA.

EU. (2000). Directive 2000/60/EC of the European Parliament and of the Council Establishing a Framework for Community Action in the Field of Water Policy (OJ L 327, 22.12.2000). 177, 199e211.

EU. (2001a). Decision No 2455/2001/EC of the European Parliament and of the Council of 20 November establishing the list of priority substances in the field of water policy and amending Directive 2000/60/EC Directive.

EU. (2001b). White Paper: Strategy for a future Chemicals Policy. Brussels, Belgium: Commission of the European Communities.

EU. (2003). Swift-Program (SS Pi-CT 2003-502 492).

EU. (2006). COM(2006) 397. Proposal for a directive of the European Parliament and of the council on environmental quality standards in the field of water policy and amending Directive 2000/60/EC.

Hynes J., Floyd S., Soini A.E., O'Connor R., and Papkovsky D.B. (2003), Fluorescence-based cell viability screening assays using water-soluble oxygen probes, *J. Biomol. Screen.*, 8, 264–272.

Hynes J., O'Riordan T C., Curtin J., Cotter T.G., and Papkovsky D.B. (2005), Fluorescence based oxygen uptake analysis in the study of metabolic responses to apoptosis induction, *J. Immunol. Methods*.

O'Mahony F.C., O'Donovan C., Hynes J., Moore T., Davenport J., and Papkovsky D.B. (2005), Optical oxygen microrespirometry as a platform for environmental toxicology and animal model studies, *Env. Sci. Technol.*, 39, 5010–5014.

O'Mahony F.C. and Papkovsky D.B. (2006), Rapid high-throughput assessment of aerobic bacteria in complex samples by fluorescence-based oxygen respirometry, *Appl. Env. Microbiol.*, 72, 1279–1287.
O'Riordan T.C., Buckley D., Ogurtsov V., O'Connor R., and Papkovsky D.B. (2000), A cell viability assay based on monitoring respiration by optical oxygen sensing, *Anal. Biochem.*, 278, 221–227.
Papkovsky D.B. (2004), Methods in optical oxygen sensing: Protocols and critical analyses, *Oxyg. Sens.*, 381, 715–735.
Papkovsky D.B. (2005), Respirometric Screening Technology (RST), *Screen. – Trends Drug Discov.*, 6, 46–47.
Sillanpaa M. and Oikari A. (1996), Assessing the impact of complexation by EDTA and DTPA on heavy metal toxicity using microtox bioassay, *Chemosphere*, 32, 1485–1497.
Vanhaecke P. and Persoone G. (1984), The arc-test: A stanardized short term routine toxicity test with *Artemia nauplii*. Methodology and evaluation, *Ecotoxicol. Test. Marine Environ.*, 2, 143–157.

Chapter 25
Omics Tools for Environmental Monitoring of Chemicals, Radiation, and Physical Stresses in *Saccharomyces Cerevisiae*

Yoshihide Tanaka[1], Tetsuji Higashi[1], Randeep Rakwal[2], Junko Shibato[2], Emiko Kitagawa[2], Satomi Murata[2], Shin-ichi Wakida[1], and Hitoshi Iwahashi[2]

Abstract The yeast *Saccharomyces cerevisiae* is one of the most characterized eucaryotes and its complete genome sequence was published in 1986. Thus, this organism is a good candidate for biological environmental monitoring. Omics (genomics, proteomics, metabolomics) technology is being applied to biological studies from prokaryotes to humans. We are applying omics technologies to environmental monitoring using yeast cells, medaka, rice, rat, and mouse. In this report, we focus on yeast omics as tools of environmental monitoring for chemicals, radiation, and physical stresses in yeast. For genomics studies, we use commercially available DNA microarrays. We analyzed the expression profiles for highly induced or repressed genes, the functional characterization of induced and repressed genes, and cluster analysis. The list of highly induced and repressed genes can help to identify candidate biomarkers and strongly induced functions; however, this may reflect only a small part of the full stress response. The functional characterization studies, on the other hand, can help elucidate the mechanisms of stress response. Cluster analysis allows comparison of different environmental stress conditions. For proteomics studies, classical two-dimensional electrophoresis and peptide sequencing or mass spectrometry were used for monitoring stress induced and modified proteins. However, protein turnover ratio and especially degradation rates were slow in yeast cells and the induction levels of proteins do not always reflect the physiological status of the cell. The role of proteomics in yeast cells must be focused on modification of proteins. For metabolomics studies, capillary electrophoresis/mass spectrometry (CE/MS) was used for the separation and identification of metabolites. This new methods have high potential for the evaluation of environmental stress.

Keywords: *Saccharomyces cerevisiae*, environmental stress, OMICS, genomics, proteomics, metablomics, DNA microarray, CE/MS

---

[1]*Human Stress Signal Research Center (HSS), National Institute of Advanced Industrial Science and Technology (AIST), 1-8-31 Midorigaoka, Ikeda, Osaka 563–8577, Japan*

[2]*HSS, AIST, Tsukuba West, 16-1 Onogawa, Tsukuba, Ibaraki 305–8569, Japan*

Y.J. Kim and U. Platt (eds.), *Advanced Environmental Monitoring*, 325–337.

25.1 Introduction

At present, more than 25 million materials are registered in the Chemical Abstracts database, and it is estimated that more than 10,000 synthetic chemicals are accumulating in the environment every year. Despite the fact that these industrial chemicals have given us numerous benefits, there is no doubt that they have damaged the environment. The chemicals being dispersed on the earth should be carefully controlled to prevent their adverse effects. In fact, many chemicals can be detected from environmental samples; however, only 10% of those chemicals can be identified by current technology (Suzuki and Utsumis 1998). Ten percent is an inadequate number to protect the environment. Furthermore, not only chemical but also physical and biological stresses including radiation, temperature, and pathogens impact humans and ecological systems. Thus, we have to understand the effects of these environmental stresses on biological systems, which necessitates monitoring these stresses.

Bioassays are used for the assessment of environmental pollution and risk assessment of environmental stresses (Celemedson et al. 1996). In bioassay systems, safety is estimated by monitoring biological responses to environmental stress. One of these studies is the Multicenter Evaluation of *in vitro*. Cytotoxicity program, organized by the Scandinavian Society for Cell Toxicology (Celemedson et al. 1996). The investigators compared LD50 data obtained *in vivo* (whole organism) and IC50 obtained *in vitro* (bioassay). They found a correlation between these parameters and defined the concept of "basal cytotoxicity". Basal cytotoxicity can be understood as the generalized toxic effect to cellular components, functions and biosynthesis that are universal to all cell lines. On the other hand, the Ames test is well known as one of the most powerful methods for monitoring the mutagenicity of environmental samples (Reifferscheid and Heil 1996). In this system, mutants of *Salmonella typhimurium* are grown in a minimum medium and mutagenicity is estimated according to the frequency of back mutation. As the frequency of back mutation is dependent on DNA damage, we can estimate the mutagenicity of chemicals or environmental stress.

An extensive literature exists on bioassay systems that include tests by Ames, Microtox, Umu, and others (Celemedson et al. 1996). Each system can be used for estimating effects by environmental stress; however, the information that can be estimated is limited to the degree of toxicity or mutagenicity. Information concerning the nature of the environmental stress remains unavailable. In addition, bioassay systems sometimes mistakenly identify natural products as the toxic substance (data not shown). Although it is important to quantify the degree of effects in the environment, information concerning the nature of stress is essential for risk assessment and prevention. Bioassay systems are required that can be used for predicting the mechanism of environmental stress.

We proposed "multiple-end-point bioassays" several years ago (Iwahashi 2000). The report introduces "multiple-end-point bioassay" systems that are based on stress sensitivities of microorganisms, responses of one kind of organism, and microarray technology. Microorganisms are screened to identify strains that are sensitive to specific stresses and the sensitivity of the isolated strain is then used for characterizing unknown chemicals or environmental samples. The "multiple-end-point bioassay" based on one kind of organisms are system using one organism and many kinds of endpoints such as growth inhibition, viability, induction of stress proteins, prion curing mutagenicity, cytoplasmic mutagenicity, and chromosomal mutagenicity. Using these endpoints we tried to characterize chemicals and environmental stresses (Iwahashi 2000). DNA microarray technology was also introduced as the candidate for the "multiple-end-point bioassay" (Iwahashi 2000).

In recent years, DNA microarray technology has developed rapidly and been widely adopted as a tool for understanding biological systems at the genomic level (Momose and Iwahashi 2001). Furthermore, this technology can be combined with proteomics and metabolomics technology. Proteomics is essentially based on the analysis of proteins using two-dimensional electrophoresis. This technology provides information on the expression levels of hundreds of proteins as well as protein modifications. Metabolomics is based on the extensive database of analysis of metabolites using NMR or CE/MS, and this is relatively new technology for biologists. A combined approach, omics technology, provides data on DNA, mRNA, proteins, and metabolites and thus allows the development of a robust "multiple-end-point bioassay". In this report, we describe the development of a combined omics approach (genomics, proteomics and metabolomics) for environmental monitoring for of chemicals, radiation, and physical stresses in yeast.

25.2 Materials and Methods

25.2.1 Strains and Growth Conditions

Saccharomyces cerevisiae S288 C (*MATαSUC2 mal mel gal2 CUP1*) was grown in YPD medium (1% Bacto Yeast Extract, 2% polypeptone, 2% glucose) at 25°C according to the procedure outlined by Kitagawa et al. (2002).

25.2.2 Stress Conditions

For chemical and radiation treatment, yeast cells growing exponentially were exposed for 2 h as follows: 35 ppm paraquat (Iwahashi 2006), saturated vitamin E (Iwahashi 2006), 0.16% supiculisporic acid (Kurita et al. 2004), 10% dimethylsulfoxide (Murata et al. 2003), 16 Gy gamma ray (Kimura et al. 2006), 25 ppm chloroacetaldehyde (Iwahashi 2006), 250 ppm capsaicin (Kitagawa et al. 2002), 5 μM thiuram (Kurita et al. 2002), 5 mM manganese chloride (Iwahashi 2006), 10 μM cadmium and 2.5 μM thiuram (Iwahashi 2006), 0.7 mM mercury(II) chloride

(Kimura et al. 2006), 0.01%, sodium dodecyl sulfate (Sirisattha et al. 2004a), ×1,500 dilution of roundup high-load (Sirisattha et al. 2004b), 15 μM cycloheximide (Iwahashi 2006), 10 mM hydrogenperoxide(H2O2 in Fig. 25.1) (Iwahashi 2006), 2 mM lead chloride (Iwahashi 2006), 1.5% pentane (Fujita et al. 2004), 5 mM thorium nitrate (Murata et al. 2006a), 400 ppm gingerol (Iwahashi 2006), 1.5 μM fluazinam (Iwahashi 2006), 5 mM 2-aminobenzimidazole (Iwahashi 2006), 0.5 mM benzpyren (Iwahashi 2006), 50 μM pentachlorophenol (Iwahashi 2006), 2 ppm zineb (Kitagawa et al. 2003), 2 ppm maneb (Kitagawa et al. 2003), 75 μM thiuram (Iwahashi 2006), 10 μM TPN (Kitagawa et al. 2003), 20 μM cadmium and 5 μM thiuram (Iwahashi 2006), 0.3 mM cadmium chloride (Momose and Iwahashi 2001), and 0.3 μM methylmercury(II) chloride (Iwahashi 2006).

For gas treatment, yeast cells growing exponentially were transferred to high pressure vessels (Iwahashi 2006) and pressured using compressed gas cylinders for 2 h as follows: 10 MPa air (Iwahashi 2006), 40 MPa nitrogen (Matsuoka et al. 2005), 0.5 MPa oxygen (Iwahashi 2006).

For physical stress treatment, yeast cells growing exponentially were frozen at −80°C for 7 days (Freeze in Fig. 25.1) (Odani et al. 2003), then treated as follows: 40 MPa at 4°C for 12 h (40 MPa 4°C in Fig. 25.1) (Iwahashi et al. 2003), 180 MPa at 4°C for 0 min (180 MPa 4°C in Fig. 25.1) (Iwahashi et al 2003), and 30 MPa 25°C for 2 h. These cells were allowed to recover for 60 min at 25–30°C. Cold shock treatment (Cold in Fig. 25.1) entailed a shift of exponentially growing yeast cells from 25 °C to 4 °C for 6 h (Iwahashi et al 2005), while pressure shock treatment (40 MPa Pressure Shock in Fig. 25.1) shifted exponentially growing yeast cells under atmosphere pressure to 40 MPa for 2 h (Iwahashi 2006), and followed by incubation under 30 MPa or 10 MPa for 16 h (30 MPa 25 °C Growth or 10 MPa 25 °C growth in Fig. 25.1) (Iwahashi 2006). Heat shock treatment was carried out by shifting exponentially growing yeast cells from 30 to 43°C for 2 h (Iwahashi et al. 1995).

For environmental samples A–E and the incinerator sample in Fig. 25.1, YPD medium were made with an environmental sample replacing the DW. These YPD media were filter sterilized (Kim et al. 2004; Murata et al. 2006b). Exponentially growing yeast cells were transferred to the YPD medium made of environmental sample.

25.2.3 DNA Microarray Analysis

Each microarray, spotted on a glass slide for hybridization with labeled mRNA probes, represented almost all ORFs of yeast (5,809~5,819 genes; depending on the lot, DNA Chip Research Inc. Yokohama, Japan). Extraction of total RNA, mRNA purification, labeling with Cy3 or Cy5, and hybridization were described previously [Kim et al. 2004; Momose and Iwahashi 2001). A Scan Array 4000 laser scanner (GSI Lunomics, Billeria, MA, USA) was used to acquire hybridization signals. Array images were analyzed with Gene Pix 4000 (Inter Medical, Nagoya, Japan). Cluster analysis of the mRNA expression profiles after the combination treatment was according to Murata et al. using the GeneSpring ver. 4.2.1 software (Silicon Genetics, CA, USA) (Momose and Iwahashi 2001; Kitagawa et al. 2002, 2003; Kurita et al. 2002, 2004; Iwahashi et al. 2003, 2005; Murata et al. 2003, 2006a, b; Odani et al. 2003; Kim et al. 2004; Sirisattha et al. 2004a, b; Matsuoka et al. 2005; Iwahashi 2006; Kimura et al. 2006).

25.2.4 *Two-Dimensional Electrophoresis*

Two-dimensional electrophoresis was carried out essentially according to O'Farrell's method (1975) and extraction of proteins, labeling with $[H^3]$ leucine, and staining with CBB were described previously (Iwahashi et al. 1995).

25.2.5 *Capillary Electrophoresis/Mass Spectrum (CE/MS) Analysis*

A Beckman P/ACE MDQ capillary electrophoresis system (Beckman Coulter, Tokyo, Japan) was connected to an Esquire 3000 plus ion trap mass spectrometer (Bruker Daltonics, Yokohama, Japan) through an electrospray ionization (ESI) source (Agilent Technologies Japan, Tokyo, Japan). Yeast extracts were prepared after the stress conditions by filtration of yeast cells (0.45 μm membrane filter), extraction of metabolites with cold methanol, and a second filtration (Microcon, 5 kDa cut-off, Millipore, Bedford, MA, USA) (Sato et al. 2004). After prefreezing the filtered solution at −80°C for 2 h, lyophilization was carried out at 25°C under vaccum (10 Pa) overnight in a lyophilizer (model FRD-MINI, Asahi Techno Glass, Chiba, Japan). The residue was dissolved in 25 μl of water/methanol (1:1, v/v). Analytical conditions were established based on a previously reported method (Soga et al. 2003; Sato et al. 2004).

25.3 Results and Discussion

25.3.1 *Genomics Technology for the Assessment of Stress Response Through mRNA Expression Levels*

Figure 25.1 shows cluster analysis of expression profiles obtained after the stress treatments (Momose and Iwahashi 2001; Kitagawa et al. 2002, 2003; Kurita et al. 2002, 2004; Iwahashi et al. 2003, 2005; Murata et al. 2003, 2006a, b; Odani et al. 2003; Kim et al. 2004; Sirisattha et al. 2004a, b; Matsuoka et al. 2005; Iwahashi 2006; Kimura et al. 2006). This calculation is based on the correlation factors between the treatments.

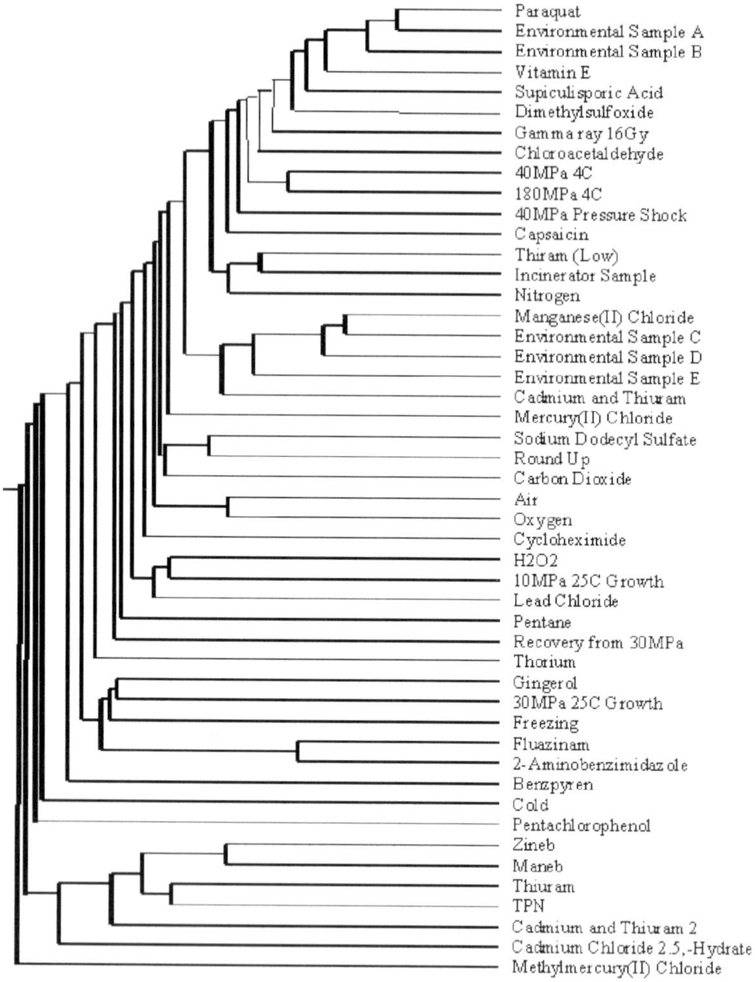

Fig. 25.1 Cluster analysis of the various environmental stress treatments to yeast cells. Each stress conditions was described previously (see text)

We may select calculation methods and the calculations were mainly based on the Euclidean distance (distance between the treatments) or Pearson CC (direction or angle between the treatments). In any calculation what we can obtain is the similarities among the expression profiles after stress treatments. Expression profiles reflect the effect of stress on cells and the effect must be specific to the stress treatments. We may speculate that clustering represents similar responses among stresses. Correlation factors for the expression profiles must be high among chemicals that cause similar damages or responses. For example, zineb, maneb, and thiuram belong to dithiocarbamate fungicides and they have similar chemical

Table 25.1 List of induced genes in the cells of recovery condition after exposure yeast cells to 40 MP of hydrostatic pressure for 16 h

Systematic name	Fold of induction	T-test P-value	Common name	Description from MIPS
YER103w	13.1	0.001	SSA4	Heat shock protein of HSP70 family, cytosolic
YFL014w	7.0	0.000	HSP12	Heat shock protein
YLR216c	5.6	0.002	CPR6	Member of the cyclophilin family
YGR142w	5.1	0.009	BTN2	Gene/protein is elevated in a btn1 mutant
YBR067c	4.8	0.018	TIP1	Esterase
YBR072w	4.5	0.038	HSP26	Heat shock protein
YGR286c	4.0	0.015	BIO2	Biotin synthetase
YHR138c	3.9	0.003		Protein involved in vacular fusion
YER142c	3.9	0.008	MAG1	3-Methyladenine DNA glycosylase
YNL274c	3.8	0.009		Putative hydroxyacid dehydrogenase
YMR002w	3.6	0.009		Unknown localised to cytoplasm and nucleus
YDR059c	3.4	0.001	UBC5	E2 Ubiquitin-conjugating enzyme
YDL100c	3.4	0.012	ARR4	Involved in resistance to heat and metal stress
YER143w	3.3	0.008	DDI1	Induced in response to DNA alkylation damage
YOR007c	3.3	0.054	SGT2	Glutamine-rich cytoplasmic protein
YJL026w	3.2	0.029	RNR2	Ribonucleoside-diphosphate reductase
YER012w	3.2	0.019	PRE1	20S Proteasome subunit C11(beta4)
YER004w	3.2	0.009		Found in Mitochondrial Proteome
YJL001w	3.2	0.004	PRE3	20S Proteasome subunit (beta1)
YDL007w	3.2	0.002	RPT2	26S Proteasome regulatory subunit
YGR037c	3.1	0.017	ACB1	Acyl-coenzyme-A-binding protein
YLR303w	3.0	0.025	MET17	O-Acetylhomoserine sulfhydrylase

structures (Kitagawa et al. 2003). Thus, these chemicals were expected to cause similar damages or similar cellular response (expression profile). As shown in Fig. 25.1, Zineb, Mneb, and Thiuram cluster and we may conclude that these chemicals cause similar damages or responses. Thus, cluster analysis allows us to understand the effect of stress on cells.

Table 25.1 shows the list of induced genes after the stress of "Recovery from 40 MPa 4C" treatment. This information is useful to understand the damage to cells by the stress treatment and the scavenging mechanism used to decrease stress. The stress of "Recovery from 40 MPa 4C" treatment seems to induce genes of *HSP*s, proteasome, and ubiquitin. We may then speculate that pressure effects protein metabolism. The list of highly induced and repressed genes can help to identify candidate biomarkers and strongly induced functions, however this may reflect only a small part of the full stress response.

Table 25.2 Functional categories of Induced genes by high pressure stress

Category and subcategory	Entry*	Induced*	Frequency*	Share*
Metabolism**				
Amino acid metabolism**	243	9	3.7	7.4
Nitrogen and sulfur metabolism	96	3	3.1	2.5
Nucleotide metabolism	227	4	1.8	3.3
Phosphate metabolism	414	14	3.4	11.5
C-compound and carbohydrate metabolism	504	14	2.8	11.5
Lipid, fatty acid and isoprenoid metabolism	272	6	2.2	4.9
Metabolism of vitamins	163	4	2.5	3.3
Secondary metabolism	77	1	1.3	0.8
Energy	365	13	3.6	10.7
Cell cycle and DNA processing	1001	14	1.4	11.5
Transcription	1063	6	0.6	4.9
Protein synthesis	476	4	0.8	3.3
Protein fate	1137	58	5.1	47.5
Protein with binding function	1034	45	4.4	36.9
Protein activity regulation	238	7	2.9	5.7
Cellular transport	1031	30	2.9	24.6
Cellular communication	234	1	0.4	0.8
Cell rescue, defense and virulence	548	32	5.8	26.2
Interaction with the cellular environment	458	8	1.7	6.6
Interaction with the environment	5	0	0.0	0.0
Transposable elements	124	0	0.0	0.0
Development (systemic)	70	1	1.4	0.8
Biogenesis of cellular components	854	11	1.3	9.0
Cell type differentiation	449	7	1.6	5.7
Unclassified proteins	2038	11	0.5	9.0

* Entry: number of genes grouped in the category or subcategory
Induced: number of genes induced by stress
Frequency: percentage of induced genes in the category
Share: percentage of induced genes of category in total induced gene number
** Capitals are functional category and lower case is subcategory

Table 25.2 shows the functional categories of induced genes after the stress of "Recovery from 40 MPa 4C" treatment. These categories were configured according to the functions of each gene by MIPS (Munich Information Center for Protein Sequences). There were 120 genes that were induced more than twofold by the "Recovery from 40 MPa 4C" treatment (Iwahashi et al. 2003) and it is not easy to factor out the meaning of the induction of 120 genes. Making a list of functional categories help us to understand the induced function for the stress treatment. Table 25.2 shows that the categories of "protein fate" and "cell rescue, defense and virulence" were significantly activated and priority of the categories of "protein fate", "cell rescue, defense and virulence", "protein with binding function", and

Table 25.3 Induced genes by stress in subcategories of proteinfate

Subcategory	Entry*	Induced*	Frequency*	Share*
Protein folding and stabilization	91	12	13.2	9.8
Protein targeting, sorting and translocation	277	8	2.9	6.6
Protein modification	606	28	4.6	23.0
Modification with fatty acids	30	0	0.0	0.0
Modification with sugar residues	68	0	0.0	0.0
Modification by phosphorylation	186	1	0.5	0.8
Modification by acetylation, deacetylation	69	0	0.0	0.0
Modification by ubiquitination, deubiquitination	77	6	7.8	4.9
Modification by ubiquitin-related proteins	20	0	0.0	0.0
Posttranslational modification of amino acids	24	1	4.2	0.8
Protein processing (proteolytic)	88	19	21.6	15.6
Assembly of protein complexes	196	7	3.6	5.7
Protein degradation	250	40	16.0	32.8
Cytoplasmic and nuclear protein degradation	186	31	16.7	25.4
Lysosomal and vacuolar protein degradation	23	2	8.7	1.6

* Entry: number of genes grouped in the category or subcategory
Induced: number of genes induced by stress
Frequency: percentage of induced genes in the category
Share: percentage of induced genes of category in total induced gene number

"cellular transport" were also significant among the induced genes. This information helps us to understand the induced functions of 120 kinds of genes and also suggests that pressure treatment affects protein metabolism and transport. We may also focus on subcategories within the main categories. The subcategories were also configured according to the functions of each gene by MIPS. Table 25.3 was focused on the subcategory in the category of "protein fate". In the "protein fate" category, it is clear that "protein degradation" is significantly activated. This suggests that high pressure stress caused protein denaturation, necessitating upregulation of protein degradation pathways.

25.3.2 *Proteomics Technology for Protein Modifications*

Figure 25.2 shows the two-dimensional (2-D) gel electrophoresis results after heat shock treatment (43°C for 2 h). The 2-D gels clearly show that the two conditions produce different patterns of protein expression. However this result came from the

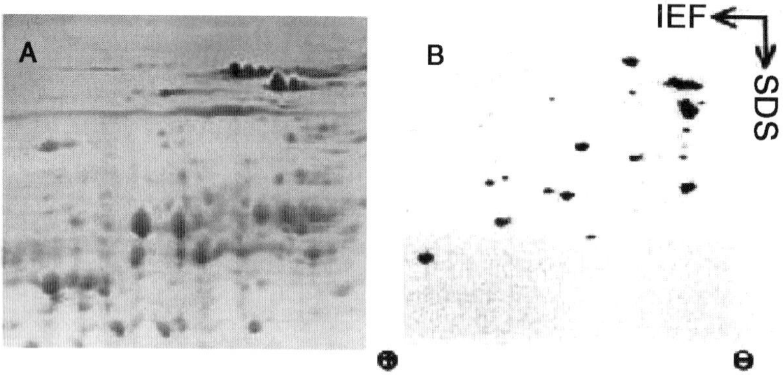

Fig. 25.2 Two-dimensional gel electrophoresis after heat shock treatment to yeast cells. Experimental conditions were described previously (Iwahashi, 2006)

same conditions of the heat shock treatment (43°C for 2h). Figure 25.2A is the CBB stained image and Fig. 25.2B is the tritium labeled autoradiography image. The CBB stained proteins represent newly synthesized protein as well as non-degraded proteins while the autoradiography shows only newly synthesized proteins. Thus, in the sample of yeast cells, especially after stress treatment, new and old proteins can be observed. The stress treatment is generally carried over a short period and the protein degradation is not complete. This suggests that proteomics methods alone are insufficient for expression analysis of yeast cells after stress conditions. The proteomics of yeast must be focused on the modification of proteins. This limitation for yeast expression analysis suggests the need for preliminary evaluation of proteomics for each organism. We previously showed that rice is an appropriate organism for proteomics analysis (Agrawal et al. 2006; Jwa et al. 2006; Kersten et al. 2006).

25.3.3 Metabolomics Technology for Understanding Flows of Metabolites

Genomics and proteomics are mainly evaluation systems for induced functions but not the products of induced functions. In contrast, metabolomics is a system for evaluating substances as the products of induced functions. Metabolomics can yield direct evidence of cellular stress. We are constructing a metabolomics system using CE/MS equipment. CE/MS was selected because this system is suitable for analysis of low molecular weight and ionic substances. The majority of metabolites are considered as ionic and small substances.

The extracted metabolites, which were injected hydrodynamically into the capillary inlet, migrated toward the MS instruments separately according to their electrophoretic mobilities. Thus we can monitor metabolites according to their ionic character and

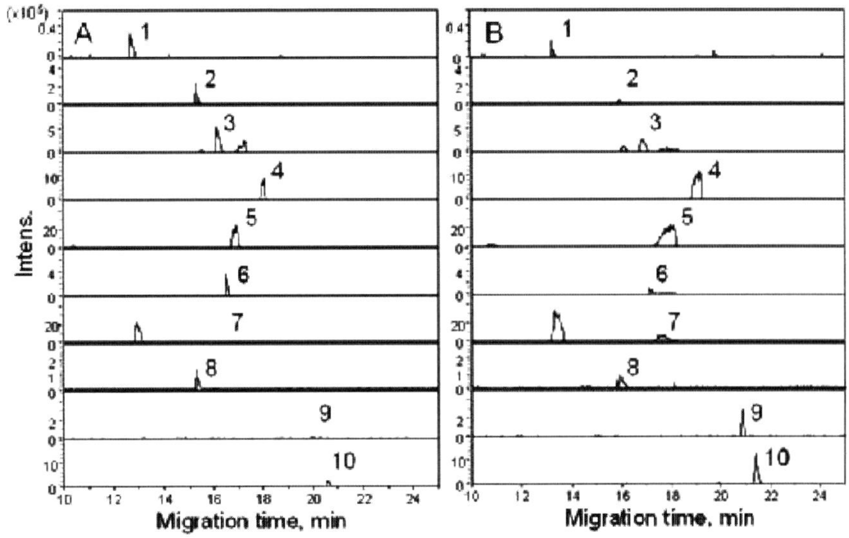

Fig. 25.3 Selected ion electropherograms of yeast extracts (A) without treatment (B) Cd stress-induced for 2h. (1) glycine (m/z=76), (2) serine (m/z=106), (3) threonine (m/z=120), (4) aspartate (m/z=134), (5) glutamate (m/z=148), (6) methionine (m/z=150), (7) O-acetyl-L-homoserine (m/z=162), (8) cystathionine (m/z=223), (9) γ-glutamylcysteine (m/z=251) (10) glutathione, reduced (m/z=308)

molecular weight. We applied 37 kinds of metabolites as candidates for analysis. These materials were selected according to the results obtained by genomics analysis. For example, we selected sulfur-containing metabolites as the stress treatment frequently induced genes related to sulfur amino acid metabolism. From group of compounds we could detect 17 metabolites in a cationic mode using a low pH electrolyte buffer. MS electropherograms of representative yeast metabolites in Cd stress-induced and control yeast extracts (2h) are shown in Fig. 25.3. Several metabolites (for example, L-homoserine and S-adenosyl-L-methionine) showed remarkable differences between control and stress-treated yeast cells. These observations were consistent with the results of genomics analysis (Momose and Iwahashi 2001). Thus, metabolomics can be combined with genomics. The needs in metabolomics are number of identified metabolites. Several dose of metabolites is not enough for the evaluation of environmental stress response to combine with genomics.

25.3.4 OMICS as the Tools of Environmental Monitoring for Chemicals, Radiation, and Physical Stresses

Now it is possible to monitor the entire yeast genome and transcriptome as well as many proteins and some kinds of metabolites. These omics systems can be combined to provide a new approach to environmental monitoring for chemicals, radiation, and physical stresses. For this integrated system to be robust we have to develop improved detection systems for modified proteins in proteomics and increase the number of detectable metabolites in metabolomics. These studies are currently underway.

References

Agrawal G.K., Jwa N.S., Iwahashi Y., Yonekura M., Iwahashi H., and Rakwal. R. (2006), Rejuvenating rice proteomics: Facts, challenges, and visions. *Proteomics,* 6, 5549–5576.

Celemedson C., et al. (44Scientists) (1996), MEIC Evaluation of acute systemic toxicology. *ALTA*, 24, 252–272.

Fujita K., Matsuyama A., Kobayashi K., and Iwahashi H. (2004),Comprehensive gene expression analysis of the response to straight-chain alcohols in *Saccharomyces cerevisiae* using cDNA microarray. *J. Appl. Microbiol.*, 97, 57–67.

Iwahashi H. (2006), Yeast genes that star in the cross protection against environmental stress, *Cryobiol. Cryotech.*, 52, 55–59.

Iwahashi H. (2000), Multiple-end-point bioassay using microorganisms. *Biotech. Biop. Eng.*, 5, 400–406.

Iwahashi H., Yang W., and Tanguay R.M. (1995), Detection and expression of the 70kDa heat shock protein ssbp1 at different temperature in the yeast *Saccharomyces cerevisiae. Biochem. Biophys. Res. Comm.*, 213, 484–489.

Iwahashi H., Shimizu H., Odani M., and Komatsu Y. (2003), Piezophysiology of genome wide gene expression levels in the yeast *Saccharomyces cerevisiae. Extremophile*, 7, 291–298.

Iwahashi H., Odani M., Ishidou E., and Kitagawa E. (2005), Adaptation of *Saccharomyces cerevisiae* to high hydrostatic pressure causing growth inhibition *FEBS Lett.*, 579, 2847–2852.

Jwa N.S., Agrawal G.K., Tomogami M., Yonekura M., Hane O., Iwahashi H., and Rakwal. R. (2006), Defense/Stress-related marker genes, proteins and secondary metabolites in defining rice self-defense mechanisms, *Plant Physiol. Biochem.*, 44, 261–273.

Kersten B., Agrawal G.K., Iwahashi H, and Rakwal R. (2006), Plant phosphoproteomics: A long road ahead. *Proteomics.*, 6, 5517–5528.

Kim H., Ishidou E., Kitagawa E., Momose Y., and Iwahashi H. (2004), Yeast DNA microarray for the evaluation of the toxicity in environmental water containing the burned ash. *Env. Monit. Assess.*, 92, 253–272.

Kimura S., Ishidou E., Kurita S., Suzuki Y., Shibato J., Rakwal, R., and Iwahashi H. (2006), DNA microarray analyses reveal a post-irradiation differential time-dependent gene expression profile in yeast cells exposed to X-rays and γ-rays. *Biochem. Biophys. Res. Comm.*, 346, 51–60.

Kitagawa E., Takahashi J., Momose Y., and Iwahashi H. (2002), The effects of the pesticide thiuram: Genome-wide screening of indicator genes by yeast DNA microarray. *Environ. Sci. Technol.*, 36, 3908–3915.

Kitagawa E., Momose Y., and Iwahashi H. (2003), Correlation of the structures of agricultural fungicides to gene expression in *Saccharomyces cerevisiae* upon exposure to toxic doses. *Environ. Sci. Technol.*, 15, 2788–2793.

Kurita S., Kitagawa E., Kim C.H., Momose Y., and Iwahashi H. (2002), Studies on the antimicrobial mechanisms of capsaicin using yeast DNA microarray. *Bios. Biotechnol. Biochem.,* 66, 532–536.

Kurita S., Shirisatta S., Kitagawa E., Momose Y., Ishigami Y., and Iwahashi H. (2004), New methods for the assessment of biological effect by surfactants using yeast DNA microarray. *J. Oleo Sci.,* 53, 387–398.

Matsuoka H., Suzuki Y., Iwahashi H., Arao T., Suzuki Y., and Tamura. K.(2005), The biological effects of high-pressure gas on the yeast transcriptome. *Braz. J. Med. Biol. Res.*, 38, 1267–72.

Momose Y., and Iwahashi H. (2001), Bioassay of cadmium using a DNA microarray: Genome-wide expression patterns of *Saccharomyces cerevisiae* response to cadmium. *Environ. Toxicol. Chem.*, 20, 2353–2360.

Murata S., Murata Y., and Iwahashi H. (2006a), Chemical toxicity of thorium in *Saccharomyces cerevisiae*. *Environ. Ecotoxicol.*, 9, 87–100.

Murata Y. Watanabe T., Sato M., Momose Y., Nakahara T., Oka S., and Iwahashi H. (2003), DMSO exposure facilitates phospholipid biosynthesis and cellular membrane proliferation in yeast cells. *J. Biol. Chem.*, 278, 33185–33193.

Murata Y., Mizukami S.M., Kitagawa E., Iwahashi H., and Takamizawa K. (2006b),The evaluation of environmental waters using yeast DNA microarray. *Chemo-Bioinform. J.*, 6, 29–46.

O'Farrell P.H. (1975), High resolution two-dimensional electrophoresis of proteins. *J. Biol. Chem.*, 250, 4007–4021.

Odani M., Komatsu Y., Oka S., and Iwahashi H. (2003), Screening of genes that respond to cryopreservation stress using yeast DNA microarray. *Cryobiology*, 47, 155–164.

Reifferscheid. G. and J. Heil (1996), Validation of the SOS/umu test using test results of 486 chemicals and comparison with the Ames test and carcinogenicity data. *Mutat. Res.*, 369, 129–145.

Sato S., Soga T., Nishioka T., and Tomita M. (2004), Simultaneous determination of the main metabolites in rice leaves using capillary electrophoresis mass spectrometry and capillary electrophoresis diode array detection. *Plant J.*, 40, 151–163.

Sirisattha S., Momose Y., Kitagawa E., and Iwahashi H. (2004a), Toxicity of anionic detergents determined by *Saccharomyces cerevisiae* microarray analysis. *Water Res.*, 38, 61–71.

Sirisattha S., Momose Y., Kitagawa E., and Iwahashi H. (2004b), Genomic profile of Roundup treatment of yeast using DNA microarray analysis. *Environ. Sci.*, 11, 313–323.

Soga T., Ohashi Y., Ueno Y., Naraoka H., Tomita M., and Nishioka T. (2003), Quantitative metabolome analysis using capillary electrophoresis mass spectrometry. *J. Proteome Res.*, 2, 488–494.

Suzuki M., and Utsumi H. (eds.) (1998), *Bioassay for the control of chemicals*. Pp. III-IV (Tokyo: Koudansya Press).

Chapter 26
Gene Expression Characteristics in the Japanese Medaka (*Oryzias latipes*) Liver after Exposure to Endocrine Disrupting Chemicals

Han Na Kim[1], Kyeong Seo Park[1], Sung Kyu Lee[2], and Man Bock Gu[*]

Abstract Endocrine disrupting chemicals (EDCs) are of concerning chemicals due to their ability to make damage or alter hormonal activities in living organisms. In this study, therefore, toxicogenomic analyses using a real time RT-PCR technique have been conducted to characterize the responses of male Japanese Medaka and to provide valuable information about the toxicological properties and their hazardous effects of three EDCs, i.e., 17-beta estradiol (E_2), nonylphenol (NP) and bisphenol A (BPA), to Japanese medaka fish. For doing that, the expression kinetics of female-related genes and other cellular toxicity representative genes, including vitellogenin (yolk protein precursor), choriogenin L (inner membrane precursor of egg), cytochrome P450 1A (CYP1A) and heat shock protein 70 (HSP70) were used. Gene expression levels at three different times after exposure with two different concentrations of EDCs were quantified by measuring messenger RNA (mRNA) concentrations in the liver extracts using the Taqman based real-time PCR method. The results showed that E_2 causes a strong estrogenic effect even at a concentration of only 1 ppb, while NP and BPA were found to cause some form of cellular toxicity, and an estrogenic effect with inducing the production of HSP 70, respectively.

Keywords: Choriogenin L, endocrine disrupting chemicals, *Oryzias latipes*, real time PCR, vitellogenin, 17-beta estradiol (E_2), bisphenol A (BPA), cytochrome P450 1A (CYP1A), heat shock protein 70 (HSP 70), nonylphenol (NP)

[1] *National Research Laboratory on Environmental Biotechnology, Gwangju Institute of Science and Technology (GIST), Gwangju 500–712, Korea*

[2] *Environmental Toxicology Devision, Korea Institute of Toxicology, 100 Jangdong, Yuseong, Daejeon, 305–343, Korea*

[*] *Corresponding author School of Life Sciences and Biotechnology, Korea University, Seoul 136–701, Korea Tel: + 82-2-3290-3417, Fax: + 82-2-928-6050, e-mail:mbgu@korea.ac.kr*

Y.J. Kim and U. Platt (eds.), *Advanced Environmental Monitoring*, 338–347.

26.1 Introduction

The use of xenotoxic chemicals has grown considerably, with approximately 100,000 compounds currently being produced on an industrial scale, while about 2000 new chemical species are introduced and released into the environment each year (Younes 1999). Some of the attributable effects of these environmental pollutants include a reduced fertility, hatchability and viability of exposed off-spring, as well as impaired hormone activity and altered sexual behavior. Such effects are thought to be long term, irreversibly affecting progeny. Among the various effects documented, endocrine disruption refers to the alteration of the normal endocrine system by chemicals that mimic hormones, or compounds that alter the synthesis, metabolism and activity of native hormones (Mclachlan 2005). Commonly reported endocrine disrupting chemicals (EDCs) in the environment that can trigger reproductive problems in fish include natural sex hormones estrogens (estradiol and estrone), and non-ionic surfactant degradation products, including nonylphenol (NP) and octylphenol (OP), and many others, such as the industrial compound bisphenol A, the contraceptive drug ethinylestradiol, fungicides, pesticides, organohalogens, heavy metals, phthalates, aromatic hydrocarbons, effluent from sewage treatment plants and run-off from animal farms (Ying and Kookana 2002).

When exposed to EDCs, fish show various toxic responses, including estrogenic potential-related effects in the liver or reproductive organs, which are the main target tissues for EDCs. Among the documented effects of EDCs, the most typical one is their ability to mimic the natural female hormone, estrogen. It is reported that the exposure of estrogenic substances can cause decreased sperm counts in males, disordered fetal development, impaired reproductive capabilities and behavior, and excessive cell proliferation promoting carcinogenesis (Birnbaum and Fenton 2003). The well-known actions of estrogenic substances are mediated through their binding to the estrogen receptor (ER) and the subsequent activation of the estrogen responsive element (ERE), which is involved in the regulation of genes (Janosek et al. 2006). It is now recognized that estrogen results in the synthesis of specific proteins required for reproduction through ERE activation. Several genes that encode proteins induced by this process are the vitellogenins (Vtg), egg yolk precursor proteins, and choriogenins (Chg), which are required for making the egg membrane (Flouriot et al. 1996; Lim et al. 1991; Murata et al. 1997). However, EDCs can also have effect on organisms via non-receptor mechanisms, including the modulation of the levels or activities of enzymes participating in biosynthesis or catabolism of estradiol, such as the cytochrome P450 family of enzymes that are involved in metabolism of chemicals (Janosek et al. 2006; Machala and Vondracek 1998). There are some studies that tested the effect of EDCs on fish using only ERE mediating gene expression, such as the ER and Vtg genes (Yamaguchi et at. 2005). However, owing to their diverse structure and elemental composition, EDCs are likely to cause broader toxic effects, not just endocrine disruption. Therefore, to investigate the effects different EDCs have on an organism fully, various biomarker genes should be utilized, including indicators of estrogenic effects.

Consequently, to better understand some of the hazardous effects caused by an exposure to three EDCs (17-beta estradiol (E_2), 4-branched nonylphenol (NP) and bisphenol A (BPA)), respectively, the expression levels of five different genes in Japanese Medaka were examined. These genes as biomarkers were selected based upon their known responses and include vitellogenin (Vtg), choriogenin-L (Chg-L), cytochrome P450 1A (CYP 1A) and heat shock protein 70 (HSP 70). Vtg and Chg-L are well known indicators for estrogenic effects. As well, cytochrome P450 1A (CYP1A) is a representative biomarker of the biotransformation and detoxification of xenobiotic compounds. In addition, heat shock protein 70 is a molecular chaperone that is synthesized under different stressful conditions. The expression levels of each biomarker after exposure to each EDC were determined by RNA quantification using the Taqman probe-based real-time PCR method (Bustin 2002).

26.2 Materials and Methods

26.2.1 Culture Method and Chemical Exposure

Medaka fish (*Oryzias latipes* d-rR strain, 5 months olds) of the orange-red variety were maintained at the Environmental Toxicology laboratory of the National Institute of Environmental Research in Korea. They were reared under constant 18-h light: 6-h dark cycles and a temperature of 25 ± 1°C. The Medaka were acclimated for a week in clean local tap water that was dechlorinated with air at least for a week prior to introducing the fish. These fish were maintained under the same light/dark cycles and a temperature range.

Male Medaka were separated from the female fish. From these, twenty male Medaka were exposed to the test chemicals in 10-L glass beakers containing 5 L of the dechlorinated local tap water. Each test group was exposed to one of the following: 1 or 100 μg/L (LC20) of 17-beta estradiol, 7.5 or 75 μg/L (LC20) of 4-branched nonylphenol or 7.5 or 75 μg/L (LC20) of bisphenol A for 4 days. All of the chemicals that were used in this study were purchased from the Sigma Aldrich Chemical Corporation (USA). These chemicals were dissolved in pure ethanol and added to the aquarium water so that the solvent concentration was 0.01% [v/v]. Groups of control fish were exposed to the same concentration of solvent alone. Air was bubbled into the water through an air stone. To avoid metabolic and microbial breakdown of the chemicals, 50% of the water was removed every day and replaced with fresh water containing the test chemical or solvent. During the chemical exposure, at each sampling time, three fish from each test and control group were randomly selected and were dissected to extract the liver. The sampling times were 1 day, 2 days and 4 days after initiating the exposure. The extracted livers were collected in a sterile 1.8 mL effendorf tube and were immediately frozen in liquid nitrogen and stored at −80°C until the RNA could be isolated, which was always less than a week after the livers were extracted.

26.2.2 Primers and Probes

Specific primers and Taqman probes for the Medaka Vtg, Chg-L, CYP1A and HSP 70 genes were designed and synthesized (Applied Biosystems, USA) taking into consideration several factors that can affect PCR efficiency, such as homology with other genes, amplicon size, melting temperature, secondary structure, and so on. The Taqman probes contain a reporter dye (6-FAM) linked to the 5′ end of the probe and a non-fluorescent quencher (NFQ) at the 3′ end of the probe. The primers and probes for each gene are listed in Table 26.1. For an endogenous control gene, the primers and probe for the 18S rRNA (pre-developed Taqman assay reagent) were purchased from Applied Biosystems (USA).

Table 26.1 PCR primers and TaqMan probes used in the analysis of the Chg-L, CYP1A, VTG and HSP 70 expression levels

Gene name	Accession number	Description	Sequence
Chg-L	AF396667	Forward primer of choriogenin L	5′-CCTGGTCTACACCTTCACTCTGA-3′
		Reverse primer of choriogenin L	5′-TGACATTCCACGATAACAACAGCTT-3′
		TaqMan probe of choriogenin L	5′-FAM-CTGGGCAGTGCCCCTGT-NFQ-3′
CYP1A	AY297923	Forward primer of cytochrome P450 1A	5′-CGCAGAAAGTTGGCCTACAGT-3′
		Reverse primer of cytochrome P450 1A	5′-TCTGCATTGCTGCCCTCTAG-3′
		TaqMan probe of cytochrome P450 1A	5′-FAM-CATTGCGCTCTTTCTC-NFQ-3′
VTG	AF268284	Forward primer of vitellogenin	5′-CACCCGTCTCTGCTGAGT-3′
		Reverse primer of vitellogenin	5′-TGAAGTGGTGAGAGCTCAAACTC-3′
		TaqMan probe of vitellogenin	5′-FAM-CATCATCGTGGATCTCTC-NFQ-3′
HSP 70	D13669	Forward primer of heat shock protein 70	5′-GGGCACGTTTTGAGGAGCTTA-3′
		Reverse primer of heat shock protein 70	5′-GAAGTGACTTCTCCACAGGATCAAG-3′
		TaqMan probe of heat shock protein 70	5′-FAM-ACGCAGACCTTTTCCG-NFQ-3′

The accession number refers to the registered sequence used from gene bank

RNA extraction and real time RT-PCR

Total RNA was isolated from the frozen livers using the RNeasy Mini kit (Qiagen, USA). The RNA quality and concentration was determined by measuring the absorption at 260 nm using a UV/VIS spectrophotometer (Perkin-Elmer Co., USA). The total RNA was diluted to the same concentration for each sample using ultra-pure nuclease-free water.

For each diluted RNA sample, 10 ng/μL was used for first-strand cDNA synthesis using a TaqMan Reverse-Transcription kit, following the RT-PCR manufacturer's two-step protocol (Applied Biosystems, USA). The conditions of the final reaction for reverse-transcription were as follows: 1× TaqMan RT buffer; 5.5 mM $MgCl_2$; 500 μM dATP, dGTP, and dCTP; 1 mM dTTP; primers (oligo $d(T)^{16}$ or random hexamer) 2.5 μM; 1.25 U/μL reverse transcriptase and 0.4 U RNase inhibitor (Applied Biosystems, USA). Reverse transcription for all samples was achieved using oligo $d(T)^{16}$ and Reverse transcriptase with the following thermal cycling conditions: 25°C for 10 min, 48°C for 30 min, and 95°C for 5 min. These cDNA samples were used for real time PCR of the five genes. On the other hand, synthesis of the cDNA from 18S rRNA was primed using random hexamer primers. The thermal cycling for the reverse transcribing of the 18S rRNA was (i) 10 min at 25°C, (ii) 60 min at 37°C and (iii) 5 min at 95°C. After reverse transcription, all cDNA samples were either used directly for real time PCR or were stored at −20°C until needed.

After the RT step, real-time PCR was performed using the DNA polymerase from the Taqman universal PCR master mix (Applied Biosystems, USA). Each target cDNA synthesized from the RNA sample was amplified using sequence-specific primers and Taqman probes. The PCR step was performed on an ABI PRISM 7000 Sequence Detection System (Applied Biosystems, USA) that has the ability to detect and record the fluorescent signals generated by the cleavage of Taqman probes. Amplification was performed with the following thermal cycling conditions: (i) 2 min at 50°C, (ii) 10 min at 95°C, (iii) 45 cycles at 95°C for 15 s and (iv) 1 min at 60°C. The results were analyzed using ABI Sequence Detector software version 1.1 (Applied Biosystems, USA).

For each target gene, samples were run with serially diluted cDNA of known quantity to construct a standard curve. The quantity of each target RNA was then determined using these standard curves based on their measured Ct (threshold cycle) values. Likewise, the level of the 18S rRNA was used as an endogenous control to normalize the variability in the RNA quality and quantity or differences in the efficiency of the PCR reaction among the samples.

26.3 Results and Discussions

The expression levels of five biomarker genes show different expression pattern after exposure to 17-beta estradiol (E_2) (Fig. 26.1). The most conspicuous results were from the vitellogenin (Vtg) and choriogenin-L (Chg-L) genes. Among other

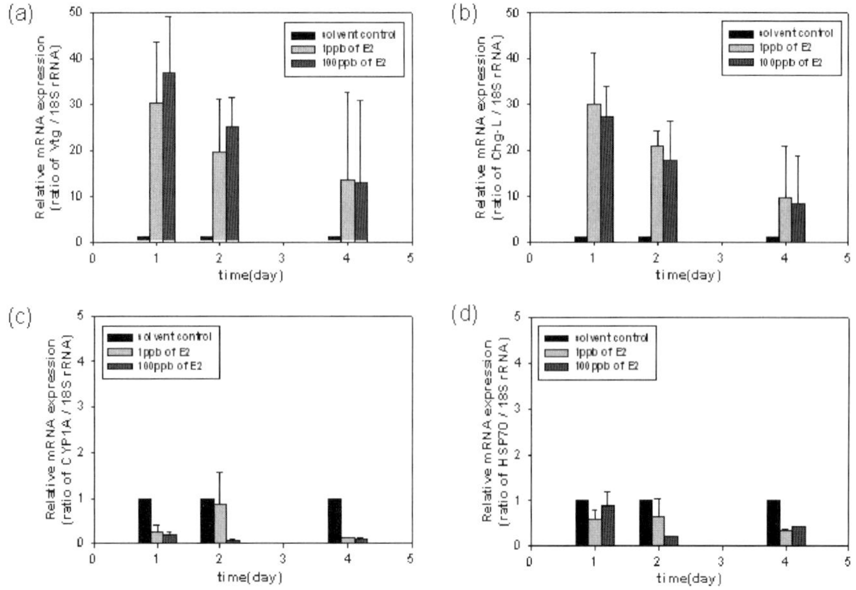

Fig. 26.1 Expression patterns of vitellogenin **(a)**, choriogenin-L **(b)**, cytochrome P450 1A **(c)** and heat shock protein 70 **(d)** in the liver of male Medaka after exposure to 1 μg/L or 100 μg/L 17-beta estradiol (black bar: solvent control, light gray: 1 μg/L, dark gray bar: 100 μg/L). All data were plotted relative to the 18S rRNA expression levels

gene expressions, CYP 1A and HSP 70 are decreased by E_2 exposure. Vtg is an egg yolk precursor protein that is synthesized primarily in the liver of females. Its expression is regulated by exposure to estrogenic substances and is normally dormant in males. In this study, a significant induction in the Vtg expression was seen after exposure to E_2 (Fig. 26.1a). Furthermore, the quantity of Vtg RNA present within the liver increased but the increasing rate decreased in a time dependent manner as the exposure progressed, with a significant difference seen only one day after the exposure was initiated. From these results, it is clear that the effect 17-beta estradiol has on the expression of Vtg is both abrupt and persistent. Changes in the Vtg expression levels were similar when the fish were exposed to either 100 or 1 μg/L of E_2. The slight difference in the fold-induction for the two concentrations implies that E_2 has a strong estrogenic effect at very low concentrations. Furthermore, these results indicate that the estrogenic response is near its maximum with only 1 μg/L E_2. This result is reasonable since hormones are normally functional at very low quantities in the organism. For example, estrogen is typically in the range of 100–150 pg per milliliter in human females (Gallicchio et al. 2005).

Chg-L is a precursor protein of the egg envelope is synthesized in the spawning female liver in response to estrogen and is found to more a sensitive gene than Chg-H for detecting estrogen (Lee et al. 2002). The Chg-L expression pattern, after exposure to E_2, is similar with that of Vtg (Fig. 26.1b). Chg-L was also highly expressed

1 day after exposure and continued to maintain a high level of expression (up to tenfold) during the test, although the increasing rate deceased as the exposure lasted. Likewise, Chg-L expression was strongly induced by 1 μg/L E_2, demonstrating the Chg-L is also sensitive to E_2.

CYP 1A is a representative biomarker for the biotransformation and detoxification of xenobiotic compounds. The results in Fig. 26.1 show that CYP 1A expression tended to decrease as time passed for both concentration of E_2 tested (except for 2 day result at lower concentration). Navas and Segner (2001) found that the ER is involved in the suppressive action of E_2 on CYP1A expression. Lower CYP1A production levels can result in deleterious side effects within the fish since the corresponding protection against harmful xenobiotics, via their metabolism by CYP 1A, would also be reduced.

HSP 70 is a molecular chaperone that is synthesized under various stressful conditions. In this study, the HSP 70 mRNA level in the Medaka liver gradually decreased after exposure to both concentrations of E_2 (Fig. 26.1d). Although these results indicate that E_2 does not cause any heat shock response within Medaka, the mechanism by which it influences many of these genes still needs to be elucidated.

The results after medaka fish were exposed to nonylphenol (NP) are shown in Fig. 26.2. Whereas the strongest response was seen with the CYP1A gene, the most sensitive responses were seen from Vtg. Furthermore, Vtg also gave the quickest

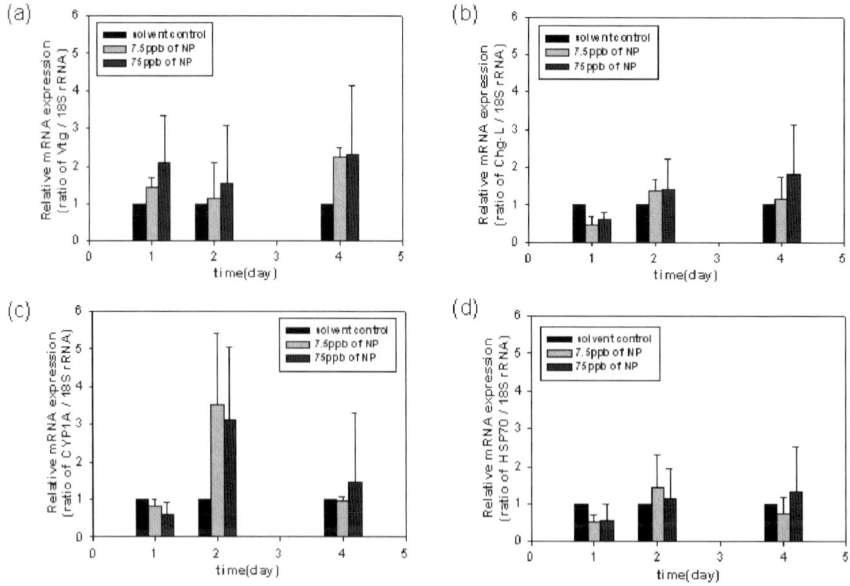

Fig. 26.2 Expression patterns of vitellogenin **(a)**, choriogenin-L **(b)**, cytochrome P450 1A **(c)** and heat shock protein 70 **(d)** in the liver of male Medaka after exposure to 7.5 μg/L or 75 μg/L nonylphenol. (black bar: solvent control, light gray: 7.5 μg/L, dark gray bar: 75 μg/L). All data were plotted relative to the 18S rRNA expression levels

response, with more than a two fold induction after only a one day exposure to 75 µg/L NP. These results demonstrate that NP has an estrogenic effect within Medaka, albeit a much weaker one than E_2. Another difference between the responses to E_2 and NP is a little response from the Chg-L gene with NP, i.e., significant responses were seen with E_2 one day after initiating the exposure, but NP had even decreasing effect on its expression at the first day. Significant increases in the CYP 1A and HSP 70 expression levels were seen after 2 days when the Medaka were exposed to 75 µg/L of NP. Induction of CYP 1A implies that NP caused some form of cellular toxicity in male Medaka since this protein has a significant role in biotransformation and/or detoxification. In HSP 70 expression, there is no significant effect on NP exposure.

The responses of male Medaka to bisphenol A (BPA) were similar with those from an exposure to NP. Vtg was induced in a dose-dependent manner, with a significant induction seen after 2 days for 75 µg/L concentration tested (Fig. 26.3). As with NP, it appears that the estrogenic effect elicited by BPA is weaker than that of E_2. The expression pattern of Chg-L is similar with those of Vtg except but the less induction level. In CYP 1A expression, there is no significant effect on BPA exposure. HSP 70 was induced only slightly by the presence of 75 µg/L BPA, but this induction was maintained, suggesting that high concentrations of BPA cause some degree of stress.

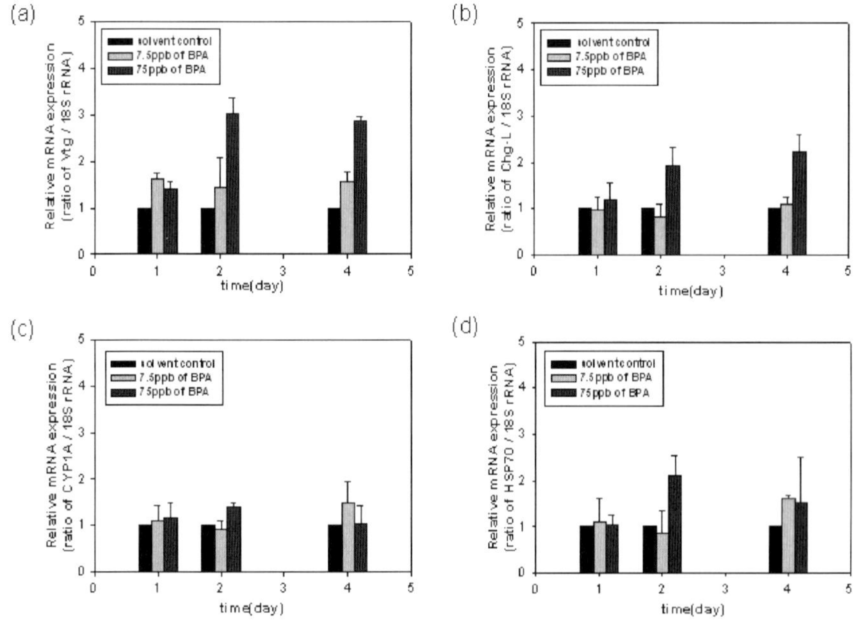

Fig. 26.3 Expression patterns of vitellogenin **(a)**, choriogenin-L **(b)**, cytochrome P450 1A **(c)** and heat shock protein 70 **(d)** in the liver of male Medaka after exposure to 7.5 µg/L or 75 µg/L of bisphenol A. (black bar: solvent control, light gray: 7.5 µg/L, dark gray bar: 75 µg/L). All data were plotted relative to the 18S rRNA expression levels

In this study, the effects of three different EDCs (17-beta estradiol, nonylphenol and bisphenol A) were analyzed by measuring the expression levels of specific biomarkers within Japanese Medaka. The biomarkers used were the genes encoding for vitellogenin, choriogenin L, cytochrome P450 1A and heat shock protein 70. Each biomarker differed in its expression pattern depending on the chemical being tested, and these differences were used to explain the characteristic toxicity signature of each compound. All three of the chemicals tested are known EDCs that have estrogenic effects within male Medaka (Yamaguchi et al. 2005). Therefore, the Vtg and Chg-L genes were selected to investigate and compare the estrogenic activities of each chemical. The results of this study clearly show that E_2 causes the strongest estrogenic effect among the compounds tested, while NP and BPA were similar in their estrogenic activities. Interestingly, exposure to E_2 leads to decreased CYP1A and HSP 70 expression, a result not seen with NP or BPA. CYP 1A is specifically responsive to NP, suggesting that NP causes some form of cellular toxicity that is modulated by CYP 1A. Similarly, HSP 70 genes were induced after exposure to NP and BPA, indicating that these compounds have some other effects on the liver cells not only just an estrogenic effect.

In summary, through real time PCR, the expression levels of five genes were quantified and it was found that each biomarker was differentially expressed according to the exposure time and concentration of the EDCs (E_2, NP and BPA) tested. From these results, the response characteristics of male Japanese Medaka were successfully investigated and analyzed.

Acknowledgements Authors are grateful to the Laboratory of Freshwater Fish Stocks within the Bioscience Center, Nagoya University (Chikusa, Nagoya, Japan) for their support in the preparation of the Japanese Medaka stocks. This work has been financially supported by the ECO project of the Korean Ministry of Environment, and authors appreciate the support.

References

Birnbaum L.S. and Fenton S.E. (2003), Cancer and developmental exposure to endocrine disruptors, *Environ. Health Persp.*, 111, 389–394.

Bustin S.A. (2002), Absolute quantification of mRNA using real-time reverse transcription polymerase chain reaction assays, *J. Mol. Endocrinol.*, 25, 169–193.

Flouriot G., Pakdel F., and Valotaire Y. (1996), Transcriptional and post-transcriptional regulation of rainbow trout estrogen receptor and vitellogenin gene expression, *Mol. Cellular Endocrinol.*, 124, 173–183.

Gallicchio L., Visvanathan K., Miller S.R., Babus J., Lewis L.M., Zacur H., et al. (2005), Body mass, estrogen levels, and hot flashes in midlife women, *Am. J. Obstet. Gynecol.*, 193, 1353–1360.

Janosek J., Hilscherova K., Blaha L., and Holoubek I. (2006), Environmental xenobiotics and nuclear receptors-interactions, effects and *in vitro* assessment. *Toxicol. In Vitro,* 20, 18–37.

Lee C., Na J.G., Lee K.C., and Park K. (2002), Choriogenin mRNA induction in male medaka, *Oryzias latipes* as a biomarker of endocrine disruption, *Aquat. Toxicol.*, 61, 233–241.

Lim E.H., Ding J.L., and Lam T.J. (1991), Estradiol-induced vitellogenin gene expression in a teleost fish, *Oreochromis aureus*. *Gen. Comp. Endocrin.*, 82, 206–214.

Machala M., and Vondracek J. (1998), Estrogenic activity of xenobiotics, *Vet. Med.*, 43, 311–317.

Mclachlan J.A. (2005), Environmental signaling: What embryos and evolution teach us about endocrine disrupting chemicals, *Endocrine Rev.*, 22 (3), 319–341.

Murata K., Sugiyama H., Yasumasu S., Iuchi I., Yasumasu I., and Yamagami K. (1997), Cloning of cDNA and estrogeninduced hepatic gene expression for choriogenin h, a precursor protein of the fish egg envelope (chorion), *Proc. Nat. Acad. Sci. U S A*, 94, 2050–2055.

Navas J.M., and Segner H. (2001), Estrogen-mediated suppression of cytochrome p4501a (cyp1a) expression in rainbow trout hepatocytes: Role of estrogen receptor. *Chem.-Biol. Interact.*, 138, 285–298.

Yamaguchi A., Ishibashi H., Kohra S., Arizono K., and Tominaga N. (2005), Short-term effects of endocrine-disrupting chemicals on the expression of estrogen- responsive genes in male medaka (*Oryzias latipes*), *Aquatic toxicology*, 72, 239–249.

Ying G.G., and Kookana R.S. (2002), Endocrine disruption: An Australian perspective. *J. Aust. Water Assoc.*, 29 (9), 42–45.

Younes M. (1999), Specific issues in health risk assessment of endocrine disrupting chemicals and international activities, *Chemosphere*, 39, 1253–1257.

Chapter 27
Optical Detection of Pathogens using Protein Chip

Jeong-Woo Choi[1,2] **and Byung-Keun Oh**[1,2]

Abstract Optical detection method based protein chips for detection of the various pathogens such as *Escherichia coli* O157:H7, *Salmonella typhimurium*, *Yersinia enterocolitica*, and *Legionella pneumophila* in contaminated environment were developed. In order to endow the orientation of antibody molecules on solid surface, protein G was introduced. Gold (Au) surface was modified with 11-mercaptoundecanoic acid (11-MUA) and the protein G was immobilized on the Au surface. And the spots of different antibodies against pathogens (*E. coli* O157:H7, *S. typhimurium*, *Y. enterocolitica*, and *L. pneumophila*) on protein G of Au surface were arrayed using a microarrayer. The responses of the various pathogens such as *E. coli* O157:H7, *S. typhimurium*, *Y. enterocolitica*, and *L. pneumophila* to the protein chip was investigated by surface plasmon resonance (SPR), fluorescence microscopy and imaging ellipsometry (IE). The lowest detection limit of the fluorescence based protein chip was 10^2 CFU/mL and the protein chip using IE could successfully detect the pathogens in concentrations varying from 10^3 to 10^7 CFU/mL.

Keywords: Fluorescence microscopy, imaging ellipsometry, protein chip, pathogen, protein G, surface plasmon resonance

27.1 Introduction

As the sequencing of the human genome project (HRP) has been finalized, biological research is entering a new era in which experimental focus will shift from identifying novel genes to determining the function of gene products

[1] *Department of Chemical and Biomolecular Engineering, Sogang University, #1 Shinsu-dong, Mapo-gu, Seoul 121–742, Korea; Tel: (+82) 2–705–8480, Fax: (+82) 2–3273–0331*

[2] *Interdisciplinary Program of Integrated Biotechnology, Sogang University, #1 Shinsu-dong, Mapo-gu, Seoul 121–742, Korea*

Y.J. Kim and U. Platt (eds.), *Advanced Environmental Monitoring*, 348–362.

(Pandey and Mann 2000). Rising to this challenge, several technologies have emerged that aim to characterize genes and/or proteins collectively rather than individually. Protein chip technology holds significant promise as a high-throughput platform for the structural and functional characterization of the components of the proteomes of humans, plants, animals, and microbes, etc. (Kelvin 2001; Wilson and Nock 2001). Protein chip is an array in which each protein occupies a defined spot on the chip. Such devices would be employed for highly parallel studies of the activities of native proteins and serve as an analytical tool somewhat analogous to DNA chips in the sense that it would be capable of monitoring protein levels in a given biological sample in a massively parallel fashion. Given the huge potential market for such devices, industrial interest in protein chip is very high. In order to develop the protein chips as the ideal proteomics-based analytical tool, it needs to be efficiently resolved the problems in the fabrication of protein chips, such as the ligand production, the proteins immobilization onto solid surface, and the monitoring of the proteins binding to the chips, etc. (Cahill 2001; Kodadek 2001).

The types of surfaces to which proteins can be immobilized fall into two categories. The first and simplest type of immobilization is physically adsorbed onto surfaces by van der Waals, hydrophobic and hydrogen-bonding interactions. The advantage of this type of immobilization is that it is very simple to perform due to not requirement of any modification of the protein. The disadvantage is that most of the immobilized protein can be inactivated due to denaturation and steric occlusion (Butler 1992). A preferred method relies on one or a small number of strong bonds between the protein and surface, leaving the protein largely unaltered, except in the vicinity of the contact point. For examples, it would be included the covalent attachment of protein, immobilization of biotinylated proteins onto streptavidin-coated surface, and immobilization of His-tagged proteins onto Ni^{2+}-chelating surface (Arenkov et al. 2000; Ruiz-Taylor et al. 2001; Zhu et al. 2001). However, due to this common methods for immobilizing proteins through (or biotin-based) interactions is by randomly conjugating lysine residues on proteins to amine-reactive surfaces (or biotinylation reagents), in order to construct the protein chips with high performance, it needs to be develop the immobilization techniques of protein in an oriented fashion (Choi et al. 2001, 2004; Vijayendran and Leckband 2001; Wilson et al. 2002). For the construction of a well-defined antibody surface, protein G, a cell wall protein found in most species of *Streptococci*, can be used as the binding material. Since protein G has a specific interaction with the F_c portion of Immunoglobulin G (IgG) (Boyle and Reis 1987), the paratope of IgG can face the opposite side of the protein G-immobilized solid support. As a result, protein G-mediated antibody immobilization can lead to a highly efficient immunoreaction (Kretschmann E. 1971; Bae et al. 2004).

In protein chip, the binding property of antigen to antibody is commonly monitored by optical detection methods. One of them is fluorescence (Angenendt et al. 2003; Choi et al. 2006; Oh et al. 2007). The assay utilizes two antibodies that simultaneously bind the same antigen: one is immobilized onto a solid surface, and the other is fluorescently labeled that can produce a fluorescent, luminescent, or colored product. Although this approach has some problems; the chemical heterogeneity of proteins makes this hopeless as a strategy for doing quantitative work because some proteins will label far more efficiently than others, and chemically labeling of proteins results in changes of their surface characteristics greatly, and a lot of labeled antibodies for all proteins is required (Kodadek T. 2001), fluorescence based detection method must be easily available in protein chip.

Otherwise there are some optical techniques with enough sensitivity to detect the binding of antigens to antibodies without the requirement of labeling process and secondary antibody labeled with a dye. Surface plasmon resonance (SPR) and imaging ellipsometry (IE) sensors have been developed to measure the binding of analytes to sensor surface, which are capable of directly detecting analytes in complex biological media with high sensitivity, with a short detection time, and with simplicity (Darren et al. 1998; Sakai et al. 1998; Bae et al. 2004; Oh et al. 2005).The SPR technique, an optical method based on the attenuation of surface plasmon generated between a metal surface and a dielectric layer, has matured to become a versatile detection tool for the study of the kinetics of receptor-ligand interaction, the adsorption of biopolymer on solid surface, peptide-antibody binding, and protein-protein interaction. IE technique is based on ellipsometry and the other optical technique involves measuring the change of the polarization state of an elliptically polarized beam reflected from thin film. It is sensitive enough to detect the adsorption of a molecular monolayer on a solid surface, such as a silicon wafer or gold surface.

Bacterial pathogens existing in contaminated environment such as *E. coli* O157: H7, *Salmonella* spp., *Yersinia* spp. and *Legionella* spp. pose a significant threat to human, animal, and agricultural health. Detection of pathogens existing in contaminated environment, therefore, is very important for public health protection (Black et al. 1978; Hussong et al. 1987; Cowden and Christie 1997; Pathirana et al. 2000; Wong et al. 2002). Consequently, considerable effort has been devoted to developing rapid, sensitive, and specific assays for these organisms. However, conventional microbiological culture methods used for the detection of micro-organism are labour-intensive and requires several days to obtain results and may be unsatisfactory to respond in a timely manner in cases of contamination. Many immunoassay techniques are widely attempted for the detection of bacteria, but they are usually expensive and require time-consuming and complex sample pretreatment procedures. For example, enzyme-linked immunosorbent assay (ELISA) is the most frequently used immunochemical approach to detect pathogens with detection limits ranging from 10^4 to 10^6 colony-forming units (CFU) per mL requiring enrichment usually for 16–24 h (De Boer and Beumer 1998; Kim et al. 1999). Therefore, alternative methods to simultaneously detect pathogens in contaminated environment with high sensitivity, with a short detection time, and with simplicity may be need.

In this study, the objective is to optically detect the various pathogens such as *Escherichia coli* O157:H7, *Salmonella typhimurium*, *Yersinia enterocolitica*, and *Legionella pneumophila* using protein chip. In order to endow the orientation of antibody molecules on solid surface, protein G was introduced. The different antibodies against pathogens (*E. coli* O157:H7, *S. typhimurium*, *Y. enterocolitica*, and *L. pneumophila*) on self-assembled protein G were selectively arrayed using a microarrayer. The responses of the each pathogen to the protein chip were investigated by SPR, fluorescence microscopy, and IE.

27.2 Experimental

27.2.1 Materials

Protein G (M.W. 22,600 Daltons) was purchased from Prozyme Inc. (USA). *S. typhimurium* (KCCM 11806) was kindly donated from the Korean Culture Center of Microorganisms (Korea). *E. coli* O157:H7 (ATCC 43895) and *Y. enterocolitica* (ATCC 700823) was kindly donated from the American Type Culture Collection (USA). *L. pneumophila* (ATCC 33154) was kindly offered from National Institute of Health in Korea. Monoclonal antibody (Mab) against *S. typhimurium* and Mab against *L. pneumophila* – fluorescein isothiocyanate (FITC) conjugate were obtained from Biogenesis, Ltd. (USA). Mab against *E. coli* O157:H7, Mab against *E. coli* O157:H7 – FITC conjugate, and Mab against *L. pneumophila* were obtained from Fitzgerald Industries International, Inc. (USA). Mab against *Y. enterocolitica,* Mab against *Y. enterocolitica* – FITC conjugate, and Mab against *Salmonella* spp. conjugate were obtained from Biodesign International, Inc. (USA). Other chemicals used in this study were obtained commercially as the reagent grade.

27.2.2 Immobilization of Antibody

BK 7 glass plate (18 mm × 18 mm, Superior, Germany) was used as the solid support and Au was sputtered to the BK 7 glass surface. Before sputtering Au, chromium (Cr) was sputtered on the glass slide to promote the adhesion of Au. The Au and the Cr film had a thickness of 43 ± 1 nm and 2 nm, respectively. The Au surface was cleaned using pirahna solution (30 vol.% H_2O_2 and 70 vol.% H_2SO_4) at 60°C for 5 min, and then rinsed with ethanol and deionized water. The self-assembled monolayer of 11-mercaptoundecanoic acid (11-MUA) on the Au surface was fabricated by submerging the prepared Au substrate into a glycerol/ethanol (1:1, v/v) solution containing 150 mM of 11-MUA for at least 12 h (Yam et al. 2001). For chemical binding between the 11-MUA adsorbed on the Au substrate and the free amine from the protein G, the carboxyl group in 11-MUA was activated by submerging the Au substrate modified with 11-MUA into a solution of 10% 1-ethyl-3-(3-dimethylaminopropyl)carbodiimide hydrochloride (EDAC) in water/ethanol (10/1, v/v) for 2 h at room temperature. The self-assembled protein G layer was fabricated by the incubation of the activated Au substrate in a solution of protein G in 10 mM phosphate buffer (PBS, pH 7.4) containing 0.14 mol/L NaCl and 0.02% (w/v) thimerosal (PBS) at room temperature for 2 h Before the immobilization of the antibody, the protein G layer by self-assembly technique on the Au substrate was blocked by inactivating the residual carboxyl group of 11-MUA with 1 M of ethanolamine. To immobilize the Mab, the protein G layer by self-assembly technique was immersed in a solution containing antibodies (50 pmol/mL Mabs) in a PBS buffer because the antibody surface loading on self-assembled protein G layer started to be saturated at 50 pmol/mL. After 4 h of incubation at 4°C, the surface was rinsed with a PBS buffer. In order to provide antigen access to the binding site of antibody by separation of antibody molecules clustered around preferred points on the surface or around other antibody molecules, Tween 20 was used.

27.2.3 Preparation of Protein Chip Based on SPR

The schematic illustration of protein chip system based on SPR was shown in Fig. 27.1. The metal coating and substrate cleaning, and the immobilization of biomolecules onto Au patterned SPR surface was performed in the same way as in above mentioned procedure. The pattern size was dia. 3 mm. The bimolecular interactions of protein chip were monitored using a SPR spectroscope (Multiskop, Optrel GbR, Germany) (Harke et al. 1997). A He-Ne laser was used as a light source to make a monochromatic light with a wavelength of 632.8 nm. The p-polarized light beam by the polarizer is used as a reference and the intensity of

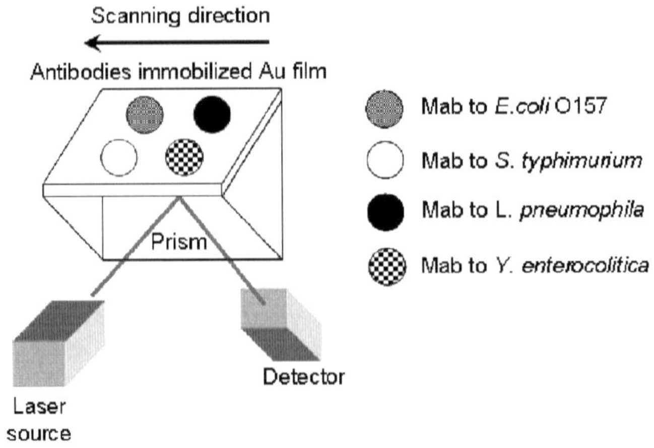

Fig. 27.1 The schematic illustration of protein chip system based on SPR

the reflected beam is measured by photo multiplier tube (PMT) sensor. A 90° glass prism (BK 7, n = 1.5168) is used as a Kretschmann ATR coupler (Kretschmann 1971). The plane face of the 90° glass prism was coupled to a BK 7 glass slide via index matching fluid. The resolution of the angle reading of the goniometer was 0.001°.

27.2.4 Preparation of Protein Chip Based on Imaging Ellipsometry

A substrate was prepared by DC magnetron sputtering of Au on a P-type Si wafer. The metal coating and substrate cleaning, and the immobilization of biomolecules were performed in the similar way as in above mentioned procedure. The Au and Cr films had thicknesses of 150 and 5 nm, respectively. The protein G solution was spotted onto 11-MUA modified Au surface using a microarrayer (NanoPlotter model 1.2, GeSiM mbH, Groβerkmannsdorf, Germany). The spotted amount per spot was 0.4 nL of a solution of 0.1 mg/mL protein G in a mixed solution of 10 mM PBS buffer and 10 vol% glycerol. The spotted substrate was incubated in a humid chamber at 4°C for at least 24 h, taking into consideration the diffusivity delay of the protein molecules, which are hindered from entering into the surface due to the viscosity of glycerol. After the incubation period, the chip was washed with PBS buffer for 20–30 min. Before the immobilization of the Mab, the residue carboxyl groups of 11-MUA on the chip were inactivated by blocking them with 3 wt% bovine serum albumin (BSA). A solution containing the Mab in PBS buffer was applied to the blocked chip. After being incubated at 4°C for 3 h, the substrate was washed with PBS buffer containing 0.1% Tween 20.

27.2.5 Preparation of Protein Chip Based on Fluorescence Image

A BK 7 type cover glass plate was used as the solid support. The metal coating and substrate cleaning, and the immobilization of biomolecules were performed in the similar way as in above mentioned procedure. The protein G was arrayed on 11-MUA modified Au surface using a microarrayer. And then the antibody molecules such as Mab against *E. coli* O157:H7, Mab against *S. typhimurium*, Mab against *Y. enterocolitica*, and Mab against *L. pneumophila* were spotted on the protein G by using the microarrayer. After incubation at 4°C for at least 24 h, the surface was washed with PBS buffer containing 0.1% Tween 20. The residue carboxyl groups of 11-MUA on protein chip were inactivated by blocking them with 3 wt % BSA.

27.2.6 Culture Condition of Pathogens

E. coli O157:H7 was cultivated in a 250 mL flask with 100 mL of medium (medium composition: pancreatic digest of casein 10 g, NaCl 5 g, yeast extract 5 g in 1 L deionized water, pH 7.0 ± 0.2 at 25°C) at 37°C with shaking at 200 rpm. *S. typhimurium* was cultivated in a 250 mL flask with 100 mL of medium (medium composition: pancreatic digest of casein 10 g, NaCl 5 g in 1 L deionized water, pH 7.4 ± 0.2 at 25°C) at 37°C with shaking at 200 rpm. *Y. enterocolitica* was cultivated in a 250 mL flask with 100 mL of medium (medium composition: pancreatic digest of casein 17 g, NaCl 5 g, papaic digest of soybean meal 3 g, K_2HPO_4 2.5 g, glucose 2.5 g in 1 L deionized water, pH 7.3 ± 0.2 at 25°C) at 37°C with shaking at 200 rpm. *L. pneumophila* was cultivated in a 250 mL flask with 100 mL of medium (medium composition: yeast extract 20 g, L-cysteine.HCl.H_2O 0.4 g and $Fe(NO_3)_3.9H_2O$ 0.1 g in 500 mL deionized water, pH 6.85–7.0 at 25°C) at 37°C under 5% CO_2 condition.

27.2.7 Imaging Ellipsometry

The imaging ellipsometry configuration, based on off-null ellipsometry, which has a component sequence of polarizer-compensator-sample-analyzer (PCSA) as shown in Fig. 27.2 (Multiskop, Optrel Gbr, Kleinmachnow, Germany), was used. For the acquisition of ellipsometric images, an objective lens (×10) was inserted between the sample and the analyzer. After the reflected beam passed through the objective and the analyzer, the intensity profile of the cross section of it was recorded in the form of an image with a resolution of 640 × 480 pixels by means of a CCD camera. The light source was a He-Ne laser beam (632.8 nm). The incident angle of the laser beam was set to 40°. The mean optical intensity (MOI) values of

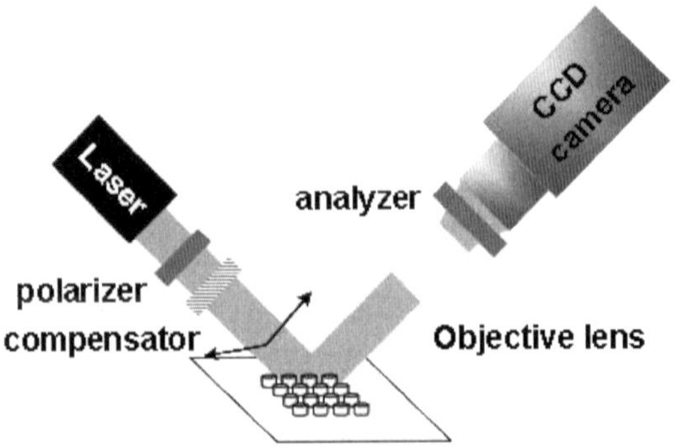

Fig. 27.2 Imaging ellipsometry system based on null-type ellipsometry

the ellipsometric images of the protein spots were calculated using the image processing software (Image-Pro, version 4.7, Media Cybernetics, Silver Spring, MD). The measurement of the ellipsometric angles, based on traditional null-ellipsometry, was performed using the PCSA null-ellipsometry system with a photodiode sensor as the optical detector.

27.3 Results and Discussion

27.3.1 Preparation of Antibody Layer

The changes of SPR curves by adsorbing 11-MUA, and by binding of protein G, and by affinity binding of antibody molecules to protein G on Au substrate in series are shown in Fig. 27.3a. As a result, the SPR angle was shifted significantly from 43.002° ± 0.02 to 43.257° ± 0.03 by the adsorption of 150 mM 11-MUA on Au surface, and from 43.257° ± 0.03 to 43.437° ± 0.03 by chemical binding between protein G (500 nM) and the activated carboxyl group of 11-MUA with EDAC, and 43.437° ± 0.03 to 43.647° ± 0.03 by affinity binding between antibody molecules and protein G on Au surface, respectively. In principle, a surface plasmon resonance is extremely sensitive to the interfacial architecture. An adsorption process leads to a shift in the plasmon resonance and allows monitoring the mass coverage at the surface with a high accuracy. (Fagerstam et al. 1992; Lundstrom 1994; Salmon et al.1997). Therefore, the shift in the SPR angle verified that thin layer of 11-MUA on Au surface was formed and protein

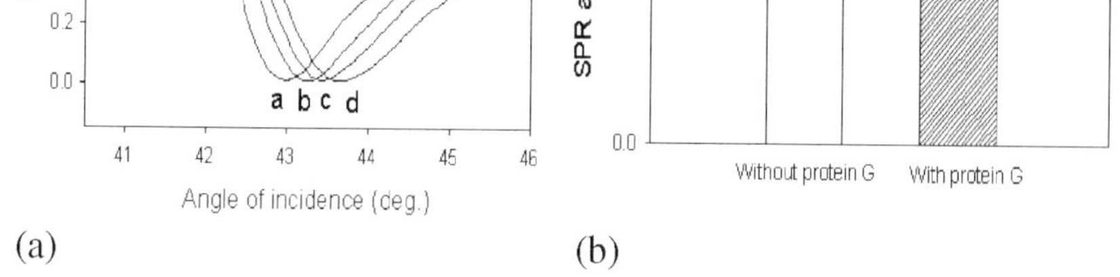

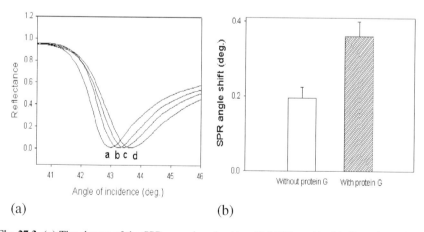

Fig. 27.3 **(a)** The change of the SPR curve by adsorbing 11-MUA, and by binding of protein G, and by affinity binding of antibody (a: bare gold, b: 11-MUA, c: protein G, d: antibody), **(b)** The effect of protein G in antibody-antigen complex formation

G molecules were well bound with 11-MUA adsorbed on Au substrate, and antibody molecules were well adsorbed on self-assembled protein G layer. From this result, it could be confirmed that the biomolecules films were formed on Au surface.

The effect of protein G about the binding interaction between antibody and antigen was investigated in comparison with shift degree of SPR angle by binding of antigen to immobilized antibody on Au substrate without/with protein G (Oh et al. 2004). The results were shown in Fig. 27.3 (b). The variation of SPR angle by the binding interaction between antibody and antigen without/with protein G is 0.195° and 0.36°, respectively. Compared with the shift degree of SPR angle by binding interaction between antibody and antigen, the shift degree of SPR angle in case of immobilized antibody on solid surface using protein G is larger than that of SPR angle in case of directly immobilized antibody on solid surface without protein G. It mean that the binding efficiency of antigen to the antibody immobilized on Au surface was improved by using protein G because the binding site of immobilized antibody on solid surface is exposed to the medium of the analytical system, since recombinant protein G used in this study has two domains that can bind to the F_c portion of IgG which is at the junction of CH_2 and CH_3 domains of the heavy chain.

27.3.2 The Protein Chip Based on SPR

The protein chip based on SPR was applied to four kinds of pathogens (ca. 10^5 CFU/mL of *E. coli* O157:H7, *S. typhimurium*, *L. pneumophila*, and *Y. enterocolitica*) in turn, and the responses of each Mab spots were analyzed with SPR (Oh et al. 2005). The response of Mab against *E. coli* O157:H7 spot of the protein chip for four pathogens was shown in Fig. 27.4a. As shown in Fig. 27.4a, the shift of SPR angle by the binding of *E. coli* O157:H7 to Mab against *E. coli* O157:H7 spot was higher than that of SPR angle by the binding of other pathogens to Mab against *E. coli* O157:H7 spot. Compared with the ratio of SPR angle shift, it was obviously observed that Mab against *E. coli* O157:H7 spot had the high selectivity with *E. coli* O157:H7 and did not react with other pathogens.

The responses of Mab against *S. typhimurium* spot, Mab against *L. pneumophila* spot, and Mab against *Y. enterocolitica* spot of protein chip for ca. 10^5 CFU/mL of three kinds of pathogens were shown in Fig. 27.4b, c, and d, respectively. As shown in Fig. 27.4b,c, and d, the shift of SPR angle by the binding of each pathogen to the corresponding Mab spots was higher than that of SPR angle by the non-specific binding of each pathogens to other Mab spots. Compared with the ratio of SPR angle shift, it was obviously observed that each Mab spots of the protein chip had the high selectivity with the corresponding pathogens. From above results, as a simultaneous detection system for the multiple pathogens, the feasibility of SPR based protein chip proposed in this study could be confirmed.

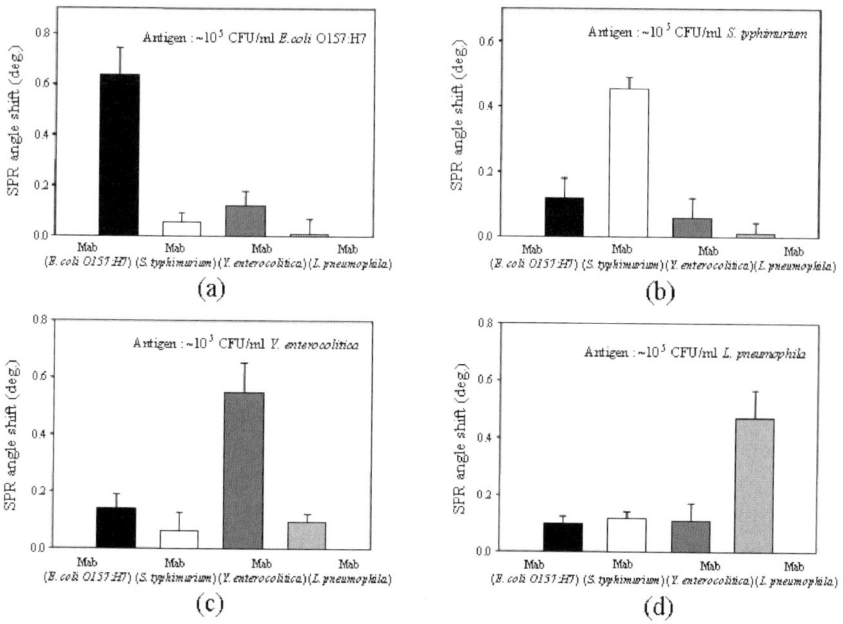

Fig. 27.4 The response of Mab against four pathogens spot of protein chip based on SPR for ca. 10^5 CFU/mL of four pathogens such as *E. coli* O157:H7, *S. typhimurium, L. pneumophila, and Y. enterocolitica* (Bar; SRP angle shift); **(a)** *E. coli* O157:H7, **(b)** S. typhimurium, **(c)** Y. enterocolitica, **(d)** L. pneumophila

27.3.3 The Response of Imaging Ellipsometry Based Protein Chip for Pathogen

The responses of the constructed protein chip for *Y. enterocolitica* were analyzed with IE. (Bae et al. 2004) On the basis of the strategy for the immobilization of antibody, the protein spots were fabricated using a microarrayer. Protein G solution was spotted onto the substrate modified with the 11-MUA layer. Following the sequence of steps that involved applying the Mab solution to the chip with protein G spots blocked with BSA, incubation, and washing with PBS buffer, the antibody array was completed. The ellipsometric images of the protein G spots blocked with BSA and the Mab (Mab to *Y. enterocolitica*) spot immobilized on the protein G spot are shown Fig. 27.5a and c, respectively. To acquire the ellipsometric images, the azimuth of the polarizer and analyzer were set to 126.0 ± 0.3° and 136.0 ± 0.3°, respectively. At the azimuth points of the polarizer and analyzer, the BSA layer region of the ellipsometric images was in the null condition. On the basis of the principle of off-null ellipsometry, the optical intensity of the light beam reflected from a film onto a solid surface is dependent on the thickness of the film, according to the Fresnel equation (Arwin et al. 2000). Therefore, the difference in optical intensity of the magnified cross section of the beam reflected from a thin film implies the local difference in the thin film thickness in the irradiated region.

In the ellipsometric images, it was found that as antibody was immobilized, the brightness of the spot image decreased. Since the azimuths of the polarizer and analyzer were set for the null condition of the BSA layer, the difference of optical intensities in the ellipsometric images corresponds to the difference between the thickness of protein spots and those of the BSA layer. The height schematics of protein G spot and Mab spot on protein G blocked with BSA are shown in Fig. 27.5b and d, respectively. This indicates that the BSA layer was denser than the protein G spot and that the difference in the surface concentration between the BSA layer and the Mab spot was reduced as result of the immobilization of the Mab onto the protein G spot.

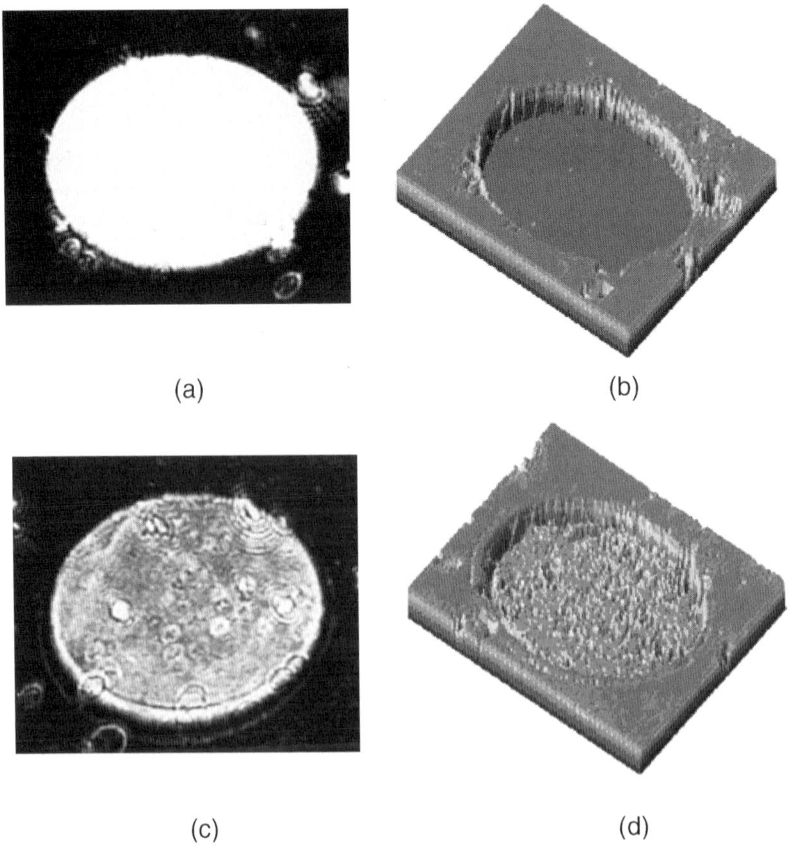

Fig. 27.5 (a) Ellipsometric image (a) and height schematic (b) of protein G spot injected by means of an inkjet-type microarrayer and ellipsometric image (c) and height schematic (d) of Mab on protein G (From Bae et al. 2004). The diameter of the spot is ~130 m

Y. enterocolitica solutions with various concentrations were applied to the completed immunosensor chips. After incubation and being washed with PBS buffer, they were dried at room temperature. It was observed that as the pathogen concentration increased, the white regions in the protein spot increased. Thus, since the region where the pathogen was bound to the substrate became thicker than the BSA layer, it was considered that the optical intensity of this region increased. The mean optical intensity (MOI) of each protein spot was calculated. The change in the MOI appeared beginning with the protein spot with a binding of 10^3 CFU/mL of pathogen, and the MOI increased up to a binding of 10^7 CFU/mL of pathogen in a manner that was proportional to the logarithm of the concentration of the pathogen. The MOI of the protein spot with a binding of 10^7 CFU/mL was almost saturated, because the output signal from an eight-bit CCD camera ranges from 0 to 255. However, a large standard deviation was observed. The reason for this was that residue salts from the PBS buffer affected the ellipsometric image. In particular, at low concentration, the signal noise due to the residue salts might result in errors in the immunosensor based on the IE. To solve this problem, it will be necessary to operate the immunosensor in an in situ flow system.

L. pneumophila and *S. typhymurium* solutions with various concentrations were also applied to the completed immunosensor chips, respectively (Bae et al. 2004, 2005). It was also observed that as the pathogen concentration increased, the white regions in the protein spot increased because the optical intensity of this region increased.

27.3.4 The Responses of Fluorescence Image Based Protein Chip for Four Pathogens

The responses of the constructed protein chip for four pathogens were analyzed with fluorescence microscopy (Oh et al. 2007). The microarray consisting of protein G onto 11-MUA modified Au surface was fabricated using a microarrayer and the images were investigated by fluorescence microscopy (Fig. 27.6A). After 24 h incubation of the microarray consisting of protein G under 4 °C condition, the sample immersed into a PBS buffer (pH 7.4) containing 3% BSA to avoid non-specific binding between 11-MUA modified Au surface and other proteins. After blocking process, antibody microarray was fabricated by using the microarrayer. After washing with PBS buffer, the antibody microarray was investigated by fluorescence microscopy. As a result, it could be confirmed that antibody microarray was well constructed onto protein G layer. The protein chip consisting of four different kinds of antibodies (Mab against *E. coli* O157:H7, Mab against *S. typhimurium*, Mab against *Y. enterocolitca*, and Mab against *L. pneumophila*) were fabricated using microarrayer, and the protein chip was applied to ca. 10^7 CFU/mL of *E. coli* O157:H7, and then in turn, it was applied to ca. 10^7 CFU/mL of *S. typhimurium*, *Y. enterocolitica*, and *L. pneumophila,* respectively. Four different kinds of pathogens such as *E. coli* O157:H7, *S. typhimurium*, *Y. enterocol-*

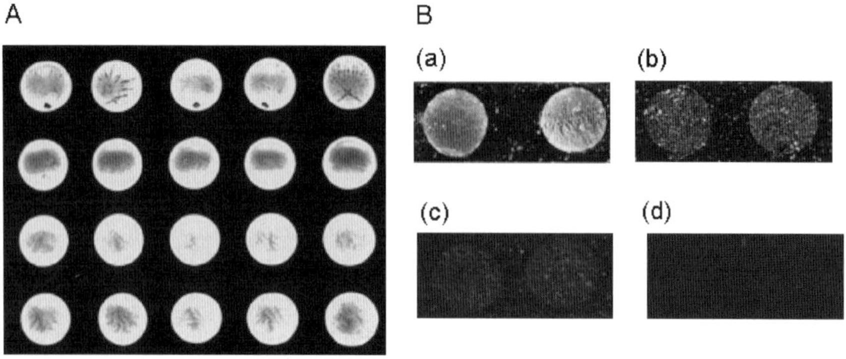

Fig. 27.6 Fluorescence images of fabricated protein chip and responses of the protein chip for *E. coli* O157:H7. **(A)** Fluorescence image of protein G microarray. **(B)** Fluorescence responses of protein chip as a function of *E. coli* O157:H7 concentration; **(a)** 10^7 CFU/mL, **(b)** 10^5 CFU/mL, **(c)** 10^2 CFU/mL, (d) 0 CFU/mL

itica, and *L. pneumophila* could be selectively detected using the constructed protein chip by fluorescence microscopy because the corresponding antigens bound to its cognate antibody spots (data not shown).

The responses of the protein chip consisting of Mab against *E. coli* O157:H7 immobilized on self-assembled protein G layer for various concentration of *E. coli* O157:H7 were investigated by using sandwich method with fluorescence microscopy. The results were shown in Fig. 27.6B. As shown in Fig. 27.6B, the fluorescence brightness and intensity of each spots increased, as the concentration of *E. coli* O157:H7 increased. The lowest detection limit of the protein chip fabricated in this study for *E. coli* O157:H7 was 10^2 CFU/mL (data not shown).

27.4 Conclusion

Optical detection method based protein chips for detection of the various pathogens such as *E. coli* O157:H7, *S. typhimurium*, *Y. enterocolitica*, and *L. pneumophila* in contaminated environment were developed. In order to endow the orientation of antibody molecules on solid surface, protein G was introduced. The protein G on Au surface modified with 11-MUA was immobilized and the spots of different antibodies against pathogens on protein G were arrayed using a microarrayer. The responses of the various pathogens in contaminated environment to the protein chip was investigated by surface plasmon resonance, fluorescence microscopy and imaging ellipsometry. The lowest detection limit of the fluorescence based protein chip was 10^2 CFU/mL and the protein chip using IE could successfully detect the pathogens in concentrations varying from 10^3 to 10^7 CFU/mL. The proposed detection technique of protein chip for various pathogens could be applied to other protein chips with a high efficiency.

Acknowledgements This work was supported by the Korea Science and Engineering Foundation (KOSEF) through the Advanced Environment Monitoring Research Center at Gwangju Institute of Science and Technology and by the Nano/Bio science & Technology Program (M10536090001–05N3609–00110) of the Ministry of Science and Technology (MOST), and by the Korea Science and Engineering Foundation (KOSEF) grant funded by the Korea government (MOST) (2006-05374).

References

Angenendt P., Glökler J., Konthur Z., Lehrach H., and Cahill D.J. (2003), 3D protein microarrays: Performing multiplex immunoassays on a single chip. *Anal. Chem.*, 75, 4368–4372.

Arenkov P., Kukhtin A., Gemmell A., Voloshchuk S., Chupeeva V., and Mirzabekov A. (2000), Protein microchips: Use for immunoassay and enzymatic reactions. *Anal. Biochem.*, 278, 123–131.

Arwin H., Welin-Klintstrom S., and Jansson R. (1993), Off-null ellipsometry revisited: Basic concentration for measuring surface concentration at solid/liquid interfaces. *J. Colloid. Interface Sci.*, 156, 377–382.

Bae Y.M., Oh B.-K., Lee W., Lee W.H., and Choi J.-W. (2004), Immunosensor for detection of *Yersinia enterocolitica* based on imaging ellipsometry, *Anal. Chem.*, 76, 1799–1803.

Bae Y.M., Oh B.-K., Lee W., Lee W.H., and Choi J.-W. (2004), Immunosensor for detection of *Legionella pneumophila* based on imaging ellipsometry, *Mater. Sci. Eng. C - Biomimetic Supramol. Syst.*, 24, 61–64.

Bae Y.M., Oh B.-K., Lee W., Lee W.H., and Choi J.-W. (2005), Study on orientation of immunoglobulin G on protein G layer, *Biosens. Bioelectron.* 21, 103–110.

Bae Y.M., Park K.-W., Oh B.-K., Lee W.H., and Choi J.-W. (2005), Immunosensor for detection of *Salmonella typhimurium* based on imaging ellipsometry, *Colloid. Surface. A: Physicochem. Eng. Aspects*, 257–258, 19–23.

Black R.E., Jackson R.J., Tsai T., Medevesky M., Shayegani M., Feeley J.C., Macleod K. L. E., and Wakelee A. M. (1978), Epidemic *Yersinia enterocolitica* infection due to contaminated chocolate milk, *N. Eng. J. Med.*, 298, 76–79.

Butler J.E., Ni L., Nessler R., Joshi K.S., Suter M., Rosenberg B., Chang J., Brown W.R., and Cantarero L. A. (1992), The physical and functional behavior of capture antibodies adsorbed on polystyrene, *J. Immunol. Methods*, 150, 77–90.

Boyle M.D.P., and Reis K. J. (1987), Bacterial Fc receptors, *Biotechnol.*, 5, 697–703.

Cahill D.J. (2001), Protein and antibody arrays and their medical applications, *J. Immunol. Methods*, 250, 81–91.

Choi J.-W., Park J.H., Lee W., Oh B.-K., Min J., and Lee W.H. (2001), Fluorescence immunoassay of HDL and LDL using protein A LB film, *J. Microbiol. Biotechnol.*, 11, 979–985.

Choi J.-W., Nam Y.S., and Fujihira M. (2004), Nanoscale fabrication of biomolecular layer and its application to biodevices. *Biotechnol. Bioprocess Eng.*, 9, 76–85.

Choi J.-W., Lee W., Oh B.-K., Lee H.-J., and Lee D.-B. (2006), Application of complement 1q for the site-selective recognition of immune complex in protein chip, *Biosens. Bioelectron.*, 22, 764–767.

Cowden J.M., and Christie P., (1997), Scottish outbreak of *Escherichia coli* O157:H7, *Health Bull.*, 55, 9–10.

Darren M.D., David C.C., Hong X.Y., and Christopher R.L. (1998), Covalent coupling of immunoglobulin G to self-assembled monolayers as a method for immobilizing the interfacial recognition layer of a surface plasmon resonance immunosensor, *Biosens. Bioelectron.*, 13, 1213–1225.

De Boer E., and Beumer R. (1998), Developments in the microbiological analysis of foods, *De Ware(n) Chem.*, 28, 3–8.

Fagerstam L.G., Frostell-Karlsson A., Karlsson R., Persson B., and Ronnberg I. (1992), Biospecific interaction analysis using SPR detection applied to kinetic binding site and concentration analysis, *J. Chromatogr.*, 597, 397–410.

Harke M., Teppner R., Schulz O.M., and Orendi H. (1997), Description of a single modular optical setup for ellipsometry, surface plasmons, waveguide modes, and their corresponding imaging technique including Brewster angle microscopy, *Rev. Sci. Instrum.*, 68, 3130–3134.

Hussong D., Colwell R.R., O'Obrien M., Weiss E., Pearson A.D., Wiever R.M., and Burge W.D. (1987), Viable *Legionella pneumophilia* not detectable by culture on agar media, *Biotechnology*, 5, 947–950.

Kelvin H.L. (2001), Proteomics: A technology-driven and technology-limited discovery science. *Trends Biotechnol.*, 19, 217–222.

Kim J.W., Jin L. Z., Cho S.H., Marquardat R.R., Frohlich A.A., and Baidoo S.K. (1999), Use of chicken egg-yolk antibodies against K88+fimbral antigen for quantitative analysis of enterotoxigenic *Escherichia coli* (ETEC) K88 by a sandwich ELSIA, *J. Sci. Food Agric.*, 79, 1513–1518.

Kodadek T. (2001), Protein microarrays: Prospects and problems. *Chem. Biol.*, 8, 105–115.

Kretschmann E. (1971), Die bestimmung optischer konstanten von metallen durch anregung von oberflachenplasmaschwingungen, *Z. Phys.*, 241, 313–324.

Lundstrom I. (1994), Real-time biospecific interaction analysis, *Biosens. Bioelectron.*, 9, 725–736.

Oh B.-K., Kim Y.K., Park K.W., Lee W.H., and Choi J.-W. (2004), Surface plasmon resonance immunosensor for the detection of *Salmonella typhimurium*, *Biosens. Bioelectron.*, 19, 1497–1504.

Oh B.-K., Lee W., Chun B.S., Bae Y.M., Lee W.H., and Choi J.-W. (2005), The fabrication of protein chip based on surface plasmon resonance for detection of pathogens, *Biosens. Bioelectron.*, 20, 1847–1850.

Oh B.-K., Kim Y.K., and Choi J.-W. (2007), The development of protein chip using protein G for the simultaneous detection of various pathogens, *Ultramicroscopy,* Submitted.

Pandey A., and Mann M. (2000), Proteomics to study genes and genomes, *Nature*, 405, 837–846.

Pathirana S.T., Barbaree J., Chin B.A., Hartell M.G., Neely W.C., and Vodyanoy V. (2000), Rapid and sensitive biosensor for Salmonella, *Biosens. Bioelectron.*, 15, 135–141.

Ruiz-Taylor L.A., Martin T.L., Zaugg F.G., Witte K., Indermuhle P., Nock S., and Wagner P. (2001), Monolayers of derivatized poly(L-lysine)-grafted poly(ethylene glycol) on metal oxides as a class of biomolecular interfaces, *Proc. Natl. Acad. Sci.*, 98, 852–857.

Sakai G., Ogata K., Uda T., Miura N., and Yamazoe N. (1998), A surface plasmon resonance-based immunosensor for highly sensitive detection of morphine, *Sensor Actuat. B,* 49, 5–12.

Salmon Z., Macleod H.A., and Tollin G., (1997), Surface plasmon resonance spectroscopy as a tool for investigating the biochemical and biophysical properties of membrane protein systems.II:Applications to biological systems. *Biochim. Biophy. Acta,* 1331, 131–152.

Vijayendran R.A., and Leckband D.E. (2001), A quantitative assessment of heterogeneity for surface-immobilized proteins, *Anal. Chem.*, 73, 471–480.

Wilson D.S., and Nock S. (2001), Functional protein microarrays, *Curr. Opin. Chem. Biol.*, 6, 81–85.

Wilson D.S., Wu J., Peluso P., and Nock S. (2002), Improved method for pepsinolysis of mouse IgG_1 molecules to $F(ab')_2$ fragments, *J. Immunol. Methods*, 260, 29–36.

Wong Y.Y., Ng S.P., Ng M.H., Si S.H., Yao S.Z., and Fung Y.S. (2002), Immunosensor for the differentiation and detection of *Salmonella* species based on a quartz crystal microbalance, *Biosens. Bioelectron.*, 17, 676–684.

Yam C.M., Zheng L., Salmain M., Pradier C.M., Marcus P., and Jaouen G. (2001), Labelling and binding of poly-(L-lysine) to functionalized gold surface Combined FT-IRRAS and XPS characterization, *Colloid. Surface. B: Biointerfaces,* 21, 317–327.

Zhu H., Bilgin M., Bangham R., Hall D., Casamayor A., Bertone P., Lan N., Jansen R., Bidlingmaier S., and Houfek T. (2001), Global analysis of protein activities using proteome chips, *Science*, 293, 2101–2105.

Chapter 28
Expression Analysis of Sex-Specific and Endocrine-Disruptors-Responsive Genes in Japanese Medaka, *Oryzias latipes*, using Oligonucleotide Microarrays

Katsuyuki Kishi[1], Emiko Kitagawa[2], Hitoshi Iwahashi[2], Tomotaka Ippongi[3], Hiroshi Kawauchi[3], Keisuke Nakazono[3], Masato Inoue[3], Hiroyoshi Ohba[3], and Yasuyuki Hayashi[3]

Abstract Gene profiling of Japanese medaka (*Oryzias latipes*) was performed using an oligonucleotide DNA microarray representing 26,689 TIGR *Oryzias latipes* Gene Indices (OLGIs). We first confirmed the high correlation coefficients (>0.94) of gene expression in individual medaka grown under the standard procedures. Secondly, we exposed male medaka to estrogenic compounds [17β-estradiol (E2), 17α-ethynylestradiol (EE2), nonylphenol (NP), octylphenol (OP), and bisphenol A (BpA)], and assessed estrogenic compounds-induced changes in mRNA expression with the medaka microarray. Histological analysis showed the production of testis-ova in the testes of male medaka exposed to the estrogenic chemicals. Microarray analysis of the chemical-treated male medaka identified estrogenic compounds-responsive OLGIs, although many of those OLGIs were not sex-specific. Based on the mRNA expression profiles, assessment of the degree of feminization/demasculinization using the combination of Pearson correlation coefficient (CC) and Euclidean distances was also attempted in order to estimate the impact of hormonally active chemicals. The calculated feminization factors indicate that E2 and EE2 treatment "weakly feminized" male medaka (~50%), while NP, OP, or BpA treatment did not significantly feminize male medaka (<6%). On the other hand, the calculated male-dysfunction factors suggest that male physiological functions were disrupted by the EE2 exposure (~50%), but not significantly by E2, NP, OP, or BpA treatments (<16%). Results demonstrate the possibility of using medaka microarrays to estimate the overall effects of hormonally active chemicals.

---

[1] *Japan Pulp & Paper Research Institute, Inc., Tokodai 5-13-11, Tsukuba, Ibaraki, 300-2635; Japan; Tel: +81-29-847-4321, Fax: +81-29-847-8923*

[2] *Human Stress Signal Research Center, National Institute of Advanced Industrial Science and Technology (AIST), Tsukuba West, Onogawa 16-1, Tsukuba 305-8569, Japan*

[3] *GeneFrontie, Corp., Nihonbashi Kayabacho 3-2-10, Chuo-ku, Tokyo, 103-0025, Japan*

Y.J. Kim and U. Platt (eds.), *Advanced Environmental Monitoring*, 363–375.

Keywords: DNA microarray, endocrine disrupter, feminization, masculinization, medaka

28.1 Introduction

Chemicals called environmental pollutants are discharged into our environment and can cause toxic effects on a variety of organisms either directly or indirectly through chemical/biochemical modification. Numerous studies have been conducted to evaluate the effects of toxicants on the survival, development, growth, reproduction and physiology of various organisms using a variety of bioassays (OECD 1998; USEPA 1993). Molecular and biochemical techniques, called biomarkers, have also been used to elucidate toxic modes of action. Moreover, a newly developed research field, toxicogenomics, the use of comprehensive gene expression changes to evaluate multiple types of toxicity, greatly expands the scope and depth of toxicological approach (Irwin et al. 2004; Snape et al. 2004). Toxicogenomic studies have been remarkably facilitated by the development of microarray technologies. DNA microarray technology burst onto the scene of molecular biological research in the mid-1990s (Ramsay 1998; Schena et al. 1998), and promises to revolutionize biological research and further our understanding of biological processes. The advent of DNA microarrays has provided a means for analyzing the expression of thousands of genes simultaneously. This advantage allows toxicologists to take a global perspective on toxicity, in that the effects of a substance on the expression of virtually all known genes can be assessed within a single microarray (Ramsay 1998).

Japanese medaka (*Oryzias latipes*), one of the focus species for eco-toxicological study, is widely accepted as an experimental model system for development, histology and sexual differentiation research (Yamamoto 1975; Ozato et al. 1992; Matsuda et al. 2002; Schartl 2004). Large-scale gene expression analyses have also been applied to medaka developmental biology (Katogi et al. 2004; Kimura et al. 2004). Recently, an oligonucleotide-based medaka DNA microarray was developed by NimbleGen Systems, Inc. (Madison, WI, USA) using the Maskless Array Synthesizer (MAS) (Singh-Gasson et al. 1999). These experimental resources and technological advances make medaka an ideal vertebrate animal model for toxicogenomic study on sexual differentiation and development.

In this study, we assessed changes in mRNA expression induced by estrogenic chemicals with the medaka DNA microarray. We first confirmed the high correlation coefficients of gene expression in individual medaka grown under the standard procedures. The average correlation coefficients for gene expression between individual mature fish were high (>0.94) for both female and male, indicating that the physiological status of medaka is highly reproducible under prescribed growth conditions. Secondly, we exposed male medaka to estrogenic chemicals; natural and synthetic estrogens: 17-estradiol (E2) and 17-ethynylestradiol (EE2); three chemicals suspected of endocrine disruptors: nonylphenol (NP), octylphenol (OP), and bisphenol A (BpA). Histological study revealed the incidence of testis-ova (oocyte formation in testes) in the exposed-medaka. Changes in the mRNA expression profile of the chemical-exposed fish were compared with the normal patterns of gene expression in male medaka in order to identify the genes responsible for feminization of male fish by the estrogenic chemicals. Furthermore, in order to estimate the physiological impact of the estrogenic chemicals, the degree of feminization/de-masculinization was also assessed based on both similarity of the mRNA expression profiles and intensities of mRNA expressions.

28.2 Materials and Methods

28.2.1 Test Organism

Japanese medaka, *Oryzias laptipes* (orange-red variety or "Himedaka"), were originally purchased from Tsuchiura Goldfish Fishery (Tsuchiura, Ibaraki, Japan) and then maintained at Japan Pulp & Paper Research Institute, Inc. for several generations. The brood stock was maintained at 24 ± 1°C in UV-disinfected, dechlorinated, carbon-treated tap water with a 16 h light-8 h dark photoperiod. The fish were fed *Artemia nauplii* (<24 h after hatching) twice daily. Medaka selected for this study were approximately 6 months post-hatch and fully mature (body wt ~400 mg; total length ~3 cm). Growth conditions followed guidelines recommended by the international toxicity test protocol (OECD 1992; http://www.env.go.jp/chemi/kagaku/). For microarray analysis, female and male fish were collected, flash-frozen in liquid nitrogen and stored at –80°C for RNA extraction.

28.2.2 Test Chemical

17-Estradiol (E2), 17-ethynylestradiol (EE2), 4-nonylphenol (NP), 4-*tert*-octylphenol (OP) and bisphenol A (BpA) were obtained from Wako Pure Chemical Industries, Ltd. (Osaka, Japan). E2/EE2, NP/OP, and BpA stock solutions of 20 µg/ml, 20 mg/ml, and 400 mg/ml, respectively, were prepared by dissolving 2 mg E2/EE2, 2 g NP/OP, and 40 g BpA, respectively, in 100 ml dimethyl sulfoxide (DMSO, Wako Pure Chemical Industries, Ltd., Osaka, Japan). Other chemicals were all analytical grade.

28.2.3 Exposure Conditions

Mature male medaka were exposed to the estrogenic chemicals for 21 days in 10-l glass chambers containing 6-l test solutions at the concentrations of 100 ng/l E2 and EE2, 100 µg/l NP and OP, and 2 mg/l BpA. Test solutions for each estrogenic chemical were prepared daily by adding 150 µl of the stock solutions of the chemicals to 30-l declorinated tap water in 40-l glass aquaria. The test solutions were delivered to each test chamber by a peristaltic pump (Cole Parmer Instrument

Co., IL, USA) at the flow rate of 1.25 l/h. The test solution in each chamber was, consequently, renewed five times daily. Two test chambers were used for each treatment group, and seven fish were placed in each chamber. Fish were maintained under a 16:8 h light:dark photoperiod and fed *A. nauplii* (<24 h after hatching) twice a day. The test chambers were cleaned once weekly, and residual bait and fences in the test chambers were removed daily. Throughout the exposure period, the DO concentration (mean ± SD) was 8.1 ± 0.1 mg/l, and pH was 7.5 ± 0.1. The water temperature in all test chambers was 24 ± 0.6°C. On the last day of exposure, three fish in each treatment group were flash-frozen in liquid nitrogen and stored at −80°C for RNA extraction. Eight fish for each treatment were also sacrificed, and their gonads were removed and subjected to histological examination. A rest of fish were also flash-frozen in liquid nitrogen and stored at −80°C.

28.2.4 Histological Examination

At the end of exposure, the gonads were removed from the sacrificed male fish and then fixed in Bouin's fixative (Muto Pure Chemicals Co., Ltd., Tokyo, Japan). The fixed gonads were embedded in paraffin blocks following standard histological procedures and cut into serial sections (5 µm thick) with a microtome (Leica Microsystems, Tokyo, Japan). The sections were stained with hematoxylin and eosin, mounted with Mount-Quick (Daido Sangyo Co., Ltd., Tokyo, Japan), and then examined under a light microscope.

28.2.5 Construction of Medaka Microarray

60-mer oligonucleotide probes for each TIGR (The Institute for Genomic Research) *Oryzias laptipes* (Japanese medaka) Gene Indices (OLGI) were designed based on the data from OLGI, updated May 17th, 2004 (release 5.0). The total number of OLGIs was 26,689, including 12,849 TCs (Tentative Consensus), 13,669 singleton ESTs (Expressed Sequence Tag), and 171 singleton ETs (Expressed Transcript), with seven probes for each OLGI. The medaka microarray was synthesized using the MAS Technology (NimbleGen Systems, Inc., Madison, WI, USA) (Singh-Gasson et al. 1999).

28.2.6 Microarray Analysis

Total RNA from whole frozen medaka was isolated using EASYPrep RNA (TaKaRa Bio Inc., Shiga, Japan) following procedures recommended by the manufacturer. Total RNA from whole frozen medaka was first converted to double-stranded cDNA, followed by the synthesis of biotin-labeled cRNA using *in vitro* transcription as described elsewhere (Eberwine et al. 1992; Nuwaysir et al. 2002). cRNA was then purified and fragmented to an average size of 50 to 200 bp. Hybridizations were performed with cRNA derived from single biosource (one-color hybridization). Hybridization, washing and scanning were carried out following standard procedures (NimbleGen Systems, Inc.). The expression level of each OLGI was calculated by averaging the intensities of signals from seven different probes. The signals between each array were normalized using RMA (Robust Multi-chip Analysis) normalization (Irizarry et al. 2003a,b,c). GeneSpring 7.1 (Agilent Technologies, CA, USA) was used for further expression analysis.

28.2.7 Calculation of Degrees of Feminization and Male Dysfunction Factors

Euclidean distances and Pearson correlation coefficients (Pearson CC) (Knudsen 2002) between gene expression data were applied in order to estimate degrees of feminization and male dysfunction. The Euclidean distance of a series of expression data $x = \{x_1, x_2, ..., x_n\}$ from origin is defined as

$$d = \sqrt{\sum_{i=1}^{n} (x_i - \underline{x})^2}$$

where $\underline{x}$ is the average values in x (in this case, the means of 26,689 OLGIs in the male, female, or chemicals-treated male groups).

The Pearson CC between two series of expression data $x = \{x_1, x_2, ..., x_n\}$ and $y = \{y_1, y_2, ..., y_n\}$ is defined as

$$\text{Pearson CC} = \cos\theta = \frac{\sum_{i=1}^{n} (x_i - \underline{x})(y_i - \underline{y})}{\sqrt{\sum_{i=1}^{n} (x_i - \underline{x})^2} \sqrt{\sum_{i=1}^{n} (y_i - \underline{y})^2}}$$

where θ is a vector angle between the two data sets, and $\underline{x}$, $\underline{y}$ is the average values in x, y, respectively (in this case, the means of 26,689 OLGIs in the male, female, or chemical-treated male groups). The Pearson CC can range between −1 and 1, with 1 indicating complete identity between the two series, 0 indicating no correlation, and −1 indicating negative correlations between all loci. Euclidean distance preserves the information about the magnitude of changes in the mRNA expression levels directly, whereas the Pearson CC reflects the similarity in expression patterns (Knudsen 2002). Thus, the Euclidean distances of the expression data sets were calculated first, and then the Pearson CC values were combined with the Euclidean distances in order to estimate the degree of feminization and male dysfunction.

28.3 Results and Discussion

28.3.1 Physiological Reproducibility of Medaka Maintained Under Standard Procedure

We first examined the variance in mRNA expression levels between individual male and female fish (Fig. 28.1). Brood stock specimens consisted of two male and two female fish. Additional three different male samples were used as the untreated controls in the estrogenic chemical exposure experiment. The correlation coefficient value was 0.98 between the two female fish and the average of correlation coefficients among male medaka was 0.94 (n=5), indicating that variances between medaka mRNA expression profiles were small. This result implies that highly reproducible physiological data can be obtained from medaka maintained under the standard conditions (OECD 1992; http://www.env.go.jp/chemi/kagaku/). The reproducibility makes it an ideal tool for evaluating medaka responses maintained in different laboratories. For example, a microarray database reflecting the physiological state of medaka raised in standard conditions could serve as the basis for GLP (Good Laboratory Practice) approval by the Japanese Ministry of the Environment. In addition, it was conceivable that only a small number (e.g. 2~3 individuals) of medaka is sufficient for microarray analyses of a particular condition, such as hormone/toxicant exposure, temperature, water quality, etc.

28.3.2 Differential Expression Profiles between Male and Female

In contrast to the high correlation coefficients between the male-male or the female-female comparisons, correlation coefficients in gene expression between individual males and females were lower (Fig. 28.1). The mean of correlation coefficients between individual males and females was 0.907. Among the 26,689

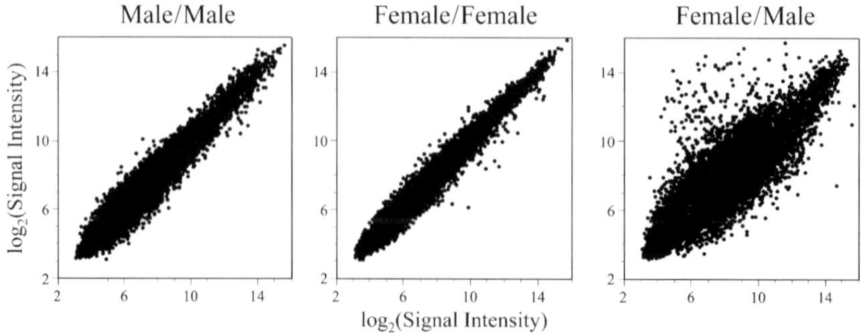

Fig. 28.1 Scatter plots of medaka microarray analyses of mRNA expression: male No. 6 vs. male No. 4 (left), female No. 6 vs. female No. 5 (center), and female No. 6 vs. male No.3 (right)

OLGIs analyzed, we identified OLGIs corresponding to genes preferentially expressed in either male or female. In the comparisons of gene expression between male and female, OLGIs that gave the ratios of averaged expression levels more than two-fold were considered as markers for male- or female-specific genes. By this criterion, 1,243 OLGIs were estimated to represent genes specifically expressed in the female, and 2,276 represented male-specific genes. The female-specific OLGIs represented sequences of previously characterized female-specific genes such as vitellogenin I/II, choriogenin H/L, estrogen receptor, ZP family genes (ZPA, ZPB, and ZPC), cyclin Bs, FIGα, and 42Sp50 (Kanamori 2000). Unlike the female-specific OLGIs, most of the male-specific OLGIs have not yet been assigned or characterized. Nonetheless, these OLGIs are good candidates for male-specific biomarkers. These results suggest that medaka microarrays can be used to distinguish physiological sex by analyzing the expression profiles of sex-specific OLGIs.

28.3.3 Gonadal Histology of Male Medaka Exposed to Estrogenic Chemicals

Estrogenic chemicals such as synthetic estrogens and alkylphenols may change the reproductive status of organisms by adversely affecting endocrine systems (Colborn et al. 1996; Oberdorster and Cheek 2001; Lathers 2002). In medaka, exposure to endogenous steroid hormones or exogenous endocrine disruptors causes inter-sex (testis-ova) or sex reversal in gonad (Yamamoto and Matsuda 1963; Kang et al. 2002a,b; Balch et al. 2004). Particularly, E2, EE2, NP, OP and BpA have been reported to induce vitellogenin and testis-ova in adult male medaka (Gronen et al. 1999; Kang et al. 2002a,b, 2003; Seki et al. 2002). Based on these previous studies, exposure concentrations of 100 ng/L E2 and EE2, 100 μg/L NP and OP, and 2 mg/L BpA were selected in this study. These exposure concentrations were used to ensure histological impairment (e.g. testis-ova formation) observed in the exposed male medaka, although some of the concentrations selected (NP, OP, and BpA) were higher than environmentally-relevant concentrations.

Histological analysis showed the formation of testis-ova in the testes of male medaka exposed to the estrogenic chemicals. Fig. 28.2 shows typical sections of the testis of the control male and the testis-ova gonad. No histologic abnormality was observed in the gonad of the control male fish. High incidences (100%, 8/8) of testis-ova were observed in 100 ng/L E2- and EE2-treated male. Four male of eight males (50%) developed testis-ova by 2 mg/l BpA-treatment, whereas testis-ova were observed in only one of 8 males (12.5%) exposed to 100 μg/l NP or OP. On the other hand, spermatozoa were still observed in the testis of the EE2-treated male (Fig. 28.2), suggesting active spermatogenesis even in the testis-ova gonads. These results suggest that the reproductive activity of the male medaka may be retained even though the estrogenic chemical treatments significantly induce female structures (oocytes) in testes of male medaka.

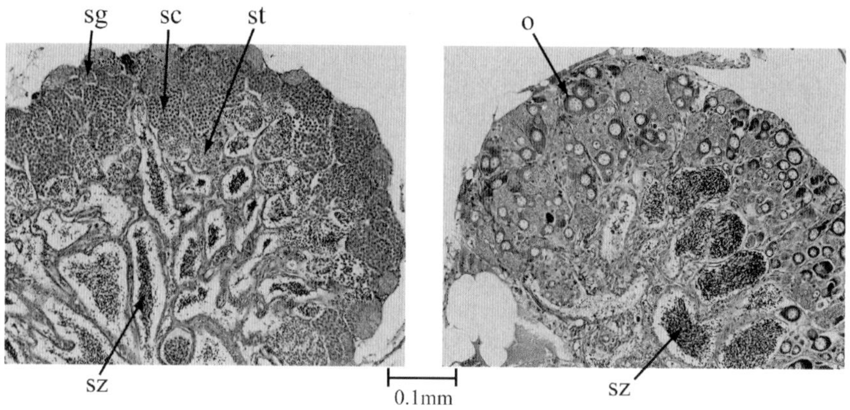

Fig. 28.2 Gonadal sections (5 μm) from male medaka. Testis of a control male (left), and testis-ova of a male exposed to 100 ng/L EE2 (right). sg: spermatogonia, sc:spermatocytes, st: spermatids, sz: spermatozoa, o:oocytes

28.3.4 Effects of Estrogenic Chemicals on Gene Expression Profiles of Male Medaka

Microarray analyses were conducted on the estrogenic chemical-treated male medaka to identify OLGIs responsive to the estrogenic chemicals. cDNAs prepared from three male medaka exposed to each chemical were applied to the microarrays. The average correlation coefficient among the three chemical-treated male medaka was more than 0.96 for each chemical, except for EE2 (the average is 0.90, varied from 0.87 to 0.97), again demonstrating the reproducibility of mRNA expression profiles. Similar expression profiles were observed between E2- and EE2-treated male, whereas NP-, OP-, and BpA-treated male medaka exhibited similar gene expression patterns (data not shown). In comparisons between the gene expression levels of control males and males exposed to each chemical, changes in averaged expression levels more than two-fold and less than 0.5-fold were considered to indicate induction and suppression, respectively. Table 28.1 summarizes the numbers of induced and suppressed OLGIs in male medaka treated with the estrogenic chemicals.

Among the female-specific OLGIs, strong induction of vitellogenins and choriogenins were observed by estrogen (E2 and EE2) (data not shown). Since these genes are not normally expressed in male fish, they are ideal biomarkers for male fish that have been exposed to exogenous estrogenic compounds (Lee et al. 2002; Sumpter and Jobling 1995). Induction of other female-specific and oocyte-specific genes, such as estrogen receptor, ZPs, cyclin Bs, and 42Sp50, was also detected in E2- and EE2-treated male medaka, suggesting that E2- and EE2-treatment activated oocyte formation in testes, supporting the histological examination in this study (Fig. 28.2). On the other hand, alkylphenol (NP, OP, and BpA)-dependent responses in gene expression of those sex-specific OLGIs (e.g. vitellogenins and choriogenins) were much weaker than that in E2- and EE2-treated male (data not shown), indicating the lower estrogenic potency of the alkylphenols.

Table 28.1 Numbers of induced/suppressed OLGIs in male medaka exposed to the estrogenic chemicals

	Induced OLGIs (treated/control > 2)			Suppressed OLGIs (treated/control < 0.5)		
Chemicals	Total	Sex-specific[1]	Non-sex-specific	Total	Sex-specific[1]	Non-sex-specific
E2 (100 ng/l)	157	105	52	106	48	58
EE2 (100 ng/l)	1705	191	1514	1451	506	945
NP (100 μg/l)	46	21	25	119	63	56
OP (100 μg/l)	204	65	139	192	23	169
BpA (2 mg/l)	68	20	48	69	46	23
Common	6	4	2	13	13	0

[1] Sex-specific OLGIs were defined as OLGIs that gave the ratios of averaged expression levels more than 2-fold in the comparisons of gene expression between male and female. By this criterion, 3,519 OLGIs were determined as sex-specific (1,243 female- and 2,276 male-specific) among 26,689 OLGIs.

Interestingly, many of OLGIs regulated by the estrogenic chemicals were non-sex-specific (Table 28.1). Their responsiveness to the estrogenic chemicals may be part of a more generalized response to the toxicity of these chemicals. Furthermore, the large portion of sex-specific OLGIs remained neither induced nor suppressed by the estrogenic chemicals, suggesting that the expression of these sex-specific genes may not be sensitive to these estrogenic chemicals. These observations indicate that estrogenic compounds do not synchronize activation of all female-specific OLGIs. Results in this study demonstrate the possibility of using the medaka microarray to distinguish hormone-responsive genes from genes responsible for other functions, such as detoxification.

28.3.5 Degrees of Feminization and Male-Dysfunction by Estrogenic Chemicals

It has been argued that elevated vitellogenin levels or induction of testis-ova may not correlate with impairment by estrogenic compounds (Yokota et al. 2001; Kang et al. 2002). The question then arises: Even though estrogenic chemicals are known to cause feminization in male medaka (Yamamoto and Matsuda 1963; Kang et al. 2002a,b), to what degree were male medaka "feminized" by estrogenic chemical-treatments? Pair-wise comparisons between the expression levels of individual OLGI sequences (e.g. vitellogenins, choriogenins, and ZPs) cannot comprehensively answer this question. Therefore, we further attempted to estimate the impact of estrogenic chemicals (the degree of feminization and male dysfunction/de-masculinization) using global mRNA expression profiling. For this purpose, we used Euclidean distance and Pearson correlation coefficient (Pearson CC) (Knudsen 2002) to compute distance and similarity, respectively, between gene expression measures.

To determine the degree of feminization or dysfunction of male, only sex specific OLGIs were used, since other OLGIs are likely to involve physiological functions or effects common to both sexes and may lead to correlations pertaining to factors other than sex-specific functions. The average normalized expression levels for each of the 3,519 sex-specific OLGIs in male, female and chemical-treated male were used for calculations.

Figure 28.3 shows the degree of feminization and the degree of male dysfunction/de-masculinization calculated using the Euclidean distances and the Pearson CCs. The feminization factors of male exposed to the estrogenic chemicals, define as the degree that the treated-male "moved" from normal male to female directions, were calculated as ~50% for the E2- and EE2-treated male medaka (Fig. 28.3), suggesting that the estrogen treatments "feminized" male medaka in some degree. While trying to measure the degree of damage on male physiological function by the chemical treatments, we asserted that the distance the treated-male migrated away from the male direction would represent the degree of male dysfunction. The "dysfunction factor" or "de-masculinization factor" for the E2- and EE2-treatment of males, estimated to be 1.3 and 49.6%, respectively, (Fig. 28.3) suggested that the EE2-treatment significantly disrupt male physiological functions whereas no serious male dysfunction was caused by the E2-treatment. On the other hand, according to the feminization/de-masculinization factor, the NP-, OP-, or BpA-treatment did not significantly

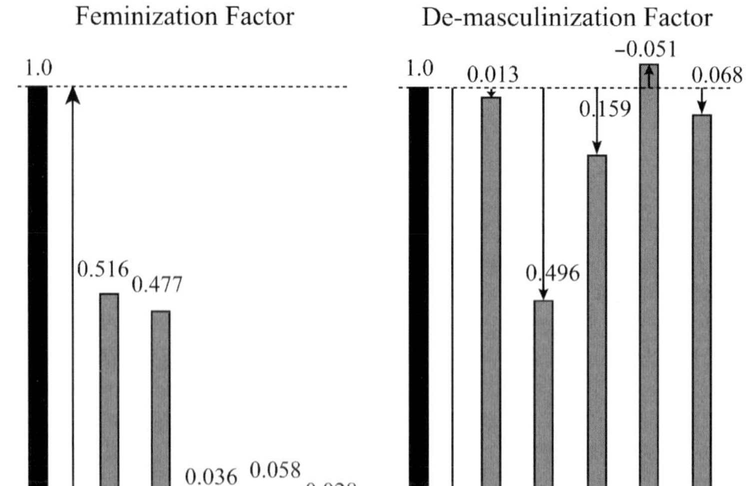

Fig. 28.3 Feminization factors (left) and male dysfunction factors (right) determined using Pearson correlation coefficients and Euclidean distances based on the expression levels of sex-specific OLGIs in each group

feminize male medaka or disrupt male physiological functions in this study (Fig. 28.3), even though some of sex-specific OLGIs were either induced or suppressed by these alkylphenols (Table 28.1).

It has been demonstrated that functional sex reversal in medaka can only be achieved by hormone exposure in a sensitive time period of early development (Yamamoto and Matsuda 1963; Yamamoto 1975; Schartl 2004). By exposing adult males to hormones, the sex of medaka cannot be functionally reversed even though the estrogen-responsive genes (e.g. vitellogenin) are dramatically induced (Sumpter and Jobling 1995; Lee et al. 2002) and the testis-ova are also observed in some cases (Yamamoto and Matsuda 1963; Kang et al. 2002; Balch et al. 2004). The calculated "feminization/de-masculinization factors" for the chemical-treated male also strongly indicate that the estrogenic chemicals treatment in this study did not sexually reverse but only altered physiology of adult male medaka in a degree depending on chemicals.

In this study, we showed the possibility to apply the global mRNA expression profiling to the estimation of the impact of hormonally active chemicals. We propose that the combination of Pearson CC and Euclidean distances could be used to calculate overall effects of hormonally active chemicals, since both the magnitude of changes in the mRNA expression levels and changes in the expression patterns are taken into account. The magnitude of changes in mRNA expression levels, however, varies significantly, so that distance might presumably be overestimated by the influence of highly expressed mRNAs. We would emphasize that although we chose whole body preparation rather than tissue specific analysis in order to establish global gene expression profiles of medaka in response to various chemical exposures, the tissue-specific analysis would probably detect more candidates for important hormone- and sex-specific gene expressions by avoiding dilution of the tissue-specific RNA by other tissues. Therefore, tissue-specific microarray analysis of hormone-dependent gene expression and weighting genes truly responsible for sexual differentiation/determination would be necessary to accurately quantify the degree of feminization or masculinization in medaka caused by hormonally active chemicals.

28.4 Conclusion

We have demonstrated that medaka microarray provided highly reproducible data to evaluate the physiological state of medaka male and female, and estrogenic chemical-responsive gene expressions in the male. The microarray approach thus identified signature gene subsets indicative of differentially expressed in normal males and females. Because those sex-specific genes must contain many feminization and male-dysfunction markers, we evaluated feminization and male-dysfunction factors of the estrogenic chemical-treated males using those gene expressions. The calculated factors seemed to be in line with the histological analysis, indicating that this could be the useful approach to assess specific hormonal activities of chemicals in medaka, and other fish models.

References

Balch G.C., Mackenzie C.A., and Metcalfe C.D. (2004), Alterations to gonadal development and reproductive success in Japanese medaka (*Oryzias latipes*) exposed to 17α-ethinylestradiol, *Environ. Toxicol. Chem.*, 23, 782–791.

Colborn T., Dumanoshi D., and Myers J.P. (1996), *Our Stolen Future*. (Dutton, New York).

Eberwine J., Yeh H., Miyashiro K., Cao Y., Nair S., Finnell R., Zettel M., and Coleman P. (1992), Analysis of gene expression in single live neurons, *Proc. Nat. Acad. Sci. U. S. A.*, 89, 3010–3014.

Gronen S., Denslow N., Manning S., Barnes S., Barnes D., and Brouwer M. (1999), Serum vitellogenin levels and reproductive impairment of male Japanese medaka (*Oryzias latipes*) exposed to 4-*tert*-octylphenol, *Env. Health Perspect.*, 107, 385–390.

Irizarry R.A., Bolstad B.M., Collin F., Cope L.M., Hobbs B., and Speed T.P. (2003a), Summaries of Affymetrix GeneChip probe level data, *Nucl. Acids Res.*, 31, e15.

Irizarry R.A., Gautier L., and Cope L.M. (2003b), *An R package for analysis of Affymetrix oligonucleotide arrays*. (Springer, Berlin).

Irizarry R.A., Hobbs B., Collin F., Beazer-Barclay Y.D., Antonellis K.J., Scherf U., and Speed T.P. (2003c), Exploration, normalization, and summaries of high density oligonucleotide array probe level data, *Biostatistics*, 4, 249–264.

Irwin R.D., Boorman G.A., Cunningham M.L., Heinloth A.N., Malarkey D.E., and Paules R.S. (2004), Application of toxicogenomics to toxicology: basic concepts in the analysis of microarray data, *Toxicol. Pathol.*, 32 (Suppl. 1), 72–83.

Kanamori A. (2000), Medaka as a model for gonadal sex differentiation in vertebrates, *Tanpakushitsu Kakusan Koso*, 45 (Suppl. 17), 2949–2953.

Kang I.J., Yokota H., Oshima Y., Tsuruda Y., Oe T., Imada N., Tadokoro H., and Honjo T. (2002a), Effects of bisphenol A on the reproduction of Japanese medaka (*Oryzias latipes*), *Env. Toxicol. Chem.*, 21, 2394–2400.

Kang I.J., Yokota H., Oshima Y., Tsuruda Y., Yamaguchi T., Maeda M., Imada N., Tadokoro H., and Honjo T. (2002b), Effect of 17β-estradiol on the reproduction of Japanese medaka (*Oryzias latipes*), *Chemosphere*, 47, 71–80.

Katogi R., Nakatani Y., Shin-i T., Kohara Y., Inohaya K., and Kudo A. (2004), Large-scale analysis of the genes involved in fin regeneration and blastema formation in the medaka, *Oryzias latipes. Mech. Dev.*, 121, 861–872.

Kimura T., Jindo T., Narita T., Naruse K., Kobayashi D., Shin I.T., Kitagawa T., Sakaguchi T., Mitani H., Shima A., Kohara Y., and Takeda H. (2004), Large-scale isolation of ESTs from medaka embryos and its application to medaka developmental genetics, *Mech. Dev.*, 121, 915–932.

Knudsen S. (2002), *A Biologist's Guide to Analysis of DNA Microarray Data*. (John Wiley & Sons, Inc., New York).

Lathers C.M. (2002), Endocrine disruptors: a new scientific role for clinical pharmacologists? Impact on human health, wildlife, and the environment, *J. Clin. Pharmacol.*, 42, 7–23.

Lee C., Na J.G., Lee K.C., and Park K. (2002), Choriogenin mRNA induction in male medaka, *Oryzias latipes* as a biomarker of endocrine disruption, *Aquat. Toxicol.*, 61, 233–241.

Matsuda M., Nagahama Y., Shinomiya A., Sato T., Matsuda C., Kobayashi T., Morrey C.E., Shibata N., Asakawa S., Shimizu N., Hori H., Hamaguchi S., and Sakaizumi M. (2002), DMY is a Y-specific DM-domain gene required for male development in the medaka fish, *Nature*, 417, 559–563.

Nuwaysir E.F., Huang W., Albert T.J., Singh J., Nuwaysir K., Pitas A., Richmond T., Gorski T., Berg J.P., Ballin J., McCormick M., Norton J., Pollock T., Sumwalt T., Butcher L., Porter D., Molla M., Hall C., Blattner F., Sussman M.R., Wallace R.L., Cerrina F., and Green R.D. (2002), Gene expression analysis using oligonucleotide arrays produced by maskless photolithography, *Genome Res.*, 12, 1749–1755.

Oberdorster E. and Cheek A.O. (2001) Gender benders at the beach: endocrine disruption in marine and estuarine organisms, *Env. Toxicol. Chem.*, 20, 23–36.

Organization for Economic Co-operation and Development (1992), *Fish, acute toxicity test, in Guidelines for Testing of Chemicals,* 203, 1–9.

Ozato K., Wakamatsu Y. and Inoue K. (1992) Medaka as a model of transgenic fish, *Mol. Mar. Biol. Biotech.,* 1, 346–354.

Ramsay G. (1998), DNA chips: state-of-the art, *Nature Biotechnology,* 16, 40–44.

Schartl M. (2004), A comparative view on sex determination in medaka, *Mech. Dev.,* 121, 639–645.

Schena M., Heller R.A., Theriault T.P., Konrad K., Lachenmeier E., and Davis R.W. (1998), Microarrays: biotechnology's discovery platform for functional genomics, *Trends Biotechnol.,* 16, 301–306.

Seki M., Yokota H., Matsubara H., Tsuruda Y., Maeda M., Tadokoro H., and Kobayashi K. (2002), Effect of ethinylestradiol on the reproduction and induction of vitellogenin and testis-ova in medaka (*Oryzias latipes*), *Env. Toxicol. Chem.,* 21, 1682–1698.

Singh-Gasson S., Green R.D., Yue Y., Nelson C., Blattner F., Sussman M.R., and Cerrina F. (1999), Maskless fabrication of light-directed oligonucleotide microarrays using a digital micromirror array, *Nat. Biotechnol.*, 17, 974–978.

Snape J.R., Maund S.J., Pickford D.B., and Hutchinson T.H. (2004), Ecotoxicogenomics: the challenge of integrating genomics into aquatic and terrestrial ecotoxicology, *Aquat. Toxicol.,* 67, 143–154.

Sumpter J.P. and Jobling S. (1995), Vitellogenesis as a biomarker for estrogenic contamination of the aquatic environment, *Env. Health Persp.*, 103 (Suppl. 7), 173–178.

U.S. Environmental Protection Agency (1993). *Methods for measuring the acute toxicity of effluents and receiving waters to freshwater and marine organisms, 4th ed.* (U.S. EPA, Cincinnati, OH (Publication No. EPA/600/4-90-027F)).

Yamamoto T. (1975). *Medaka (Killifish): Biology and Strains.* (Keigaku, Tokyo).

Yamamoto T.O. and Matsuda N. (1963), Effects of estradiol, stilbestrol and some alkyl-carbonyl androstanes upon sex differentiation in the medaka, *Orvzias latipes. Gen. Comp. Endocrinol.,* 3, 101–110.

Yokota H., Seki M., Maeda M., Oshima Y., Tadokoro H., Honjo T., and Kobayashi K. (2001), Life-cycle toxicity of 4-nonylphenol to medaka (*Oryzias latipes*), *Env. Toxicol. Chem.,* 20, 2552–2560.

Chapter 29
Assessment of the Hazard Potential of Environmental Chemicals by Quantifying Fish Behaviour

Daniela Baganz and Georg Staaks

Abstract Using the spontaneous locomotor behaviour of fish as a toxicological parameter, sublethal effects to the naturally occurring cyanotoxin microcystin-LR (MC-LR) and a characteristic man-made chemical 2.4.4′-trichlorobiphenyl (PCB 28) were investigated under laboratory conditions. Swimming activity of two fish species (*Danio rerio* and *Leucaspius delineatus*) was monitored continuously by using an automated video-monitoring and object-tracing system. For analysing cyclic aspects the basic behavioural analyses were combined with chronobiological procedures such as power spectral analysis. Using these methods it was shown that dissolved MC-LR concentrations between 0.5 and 50 μg l^{-1} and PCB 28 concentrations at 100 and 150 μg l^{-1} acted as stressors and caused significant changes in the behaviour and circadian activity rhythms of *Danio rerio* as well as *Leucaspius delineatus*. For both species elevated concentrations of the stressors led to a reduction of their activity. It was proved that the basic behavioural analyses combined with chronobiological procedures could be valuable tools for the study of stressful or even harmful environmental factors in the field of ecotoxicology as well as for biomonitoring. Some findings of this study build the basis for the development of a new low-budget fish biomonitoring system for drinking water protection.

Keywords: Behavioural ecotoxicology, fish, chemical stressors, early warning systems of drinking water quality

Department of Biology and Ecology of Fishes, Leibniz-Institute of Freshwater Ecology and Inland Fisheries, Berlin, Germany

Y.J. Kim and U. Platt (eds.), *Advanced Environmental Monitoring*, 376–389.

29.1 Introduction

Behaviour is the result of the interaction(s) of an organism with its environment. Thus it represents the integration of underlying physiological processes with the environmental stimuli that trigger them and the evolutionary forces that have shaped them and continue to do so (Grue et al. 2002). Behavioural tests addressing whole organism-level effects provide primary signalling about a wide range of toxic compounds in water (e.g. Reide and Siegmund 1989; Boujard and Leatherland 1992; Spieser et al. 2000; Campbell et al. 2002; Schmidt et al. 2004; Baganz et al. 2004).

In evaluating the impact of stressors in freshwater systems, fish have a special importance because they are situated at the end of the aquatic food chain, and thus may also indicate a contamination with persistent pollutants at lower trophic levels. Therefore research on fish behaviour as an indicator of toxic effects is currently receiving more and more attention (Chon et al. 2002).

To analyse pollutant-induced effects on fish behaviour, we investigated the cyanobacteria toxin microcystin-LR (MC-LR) and a trichlorobiphenyl (PCB 28) at sub-lethal levels. Both substances are widespread in the aquatic environment, but there is rather little knowledge about their impact on fish behaviour.

Microcystins are cyanobacterial hepatotoxins that are produced by some cyanobacterial genera, e.g., *Microcystis, Anabaena, Oscillatoria, Nostoc, and Anabaenopsis*. The acute hepatotoxic symptoms that result from microcystin exposure are generally caused by binding and inhibiting the serine/threonine protein phosphatase types 1 and 2A (MacKintosh et al. 1990, 1995; Solter et al. 1998).

The ortho-substituted congener (non-coplanar PCB) 2,4,4′ trichlorobiphenyl (PCB 28; Ballschmiter and Zell 1980) is one of 7 PCB congeners which has been identified by the International Council for the Exploration of the Seas (ICES) as markers of the degree of contamination (Garritano et al. 2006).

Potential species-specific reactions to chemical stressors were regarded by choosing two different fish species: the tropical species *Danio rerio* (zebrafish) that is a model system for integrative physiology and toxicology and the temperate species *Leucaspius delineatus* (sunbleak).

Because all behavioural reactions of organisms are essentially coupled with natural processes and physiological reactions that mainly have rhythmic components, even behavioural parameters normally occur in rhythmic structures. To consider these cyclic aspects the basic behavioural analyses were combined with chronobiological procedures such as time series analysis and power spectral analysis. Analysis of the degree of synchronisation between activity rhythms and their zeitgeber (e.g. by power spectral analysis, effects of zeitgeber) as well as the quantification of the harmonic frequency structure of activity rhythms (by power spectral analysis) proved to be good indicators for environmental changes, corresponding to the findings of Siegmund and Biermann (1989, 1990); Scheibe et al. (1999) and Baganz et al. (2005).

This study is engaged in the still emerging field of behavioural ecotoxicology which integrates the three different disciplines: ethology, toxicology and ecology (Dell'Omo 2002). Furthermore, the methods of behavioural analyses used to indicate the presence of a stressor are related to the field of biomonitoring. Many biological early warning systems (BEWS) that evaluate the physiological and behavioural responses of whole organisms to water quality have been developed in recent years (Van der Schalie et al. 2001). BEWS have proved to be very useful for continuous monitoring of (industrial) effluents, water intake, and river or seawater quality control (Se Zwart et al. 1995).

29.2 Materials and Methods

For both species (*Danio rerio* and *Leucaspius delineatus*), 6 schools of seven adult individuals each were kept in 15-litre glass aquaria with a swimming space of 40 × 25 × 15 cm. For *Danio rerio* one school consisted of three females and four males. For *Leucaspius delineatus* the sex ratio could not be ascertained *in vivo*. Mean total length, mean body mass and age of individuals are shown in Table 29.1.

The experiments aimed to maintain very constant external conditions concerning water quality parameters, artificial illumination in a distinct time regime, avoidance of optical or visual perturbations and exclusion of noise and vibrations. Feeding, temperature and artificial light/dark rhythms were constant and automatically controlled. Fish were exposed to a 12:12 h light/dark rhythm (without any natural light). The animals were automatically fed with TetraMin flakes twice a day at a ratio of 3% body mass per day, 3 and 7 h after light-on.

For the experiments, aerated drinking water ("Berlinwasser Holding Friedrichshagen") was used. The physico-chemical composition of the water is listed in Table 29.2. Basic water quality parameters of pH, oxygen and ammonium were measured both in storage tanks and aquaria once a day. The physico-chemical parameters of the used aquarium water were constant over the exposure period and within the normal physiological ranges for fish (Schäperclaus 1991; Schreckenbach et al. 2001).

After 3 weeks of acclimatisation to the test conditions the behaviour of all fish groups was recorded under standard test conditions. Thereafter, four groups of both species were exposed to the test substances MC-LR or PCB 28. For the experiments with MC-LR both fish species were exposed to four different (nominal) concentrations of MC-LR: 0.5, 5 and 15 μg l^{-1} for a period of 17 days each and 50 μg l^{-1} for

Table 29.1 Mean total length (TL), mean body mass (BM) and age of *Danio rerio* and *Leucaspius delineatus* for the tests with MC-LR and PCB 28

	Danio rerio			Leucaspius delineatus		
	TL [cm]	BM [g]	Age [d][1]	TL [cm]	BM [g]	Age [d][1]
MC-LR	3.45 ± 0.26	0.25 ± 0.02	160–171	4.55 ± 0.22	0.30 ± 0.02	170–185
PCB 28	3.47 ± 0.25	0.26 ± 0.03	160–171	4.57 ± 0.21	0.31 ± 0.03	170–185

[1] Age of the individuals at the start of each experiment

Table 29.2 Physico-chemical parameters of the aquarium water for the experiments with *Danio rerio* and *Leucaspius delineatus* under the influence of MC-LR and PCB 28

Parameter	Unit	Measured value
Temperature	°C	26 ± 0.5[a] 20 ± 0.5[b]
Conductivity	$\mu S\ cm^{-1}$	720 ± 10
pH- value		7.5–7.7
Total hardness	°dH	16.5 ± 0.5
Total organic carbon (TOC)	$mg\ l^{-1}$	4.1 ± 0.4
Oxygen	$mg\ l^{-1}$	7.7 ± 0.2[a] 8.2 ± 0.3[b]
Ammonium	$mg\ l^{-1}$	<0.5
Iron	$mg\ l^{-1}$	<0.03
Nitrate	$mg\ l\text{-}1^{-1}$	5 ± 0.2
Nitrite	$mg\ l^{-1}$	<0.03

[a] For the experiments with *Danio rerio*
[b] For the experiments with *Leucaspius delineatus*

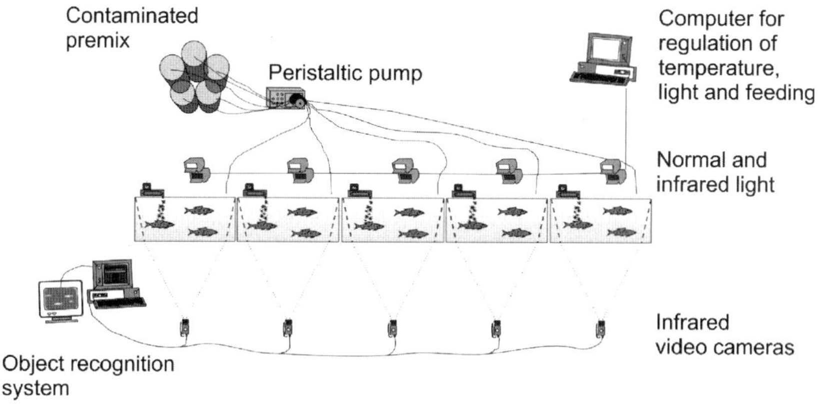

Fig. 29.1 Scheme of the equipment of the BehavioQuant system

a period of 6 days. For the experiments with PCB 28, four groups of fish were exposed to two different (nominal) concentrations of PCB 28 (duplicates): 100 and 150 $\mu g\ l^{-1}$ for a period of 8 days.

The fish activity was monitored continuously with the automated video processing system BehavioQuant (Spieser et al. 2000). The experimental design is shown in Fig. 29.1. Fish were observed by video cameras, one in front of each tank, which were able to handle normal as well as infrared light, enabling continuous observation even during the night. The positions of the untagged fish were recorded in a two-dimensional area, data were digitised and paths of individual fish afterwards tracked by the object recognition software. Thus it was possible to reconstruct the real movements of every fish in the school.

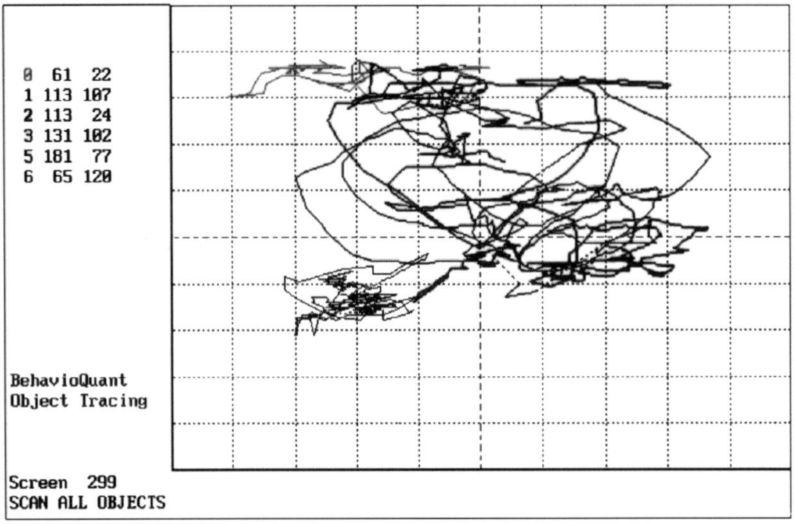

Fig. 29.2 Screen shot of the movement tracks of one fish group during one measuring interval of 2 minutes. The different lines represent the single individuals

The screen shot (Fig. 29.2) shows an example of the movement tracks of one fish school. Video was filmed at a frequency of 25 frames per second, and overall there were 69 measuring cycles per day, i.e. 3 cycles per hour. Each single measuring cycle lasted two minutes. Each picture of the experimental chamber was in real time compared point-by-point with a background reference picture. The x-y positions of recognized objects were written to disk for each measuring interval. The raw data were pre-processed and converted into tables which contained the behavioural parameter values: the motility is the swimming velocity in video-pixels per second and the turns are the number of changes of the direction per second. Motility and turns characterised the swimming activity.

29.2.1 Statistics and Calculations

29.2.1.1 Mean Motility Analysis

Mean motility was analysed over the whole measured time per day (23 h d^{-1}) and over the whole exposure period. The time interval of three hours after the onset of light was specifically regarded, whereby the data were averaged per hour. On these bases the results of the exposed groups were statistically compared with those of the controls reflecting the same time period.

Statistical analysis of results was performed in SPSS 11.5. using its advanced models one-factorial ANOVA procedure. Homogeneity of variances tested by Levenes test proved that we could not assume equal variances for all of the groups. Following this, we used the Welch statistic instead of the F statistic as well as the Dunnett's T3 post hoc test for comparison of groups. This pair-wise comparison test based on the Studentized maximum modulus is appropriate when variances are unequal.

29.2.1.2 Effects of Zeitgeber

Effects of zeitgeber (time trigger) that were calculated as a quotient of the motility during the light phase and the overall motility during the light and dark phase allow a distinction between diurnal and nocturnal activity rhythms of the test species. Values between 0 and 0.5 indicated that the animals were nocturnally active and values between 0.5 and 1 that they were diurnally active. The student's t-Test of SPSS 11.5. was used to compare the group values to 0.5 indicating statistical significance of diurnality or nocturnality of fish. Generally significant differences were accepted at $p < 0.05$. In all figures and tables a significance level of

1. $p < 0.05$ is indicated by one asterisk,
2. $p < 0.01$ is indicated by two asterisks, and
3. $p < 0.005$ is indicated by three asterisks.

29.2.1.3 Power Spectral Analysis

A power spectral analysis of the motility, which is a Fourier transformed autocorrelation function, was used to quantify the harmonic frequency structure of the activity rhythms of *Danio rerio* and *Leucaspius delineatus*. The calculations were performed by means of the program "Zeit" the application of which is described in Scheibe et al. (1999). Periodic frequencies which explain a significant proportion of the total variation of the original data series and which are furthermore harmonic to the circadian period were ascertained. Periods are called harmonic in a chronobiological context if their lengths are integer dividers of 24 h. All periods of the power spectra were tested for significance by the integrated function of the program "Zeit" (see Scheibe et al. 1999).

29.3 Results and Discussion

It was shown that dissolved MC-LR concentrations between 0.5 and 50 μg l^{-1} and PCB 28 concentrations at 100 and 150 μg l^{-1} acted as stressors and caused significant changes in the behaviour and circadian activity rhythms of *Danio rerio* as well as *Leucaspius delineatus*. So for both species elevated concentrations of the stressors (MC-LR and PCB 28) led to a reduction in their activity. This reduced reaction of fish to light under chemical stress conditions is exemplarily illustrated by the

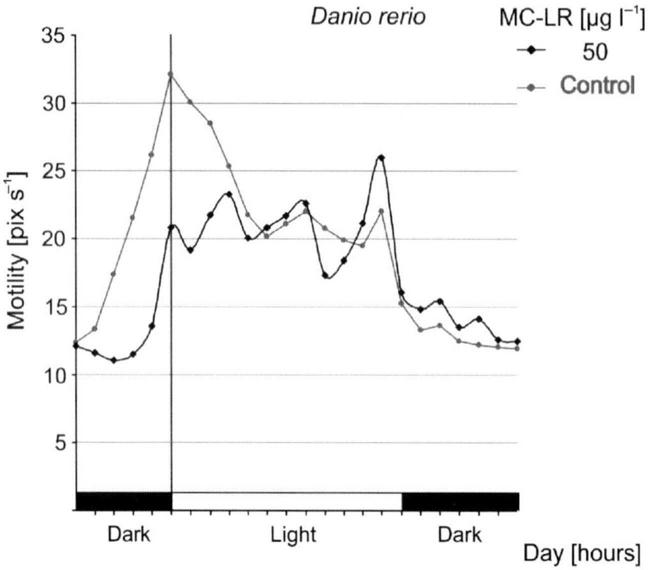

Fig. 29.3 Smoothed curve of average motility of *Danio rerio* over the whole period of exposure to MC-LR (50 µg l^{-1}) compared to the control

daily activity curves for *Danio rerio* (Fig. 29.3). Further evaluation focused on the time interval of three hours after the onset of light, because the changes between exposure groups and controls were especially evident during this period of the day (Fig. 29.4 and Fig. 29.5). This was the time period of the daily mating and spawning behaviour of the species *Danio rerio*.

The changes in the time interval of three hours after the onset of light were statistically significant for the lowest MC-LR concentration of 0.5 µg l^{-1} which led to an increase in motility and for the highest MC-LR concentration of 50 µg l^{-1} which led to a decrease in motility (Fig. 29.4). Under the influence of PCB 28 the motility of *Danio rerio* as well as *Leucaspius delineatus* was significantly reduced for both test concentrations of 100 µg l^{-1} and 150 µg l^{-1} (Fig. 29.5).

Furthermore under the influence of the MC-LR and PCB 28 a degree of desynchronisation of the activity to the zeitgeber light was brought about in both fish species. This was indicated by the effects of zeitgeber that allow a distinction between diurnal and nocturnal activity rhythms of the test species (Figs. 29.6 and 29.7). For *Danio rerio* exposed to MC-LR a phase delay occurred, whereby at all concentrations these changes could only be registered during the light phase. Therefore, *Danio rerio* remained diurnally active. In contrast, the phase of *Leucaspius delineatus* advanced, whereby this shift was so drastic that a phase reverse occurred, and this species became significantly nocturnal.

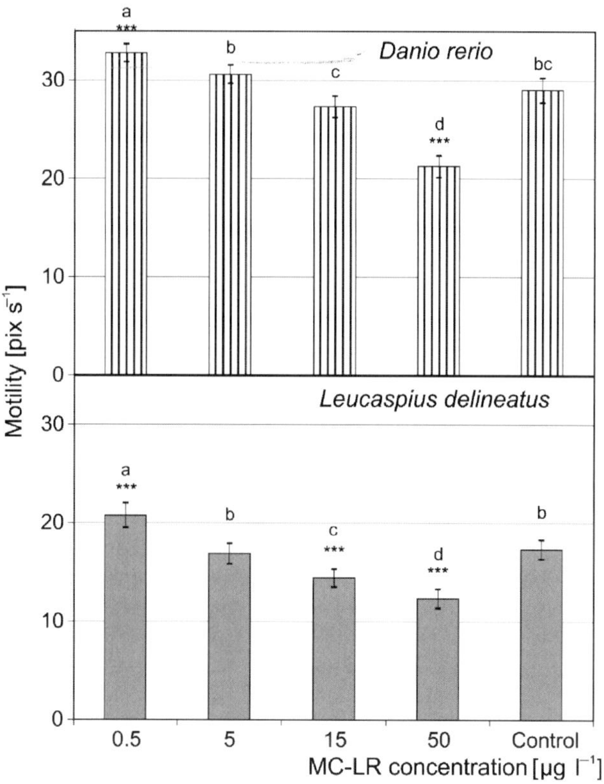

Fig. 29.4 Average motility of *Danio rerio* and *Leucaspius delineatus* during the time interval of 3 hours after the onset of light over the whole period of exposure to MC-LR. Significant differences of the exposure groups compared with those of the controls are indicated by the asterisks. Different letters indicate significantly different groups. Means and 95% confidence intervals are shown

Influenced by PCB 28 a degree of desynchronisation to the light/dark change, which led to a phase advance, was only found for *Leucaspius delineatus*. Both species remained diurnally active, whereby the values of the effects of zeitgeber were significantly reduced for *Danio rerio* as well as *Leucaspius delineatus*.

To evaluate further cyclic aspects, the basic behavioural analyses were combined with chronobiological procedures such as power spectral analysis of the motility data. The power spectral analysis with their amplitude coefficients showed that the dominance of the circadian rhythmic peak (of 24 h) was reduced under the influence of chemical stressors. For *Danio rerio* the explained proportion of the 24-hour period increased at lower MC-LR concentrations of 0.5 and 5 μg l^{-1} and decreased at elevated MC-LR concentrations of 15 and 50 μg l^{-1} (Fig. 29.8).

For *Leucaspius delineatus* at elevated MC-LR concentrations of 5 μg l^{-1}, 15 μg l^{-1} and 50 μg l^{-1} the dominance of circadian rhythms ($\tau = 24$ h) was clearly reduced and

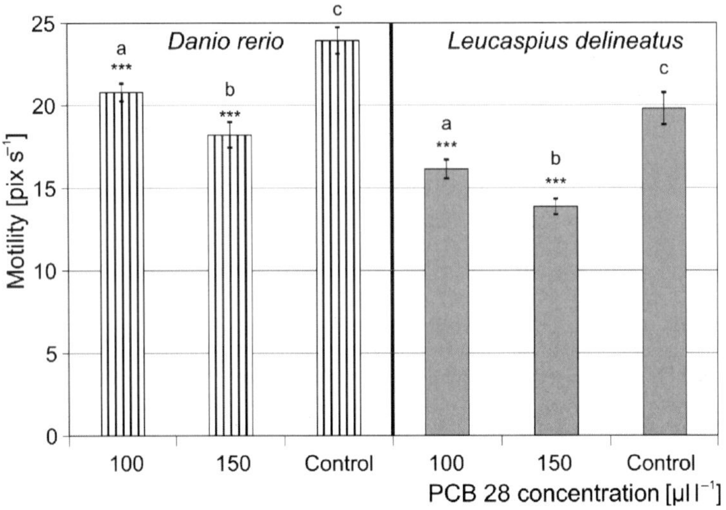

Fig. 29.5 Average motility of *Danio rerio* and *Leucaspius delineatus* during the time interval of 3 hours after the onset of light over the whole period of exposure to PCB 28. Significant differences of the exposure groups compared with those of the controls are indicated by the asterisks. Different letters indicate significantly different groups. Means and 95% confidence intervals are shown

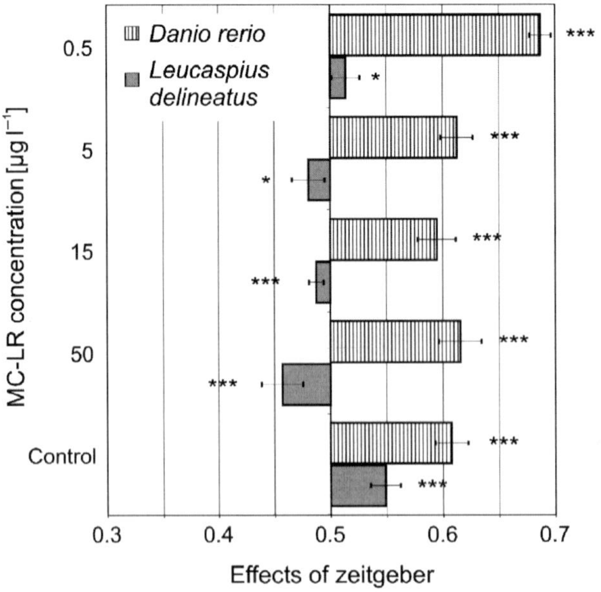

Fig. 29.6 Effects of zeitgeber with 95% confidence intervals for *Danio rerio* and *Leucaspius delineatus* exposed by MC-LR. Asterisks indicate significant differences from the value of 0.5. At values between 0 and 0.5, the animals are nocturnally active and between 0.5 and 1 they are diurnally active. Means and 95% confidence intervals are shown

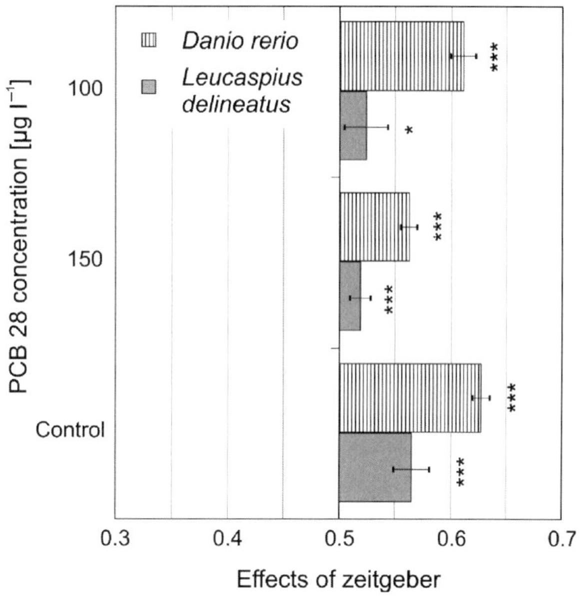

Fig. 29.7 Effects of zeitgeber with 95% confidence intervals for *Danio rerio* and *Leucaspius delineatus* exposed by PCB 28. Asterisks indicate significant differences from the value of 0.5. At values between 0 and 0.5, the animals are nocturnally active and between 0.5 and 1 they are diurnally active. Means and 95% confidence intervals are shown

simultaneously the proportion of a harmonic oscillation with a 12-hour rhythm increased (Fig. 29.8).

The findings in this study indicate that the non-invasive automatic registration of activity data is a suitable approach to more sensitive ecotoxicological research methods and practicable for a range of further applications such as biomonitoring systems. Automated biomonitoring or biological early warning systems are defined as systems that detect toxic conditions on a continuous basis in whole organisms (Butterworth et al. 2000). In contrast to physico-chemical analyses biomonitors facilitate an unspecific indication of pollutants including synergistical and antagonistical effects in water monitoring. Using behavioural endpoints for biomonitoring has the advantages that

1. their high sensitivity is comparable with other toxicological tests, e.g., enzymatic tests,
2. they have the capability of an online monitoring process without disturbing the test organisms and
3. organisms respond with behavioural changes within short time periods (Blübaum-Gronau et al. 2000).

Since critical concentrations at a very low level can be measured and potential dose-effect relationships can be registered, biomonitoring using fish behaviour is about to become a standard. Some research has to be done along this path to simplify the modelling of answer reactions to make its detection more reliable and more independent of the kind of reaction. So, the results of this study show that in some cases it is necessary to register the absolute deviation from standard values

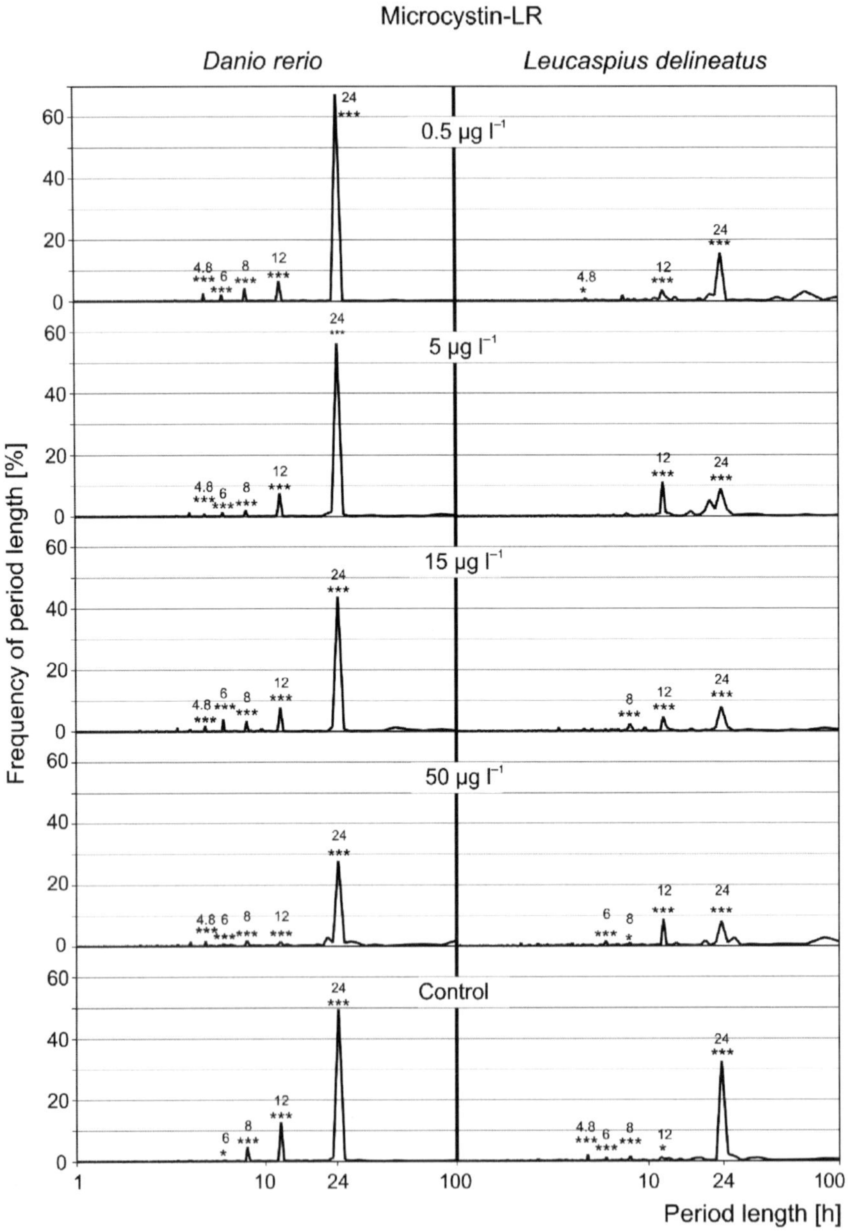

Fig. 29.8 Power spectrum of motility of *Danio rerio* and *Leucaspius delineatus* exposed to MC-LR

independent on their direction (e.g. if there is an increase or a decrease in activity and/or in the calculated values of rhythmical parameters).

The next stage of the work is the development of a low-budget biomonitoring system for drinking water protection, which is already in progress. This project is

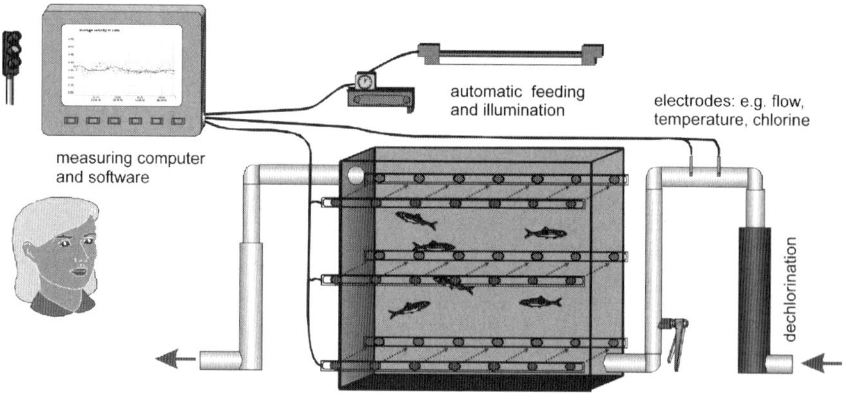

Fig. 29.9 Scheme of the equipment of the low-budget biomonitoring system

an integrated part of the EU funded Project "TECHNEAU"[1]. The main application of the system will consist of a continuous monitoring of drinking water in a distribution system to prevent terrorist attacks or major accidents, which can severely affect the water quality and thus potentially have a high impact on human health.

For this purpose the system needs to

1. respond quickly to changes in water quality, because quick action may be necessary to prevent contaminated water from reaching the public,
2. produce reliable results (no false alarms), because the drinking water supply must not be shut down for no good reason.

To be accepted in practice and affordable for small scale water suppliers and for enterprises in developing countries the system must:

be cheap,
require almost no maintenance at all (so must be robust),
be operated by remote control,
not need highly skilled personnel to interpret the data from the sensor.

This requires this system to be produced with minimal expenditure on hard- and software by using photoelectric barriers. So, instead of the highly sophisticated video-processing system used in the study presented, a photoelectric barrier system have to be applied (Fig. 29.9). Similar to both systems is the utilisation of fish behaviour as a parameter to indicate water pollution.

Through financing the integrated project "TECHNEAU", the European Commission has decided to stimulate the development and application of innovative and cost-effective European strategies and technologies for safe drinking water supply and the methods of behavioural biomonitoring can significantly contribute to this aim.

[1] TECHNEAU: Technology Enabled Universal Access to Safe Water (http://www.techneau.org)

References

Baganz D., Staaks G., Pflugmacher S., and Steinberg C. (2004), A Comparative Study on Microcystin-LR induced behavioural changes of two fish species (*Danio rerio* and *Leucaspius delineatus*). *Environ. Toxicol.*, 19, 564–570.

Baganz D., Siegmund R., Staaks G., Pflugmacher S., and Steinberg C. (2005), Temporal pattern in swimming activity of two fish species (*Danio rerio* and *Leucaspius delineatus*) under chemical stress conditions. *Biol. Rhythm. Res.*, 36(3), 263–276.

Ballschmiter K., and Zell M. (1980), Analysis of polychlorinated biphenyls (PCB) by glass capillary gas chromatography. *Fresen. J. Anal. Chem.*, 302, 20–31.

Blübaum-Gronau E., Hoffmann M., Spieser O.H., and . Scholz W. (2000), Continuous water monitoring. In Butterworth F.M., Gunatilaka A.,, and Gonseblatt M.E., (Eds.), *Biomonitors and Biomarkers Biomarkers as Indicators of Environmental Change* (pp. 123–141). Kluwer Publ., New York, 2, ISBN 0-306-46387-3.

Butterworth F.M., Gunatilaka A., and Gonsebatt M.E. (Eds.) (2000). *Biomonitors and Biomarkers as Indicators of Environmental change.* (Kluwer Academic/Plenum Publishers. New York, 2).

Boujard T., and Leatherland J.F. (1992), Circadian rhythms and feeding time in fishes. *Environ. Biol. Fish.*, 35, 109–131.

Campbell H.A., Handy R.D., and Sims D.W. (2002), Increased metabolic cost of swimming and consequent alterations to circadian activity in rainbow trout (*Oncorhynchus mykiss*) exposed to dietary copper. *Can. J. Fish. Aquat. Sci.*, 59, 768–777.

Chon T.S., Kwak I.S., Song M.Y., Ji C.W., Kim C.K., Cha E.Y., Koh S.C., Kim J.S., Leem J.B., and Lee S.K. (2002), Self-organizing Mapping on response behavior of indicator species exposed to toxic chemicals for developing automatic bio-monitoring system in aquatic environment. *Int. J. Ecol. Model. (SCI).*

Dell'Omo G. (Ed.) (2002). *Behavioural Ecotoxicology.* (John Wiley and Sons, LTD, ISBN 0-471-96852-8).

Garritano S., Pinto B., Calderisi M., Cirillo T., Amodio-Cocchieri R., and Reali D. (2006, March 30), Estrogen-like activity of seafood related to environmental chemical contaminants. *Environ Health.*, 5:9, from http://www.ehjournal.net/content/5/1/9.

Grue C.E., Gardner S.C., and Gibert P.L. (2002), On the Significance of Pollutant-induced Alterations in the Behaviour of Fish and Wildlife. In Dell'Omo G. (Ed.), *Behavioural Ecotoxicology* (pp. 1–90). John Wiley and Sons, LTD. ISBN 0-471-96852-8.

MacKintosh C., Beattie K.A., Klumpp S., Cohen P., and Codd G.A. (1990), Cyanobacterial microcystin-LR is a potent and specific inhibitor of protein phosphatases 1 and 2 A from both mammals and higher plants. *FEBS Lett.*, 264, 187–192.

MacKintosh R.W., Dalby K.N., Campbell D.G., Cohen P.T., Cohen P., and MacKintosh C. (1995), The cyanobacterial toxin microcystin binds covalently to cysteine-273 on protein phosphatase 1. *FEBS Lett.*, 371, 236–240.

Reide M., and Siegmund R. (1989), Time pattern analysis of swimming activity and heart rate under the influence of chloramphenicol in carp (*Cyprinus carpio* L.) and rainbow trout (*Salmo gairdneri* R.). *Aquaculture*, 80, 315–324.

Schäperclaus W. (Ed.) (1991), *Fish Diseases.* (5th corr., rev. and subst. enl. ed. Oxonian Press, New Delhi, ISBN 81-7087-057-7).

Scheibe K. M., Berger A., Langbein J., Streich W.J., and Eichhorn K. (1999), Comparative analysis of ultradian and circadian behavioural rhythms for diagnosis of biorhythmic state of animals. *Biol. Rhythm. Res.*, 30, 216–233.

Schreckenbach K., Knösche R., and Ebert K. (2001), Nutrient and energy content of freshwater fishes. *J. Appl Ichthyol.*, 17, 1–3.

Se Zwart D., Kramer K.J.M., and Jenner H.A. (1995), Practical experiences with the biological early warning system Mosselmonitor. *Environ. Toxicol. Water Qual.*, 10(4), 237–247.

Schmidt K., Steinberg C.E.W., Pflugmacher S., and Staaks G. (2004), Xenobiotic substances such as PCB mixtures (Arochlor 1254) and TBT can influence swimming behavior and biotransformation activity (GST) of carp (*Cyprinus carpio*). *Environ. Toxicol.*, 19(5), 460–470.

Siegmund R., and Biermann K. (1989), Chronobiological parameters as indicators of environmental alteration on fish. *Chronobiologia* 16, 181.

Siegmund R., and Biermann K. (1990), Chronobiological parameters as indicators of a disturbed organism-environmental relation in fish. *Prog. Clin. Biol. Res.*, 341B, 627–637.

Solter P.F., Wollenberg G.K., Huang X., Chu F.S., and Runnegar M.T. (1998), Prolonged sublethal exposure to the protein phosphatase inhibitor microcystin-LR results in multiple dose-dependent hepatotoxic effects. *Toxicol. Sci.*, 44, 87–96.

Spieser O.H., Schwaiger J., Ferling H., and Negele R.D. (2000), An introduction to behavioral monitoring-effects of nonylphenol and ethinylestradiol on swimming behavior of juvenile carp. In Butterworth F.M., Gunatilaka A., Gonsebatt M.E. (Eds.), *Biomonitors and Biomarkers as Indicators of Environmental change* (pp 93–112). Kluwer Publ., New York, 2, ISBN 0-306-46387-3.

Van der Schalie W.H., Shedd T.R., Knechtges P.L., and Widder M.W. (2001), Using higher organisms in biological early warning systems for real-time toxicity detection. *Biosensors and Bioelectronics*, 16(7–8), 457–465.

Chapter 30
Biomonitoring Studies Performed with European Eel Populations from the Estuaries of Minho, Lima and Douro Rivers (NW Portugal)

Carlos Gravato[1,3], Melissa Faria[1], Anabela Alves[1], Joana Santos[1], and Lúcia Guilhermino[1,2]

Abstract Contaminants' presence in the aquatic environment is relevant for the disturbance of the European stocks of diadromic species. The main goal of this study was to compare the biotransformation and oxidant/antioxidant *status* of yellow eel (*Anguilla anguilla*) populations from the estuaries of Minho (reference), Lima and Douro (contaminated) Rivers. Comparatively to the values determined in eels from the reference estuary, low total glutathione and reduced glutathione levels associated with high lipid peroxidation levels and benzo(a)pyrene-type metabolites' concentrations were found in liver from eels collected in the estuary of Lima river. Eels from Douro estuary showed high liver ethoxyresorufin-*O*-deethylase, catalase, glutathione peroxidase, total glutathione, reduced glutathione and oxidized glutathione levels associated with low lipid peroxidation and benzo(a)pyrene-type metabolites relatively to fish from the reference estuary. The pollution present in the estuaries of Lima and Douro Rivers is causing alterations on biotransformation and antioxidant stress parameters. In addition, Lima estuary eels are exposed to polycyclic aromatic hydrocarbons as indicated by the high levels of metabolites found. Since polycyclic aromatic hydrocarbons interfere with reproductive parameters and increased cytochrome P450 1A1 activity (as found in eels from the Douro estuary) interfere with reproduction, the exposure of eels to pollution in Lima and Douro estuaries may be decreasing their reproductive potential. In addition, energy to face chemical stress may be allocated from processes such as growth and weight increase that are factors determinant for the success of the long migration to the reproduction area. Therefore, pollution may be decreasing the contribution of these populations to the species evolution.

Keywords: Biomonitoring, biomarkers, biotransformation, oxidative stress, *Anguilla anguilla*

---

[1] *CIMAR-LA/CIIMAR – Centro Interdisciplinar de Investigação Marinha e Ambiental, Laboratório de Ecotoxicologia, Universidade do Porto, Rua dos Bragas, 177, 4050–123 Porto, Portugal*

[2] *ICBAS – Instituto de Ciências Biomédicas de Abel Salazar, Universidade do Porto, Departamento de Estudos de Populações, Laboratório de Ecotoxicologia, Largo Professor Abel Salazar 2, 4099–003, Porto, Portugal*

[3] *Departamento de Biologia, Universidade de Aveiro, 3810–193 Aveiro, Portugal*

Y.J. Kim and U. Platt (eds.), *Advanced Environmental Monitoring*, 390–401.

30.1 Introduction

Anthropogenic activities are the main factor leading to the contamination of marine environments by complex mixtures of pollutants. Estuarine ecosystems are the major sink for many of the potentially hazardous chemical contaminants due to their proximity to urbanized areas. The use of biomarkers to assess the effects of pollutants in wildlife has been recommended for biomonitoring programs in addition to chemical analyses (Cajaraville et al. 2000; Solé 2000). Biomonitoring programmes, using fish collected from estuaries contaminated by polycyclic aromatic hydrocarbons (PAHs), metals and endocrine disruptors (Law et al. 1997; Matthiessen et al. 1998; Woodhead et al. 1999), have revealed an increase of the levels of several key biomarkers, including ethoxyresorufin-*O*-deethylase (EROD) activity (Kirby et al. 1999), bile metabolites (Ruddock et al. 2003) and DNA damage (Lyons et al. 2004). The levels of antioxidants have been also proposed as biomarkers due to their importance in the protection against oxyradicals (Orbea et al. 2002; Regoli et al. 2002). Those responses, such as non-enzymatic and enzymatic antioxidants, are part of the defence systems that prevent the oxidative damage of macromolecules resulting from an increment of reactive oxygen species (ROS) production induced by pollutants (Livingstone et al. 1990; Winston and Di Giulio 1991; Lemaire and Livingstone 1993; Filho 1996).

The eel *Anguilla anguilla* is widely distributed in European estuaries, and brackish and freshwater basins. This organism is a bottom-dwelling species with territorial behaviour and is generally immature in European waters (Colombo and Grandi 1995). Eels can bioaccumulate elevated concentrations of pollutants due to their high lipid content and longevity. Several field and laboratory studies have been performed to characterize the main responses induced by pollutants in eels (Pacheco and Santos 1997; Fenet et al. 1998; Roche et al. 2000; Schlezinger and Stegeman 2000; Livingstone et al. 2000; Doyotte et al. 2001; Peña-Llopis et al. 2001; Peters et al. 2001; Langston et al. 2002; Regoli et al. 2003; Buet et al. 2006). After the continental phase of their life cycle (6–12 and 9–20 years for males and females, respectively), eels migrate to the Sargasso Sea where they reproduce and die (Colombo and Grandi 1995). The presence of environmental contaminants in the ecosystems where the continental phase occurs (e.g. estuaries) may be relevant for the disturbance of the European stocks of this species, since it may affect their health condition and, thus, to compromise their survival to the long term migration to the reproduction area.

The main goal of this research work was to compare the biotransformation and oxidant/antioxidant status of yellow eel populations from the estuaries of Minho (reference site), Lima and Douro (contaminated sites) Rivers. Liver phase I and II biotransformation were assessed as ethoxyresorufin-*O*-deethylase (EROD) and glutathione *S*-transferase (GST) activities, respectively. The benzo(a)pyrene (B(a)P)-type metabolites were determined in bile and liver cytosol. Non-enzymatic antioxidant status was determined as reduced glutathione (GSH), oxidized glutathione (GSSG), total glutathione (TG) levels and GSH/GSSG ratio. Liver antioxidant enzymes were determined as glutathione peroxidase (GPx), glutathione reductase (GR), superoxide dismutase (SOD) and catalase (CAT) activities. Lipid peroxidation (LPO) was also determined in eel's liver.

30.2 Material and Methods

The estuary of Minho River located on the border of Portugal and Spain is considered as one of the least contaminated estuaries along the Portuguese coast and several studies have been performed using it as a reference area (Castro et al. 2004; Monteiro et al. 2005; Rodrigues et al. 2006). The Lima River is subjected to contaminants resulting from harbour activities (Viana do Castelo harbour), discharges of urban origin, released without prior treatment by some municipalities (INAG 2000; Cairrão et al. 2004) and from a paper mill industry. The domestic sewage of over one million inhabitants and industrial effluents are still discharged, mostly without treatment, directly into the Douro estuary and its tributaries (Mucha et al. 2004; Ferreira et al. 2004, 2006).

Sampling activities were carried out in the estuaries of Minho, Lima and Douro Rivers (Northwest of Portugal) during November 2005. Physical and chemical parameters determined in the three estuaries were: 10.5°C, 7.5 ‰, pH 7.9, dissolved oxygen (DO) 9.3 mg/L and 12.7 μS^{-1} in Minho; 12.0°C, 5.3 ‰, pH 6.4, DO 9.1 mg/L and 11.7 μS^{-1} in Lima; 11.8°C, 13.5 ‰, pH 8.3, DO 11.2 mg/L and 22.6 μS^{-1} in Douro.

Eels were measured, weighed and sacrificed by cervical severance. Analyzed specimens did not exhibit significant size differences with means of wet body weight (g) and length (cm) of 62 ± 7 and 32 ± 1 in Minho (n = 25), 71 ± 9 and 34 ± 1 in Lima (n = 25), 84 ± 12 and 35 ± 1 in Douro (n = 36). Liver and bile were rapidly isolated, frozen and maintained at -80°C until processed for analysis.

Livers were systematically divided into three parts:

1. One part of each liver was homogenized (1:5) in 50 mM Tris-HCl, pH 7.4 containing 0.15 M KCl. Liver microsomes were prepared according to the methods previously described and EROD activity was quantified (Burke and Mayer 1974; Stegeman and Kloepper-Sams 1987).
2. Another portion of each liver was homogenized (1:10) in 0.1 M K-phosphate buffer (pH 7.4). Part of this liver homogenate was used to determine the extent of endogenous LPO by measuring the thiobarbituric acid reactive substances (TBARS) according to Ohkawa (1979) and Bird and Draper (1984), with the adaptations described by Filho et al. (2001) and Torres et al. (2002). The remaining liver homogenate was centrifuged for 20 min at 12,000 rpm (4°C) to obtain the post-mitochondrial supernatant (PMS). SOD activity was determined in PMS (Flohé and Ötting 1984). GPx activity was determined in PMS by measuring the decrease in NADPH at 340 nm and using H_2O_2 as substrate (Mohandas et al. 1984). GR activity was assayed in PMS according to Cribb et al. (1989). CAT activity was determined in PMS and represents the H_2O_2 consumption obtained at 240 nm in the presence of H_2O_2 (Clairborne 1985).

TG content (GSH+GSSG) and oxidized glutathione (GSSG) were determined in PMS at 412 nm, using a recycling reaction of GSH with 5,5'-dithiobis (2-nitrobenzoic acid) (DTNB) in the presence of GR excess (Tietze 1969; Baker et al. 1990). 2-Vinyl-pyridine was used to conjugate GSH for the GSSG determination (Griffith 1980). GSH was calculated by subtracting GSSG from the TG levels. GSH/GSSG ratios were expressed according to Peña-Llopis et al. (2001);

3. The remaining part of each liver was homogenized in 1 mL 0.1 M K-phosphate buffer (pH 6.5) and centrifuged during 30 min at 9000 g (4°C). The supernatant was further diluted in the same buffer to a final concentration of 0.5 mg/mL and GST activity was determined following the conjugation of GSH with 1-chloro-2,4-dinitrobenzene (CDNB) at 340 nm (Habig et al. 1974).

The protein concentration of liver supernatants was determined according to the Bradford method, whereas bile samples were diluted with 10 vol of water to give an initial bile solution that was assayed for protein content (Bradford 1976).

Biliary and liver PMS metabolites are reported on the basis of mg protein as previously adapted by Gagnon and Holdway (2000). The initial bile solution and the liver PMS were further diluted in methanol 50% to 1:200. Fluorescent readings were made at 380/430 nm for benzo(a)pyrene (B(a)P)-type metabolites (Lin et al. 1996; Aas et al. 1998; Gagnon and Holdway 2000), using B(a)P as a reference standard.

SIGMASTAT 2.03 software was used for statistical analysis. The experimental data were first tested for normality and homogeneity of variance. Then, one-way ANOVA followed by Tukey test (Zar 1996) were used to compare different treatments. Differences between means were considered significant when $p < 0.05$.

30.3 Results

The results showed that liver EROD activity was significantly higher (~330%) in eels from the Douro estuary than in eels collected in the reference and in Lima estuaries (Fig. 30.1A). Liver phase II biotransformation, assessed as GST activity, was significantly induced in eels collected in the estuaries of Lima (~41%) and Douro (~27%) Rivers, compared to the reference site (Fig. 30.1B). In addition, BaP-type metabolites in bile were not significantly different in the 3 estuaries, despite their slight lower amount in the bile of eels from Douro (Fig. 30.1C). However, BaP metabolites level was significantly increased in the liver of eels collected in the Lima estuary (~65%), whereas it was significantly decreased (~30%) in the liver of the eels collected in the Douro estuary, compared to the reference site (Fig. 30.1D).

The non-enzymatic antioxidant parameters, namely TG (~27%), GSH (~65%) and GSH/GSSG ratio (~63%), were significantly decreased in the liver of the eels collected in the Lima estuary compared to the reference site (Fig. 30.2A,B,D). Eels from the Douro estuary exhibited a significantly higher GSSG content (~20%) and slightly increased levels of TG (~26%) and GSH (~34%), as well as GSH/GSSG ratio (~10%), when compared to Minho estuary eels (Fig. 30.2).

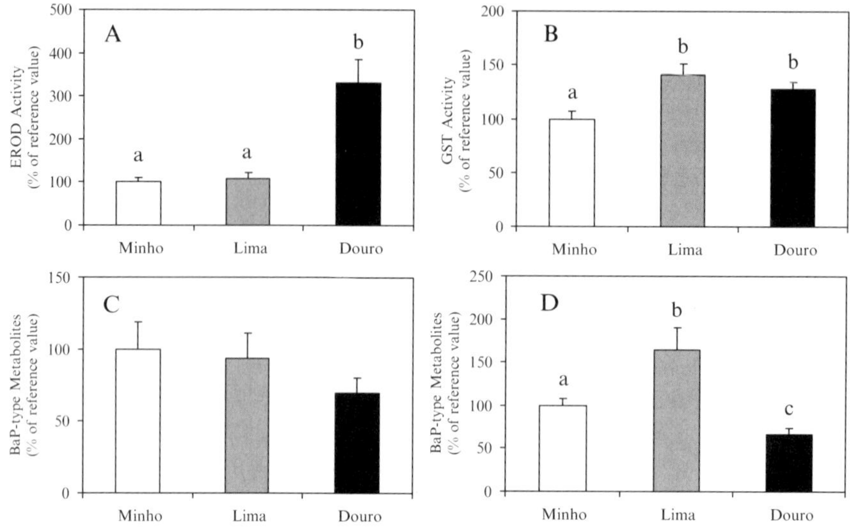

Fig. 30.1 Liver EROD activity (**A**), liver GST activity (**B**), and BaP-type metabolites in the bile (**C**) and liver (**D**) determined in eels from the estuaries of rivers Minho (reference site), Lima and Douro (contaminated sites). Results are presented as the percentage of reference site mean levels (EROD: 39.3 ± 11.1 pmol/min/mg protein; GST: 112.5 ± 7.8 nmol/min/mg protein; BaP-type metabolites in the bile: 74.3 ± 14.1 ng BaP/mg protein; BaP-type metabolites in the liver: 21.0 ± 1.6 ng BaP/mg protein) with corresponding S.E.M. bars. Different characters indicate statistical significant differences among sampling sites as indicated by Tukey honestly significant difference multiple-comparison test ($p < 0.05$) for each parameter

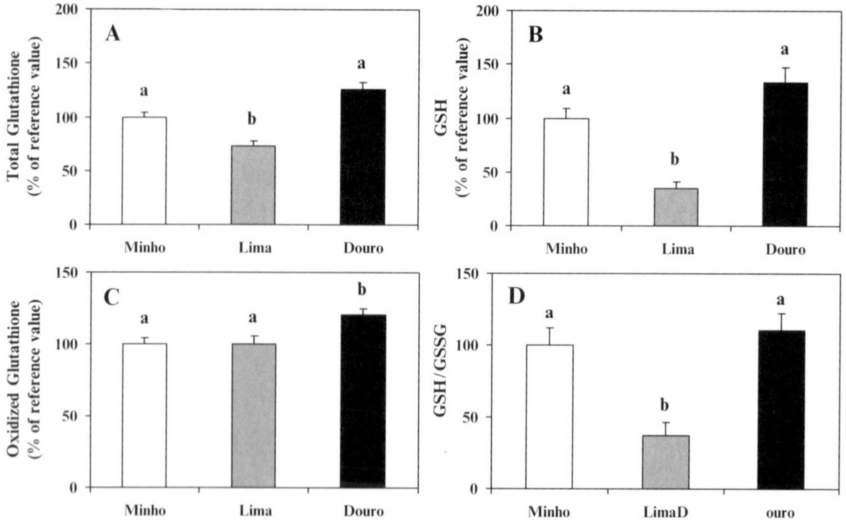

Fig. 30.2 Liver TG (**A**), GSH (**B**), GSSG (**C**) and GSH/GSSG ratio (**D**) determined in eels from the estuaries of rivers Minho (reference site), Lima and Douro (contaminated sites). Results are presented as the percentage of reference site mean levels (TG: 0.88 ± 0.04 nmol/min/mg protein; GSH: 0.34 ± 0.03 nmol/min/mg protein; GSSG: 0.25 ± 0.01 nmol/min/mg protein; GSH/GSSG: 1.53 ± 0.18) with corresponding S.E.M. bars. Different characters indicate statistical significant differences among sampling sites as indicated by Tukey honestly significant difference multiple-comparison test ($p < 0.05$) for each parameter

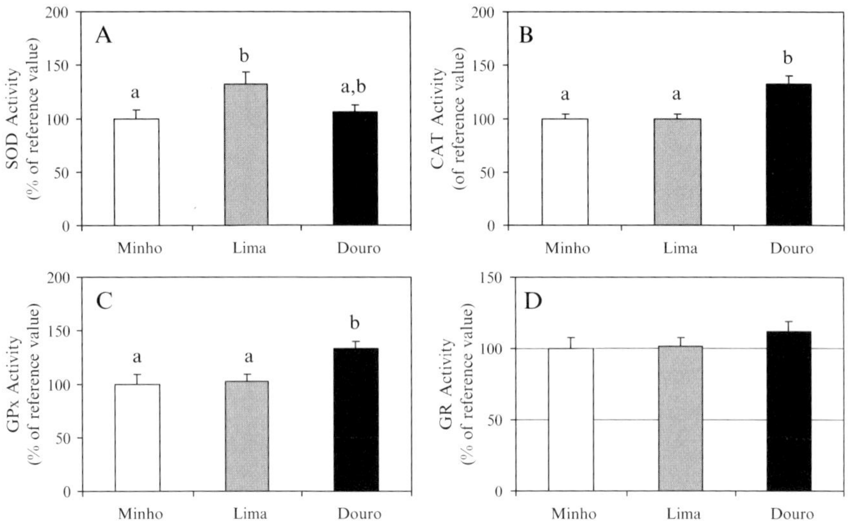

Fig. 30.3 Liver SOD (**A**), CAT (**B**), GPx (**C**) and GR (**D**) activities determined in eels from the estuaries of rivers Minho (reference site), Lima and Douro (contaminated sites). Results are presented as the percentage of reference site mean levels (SOD: 1.1 ± 0.1 U/mg protein; CAT: 16.8 ± 0.8 μmol/min/mg protein; GPx: 1.8 ± 0.2 nmol/min/mg protein; GR: 3.8 ± 0.3 nmol/min/mg protein) with corresponding S.E.M. bars. Different characters indicate statistical significant differences among sampling sites as indicated by Tukey honestly significant difference multiple-comparison test ($p < 0.05$) for each parameter

SOD activity was significantly higher (~32%) in the liver of eels collected in the Lima estuary than in eels from the estuary of Minho River (Fig. 30.3A). Liver CAT (~32%) and GPx (~33%) activities were significantly higher in eels from the Douro estuary than in eels collected in the estuaries of Lima and Minho Rivers (Fig. 30.3B, C). Although not significant, the eels from Douro estuary showed a slightly higher GR activity (~12%) than eels from Lima and Minho estuaries (Fig. 30.3D).

The oxidative damage, assessed as LPO, was significantly higher in the liver of eels from the estuary of Lima River than in the liver of eels collected in the estuaries of Minho and Douro Rivers (Fig. 30.4). Moreover, LPO was significantly lower in the liver of eels from the estuary of the Douro River than in eels from the estuary of Minho River (Fig. 30.4).

30.4 Discussion

In the present study, induced EROD activity was found in eels from the estuary of Douro River that is polluted with xenobiotics both from urban and industrial sources (Ferreira et al. 2004, 2006), therefore, in good agreement with previous studies reporting an induced liver EROD activity in mullets from this estuary

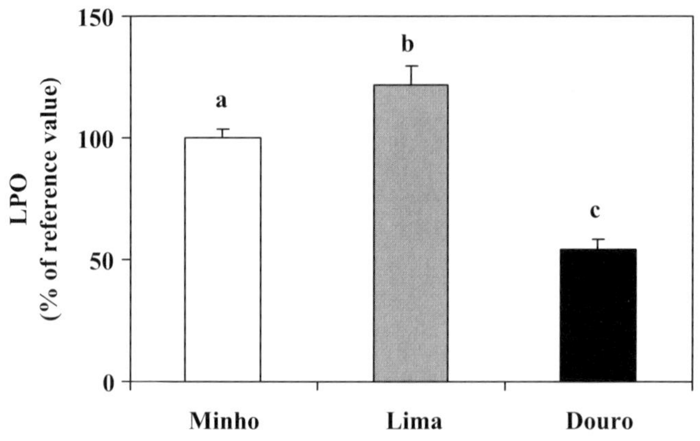

Fig. 30.4 Liver LPO determined in eels from the estuaries of rivers Minho (reference site), Lima and Douro (contaminated sites). Results are presented as the percentage of reference site mean levels (LPO: 92.4 ± 3.0 nmol/g wt) with corresponding S.E.M. bars. Different characters indicate statistical significant differences among sampling sites as indicated by Tukey honestly significant difference multiple-comparison test ($p < 0.05$) for each parameter

(Ferreira et al. 2004, 2006). EROD activity increase was previously observed in several fish species after exposure to organic pollutants, such as PAHs, causing very strong increases in CYP1A catalytic activities (Van der Oost et al. 2003). The present study demonstrated not only a significant liver EROD induction in eels from the Douro estuary, but also a significant increase in liver GST activity. This phase II biotransformation induction in eels was associated with low BaP–type metabolites in bile and liver of eels from Douro suggesting that chemicals other than PAHs are responsible for the induction found in eels from this estuary.

Previous studies showed that the hepatic cytochrome P450 1A activity was inversely related to egg viability, fertilization success and to the successful development from fertilization through hatching, demonstrating that this biotransformation activity interfere with the ability to regulate sex steroids (Spies et al. 1988). Therefore, it may be hypothesized that EROD induction found in eels from Douro River estuary may have negative effects on sexual maturation of the individuals. In addition, eels from this estuary showed high CAT, GPx, TG, GSH and GSSG levels that were associated with low LPO and B(a)P-type metabolites either in the bile or in the liver. High liver biotransformation and antioxidant capacity in eels from the estuary of the Douro River might be important to their adaptation to the presence of contaminants in this estuary. It has been proposed that animals inhabiting polluted environments can develop some adaptations or compensatory mechanisms (Regoli and Principato 1995). Chronically intoxicated fish can, for example, reduce the inflow of toxicants by building up morphological barriers and they may activate mechanisms of detoxification (Lindström-Seppä and Pesonen 1986; Andersson et al. 1988), which make them more resistant to toxicants (Bucher and Hofer 1990). However, the mechanisms used to face chemical stress require energy that could have been invested in growth and weight increase, factors that are determinant for the success of the long migration until the reproduction area (Sargasso Sea).

Eels from the estuary of Lima River showed low liver TG, GSH, GSH/GSSG and LPO levels associated with high B(a)P-type metabolites liver concentrations. These results suggest an impairment of their detoxification and anti-oxidative defenses and the occurrence of lipid peroxidation. The low levels of GSH found may be related with the low EROD activity found in these animals. The consumption of GSH due to the scavenging of oxyradicals or as a cofactor (GPx and GST activities) represents a drain on intracellular reducing equivalents (Winston and Di Giulio 1991). The tissue thiol status modulates CYP 1A gene expression and catalytic activity (Otto et al. 1997), suggesting that the low liver EROD activity observed in eels from the estuary of Lima river can be a consequence of GSH depletion. Overall, these results indicate that eels from Lima River estuary are under oxidative stress probably due to long-term PAHs exposure. Fish vitellogenesis can be affected by pollutants with know affinity for the estrogenic receptor, such as PAHs (Nicolas 1998). Since, eels from Lima estuary are exposed to PAHs as indicated by the high levels of liver bile metabolites found, negative effects on sexual development may be expected. In addition, oxidative stress may also interfere with the health condition of the individuals and, thus, to decrease their probability of survive during the long migration to reproduce. In fact, previous studies demonstrated clear relationships between biochemical alterations induced by environmental pollution (PAHs) and fish diseases, such as preneoplastic hepatic lesions, hepatic neoplasms and skin diseases (Myers et al. 1994; Vethaak et al. 1996). Winston and Di Giulio (1991) observed elevated rates of idiopathic lesions and neoplasia among fish inhabiting polluted environments that were related to the increased oxidative stress associated with pollutant exposure.

In conclusion, the findings of this study suggest that the exposure of eels to pollution during their development in the estuaries of Lima and Douro Rivers is inducing changes on biomarkers involved in physiological functions determinant for the survival and performance of the eels, namely, biotransformation enzymes and anti-oxidative stress defences. Studies from several authors suggest that these alterations may have negative effects on sexual development. In addition, mechanisms used to face chemical stress need energy that is probably allocated from other functions such as tissue repair, growth and weight increase which are determinant for a successful migration until the reproduction area. Therefore, exposure to pollution due to their continental phase may reduce the contribution of Lima and Douro eel populations for the species evolution. More studies are required to relate the alterations found in biomarkers with possible effects on reproductive parameters and health condition of the animals.

Acknowledgements The authors express their gratitude to Dr. C. Antunes, A. Oliveira and to I. Cunha for their help during the capture of the eels. This study was funded by "Fundação para a Cíência e a Tecnologia" of Portugal (FCT) (project EELEANORA, POCTI/BSE/47918/2002; Post-Doc grant to C. Gravato, POCTI/BPD/21070/2004) and by EU FEDER funds.

References

Aas E., Beyer J., and Goksoyr A. (1998), PAH in fish bile detected by fixed wavelength fluorescence. *Mar. Environ. Res.*, 46, 225–228.

Andersson T., Förlin L., Härdig J., and Larsson A. (1988), Physiological disturbances in fish living in coastal waters polluted with bleached kraft pulp mill effluents. *Can. J. Fish. Aquat. Sci.*, 45, 1525–1536.

Baker M.A., Cerniglia G.J., and Zaman A. (1990), Microtiter plate assay for the measurement of glutathione and glutathione disulfide in large numbers of biological samples. *Ann.. Biochem.*, 190, 360–365.

Bird R.P., and Draper A.H. (1984), Comparative studies on different methods of malondyhaldehyde determination. *Methods Enzymol.*, 90, 105–110.

Bradford M. (1976), A rapid and sensitive method for the quantification of microgram quantities of protein utilizing the principle of protein-dye binding. *Ann. Biochem.*, 72, 248–254.

Bucher F., and Hofer R. (1990), Effects of domestic wastewater on serum enzyme activities of brown trout (*Salmo trutta*). *Comp. Biochem. Physiol.*, 97, 381–385.

Buet A., Banas D., Vollaire Y., Coulet E., and Roche H. (2006), Biomarker responses in European eel (*Anguilla anguilla*) exposed to persistent organic pollutants. A field study in the Vaccarès lagoon (Camargue, France). *Chemosphere*, 65, 1846–1858.

Burke M.D., and Mayer R.T. (1974), Ethoxyresorufin: Direct fluorimetric assay of a microsomal-*O*-deethylation which is preferentially inducible by 3-methylcholantrene. *Drug. Metab. Dispos.*, 2, 583–588.

Cairrão E., Couderchet M., Soares A.M.V.M., and Guilhermino L. (2004), Glutathione-*S*-transferase activity of *Fucus* spp. as a biomarker of environmental contamination. *Aquat. Toxicol.*, 70, 277–286.

Cajaraville M.P., Bebianno M.J., Blasco J., Porte C., Sarasquete C., and Viarengo A. (2000), The use of biomarkers to assess the impact of pollution in coastal environments of the Iberian Peninsula: A practical approach. *Sci. Total Environ.*, 247, 201–212.

Castro M., Santos M.M., Monteiro N.M., and Vieira N. (2004), Measuring lysosomal stability as an effective tool for marine coastal environmental monitoring. *Mar. Environ. Res.*, 58, 741–745.

Clairborne A. (1985), Catalase activity. In: Greenwald R.A. (Ed.). *CRC Handbook of Methods in Oxygen Radical Research* (pp. 283–284). Boca Raton, FL: CRC Press.

Colombo G., and Grandi G. (1995), Sex differentiation in the European eel: Histological analysis of effects of sex steroids on the gonad. *J. Fish Biol.*, 47, 394–413.

Cribb A.E., Leeder J.S., and Spielberg S.P. (1989), Use of a microplate reader in an assay of glutathione reductase using 5,5'-dithiobis(2-nitrobenzoic acid). *Ann. Biochem.*, 183, 195–196.

Doyotte A., Mitchelmore C.L., Rinisz D., McEvoy J., Livingstone D.R., and Peters L.D. (2001), Hepatic 7-ethoxyresorufin *O*-deethylase activity in the eel (*Anguilla anguilla*) from the Thames estuary and comparisons with other United Kingdom estuaries. *Mar. Pollut. Bull.*, 42, 1313–1322.

Fenet H., Casellas C., and Bontoux J. (1998), Laboratory and field-caging studies on hepatic enzymatic activities in European eel and rainbow trout. *Ecotox. Environ. Saf.*, 40, 137–143.

Ferreira M., Antunes P., Gil O., Vale C., and Reis-Henriques M.A. (2004), Organochlorine contaminants in flounder (*Platichthys flesus*) and mullet (*Mugil cephalus*) from Douro estuary, and their use as sentinel species for environment monitoring. *Aquat. Toxicol.*, 69, 347–357.

Ferreira M., Moradas-Ferreira P., and Reis-Henriques M.A. (2006), The effect of long-term depuration on phase I and phase II biotransformation in mullets (*Mugil cephalus*) chronically exposed to pollutants in River Douro Estuary, Portugal. *Mar. Environ. Res.*, 61, 326–338.

Filho D.W. (1996), Fish antioxidant defences—A comparative approach. *Braz. J. Med. Biol. Res.*, 29, 1735–1742.

Filho D.W., Tribess T., Gáspari C., Cláudio F.D., Torres M.A., and Magalhães A.R.M. (2001), Seasonal changes in antioxidant defenses of the digestive gland of the brown mussel (*Perna perna*). *Aquaculture*, 203, 149–158.

Flohé L., and Ötting F. (1984), Superoxide dismutase assays. *Method. Enzymol.*, 105, 93–104.

Gagnon M.M., and Holdway D.A. (2000), EROD induction and biliary metabolite excretion following exposure to the water accommodated fraction of crude oil and to chemically dispersed crude oil. *Arch. Environ. Contam. Toxicol.*, 38, 70–77.

Griffith O.W. (1980), Determination of glutathione and glutathione disulfide using glutathione reductase and 2-vinyl-pyridine. *Ann. Biochem.*, 106, 207–212.

Habig W.H., Pabst M.J., and Jakoby W.B. (1974), Glutathione-*S*-transferases, the first enzymatic step in mercapturic acid formation. *J. Biol. Chem.*, 249, 7130–7139.

INAG (2000). Planos das bacias Hidrográficas dos rios Luso-Espanhois-Síntese. Caracterização e Diagnóstico. Instituto da Água, Direcção de serviços de recursos Hídricos, Divisão de recursos Subterrâneos, pp. 398.

Kirby M.F., Matthiessen P., Neall P., Tylor T., Allchin C.R., Kelly C.A., Maxwell D.L., and Thain J.E. (1999), Hepatic EROD activity in flounder (Platichthys flesus) as an indicator of contaminant exposure in English estuaries. *Mar. Pollut. Bull.*, 38, 676–686.

Langston W.J., Chasman B.S., Burt G.R., Pope N.D., and McEvoy J. (2002), Metallothionein in liver of eels *Anguilla anguilla* from the Thames estuary: An indicator of environmental quality? *Mar. Environ. Res.*, 53, 263–293.

Law R.J., Dawes V.J., Woodhead R.J., and Matthiessen P. (1997), Polycyclic aromatic hydrocarbons (PAH) in seawater around England and Wales. *Mar. Pollut. Bull.*, 34, 306–322.

Lemaire P., and Livingstone D.R. (1993), Pro-oxidant/antioxidant processes and organic xenobiotics interactions in marine organisms, in particular the flounder *Platichthys flesus* and mussels *Mytilus edulis*. *Trend. Comp. Biochem. Physiol.*, 1, 1119–1150.

Lin E.L.C., Cormier S.M., and Torsella J.A. (1996), Fish biliary polycyclic aromatic hydrocarbon metabolites estimated by fixed-wavelength fluorescence: Comparison with HPLC-fluorescent detection. *Ecotoxicol. Environ. Saf.*, 35, 16–23.

Lindström-Seppä P., and Pesonen M. (1986), Biotransformation enzymes in fish as tools for biomonitoring the aquatic environment. *Acta Biol. Hung.*, 37, 85–95.

Livingstone D.R., Mitchelmore C.L., Peters L.D., O'Hara S.C., Shaw J.P., Chesman B.S., Doyotte A., McEvoy J., Ronisz D., Larsson D.G., and Forlin L. (2000), Development of hepatic CYP1A and blood vitellogenin in eel (*Anguilla anguilla*) for use as biomarkers in the Thames Estuary, UK. *Mar. Environ. Res.*, 50, 367–371.

Livingstone D.R., Garcia Martinez P., Michel X., Narbonne J.F., O'Hara S., Ribera D., and Winston G.W. (1990), Oxyradical generation as a pollution-mediated mechanism of toxicity in the common mussel, *Mytilus edulis* L., and other molluscs. *Funct. Ecol.*, 4, 415–424.

Lyons B.P., Stentiford G.D., Green M., Bignell J., Bateman K., Feist S.W., Goodsir F., Reynolds W.J., and Thain J.E. (2004), DNA adduct analysis and histopathological biomarkers in European flounder (*Platichthys flesus*) sampled from UK estuaries. *Mutat. Res.*, 552, 177–186.

Matthiessen P., Bifield S., Jarret F., Kirby M.F., Law R.J., McMinn W.R., Sheahan D.A., Thain J.E., and Whale G.F. (1998), An assessment of sediment toxicity in the River Tyne estuary, UK, by means of bioassays. *Mar Environ. Res.*, 45, 1–15.

Mohandas J., Marshall J.J., Duggins G.G., Horvath J.S., and Tiller D. (1984), Differential distribution of glutathione and glutathione related enzymes in rabbit kidney. Possible implications in analgesic neuropathy. *Cancer Res.*, 44, 5086–5091.

Monteiro M., Quintaneiro C., Morgado F., Soares A.M.V.M., and Guilhermino L. (2005), Characterization of the cholinesterases present in head tissues of the estuarine fish *Pomatoschistus microps*: Application to biomonitoring. *Ecotox. Environ. Saf.*, 62, 341–347.

Mucha A.P., Bordalo A.A., and Vasconcelos M.T.S.D. (2004), Sediment quality in the Douro river estuary based on trace metal contents, macrobenthic community and elutriate sediment toxicity test (ESTT). *J. Environ. Monit.*, 6, 585–592.

Myers M.S., Stehr C.M., Olsen O.P., Johnson L.L., McBain B.B., Chan S.L., and Varanasi U. (1994), Relationships between toxicopathic hepatic lesions and exposure to chemical contaminants in English sole (*Pleuronectus vetulus*), starry flounder (*Platichthys stellatus*), and white croaker (*Genyonemus lineatus*) from selected marine sites on the Pacific coast, USA. *Environ. Health Perspect.*, 102, 200–215.

Nicolas J.M. (1998), Vitellogenesis in fish and the effects of polycyclic aromatic hydrocarbon contaminants. *Aquat. Toxicol.*, 45, 77–90.

Ohkawa H. (1979), Assay for lipid peroxides in animal tissues by thiobarbituric acid reaction. *Anal. Biochem.*, 95, 351–358.

Orbea A., Ortiz-Zarragoitia M., Sole M., Porte C., and Cajaraville M.P. (2002), Antioxidant enzymes and peroxisome proliferation in relation to contaminant body bordens of PAHs and PCBs in bivalve molluscs, crabs and fish from the Urdaibai and Plentzia estuaries (Bay of Biscay). *Aquat. Toxicol.*, 58, 75–98.

Otto D.M.E., Sen C.K., Casley W.L., and Moon T.W. (1997), Regulation of 3,3',4,4'-tetrachlorobiphenyl induced cytochrome P450 metabolism by thiols in tissues of rainbow trout. *Comp. Biochem. Physiol.*, 117, 29–309.

Pacheco M., and Santos M.A. (1997), Induction of EROD activity and genotoxic effects by polycyclic aromatic hydrocarbons and resin acids on the juvenile eel (*Anguilla anguilla* L.). *Ecotox. Environ. Saf.*, 38, 252–259.

Peña-Llopis S., Pena J.B., Sancho E., Fernández-Vega C., and Ferrando M.D. (2001), Glutathione-dependent resistance of the European eel *Anguilla anguilla* to the herbicide molinate. *Chemosphere*, 45, 671–681.

Peters L.D., Doyotte A., Mitchelmore C.L., McEvoy J., and Livingstone D.R. (2001), Seasonal variation and estradiol-dependent elevation of Thames estuary eel *Anguilla anguilla* plasma vitellogenin levels and comparisons with other United Kingdom estuaries. *Sci. Total Environ.*, 279, 137–150.

Regoli F., and Principato G. (1995), Glutathione, glutathione-dependent and antioxidant enzymes in mussels, *Mytilus galloprovincialis*, exposed to metals under field and laboratory conditions: Implications for the use of biochemical biomarkers. *Aquat. Toxicol.*, 31, 143–164.

Regoli F., Pellegrini D., Winston G.W., Gorbi S., Giuliani S., Virno-Lamberti C., and Bompadre S. (2002), Application of biomarkers for assessing the biological impact of dredged materials in the Mediterranean: The relationship between antioxidant responses and susceptibility to oxidative stress in the red mullet (*Mullus barbatus*). *Mar. Pollut. Bull.*, 44, 912–922.

Regoli F., Winston G.W., Gorbi S., Frenzilli G., Nigro M., Corsi I., and Focardi S. (2003), Integrating enzymatic responses to organic chemical exposure with total oxyradical absorbing capacity and DNA damage in the European eel *Anguilla anguilla. Environ. Toxicol. Chem.*, 22, 2120–2129.

Roche H., Buet A., Jonot O., and Ramade F. (2000), Organochlorine residues in European eel (*Anguilla anguilla*), crucian carp (*Carassius carassius*) and catfish (*Ictalurus nebulosus*) from Vaccarès lagoon (French National Nature reserve of Camargue)—Effects on some physiological parameters. *Aquat. Toxicol.*, 48, 443–459.

Rodrigues P., Reis-Henriques M.A., Campos J., and Santos M.M. (2006), Urogenital papilla feminization in male *Pomatoschistus minutus* from two estuaries in northwestern Iberian Península. *Mar. Environ. Res.*, 62, 258–262.

Ruddock P.J., Bird D.J., McEvoy J., and Peters L.D. (2003), Bile metabolites of polycyclic aromatic hydrocarbons (PAHs) in European eels *Anguilla anguilla* from United Kingdom estuaries. *Sci. Total Environ.*, 301, 105–117.

Schlezinger J.J., and Stegeman J.J. (2000), Induction of cytochrome P450 1A in the American eel by model halogenated and non-halogenated aryl hydrocarbon receptor agonists. *Aquat. Toxicol.*, 50, 375–386.

Solé M. (2000), Assessment of the results of chemical analyses combined with the biological effects of organic pollutants on mussels. *Trend. Anal. Chem.*, 19, 1–9.

Spies R.B., Rice D.W., and Jr, Felton J. (1988), Effects of organic contaminants on reproduction of the starry flounder *Platichthys stellatus* in San Francisco Bay, I. Hepatic contamination and mixed-function oxidase (MFO) activity during the reproductive season. *Mar. Biol.*, 98, 181–189.

Stegeman J.J., and Kloepper-Sams P.J. (1987), Cytochrome P-450 enzymes and monooxygenase activity in aquatic animals. *Environ. Health Perspect.*, 71, 87–95.

Tietze F. (1969), Enzymic method for quantitative determination of nanogram amounts of total and oxidized glutathione. *Ann. Biochem.*, 27, 502–522.

Torres M.A., Testa C.P., Gáspari C., Masutti M.B., Panitz C.M.N., Curi-Pedrosa R., Almeida E.A., Di Mascio P., and Filho D.W. (2002), Oxidative stress in the mussel *Mytella guyanensis* from polluted mangroves on Santa Catarina Island, Brazil. *Mar. Pollut. Bull.*, 44, 923–932.

Van der Oost R., Beyer J., and Vermeulen N.P.E. (2003), Fish bioaccumulation and biomarkers in environmental risk assessment: A review. *Environ. Toxicol. Pharm.*, 13, 57–149.

Vethaak A.D., Jol J.G., Meijboom A., Eggens M.L., ap Reinallt T., Westen P.W., Van de Zande T., Bergman A., Dankens N., Ariese F., Baan R.A., Everts J.M., Opperhuizen A., and Marquenie J.M. (1996), Skin and liver diseases induced in flounder (*Platichthys flesus*) after long-term exposure to contaminated sediments in large-scale mesocosms. *Environ. Health Perspect.,* 104, 1218–1229.

Winston G.W., and Di Giulio R.T. (1991), Prooxidant and antioxidant mechanisms in aquatic organisms. *Aquat. Toxicol.*, 19, 137–161.

Woodhead R.J., Law R.J., and Matthiessen P. (1999), Polycyclic aromatic hydrocarbons (PAH) in surface sediments around England and wales and their possible biological significance. *Mar. Pollut. Bull.*, 38, 773–779.

Zar J.H. (1996), *Biostatistical Analysis*. Third Edition. USA: Prentice Hall International, Inc.

Chapter 31
In Vitro Testing of Inhalable Fly Ash at the Air Liquid Interface

Sonja Mülhopt[1], Hanns-Rudolf Paur[1], Silvia Diabaté[2], and Harald F. Krug[2]

Abstract The aim of this study is to analyse the toxicological potential of fine and ultrafine particles from industrial combustion processes using a biotest. This biotest is performed by near-realistic exposure of cultivated lung cells at the air-liquid interface and analysing the biological responses. Important steps in this work are to develop the exposure system for the use at industrial particle sources, to provide reproducible deposition conditions for submicron particles and to validate the exposure protocol for the bioassay. The presented technique maintains the viability of the cells but is sensitive for inflammatory effects. Exposure experiments with the ultrafine fraction of fly ash from a municipal waste incinerator have shown an increased release of IL-8 as a function of exposure time and dose. The presented exposure method and the lung specific bioassay seem to be an appropriate model to simulate the inhalation of particulate air pollution and to screen the biological effects of particulate emissions from different sources.

Keywords: Aerosol, bioassay, PM10, toxicity, ultrafine particles

31.1 Introduction

Particle emissions from industry and traffic are important sources of anthropogenic airborne particles in the environment. In recent years epidemiological studies have shown that there is an association between the concentration of fine particles (PM_{10}, $PM_{2.5}$) in the atmosphere and the rate of mortality or morbidity due to respiratory and cardiovascular diseases (Peters et al., 1997; Wichmann et al.,

[1] *Forschungszentrum Karlsruhe, Institute for Technical Chemistry, Thermal Waste Treatment Division, Hermann-von-Helmholtz-Platz 1, 76344 Eggenstein – Leopoldshafen, Germany; Tel: + 49 7247 82 3807, Fax: + 49 7247 82 4332*

[2] *Forschungszentrum Karlsruhe, Institute for Toxicology and Genetics, Hermann-von-Helmholtz-Platz 1, 76344 Eggenstein – Leopoldshafen, Germany*

Y.J. Kim and U. Platt (eds.), *Advanced Environmental Monitoring*, 402–414.

2000a). The causes of the toxicological effects of fine particles to the human organism are largely unknown. Besides their chemical composition, the particle number and the surface of the particles seem to be of particular importance for the effects. So not only particle mass but more the number concentration seems to be a measure for the toxic potential of an inhaled aerosol (Wichmann et al., 2000b). For the quantitative assessment of the toxicity of industrial submicron particles the dose-response relationship is essential. The long term objective of this project is to use the toxicological results on relevant particle parameters for the developments of emission control strategies which will reduce the aerosol components responsible for the health effects in the population. Fly ash from a municipal waste incinerator plant was selected to study *in vitro* its responses in target cells of the lung.

Inhalation experiments with animals (in vivo studies) provide the most comprehensive data on the toxicological potential of an aerosol; however, they are very complex and expensive. *In vitro* exposure of cultivated lung cells allows a series of experiments in short time periods under reproducible conditions (Diabaté et al., 2002; Voelkel et al., 2003; Wottrich et al., 2004).

In vitro exposure of cells usually takes place in culture medium in which the particles are resuspended. However resuspension in a liquid changes the properties of the particles, moreover the dose is not applied directly to the cells. The amount of particles which finally reach the cell culture remains unknown unless radioactive labelled particles are used and quantified after the exposure. Therefore a newly developed technique to study the biological effects of an aerosol was applied, in which the particles directly contact the air-liquid interface of the cells (Aufderheide et al., 2005). An exposure at the air-liquid interface is technically challenging, however it resembles the situation in the lung during inhalation of atmospheric particulate matter.

The aim of this study is to analyse the toxicological potential of fine and ultrafine particles from industrial combustion processes using a biotest. This biotest is performed by near-realistic exposure of cultivated lung cells at the air-liquid interface and analysing the biological responses. Important steps in this work were to develop the exposure system for the use at industrial particle sources, to reproducibly generate a test aerosol, to determine the deposition efficiency of the aerosol and to provide an exposure protocol for the bioassay.

31.2 Experimental Procedures

31.2.1 Aerosol Generation

Particles are generated in the flow reactor AEOLA (Fig. 31.1). AEOLA (**AE**r**O**sol **LA**boratory facility) is a pilot plant for the generation of dry aerosols in a wide range of concentrations. Ambient air or flue gas from a light-oil burner is used as carrier gas. Relevant concentrations of gaseous trace substances (SO_2, NO_x, dioxin,

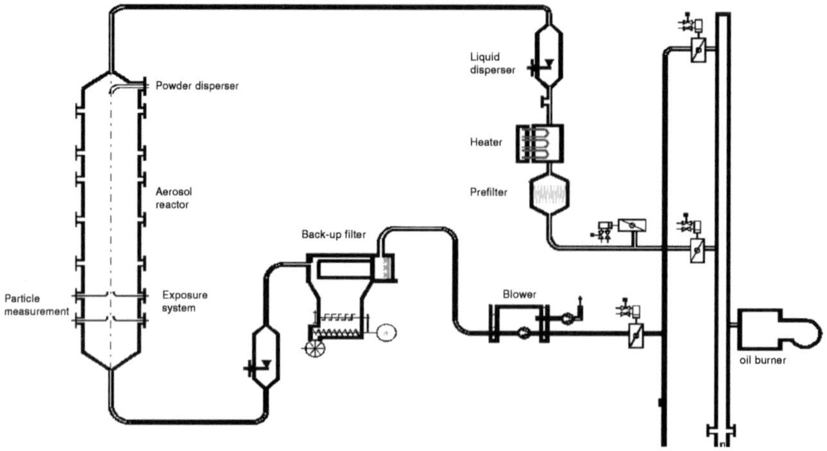

Fig. 31.1 Schematic representation of the AEOLA aerosol laboratory

etc.) may be added to the gas. At flow rates of up to 800 Nm³/h and temperatures from 20 to 200°C, gas and particle concentrations can be varied over wide ranges. The carrier gas first flows through a HEPA filter. This filter ensures that the gas is particle-free and later measurements are not falsified by impurities. Further downstream is an air heater, which allows heating the gas to the desired temperature. Fluids like fluorescein-sodium solution or organics can be sprayed into the heated gas stream via a two-phase nozzle (Schlick, Typ S4) driven with 2 bar and 1.2 l/h fluid. Dry powders are suspended by an aerosol generator using a rotating brush. For the exposure experiments AEOLA is operated at a volume flow of 500 Nm³/h and a temperature of 80°C. Stability of total number and mass concentration of the generated aerosol was controlled using an optical particle counter (scattered light analyser PCS 2000, PALAS, Karlsruhe).

The aerosol samples are taken at the lower part of an 8 m high flow reactor which produces constant and homogeneous aerosol flow. Afterwards the off gas passes a bag filter and the blower before leaving to a stack.

For the bioassay the fine fraction of fly ash from a municipal waste incinerator is dispersed in the upper part of the reactor as described above. The fly ash was collected in a municipal waste incinerator and the fraction (MAF02) below 20 μm was prepared by air separation (Fig. 31.2).

The chemical composition of this fly ash is determined by Total Reflection X-Ray Fluorescence Analysis, ion chromatography and elemental analysis. Main components are nearly 70% salts, metals and heavy metals and approximately 1% carbon (Table 31.1). Birnbaum et al. (1996) reported a density similar to quartz of about 2.2 g/cm³ for complete fly ash. For the fine fraction below 20 μm we determined the density to 2.7 g/cm³ by a helium pyknometer (Micromeritics, average of 10 measurements). By the same method the density of solid fluorescein sodium is determined to 1.5 g/cm³.

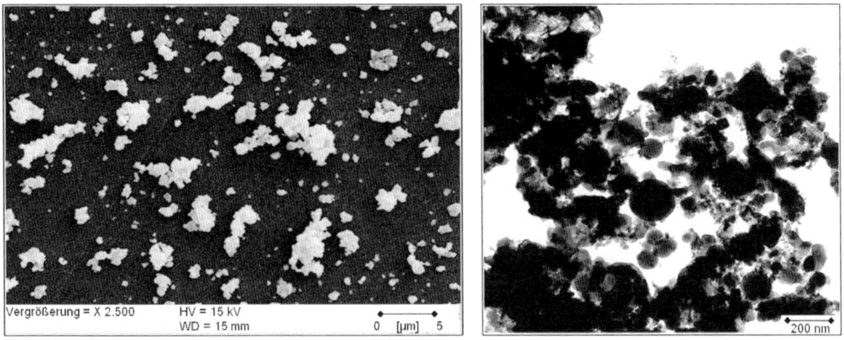

Fig. 31.2 Left: Scanning electron microscopy of a filter sample of fly ash. Right: Transmission electron microscopy of the insoluble fraction of fly ash

Table 31.1 Main components of the fine fraction of fly ash in percent by weight

Al	Mg	Na	Si	K	Ca	Fe	Cl	S
1.2	0.5	7.3	3.6	12.0	6.4	0.9	6.5	6.0

31.2.2 Aerosol Measurement

The resuspended aerosol is described by its number size distribution (Fig. 31.3) that is determined using a scanning mobility analyser SMPS 3071 (TSI). This instrument is also used to determine the grade efficiency by measuring the size distribution upstream and downstream of the exposure chamber (Fig. 31.4). The size distributions and the calculated grade efficiencies correspond to the gas composition of mobility analyser and are not corrected for particle growth in humid air. At relative humidity of 85% as used in this study this factor was determined to 1.3 in previous experiments. Similar data are reported for sodium sulphate (Mätzing et al., 1996) and for other combustion aerosols (Weingartner et al., 1995). For the scanning electron microscopy Nuclepore filters are sampled according to VDI 2066 by drawing the aerosol for a defined time at a defined volume flow through the filter and correcting the volume to normal conditions.

31.2.3 Sampling System and Exposure

For the *in vitro* exposure of cultured pulmonary cell lines to ultra fine particles a sampling system which simulates the human respiratory tract is required. The exposure of the cell cultures takes place in the exposure unit CULTEX which is integrated in the developed exposure system (Mülhopt et al., 2004a + b). The flow chart is shown in Fig. 31.4.

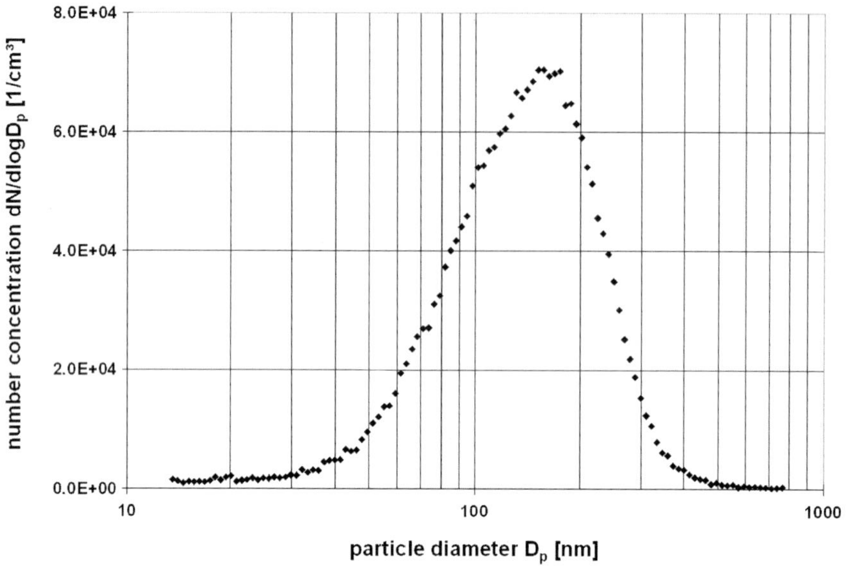

Fig. 31.3 Size distribution of the fine fraction of fly ash (MAF02) suspended in air, sampled from AEOLA, and determined by the Scanning Mobility Particle Sizer SMPS (TSI), (X_{modal} = 165 nm; σ_g = 1.65)

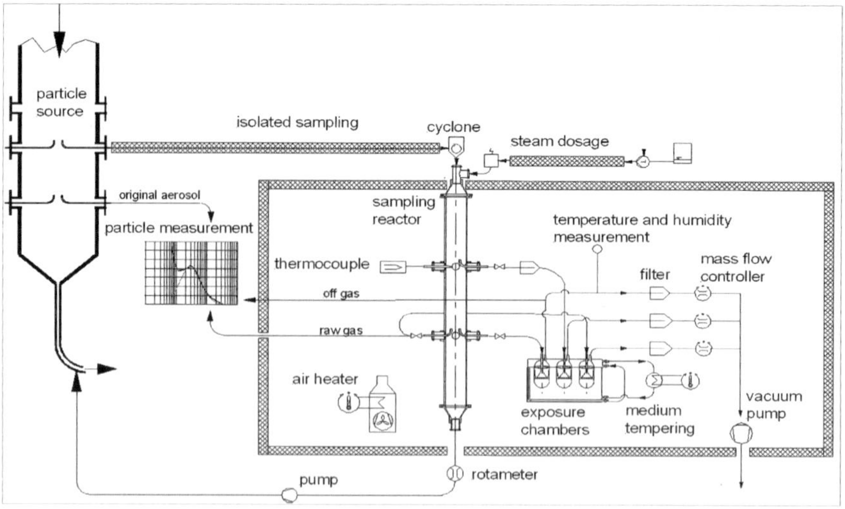

Fig. 31.4 Flow chart of the exposure system installed in the AEOLA aerosol laboratory

The main components in the sampling and exposure system are:

1. *Sampling of the aerosol* from the AEOLA reactor at a flow rate of 2 m^3/h.
2. *Removal of coarse particles* by a cyclone (Stage 1 of the Sierra Stack 5 Cascade Cyclone) with a cut off diameter of ~1 µm. Thus, individual large particles are removed that would contribute considerably to the deposited mass and therefore to the dosimetry. Additionally, large particles would probably disturb the bioassay.
3. *Humidification of the aerosol:* To avoid drying of the cell cultures the aerosol is humidified to conditions corresponding to the lower region of human lungs. Therefore, water vapor is injected at a rate of 80 g/h into the humidification reactor downstream of the cyclone to achieve a relative humidity of >85% at 37 ± 2°C.
4. *Exposure of the lung cells:* The exposure units CULTEX (Vitrocell, Gutach) (Aufderheide et al., 2000) consist of an inlet system into which the aerosol flows, and the lower unit that contains the cell cultures. The aerosol is directed via concentrical steel funnels into the center above the cell layer and directed to the outside (Fig. 31.5). It leaves the exposure chamber and enters the off gas line. For each cell culture an extra aerosol sample is drawn from the reactor. The advantage of this separation of aerosol flows is to avoid fluctuations by linking the chambers and particle losses due to dividing the flows. Each aerosol flow is regulated by its own mass flow controller downstream of the exposure chambers. The separation is also necessary to have the possibility of filtering one or two aerosol flows before entering the exposure chamber so there are simultaneous exposures of cell cultures to particles and to particle free tracer gas. The lower unit of the exposure system accommodates the Transwell inserts (Corning, Wiesbaden) with the cell culture. The cell cultures adhere to the membrane of Transwell-Clear inserts with a diameter of 24 mm and 0.4 µm pores (catalogue number 3450). The cell layer is wetted by the nutrient medium (RPMI 1640, Invitrogen, Karlsruhe) supplemented with 10 mM HEPES (4-(2-hydroxyethyl)piperazine-1-ethanesulfonic acid) from below. The cell cultures used are kept at 37±2 °C by circulating warmed water in the housing of the exposure unit.

Fig. 31.5 Left: Exposure chamber with three cell culture inserts. Right: 3D image of the aerosol inlet above a Transwell membrane insert with cell culture

31.2.4 Determination of the Deposited Dose

To determine the deposited mass on the surface of a cell culture a dosimetric procedure is developed. An aerosol of fluorescein-sodium is generated in the AEOLA flow reactor as described above. The size fraction below 1 µm is separated by the cyclone and the membrane surface without cells is exposed to the aerosol for 1 to 2 hours. The polyethylene membranes are cut out of the Transwell-clear inserts by a scalpel and extracted with 10 ml of distilled water in an ultrasonic bath for 10 min at room temperature. Depending on the concentration, the solution is diluted such that spectroscopy could be performed in the range of the calibration curve with known concentrations of fluorescein sodium. The deposited mass is calculated from the fluorescence intensity of the solution measured with a fluorescence spectrometer (Aminco Bowman Series 2 Luminescence Spectrometer) at the wavelength of 509 nm. The fluorescence intensity of fluorescein-sodium standards exhibits a linear behaviour in the concentration range of 0.01 – 0.1 µg/ml.

31.2.5 Biotest

For the biotest, a co-culture system of BEAS-2B, immortalized cells from normal human bronchial epithelium (ATCC, Rockville, MD) and differentiated THP-1 macrophages, was seeded onto porous Transwell-clear membranes. At the time of the experiments approximately 1×10^6 cells covered the membrane surface area of 4.5 cm^2. The confluent monolayer is exposed to filtered air or to aerosol at the air-liquid interface having access to medium through the pores (0.4 µm in diameter). These samples were compared to controls kept in the laboratory.

After a post-incubation period of 20 h, the medium is analyzed for lactate dehydrogenase (LDH) (Roche Mannheim) and interleukin-8 (IL-8) release (ELISA kit from BD Pharmingen, Heidelberg). The cells are tested for viability with alamarBlue (Serotec, Düsseldorf), a reagent to determine the relative metabolic activity of the cells (Diabaté et al., 2006). The values are reported as arithmetic mean values with the error bars representing the standard error of the mean (s.e.m.) of independent experiments with several samples each as indicated in the figure captions.

31.3 Results and Discussion

31.3.1 Characterization of the Exposure System

To get reproducible data of biological responses all details of the exposure system were characterized. For this purpose we used the dosimetry method described above to optimize and characterize the system for parameters like deposition

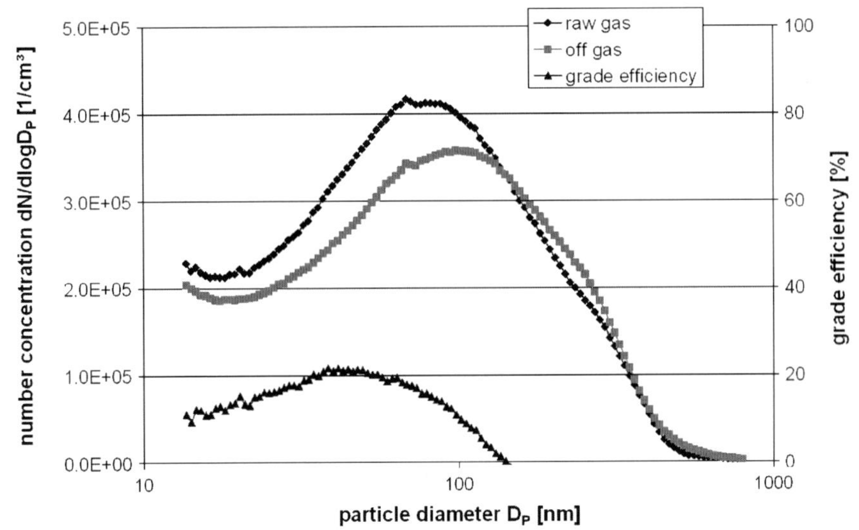

Fig. 31.6 Mean size distribution and grade efficiencies for an exposure of cell cultures (A549 cells, r.h. 85%) to fluorescein-sodium particles, mean values, n(raw gas) = 5, standard deviation 20%, n(offgas) = 20, standard deviation 38%

efficiency, reproducibility, deposited mass per membrane and grade efficiency for the model aerosol of fluorescein-sodium particles.

To describe the deposition behaviour of the whole exposure system the grade efficiency is determined by measuring the size distribution upstream and downstream of the exposure chambers containing cell cultures with SMPS (Fig. 31.6). Measurement of raw gas takes place in the reactor after cyclone and humidification, the measurement of the clean gas was full flow after the CULTEX chamber. The total number concentration amounts 4.1E + 05 ± 15% in the raw gas and 3.8E + 05 ± 30%in the off gas over the measurement range of 13 – 800 nm. That means 10% of the particles remain in the exposure system.

For determining the grade efficiency a flow rate of 300 ml/min is used because of the defined aerosol inlet flow of the SMPS. We assume there is nearly no difference to the flow rate of 100 ml/min used for the exposure of cell cultures because in both cases the gas velocities above the membrane are below 0.012 m/s. Out of this and the low particle diameter the main deposition mechanism to expect above the cell culture is diffusion. In the exhaust part of the exposure chamber small diameters and sharp changes in flow direction may create other deposition mechanism like impaction.

The deposition efficiency on single membranes is determined from the deposited mass by fluorescence spectrometry as described above. For a 1 h exposure at 300 ml/min, the deposited mass was 3.9 μg per 24 mm membrane with a standard deviation of 11%. This corresponds to 2.3% of the total mass that passed the membrane.

In the results of the grade efficiency the influence of the geometry of the system is shown. Losses in the off gas part of the exposure system and in the tubes of the sampling is part of the grade efficiency. Due to this the deposition efficiency measured by the fluorescein-sodium dosimetry is much lower than the efficiency calculated from the grade efficiency.

31.3.2 Exposure of Human Lung Cells Towards Fly Ash

For the exposure experiments with fly ash during the exposure of the cell cultures a similar behaviour as determined for fluorescein sodium particles is observed. This indicates that the dose on a mass basis was extremely low. While exposure experiments the total number concentration amounts to 4E + 04 ± 25% $1/cm^3$. Based on the deposition rate of 2.3% of mass determined for fluorescein-sodium the deposited mass of fly ash per membrane can be estimated. The basis for this estimation is the similarity of the aerosols in structure (both are salts), size distribution, number concentration and deposition behaviour. So the deposited mass is approximately 0.09 μg per membrane and hour. Similar results have been reported for the exposure of A549 cells to carbonaceous particles by Bitterle et al. (2006). The 1-hour dose per unit of area of cell culture is in the range of 20 $ng/cm^{2*}h$. This concentration of 20 ng/cm^2 corresponds to about the 20 fold of a 24 hour-dose of PM_{10} inhaled by an adult at a polluted urban site (100 $\mu g/m^3$).

A co-culture of BEAS-2B cells with characteristics of the human bronchial epithelium and THP-1 macrophages was used as target cells for the exposure experiments. Prior studies revealed that its response to fly ash was most sensitive compared to other cell lines. However, to detect an effect due to deposited fly ash it is necessary to demonstrate that the cells did not suffer from the exposure conditions themselves. Comparably to in vivo conditions, where bronchial epithelial cells are directly exposed to inhaled air, it is of essential importance for the viability of the cells not to dry out. Therefore, the evaporation of water by the gas flow must be replaced by moistening through the membranes. The influence of the volume flow is shown by passing filtered humidified air at different flow rates over the cells.

The viability of the cells, which was indicated by reduction of the alamarBlue reagent, is analysed 20 h after exposure. No difference is observed between the submerged controls which are kept unchanged in the incubator and the control cells which are kept at air-liquid conditions without flow in normal air at 37°C (Fig. 31.7).

The viability of the cells which are transported to the AEOLA and exposed to filtered air decreased with increasing flow rate. For exposure with aerosol we decided to employ a flow rate of 100 ml/min per membrane. At 100 l/min, the viability of the cells is not affected up to an exposure time of 6 h to filtered air or MAF02 aerosol (Fig. 31.8). According to this for the toxicity tests the flow rate of 100 ml/min is used. As a marker for the inflammatory response the release of the

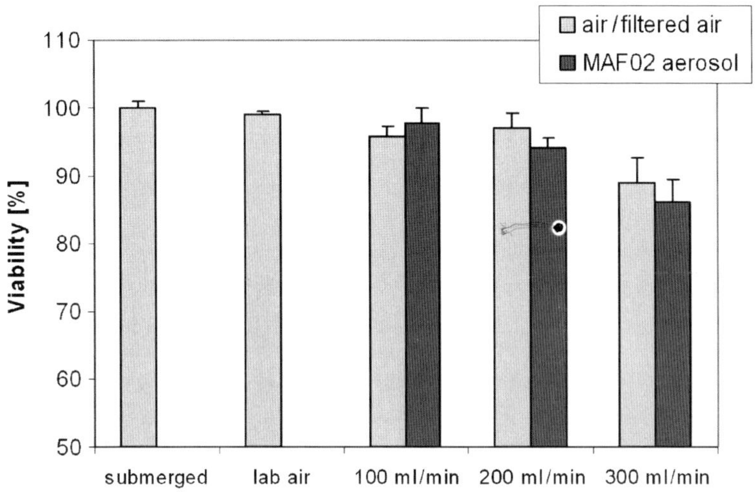

Fig. 31.7 Viability of co-cultures of BEAS-2B and THP-1 cells exposed at the air-liquid interface as a function of the flow rate (r.h. = 85%, t = 1 h, MAF02 = submicron fraction of fly ash)

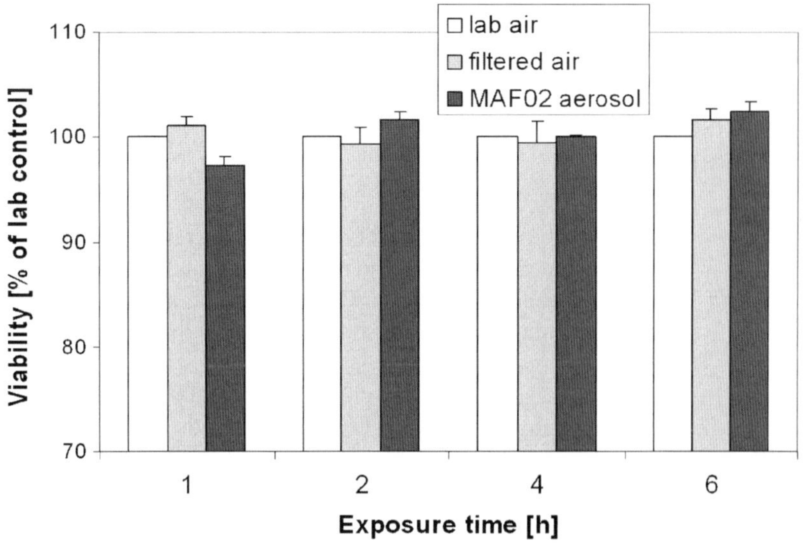

Fig. 31.8 Viability of co-cultures of BEAS-2B and THP-1 cells as a function of the exposure time (flow rate 100 ml/min per membrane, r.h. = 85%, MAF02 = submicron fraction of fly ash)

protein interleukin-8 is determined. The release of interleukin-8 increases in dependence of the exposure time (Fig. 31.9). As the deposited mass of fly ash is constant during the experiment this correlates with the dose.

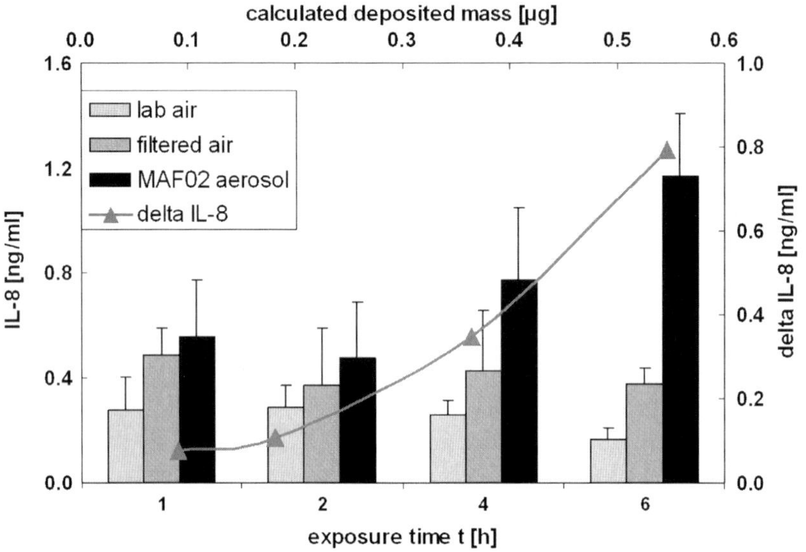

Fig. 31.9 Effects of exposure to filtered air and the submicron fraction of fly ash aerosol on the IL-8 release of human lung cells (number of experiments $n = 3$ for $t = 2, 4, 6\,h$ and $n = 2$ for $t = 1\,h$, number of samples in each experiment $m = 1$ for lab air, $m = 2$ for filtered air, $m = 3$ for MAF02 aerosol) (co-culture of BEAS-2B with THP-1 cells, r.h. = 85%, particle number concentration $3.58\,E + 04\ 1/cm^3$) columns to left y-axis, curve to upper x-axis and right y-axis

31.4 Summary

An exposure system and a protocol was developed and characterized to measure the dose response relationship between submicron particles and the inflammatory response of human lung cells. The exposure system can be applied to measurements up to six hours at industrial particle sources to assess the risk of submicron particles.

The exposure system was applied for in vitro testing of submicron fly ash from a municipal waste incinerator. The IL-8 release of lung cells exposed to this fly ash increased by a factor of 4 compared to the controls. This demonstrates the potential of submicron fly ash to trigger the inflammation of lung cells. IL-8 is a key biomarker of lung cells exposed to airborne pollutants and an indicator of a pro-inflammatory response which may contribute to lung inflammation (Steerenberg et al., 1998; Monn et al., 1999; Diabaté et al., 2002). A similar exposure system has also been employed by other research groups e.g. to study the effects of diesel exhaust particles (Cheng et al., 2003) or metallic nanoparticles (Cheng 2004) on the IL-8 release from A549 cells which have been exposed at the air-liquid interface.

The advantage of aerosol exposure at the air-liquid layer compared to submerged exposure is the direct contact of the aerosol with the target cells without further treatment of the particles such as deposition on filters and resuspension. The results of this study show that an *in vitro* exposure of pulmonary cells at the air-liquid interface can be conducted by the way described under methods without loss of cell viability. That means transport and exposure to low flow rates has no effect on the viability of the cells.

This exposure technique can therefore be used for screening the toxicological potential of unknown aerosols in order to identify potential emitters of health relevant particles. Furthermore, new developments in the technology of the combustion process and flue gas cleaning can directly be evaluated.

Acknowledgements We thank H. Fischer for taking the SEM images, F. Seidenstricker and M. Hauser for useful help in the laboratory. The financial support of BWPLUS (Baden-Württemberg Research Program Securing a Sustainable Living Environment) is appreciated.

References

Aufderheide M., and Moor U. (2000), CULTEX—An alternative technique for cultivation and exposure of cells of the respiratory tract to airborne pollutants at the air/liquid interface. *Exp. Toxicol. Pathol.,* 52, 265–270

Aufderheide M. (2005), Direct exposure methods for testing native atmospheres. *Exp. Toxicol. Pathol.*, 57, Supplement 1, 213–226

Birnbaum L., Richers U., and Koeppel W. (1996), Untersuchung der physikalisch/chemischen Eigenschaften von Filterstäuben aus Müllverbrennungsanlagen (MVA). *Wissenschaftliche Berichte/Forschungszentrum Karlsruhe*, FZKA-5693

Bitterle E., Karg E., Schroeppel A., Kreyling W. G., Tippe A., Ferron G. A., Schmid O., Heyder J., Maier K. L., and Hofer T. (2006), Dose-controlled exposure of A549 epithelial cells at the air-liquid interface to airborne ultrafine carbonaceous particles. *Chemosphere*, 65, (10), 1784–1790.

Cheng M. D. (2004), Effects of nanophase materials (≤20 nm) on biological responses. *J. Env. Sci. Health - Part A Toxic/Hazard. Subst. Env. Eng.,* 39, (10), 2691–2705

Cheng M. D., Malone B., and Storey J. M. E. (2003), Monitoring cellular responses of engine-emitted particles by using a direct air-cell interface deposition technique. *Chemosphere,* 53, 237–243.

Diabaté S., Mülhopt S., Paur H.-R., and Krug H. F. (2002), Pro-inflammatory effects in lung cells after exposure to fly ash aerosol via the atmosphere or the liquid phase. *Ann. Occup. Hyg.,* 46, 382–385.

Diabaté S., Mülhopt S., Paur H.-R., and Krug H. F. (2007). Responses of human lung cells after exposure to ultrafine particles of incinerator fly ash at the air-liquid interface, submitted

Mätzing H., Baumann W., and Paur H.-R. (1996), Bimodal aerosol coagulation with simultaneous condensation/evaporation, *J. Aerosol Sci.*, 27, Supplement 1, S363–S364

Monn C., and Becker S. (1999), Cytotoxicity and induction of proinflammatory cytokines from human monocytes exposed to fine (PM2.5) and coarse particles (PM10–2.5) in outdoor and indoor air. *Toxicol. App. Pharmacol.*, 155, 245–252.

Mülhopt S., Seifert H., and Paur H.-R. (2004a, June), Exposure technique for a lung specific bioassay for the assessment of industrial ultra fine particle-emissions. (Paper presented at the 7th International Conference on Nanostructured Materials, Wiesbaden; Germany)

Mülhopt S., Paur H-R., and Seifert H. (2004b), Expositionsverfahren für einen lungen-spezifischen Bioassay zur Bewertung industrieller Feinstpartikel-Emissionen. *BWPLUS Report* 2004, Retrieved from http://www.bwplus.fzk.de/berichte/SBer/BWB21018SBer.pdf

Peters A., Wichmann H. E., Tuch T., Heinrich J., and Heyder J. (1997), Respiratory effects are associated with the number of ultra-fine particles. *Am. J. Resp. Crit. Care Med.,* 155, 1376–1383

Steerenberg P. A., Zonnenberg J. A., Dormans J. A., Joon P. N., Wouters I. M., van Bree L., Scheepers P. T., and Van Loveren H. (1998), Diesel exhaust particles induced release of interleukin 6 and 8 by (primed) human bronchial epithelial cells (BEAS 2B) in vitro. *Exp. Lung Res.*, 24, 85–100.

Voelkel K., Krug H. F., and Diabaté S. (2003), Formation of reactive oxygen species in rat epithelial cells upon stimulation with fly ash. *J. Biosci.,* 28, 51–55.

Weingartner E., Baltensperger U., and Burtscher H. (1995), Growth and structural changes of combustion aerosols at high relative humidity. *J. Aerosol Sci.*, 26, Supplement 1, S667-S668

Wichmann H. E., and Peters A. (2000a), Epidemiological evidence of the effects of ultrafine particle exposure. *Phil. Trans.: Math. Phys. Eng. Sci. (Series A),* 358, (1775), 2751–2769.

Wichmann H. E., Spix C., Tuch T., Wölke G., Peters A., Heinrich J., Kreyling W. G., and Heyder J. (2000b), Daily Mortality and Fine and Ultrafine Particles in Erfurt, Germany, Part I: Role of Particle Number and Particle Mass. *Research Report Number* 98, Health Effects Institute

Wottrich R., Diabaté S., and Krug H. F. (2004), Biological effects of ultrafine model particles in human macrophages and epithelial cells in mono- and co-culture. *Int. J. Hyg. Env. Health,* 207, (4), 353–361.

List of Abbreviations

ACE 2	Second Aerosol Characterization Experiment
DIN	Deutsche Industrie Norm (German industrial standard)
DOC	Dissolved organic Carbon
DRI	Desert Research Institute
DT	DUSTTRAK
EARLINET	European Aerosol Research Lidar Network
INDOEX	Indian Ocean Experiment
LIF	Laser induced fluorescence
LLG	Laser-Laboratorium Göttingen
MODIS	Moderate Resolution Imaging Spectroradiometer
MPL	Micro-pulse lidar
Nd:YAG	Neodymium: Yttrium Aluminum Garnet
NOM	Natural organic matter
OP-FTIR	Open path-Fourier transform infrared spectrometer
OP-LT	Opwen path-laser transmissometer
ORS	Optical remote sensing
PAH	Polycyclic aromatic hydrocarbon
PM	Particulate matter
TOMS	Total Ozone Mapping Spectrometer
UV	Ultra-violet

Index

A
Air pollution monitoring, 3–17
Absorbing, 9, 54, 144, 146, 162, 163, 169, 170, 175, 176, 179, 184, 187, 188
Absorption, 5, 7–9, 11, 14, 22, 39–41, 50–59
Acclimatization, 378
ACE 2, 161–163
Acute toxicity
  mortality, 249–251, 255, 314, 402
ADEOS II, 218
Aerosol Particle Seizer, 209
Aerosol Robotic Network (AERONET)
  sun/sky photometer, 180
Aerosol studies using polarization, 137
Aerosol sulfate, 91
Aerosol type, 136, 137, 141, 158, 161, 162, 164, 191, 193, 194
Aerosols
  coarse mode, 102, 122, 126, 157
  desert dust, 184, 218
  elastic-Raman lidar, 180
  fine mode, 187, 188
  optical parameters, 223
  properties of, 15, 108, 112, 145, 149, 155, 156, 168, 173, 180, 188, 403
Agent stimulant
  erwinia herbicola, 205, 210, 214
  ovalbumin, 210, 212
Air liquid interface, 403, 411–413
Air pollutants of SO_2, NO_x, O_3, 5, 46, 91, 103
Air pollution, 5, 91, 92, 97, 100, 101, 103, 138, 190
Airborne gaseous, 90, 91
Aircraft measurements, 90–93, 97, 103
Angstrom coefficient, 186–188
Ångström exponent, 157–159, 162, 163
Anguilla, 391
Animal based testing, 323
Anthropogenic, 38, 70, 78, 91, 158, 160–163, 179, 205, 391, 402
AOT, 186–188, 191–195, 198, 199, 218–220, 222–227
Apportionment factor, 148, 149
Aquatic colloids in nature, 234–237
Arctic, 137–139, 161
Assessment of "good ecological status", 297, 301, 308
Atmospheric particles, 168
Avoidance behavior test, 251

B
Back blast, 144–149, 151, 153
Background Asian dust, 122, 123, 126–128, 130–133
Background KOSA, 122, 129, 130
Backscatter coefficient, 159, 160, 181, 183–185
Back trajectories, 180, 181, 183, 184, 186
BAER, 191, 193, 195, 219–222
Balloon-borne measurement, 123, 124, 126, 128, 129, 131
Behavioural ecotoxicology, 377
17-Beta estradiol (E_2), 340, 342, 343, 346
Bile metabolites, 391, 397
Bioaerosol threat, 204
Bioassay, 249
Bioconcentration, 254
Biogenic, 138
Biological agent
  bacillus anthracis, 210, 213
  yersinia pestis, 210, 214
Biomarkers, 298–301, 308, 331, 340, 346, 364, 369, 370, 391, 397
Biomonitoring studies, 390–397
Biosensor, 298–301, 306, 308, 309
Biotic ligand model (BLM), 255

Biotransformation, 301, 340, 344, 345, 391, 393, 396, 397
Bisphenol A (BPA), 304, 308, 339, 340, 345, 346, 364, 365
Bradbury-Nielson gate, 74
Bromate, 272, 274–280, 292

C
Calibration curve, 150, 177, 408
Carbon dioxide, 78, 79
Cavity enhanced spectroscopy, 52
CCF (Concordance Correlation Factor), 28, 29, 33
CE/MS, 327, 329, 334
Chemical ionisation reaction, 67, 72
Chemical Ionisation Reaction (CIR-MS) Mass Spectrometry, 67
Chemical sensing of trace gases, 50
Chemical stress, 377, 381, 383, 396, 397
Chlorine monoxide, 38
Choriogenin L, 340–346
Chronic toxicity
  growth, 299, 315, 364, 365, 397
  reproduction, 249–251, 299, 313, 364
Circadian activity, 381
Climate, 78, 91, 117, 122, 131, 136, 156, 158, 179, 180, 218, 219, 340
Complex refractive index, 144, 146, 148, 157, 169, 171–173, 175–177, 181
Continuous monitoring, 38, 79, 83, 84, 168, 272, 378, 387
Cyanobacteria, 377
Cytochrome P450, 339–341, 343, 344, 346, 396
Cytochrome P450 1A (CYP1A), 340, 341, 343–346, 396

D
Danio rerio, 377–379, 381–386
Dark vegetation, 221
Depolarization, 137–142, 157–159, 181, 183–185, 187, 188
Depth, 75, 94, 145, 152, 161, 162, 206, 214, 256, 302, 364
Discrete ordinate radiative transfer (DISORT), 222
Dispersion plume, 37, 107, 117
DNA microarray, 327, 328, 364
DOAS, 8, 9, 11–14, 16, 22, 39–47, 108–112, 116
DOC, 235, 236, 254, 255, 261–269, 283, 284, 286–291
Douro river, 395, 396
Drinking water, 260, 261, 263, 264, 267, 269, 271, 272, 278, 282, 294, 301–303, 306, 308, 378, 386, 387
Drinking water quality, 376
Drinking water treatment, 260, 261, 263, 267, 269, 278
Drug exposure, 297
DT, 144, 147–149
Dual wavelength, 170, 171
Dust, 11, 91, 122–124, 126–133, 137, 144–149
DUSTTRAK, 144

E
EARLINET, 156, 161, 162, 180
Early warning system, 378, 385
Earthworm toxicity test
  Eisenia andrei, 250
  Eisenia fetida, 249, 250
Eco-toxicological classification in sediment, 249, 297, 298, 306
Eels, 391–397
Effect assessment, 297
Effective radius, 157, 158, 160, 163, 191, 192, 197
Effluent organic matter, 259–269
Elastic scattering, 209
Emission, 6, 8, 10, 21–23, 25, 27, 29, 30–32, 34, 38, 47, 78, 79, 83, 86–88, 91
Emission sources
  area, 21–23, 25, 28, 79, 97, 108, 131–133, 206, 220
  fugitive, 21–23, 25, 28, 29, 143, 143–154
Emission strength, 47, 79, 86–88, 133
Endocrine disrupting chemicals, 339
Endocrine disruptor, 299, 364, 369, 391
Endocrine effects, 298, 301, 304, 306
Environmental stress, 298, 326, 327, 330, 335
Enzymatic antioxidants, 391
EOS/AQUA, 218
EOS/Terra, 218
Estuaries, 391–397
Euclidean distance, 330, 367, 371–373
European Aerosol Research Lidar Network (EARLINET), 156, 180
Exposure, 56, 250, 252, 254, 298, 301, 308, 315, 318, 319, 322, 331, 339, 340, 342
External cavity quantum cascade laser (EC-QCL), 53, 54, 56
Extinction, 10, 11, 108, 141, 144–146, 148, 149, 152, 157, 158, 181, 183–185, 187, 192

Extinction coefficient, 11, 145, 157, 181, 184, 185
Extinction efficiency, 144–146
Extinction spectrum, 145, 146

F
Feminization, 365, 367, 371–373
Final remarks, 246
Fish, 301, 308, 339, 340, 343, 344, 363, 365, 366, 368–370, 377–382, 385, 387, 391, 396, 397
Fluorescence, 6, 8, 93, 94, 204–207, 209–212, 214, 241–243, 261, 263, 265, 266, 274, 283, 285–287
Fluorescence based oxygen sensing, 313
Fluorescence microscopy, 351, 359, 360
Forest-fire smoke, 158–164
Formaldehyde (H_2CO) detection, 60
Free troposphere, 122, 123, 126–131, 133, 155, 156, 161, 162, 188

G
Generation of aquatic colloid-borne actinides, 240–244
Genomics, 327, 329, 334, 335, 364
Glutathione, 335, 391–394

H
Hadamard transform, 65, 72, 73
Haze, 11, 13, 137–139, 161, 225
Heat shock protein 70 (HSP 70), 340, 341, 343–346
Hydroxyl radical, 272

I
Ice fog, 138–140
Imaging ellipsometry, 350, 353, 354, 357
In vitro testing, 402–412
INDOEX, 161–163
Integrated Cavity Output Spectroscopy (ICOS), 52, 58
Intensified Charge Couple Device (ICCD), 203
Interband cascade lasers, 51
Intercontinental, 158
Interferometer, 114, 115
Introduction, 11, 250
Inversion, 124, 144–146, 148, 152, 157, 169, 181, 183

L
Laser absorption spectroscopy (LAS), 50–52, 58
Laser Induced Fluorescence (LIF), 8, 203–206, 210, 212, 283, 287, 288
Leucaspius delineatus, 376–378, 381–383
Lidar
  backscatter coefficient, 183, 185
  depolarization ratio, 183, 184, 188
  elastic-Raman lidar, 180
  extinction coefficient, 181, 184
  lidar equation, 205, 206
  lidar ratio, 158, 181, 183, 184
Lima river, 390, 392, 395, 397
Linear mixing model (LMM), 217, 221–223
Lipid peroxidation, 390, 392, 397
Long-range trans boundary, 90–92, 104
Long-range transport, 40, 91, 104, 122, 129, 130, 133, 180
Look-Up Table (LUT), 222
LT, 144

M
Manzanillo Mexico, 111, 116
Marine, 44, 161, 162, 163, 218, 252, 314, 323, 391
Masculinization, 373
Mass concentration, 90, 93, 102, 104, 122, 126, 128, 129, 133, 144, 147, 149, 152, 154, 404
Mass factor, 149
Mass flux of Asian dust particle, 12, 122, 126–128
Mass spectrometry, 7, 50, 325
Medaka, 325, 340, 344, 345, 363–366, 368, 369
MERIS, 191, 193, 195, 197, 200, 201
Metablomics, 325, 327, 334–336
Microcystin-LR, 376, 377
Microphysical, 155, 156, 157, 158, 163, 180, 184, 186, 188
Micro-pulse lidar (MPL), 143, 144, 151, 153
Mie, 10, 11, 43, 144, 145, 168, 172
Migration of colloid-borne actinides, 234, 244–246
Minho river, 392, 395
Minimum reflectance technique (MRT), 217, 222, 228
Model
  dispersion, 78, 107, 117
  radiative transfer, 39, 43, 112, 113
Multi-angle Imaging Spectro-Radiometer (MISR), 218
Multi-channel reflectance, 72
Multi-spectral, 143, 144, 154, 205
Multivariate analysis, 205, 214
Multiwavelength, 155, 156, 163, 164

N
Natural organic matter, 259–269, 286
Network, 79, 156, 180, 190, 199, 200, 204, 223
Neutral red retention time (NRRT)
lysosomal membrane stability, 252
Nitric oxide (NO) detection, 5, 7, 17
NNLS (Non-negative Least Squares), 24
NO^+, 65–68
NO_2 emissions, 6
Non-enzymatic Antioxidants, 391, 393
Nonylphenol (NP), 304, 338, 339, 340, 344, 363, 364

O
OH•, 5, 13, 14, 17, 39, 272, 277, 278
OMICS, 325–336
One compartment model, 253–255
On-line process control, 282
Open-path
FTIR (Fourier Transform Infrared), 22, 144
TDLAS (Tunable Diode Laser Absorption Spectroscopy), 22
Optical, 11, 23, 52, 58, 93, 109, 112, 114, 116, 123
Optical detection method, 348–361
Optical measurement, 167–177
Optical oxygen respirometry, 313
Optical particle spectrometer, 168, 169–172, 176
Optical Remote Sensing (ORS)
DIAL (Differential Absorption LIDAR), 22
OP-FTIR (Open-Path Fourier Transform Infrared), 22
TDLAS (Tunable Diode Laser Absorption Spectroscopy), 22
UV-DOAS (Ultra-Violet Differential Optical Absorption Spectroscopy), 22
Optical thickness, 181, 186, 190, 220
Optimization, 22, 175, 272, 278
Oryzias latipes, 338, 340, 363, 364
OTM-10 (Other Test Method 10), 35
Overlap, 27, 138, 157, 206
Oxidative damage, 391, 395
Oxygen consumption assay, 313
Oxygenated volatile organic compounds, 64
Ozone, 5, 10, 14, 17, 38, 56, 68, 93, 94, 102, 218
Ozone/Hydrogen peroxide, 272

P
Particle, 11, 80, 93, 102, 104, 109, 122–124, 126, 128
Particulate Matter (PM)
PM10, 143
Pathogen, 271, 326, 350, 351, 354, 356–360
Paul Scherrer Institut, 71
PCB 28, 376, 378, 379, 381–383
Pearson correlation coefficient, 28, 29, 363, 367, 371
Photoacoustic spectroscopy (PAS), 53–57
Photoelectric barrier system, 387
α-Pinene, 71
Polarization and Directionality of the Earth's Reflectance (POLDER), 218
Pollution, 4, 5, 45, 65, 78–84, 88
Polycyclic aromatic hydrocarbons, 283, 391
Power plant, 38, 39, 40, 45–7, 108, 110, 113, 115
Power spectral analysis, 377, 381, 383
Profile, 22, 23, 26, 29, 30, 52, 53, 70, 71, 94, 110
Protein chip, 348–359
Protein G, 349, 351, 352, 353, 355–360
Proteomics, 309, 327, 333, 334, 336
Proton Transfer Reaction Mass Spectrometry (PTR-MS), 64, 65
Provenance of aquatic colloids, 234, 239, 240
Pseudo Random Binary Sequence, 72–74

Q
Quantification of aquatic colloids by laser-induced breakdown detection (LIBD), 237–239
Quantum cascade lasers, 52, 53
Quartz enhanced photoacoustic spectroscopy (QEPAS), 54–57

R
Radial Plume Mapping (RPM)
1D-RPM, 29
Horizontal (HRPM), 23–25
Vertical (VRPM), 25–29
Raman lidar, 140, 155–164, 184, 204
Raman signal, 157, 286
Reactive oxygen species, 391
Reagent ion, 66, 67
Real time measurements, 13, 38, 39
Real time PCR, 340, 342, 346
Refractive index, 144, 145, 148, 149, 157, 168–174, 176, 177, 181
Remote sensing, 3, 6, 7, 17, 38, 39, 47, 137, 218, 219
Rotating Shadowband Radiometer (RSR), 222, 223

S
Saccharomyces cerevisiae, 325–336
Sahara, 159, 161, 162, 180, 181, 186–188

SAPHIR, 68, 69
Satellite images, 45, 181–184, 188
SBFM (Smooth Basis Function Minimization), 26
Scattering
  elastic scattering, 209
  raman scattering, 204
Schematic, 358
Sea-viewing Wide Field-of-view Sensor (SeaWiFS), 193, 218
Secondary organic aerosol, 71, 72
Self-assembly method, 352
SIGIS (Scanning Infrared Gas Imaging System), 114, 115
Single-scattering albedo, 157, 158, 163, 164, 220
Size distribution, 10, 124, 129, 136, 144, 145, 157, 181, 183, 191–193, 218, 405, 410
Size measurement, 168–170, 172
Smoke, 139, 140, 149, 158, 159, 161, 162, 164, 218
SO_2 emissions, 91
Species-specific reactions, 377
Spectral, 10, 11, 31, 39–41, 51, 53, 55, 60, 108, 109, 112
Spectral signature, 204, 205, 207, 210, 211, 213, 214
Spectrometer, 16, 53, 58, 65, 66, 72, 109, 112, 113, 144, 168, 172–177, 207–209, 218, 286, 408
Spectroscopy, 8–13, 39, 52, 58, 263, 286, 408
Spectrum
  background, 110
  dark, 109
  differential, 108–110
  reference, 109, 110, 114, 115
SSE (Sum of Squared Errors), 26, 27, 30,
Standoff detection, 204
Stochastic procedure, 169
Sun photometer, 156
Surface plasmon resonance, 350, 355, 360
Swimming activity, 380

T
Testis ova, 364, 369, 371, 373
Time series analysis, 377
TOA reflectance, 220, 222
Toxicity, 56, 234, 248–252, 254–256, 298, 308, 312–316, 318–320
Toxicity testing, 314, 315
Toxicokinetics
  elimination rate, 254
  uptake rate, 254
Toxin, 212, 377
Trace gas, 3–14, 16, 17, 41, 46, 50–61
Trajectory, 78, 83, 88, 105, 109, 110, 158, 164
Transmissometer, 143
Transport, 5, 9, 16, 40, 45, 47, 78, 88, 91, 92, 97, 100
1,3,5-Trimethylbenzene, 71

U
Ultrafine particle, 403

V
Video-monitoring, 376
Vitellogenin, 308, 339, 340, 342, 346, 369, 371
Volatile organic compound, 64, 65
Volcanic, 38, 109, 140, 141